Hypothalamic Regulation of Obesity

Hypothalamic Regulation of Obesity

Editors

Rosalía Rodríguez-Rodríguez
Cristina Miralpeix

MDPI • Basel • Beijing • Wuhan • Barcelona • Belgrade • Manchester • Tokyo • Cluj • Tianjin

Editors
Rosalía Rodríguez-Rodríguez
Basic Sciences Department
Universitat Internacional de Catalunya
Sant Cugat del Vallés (Barcelona)
Spain

Cristina Miralpeix
INSERM Neurocentre Magendie
University of Bordeaux
Bordeaux
France

Editorial Office
MDPI
St. Alban-Anlage 66
4052 Basel, Switzerland

This is a reprint of articles from the Special Issue published online in the open access journal *International Journal of Molecular Sciences* (ISSN 1422-0067) (available at: www.mdpi.com/journal/ijms/special_issues/hypothalamic_obesity).

For citation purposes, cite each article independently as indicated on the article page online and as indicated below:

LastName, A.A.; LastName, B.B.; LastName, C.C. Article Title. *Journal Name* **Year**, *Volume Number*, Page Range.

ISBN 978-3-0365-7393-9 (Hbk)
ISBN 978-3-0365-7392-2 (PDF)

Contents

About the Editors

Rosalía Rodríguez-Rodríguez

Rosalia Rodriguez-Rodriguez has a PhD in Pharmacology and is an established researcher, associate Professor, co-IP of the Neurolipid Group, and Head of Biomedicine at UIC Barcelona. She has more than 15 years of research expertise in the study of obesity and related complications in cellular and animal models at national and international institutions (Aarhus University, Oxford University, University of Bath). Her current research involves the study of the molecular mechanisms in the hypothalamus underlying obesity and the generation of drug delivery systems based in nanomedicines to target brain cells and manage obesity. She has been the PI of 5 national/international research projects and has participated in more than 15 scientific projects, and in all of them, she has shown passion for and commitment to the project's progression and future steps. She has 2 international patents (1 exploited), has authored 67 publications (H index 28), has supervised 9 predoctoral students, and is the associate journal Editor and Expert Evaluator of national and international research institutions.

Cristina Miralpeix

Dr Cristina Miralpeix gained a PhD in Biomedicine in 2019 and is currently a postdoctoral researcher in the Energy Balance and Obesity group led by Dr Daniela Cota at Neurocentre Magendie. Her main interest is the development of obesity and the crosstalk between peripheral metabolism and brain circuits. She has particular interest and expertise in the neuronal pathways that regulate feeding behaviour in animal models of obesity and also in the endocannabinoid system. She has been awarded several grants, one of them being a postdoc grant from the Fondation Recherche Médical for 3 years.

Preface to "Hypothalamic Regulation of Obesity"

The obesity epidemic is a major socioeconomic problem that urgently necessitates better understanding of the mechanisms that mediate the imbalance between food intake and energy expenditure and of obesity-related metabolic and cardiovascular complications. Gaining insight into the cellular basis of obesity could lay the foundations for the development of new therapeutic strategies.

In the last few decades, it has been demonstrated that the hypothalamus regulates energy homeostasis. The hypothalamus contains hormonal- and nutrient-sensing nuclei that organize central and peripheral responses to maintain normal body weight, food intake, energy expenditure and nutrient partitioning. Within the hypothalamus, numerous specialized neuronal populations are connected to each other and to various extrahypothalamic brain regions to coordinate energy homeostasis. Evidence also suggests non-neuronal populations, such as astrocytes, participate, and interesting interplay may exist between astrocytes and hypothalamic neurons, whose disruption leads to insulin resistance and obesity.

This reprint reports on the most recent insights into the hypothalamic neuronal and non-neuronal pathways involved in obesity development. Novel findings related to key systems, such as endocannabinoid and melanocortin-related pathways, in the hypothalamus are discussed. In addition, the emerging contribution of the crosstalk between the hypothalamus and peripheral tissues, such as gut and adipose tissue, in obesity is a promising topic. Therefore, here, we present an overview of the most recent mechanisms underlying the hypothalamic regulation of obesity and the impact of these investigations in the prevention and treatment of human obesity for the scientific community and doctors. We thank the authors who have contributed their scientific findings to this reprint.

Rosalía Rodríguez-Rodríguez and Cristina Miralpeix
Editors

International Journal of
Molecular Sciences

MDPI

Editorial

Hypothalamic Regulation of Obesity

Rosalía Rodríguez-Rodríguez [1,*] and Cristina Miralpeix [2]

1 Basic Sciences Department, Faculty of Medicine and Health Sciences, Universitat Internacional de Catalunya, 08195 Sant Cugat del Vallès, Spain
2 INSERM, Neurocentre Magendie, U1215, University of Bordeaux, 3300 Bordeaux, France; cristina.miralpeix@inserm.fr
* Correspondence: rrodriguez@uic.es; Tel.: +34-935-042-002

Citation: Rodríguez-Rodríguez, R.; Miralpeix, C. Hypothalamic Regulation of Obesity. *Int. J. Mol. Sci.* **2021**, *22*, 13459. https://doi.org/10.3390/ijms222413459

Received: 9 December 2021
Accepted: 13 December 2021
Published: 15 December 2021

Publisher's Note: MDPI stays neutral with regard to jurisdictional claims in published maps and institutional affiliations.

Obesity has now reached pandemic proportions and represents a major socioeconomic and health problem in our societies. Up to now, obesity has been considered a medical issue only in high-income countries, but this disease is now dramatically on the rise in low- and middle-income countries, particularly in urban settings, affecting both adults and the pediatric population [1].

The simplest definition of overweight and obesity is an excessive accumulation of fat due to an imbalance between the energy in and the energy out, and this adiposity represents a risk to health [1]. But the whole picture is more complex than this, and obesity is a marker of a serious metabolic dysregulation that involves several diseases such as type 2 diabetes, cancer, cardiovascular pathologies, and most recently for COVID-19 infection [1,2]. Since the current approaches to combatting obesity and its complications have limited clinical effectiveness, gaining insight into the cellular and molecular basis of obesity could lay the foundations for the development of new strategies to prevent metabolic disruption and to treat this somewhat unaddressed medical issue.

In the last few decades, it has been strongly demonstrated that the hypothalamus is the master regulator of energy homeostasis. The hypothalamus contains hormone- and nutrient-sensing nuclei that organize central and peripheral responses for maintaining normal body weight, food intake, energy expenditure, and nutrient partitioning. Within the hypothalamus, specialized sub-populations of neurons are connected to each other and to various extrahypothalamic regions to coordinate energy balance. Evidence has also highlighted the participation of non-neuronal populations (i.e., microglia and astrocytes), and even the interesting cross-talk between these types of brain cells, whose disruption leads to insulin resistance and obesity. In this Special Issue, we report on the most recent insights into the hypothalamic circuitries and pathways involving neurons [3–7], astrocytes [8,9], and microglia [10,11] in obesity development and associated complications. The emerging contribution of astrocyte–neuron [9] and microglia–neuron cross-talks [12] in the hypothalamus and the contribution of microbiota and the gut-brain axis controlling food intake and energy homeostasis [13] are also presented in this Special Issue (Figure 1).

Regarding the importance of hypothalamic targets in obesity development, Quiñones et al. [3] elegantly review the most relevant studies focusing on the deacetylases sirtuins in hypothalamic neurons, SIRT1 and SIRT6, as multifaceted mediators of energy metabolism, controlling processes such as food intake, food preference, puberty, body weight, adiposity, and glucose and insulin homeostasis. Importantly, the complexity of the hypothalamic nuclei in obesity development is supported by the fact that the effect of SIRT1 on energy balance depends on the neuronal type where it is acting. For instance, lacking SIRT1 or SIRT6 in POMC neurons affects energy expenditure and adiposity; however, SIRT1 in AgRP neurons specifically affects eating behavior, and in oxytocin neurons in the PVN controls diet preference. In addition to sirtuins, Fosch et al. [4] provide new insights into other targets in SF1 neurons of the ventromedial hypothalamus that play a critical role not only in body weight control and adiposity, but also in glucose tolerance and insulin and leptin sensitivity, with minimal effects on food intake and differential actions between male and female mice. The authors also report how nutrient sensors such as AMPK and SIRT1, glutamatergic transmission, synaptic receptors, and mediators of autophagy in SF1 neurons are promising therapeutic targets against obesity and diabetes, reinforcing the "central role" of the hypothalamus in the control of peripheral metabolism beyond the regulation of feeding behavior [4]. In this sense, Fukumura et al. [5] found new evidence in this special issue about the effects of a novel hypothalamic small protein, named neurosecretory protein GL (NPGL), on glucose and insulin homeostasis. In this study, overexpression of *Npgl* in the mediobasal hypothalamus of mice improved glucose tolerance and attenuated insulin resistance and hyperglycemia under high-fat diet (HFD) exposure, without significant changes in food intake. The concomitant increased mRNA expression levels of galanin (a neuropeptide that co-localizes with NPGL in arcuate neurons and regulates glucose homeostasis) suggests that both peptides could regulate each other for these actions [5].

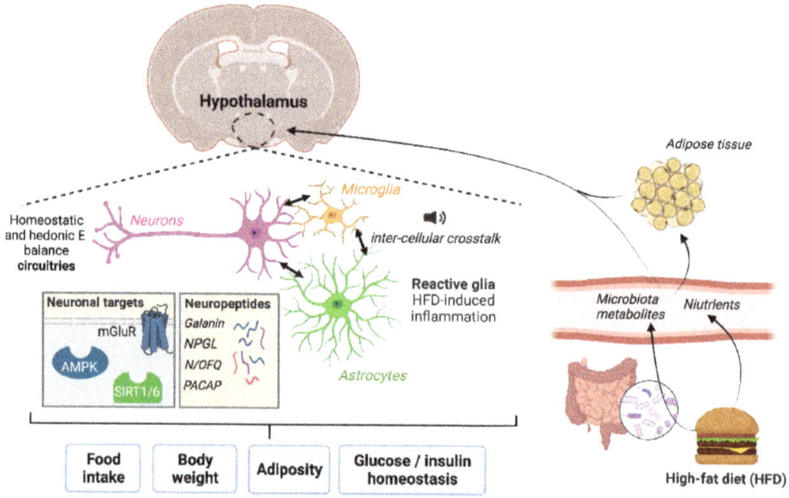

Figure 1. In this Special Issue, we report on the most recent insights into the neuronal targets and neuropeptides of the hypothalamic circuitries involved in obesity development and progression. In addition, the emerging contribution of astrocyte-neuron and microglia-neuron cross-talks in the hypothalamus in response to a high-fat diet, and the implication of the gut-brain axis controlling food intake and energy homeostasis are also presented

When investigating the hypothalamic regulation of obesity, in addition to the identification of potential targets mediating the homeostatic circuits, it is also important to explore the hedonic circuits mediating energy balance, identifying the salient neural, hormonal, and humoral components involved, as remarkably reviewed by Gastelum et al. [7]. They explore the sexual dimorphism on these circuits, paying special attention to the role of two emerging neuropeptides, the nociception/orphanin FQ and the pituitary adenylate cyclase-activating polypeptide (PACAP) in the neuronal activity regulation in positive and negative energy states. Jang et al. [6] also investigate leptin-mediated feeding circuits in hypothalamic neurons, identifying the expression of the angiopoietin-like growth factor (AGF) in POMC neurons of the arcuate nucleus as a downstream factor involved in leptin signaling in the hypothalamus.

Hypothalamic circuits controlling energy balance in response to feeding are also mediated by the gut-to-brain nutrient signaling, whose disruption leads to obesity. As extensively reported by Romani-Perez et al. [13], aberrant feeding patterns or unhealthy diets might alter gut microbiota–diet interactions and modify nutrient-sensing information from the gut to the hypothalamus, impairing energy homeostasis. They identify microbiome-based strategies to improve the gut-brain axis function and hence combat obesity.

An already well-established concept in the neurobiology of obesity is that neurons are not the only characters playing a main role in the response processing and the transmission of information. This is also true in the hypothalamus, where a maladaptive interaction between neurons and their surrounding microglia and astrocytes likely contribute to the development and progression of obesity [9,12]. The perspective elegantly presented by Leon et al. [12] discusses recent evidence that sheds light on the cellular and molecular mechanisms underlying microglia–neuron communication in hypothalamic circuits that are crucial for body weight and food-intake control in response to HFD exposure. Interestingly, microglia, as key players in the immunological response, are sensitive to fuel substrate and can influence neuronal activity in part via cytokines. However, the maladaptive microglia–neuron cross-talk observed in response to overnutrition seems to involve synergy with astrocytes, which are key components of the tripartite synapse, as discussed below. In addition to these cellular interactions, Mendes et al. [10] examine the subtypes of microglia that may be involved in HFD-induced hypothalamic inflammation in the early stages of the disease. They also remarkably discuss which models can be useful for obtaining the most reliable data when exploring distinct subsets of microglia, and clarify the changes in hypothalamic microglia signature in response to HFD-induced obesity based on novel transcriptomic analysis. The importance of microglia activation in the hypothalamus in energy balance has also been reported by Lopez-Gambero et al. [11]. They found that glial activation and metabolic dysfunction in the hypothalamus of a mouse model of Alzheimer's disease (5XFAD mouse) leads a negative energy balance and lower insulin and related hormones in plasma, contributing to the development of the pathology. Interestingly, this phenotype was more marked in female than male mice, indicating useful information for early detection of the disease and sexual dimorphism in the pathophysiology of Alzheimer's disease.

The function of hypothalamic astrocytes and the interplay between astrocytes and neurons are also becoming very important in the study of obesity development. In this special issue, Song et al. [8] show the ability of the adipose tissue-derived hormone adiponectin to control nutrient metabolism in hypothalamic astrocytes by enhancing glucose uptake, glycolytic activity, and lactate and ketone body production. In this study, central administration of adiponectin increases the number of astrocytes in the hypothalamus of mice. The most relevant findings regarding molecular mechanisms by which hypothalamic astrocytes are involved in the pathogenesis of obesity have been well reviewed by Gonzalez-Garcia et al. [9]. They expose the idea of how a disruption in the astrocyte–neuron communication within the hypothalamus is altered by exposure to obesogenic insults and the impact this has on obesity development. In addition, they point out the emerging idea that astrocytes are heterogeneous and that there might exist subpopulations that contribute to specific actions. Therefore, the identification of specific hypothalamic astrocyte subpopulations might be critical for the pathophysiology of obesity [9]. The study of the potential role of other brain cells and how they are interconnected with glia and neuronal cells in the hypothalamic regulation of obesity is also promising (e.g., tanycytes and endothelial cells) [12].

Therefore, the manuscripts published in this Special Issue provide new insights into the hypothalamic neuronal and non-neuronal pathways involved in the pathophysiology and development of obesity and related diseases such as diabetes (Figure 1). Via four original papers, six review manuscripts, and one perspective, this Special Issue will substantially contribute to the identification of novel therapeutic targets in the hypothalamus and novel approaches to combatting the progression of human obesity.

Author Contributions: All authors (R.R.-R. and C.M.) have participated in writing and editing. All authors have read and agreed to the published version of the manuscript.

Funding: This work was supported by the Spanish Ministerio de Economía, Industria y Competitividad (MINECO), Agencia Estatal de Investigación (AEI), and Fondo Europeo de Desarrollo Regional (FEDER) (Grants PCI2018-092997/AEI to R.R.-R.; PID2020-114953RB-C22 to R.R.-R. by MCIN/AEI/10.13039/501100011033).

Institutional Review Board Statement: Not applicable.

Conflicts of Interest: The authors declare no conflict of interest.

References

1. World Health Organization. Obesity and Overweight. Available online: https://www.who.int/news-room/fact-sheets/detail/obesity-and-overweight (accessed on 19 April 2021).
2. World Health Organization. Diabetes. Available online: https://www.who.int/news-room/fact-sheets/detail/diabetes (accessed on 19 April 2021).
3. Quiñones, M.; Martínez-Grobas, E.; Fernø, J.; Pérez-Lois, R.; Seoane, L.M.; Al Massadi, O. Hypothalamic Actions of SIRT1 and SIRT6 on Energy Balance. *Int. J. Mol. Sci.* **2021**, *22*, 1430. [CrossRef]
4. Fosch, A.; Zagmutt, S.; Casals, N.; Rodríguez-Rodríguez, R. New Insights of SF1 Neurons in Hypothalamic Regulation of Obesity and Diabetes. *Int. J. Mol. Sci.* **2021**, *22*, 6186. [CrossRef]
5. Fukumura, K.; Narimatsu, Y.; Moriwaki, S.; Iwakoshi-Ukena, E.; Furumitsu, M.; Ukena, K. Effects of Overexpression of Neurosecretory Protein GL-Precursor Gene on Glucose Homeostasis and Insulin Sensitivity in Mice. *Int. J. Mol. Sci.* **2021**, *22*, 4681. [CrossRef] [PubMed]
6. Jang, Y.; Heo, J.Y.; Lee, M.J.; Zhu, J.; Seo, C.; Go, D.H.; Yoon, S.K.; Yukari, D.; Oike, Y.; Sohn, J.-W.; et al. Angiopoietin-Like Growth Factor Involved in Leptin Signaling in the Hypothalamus. *Int. J. Mol. Sci.* **2021**, *22*, 3443. [CrossRef] [PubMed]
7. Gastelum, C.; Perez, L.; Hernandez, J.; Le, N.; Vahrson, I.; Sayers, S.; Wagner, E.J. Adaptive Changes in the Central Control of Energy Homeostasis Occur in Response to Variations in Energy Status. *Int. J. Mol. Sci.* **2021**, *22*, 2728. [CrossRef]
8. Song, N.; Jeong, D.Y.; Tu, T.H.; Park, B.S.; Yang, H.R.; Kim, Y.J.; Kim, J.K.; Park, J.T.; Yeh, J.-Y.; Yang, S.; et al. Adiponectin Controls Nutrient Availability in Hypothalamic Astrocytes. *Int. J. Mol. Sci.* **2021**, *22*, 1587. [CrossRef]
9. González-García, I.; García-Cáceres, C. Hypothalamic Astrocytes as a Specialized and Responsive Cell Population in Obesity. *Int. J. Mol. Sci.* **2021**, *22*, 6176. [CrossRef] [PubMed]
10. Mendes, N.F.; Jara, C.P.; Zanesco, A.M.; de Araújo, E.P. Hypothalamic Microglial Heterogeneity and Signature under High Fat Diet–Induced Inflammation. *Int. J. Mol. Sci.* **2021**, *22*, 2256. [CrossRef] [PubMed]
11. López-Gambero, A.J.; Rosell-Valle, C.; Medina-Vera, D.; Navarro, J.A.; Vargas, A.; Rivera, P.; Sanjuan, C.; Rodríguez de Fonseca, F.; Suárez, J. A Negative Energy Balance Is Associated with Metabolic Dysfunctions in the Hypothalamus of a Humanized Preclinical Model of Alzheimer's Disease, the 5XFAD Mouse. *Int. J. Mol. Sci.* **2021**, *22*, 5365. [CrossRef] [PubMed]
12. Léon, S.; Nadjar, A.; Quarta, C. Microglia–Neuron Crosstalk in Obesity: Melodious Interaction or Kiss of Death? *Int. J. Mol. Sci.* **2021**, *22*, 5243. [CrossRef] [PubMed]
13. Romaní-Pérez, M.; Bullich-Vilarrubias, C.; López-Almela, I.; Liébana-García, R.; Olivares, M.; Sanz, Y. The Microbiota and the Gut–Brain Axis in Controlling Food Intake and Energy Homeostasis. *Int. J. Mol. Sci.* **2021**, *22*, 5830. [CrossRef] [PubMed]

International Journal of *Molecular Sciences*

MDPI

Perspective

Microglia–Neuron Crosstalk in Obesity: Melodious Interaction or Kiss of Death?

Stéphane Léon, Agnès Nadjar and Carmelo Quarta *

Neurocentre Magendie, Physiopathologie de la Plasticité Neuronale, INSERM U1215, University of Bordeaux, 33000 Bordeaux, France; stephane.leon@inserm.fr (S.L.); agnes.nadjar@inserm.fr (A.N.)
* Correspondence: carmelo.quarta@inserm.fr

Abstract: Diet-induced obesity can originate from the dysregulated activity of hypothalamic neuronal circuits, which are critical for the regulation of body weight and food intake. The exact mechanisms underlying such neuronal defects are not yet fully understood, but a maladaptive cross-talk between neurons and surrounding microglial is likely to be a contributing factor. Functional and anatomical connections between microglia and hypothalamic neuronal cells are at the core of how the brain orchestrates changes in the body's metabolic needs. However, such a melodious interaction may become maladaptive in response to prolonged diet-induced metabolic stress, thereby causing over-feeding, body weight gain, and systemic metabolic perturbations. From this perspective, we critically discuss emerging molecular and cellular underpinnings of microglia–neuron communication in the hypothalamic neuronal circuits implicated in energy balance regulation. We explore whether changes in this intercellular dialogue induced by metabolic stress may serve as a protective neuronal mechanism or contribute to disease establishment and progression. Our analysis provides a framework for future mechanistic studies that will facilitate progress into both the etiology and treatments of metabolic disorders.

Keywords: microglia–neuron communication; hypothalamus; obesity; metabolism

Citation: Léon, S.; Nadjar, A.; Quarta, C. Microglia–Neuron Crosstalk in Obesity: Melodious Interaction or Kiss of Death? *Int. J. Mol. Sci.* **2021**, 22, 5243. https://doi.org/10.3390/ijms22105243

Academic Editors: Rosalía Rodríguez-Rodríguez and Cristina Miralpeix

Received: 22 April 2021
Accepted: 13 May 2021
Published: 15 May 2021

1. Introduction

Over the last 25 years, since the discovery of the hormone leptin [1], enormous advancements have been made in understanding how the brain directs changes in feeding behaviors and systemic metabolic pathways to maintain the body's metabolic needs [2]. This advancement has initiated a consensus on a "brain-centric" view, which now considers that obesity does not only result from the dysfunctional activity of peripheral organs, such as the adipose tissue, but can also originate from the brain [3].

Large-scale genomic studies have demonstrated that body weight excess and the associated systemic metabolic defects can derive from spontaneous mutations that involve genes that are expressed in the central nervous system (CNS) [3]. Moreover, multiple lines of evidence have now demonstrated that hypercaloric diet feeding can trigger obesity and multiple metabolic comorbidities via central mechanisms of action [4–7].

Most of these mechanisms involve changes in the activity of neuronal networks located in the hypothalamus that influence fuel (food) ingestion, dissipation, or storage [8–10]. Brain-to-periphery axes controlled by this brain region allow maintenance of constant levels of energy reserves, or the so-called condition of energy balance. The melanocortin circuit, which is formed by different neuronal populations that respectively express the peptidergic precursor pro-opio-melanocortin (POMC) or the neuropeptide agouti-related protein (AgRP), plays a crucial role. When the synaptic plasticity of this circuit is impaired—for instance, in response to prolonged feeding with hypercaloric diets—this neurobiological defect can lead to energy balance dysregulation [5,6], and, therefore, to the establishment or the progression of diet-induced obesity (DIO) [7,11].

The pivotal role played by the hypothalamus in the etiology of obesity and its associated sequelae may not seem surprising given that this brain area has long been known to influence food intake and energy handling. More surprisingly, however, non-neuronal cells may be the key piece of the puzzle, at least based on more recent information.

Astrocytes have classically been known to participate in the so-called "tripartite synapse", e.g., the integrated functional unit whereby these glial cells communicate with two or more neurons to accommodate their synaptic transmission [12]. Within this cell-to-cell process of communication, however, microglial cells are emerging as key players given their influence on the formation, function, plasticity, and elimination of synapses [13,14].

During overfeeding, microglial cells accumulate in the hypothalamus [15] ("microgliosis"), undergo morphological activation, and augment the production of proinflammatory cytokines and other neuronally secreted factors that potentially interfere with synaptic transmission [16]. These changes seem to occur specifically in the hypothalamus, but not in other brain areas [16], suggesting that a "healthy" microglial–neuronal crosstalk may be crucial for hypothalamus-based regulation of energy balance, whereas a defective process of intercellular communication may promote neuronal dysfunction, energy imbalance, and ultimately obesity.

Our understanding of the role of microglia–neuron interactions in modulating neuronal function in extrahypothalamic brain areas has significantly improved over the last decade [17]. However, the molecular and cellular underpinnings of such intercellular dialogue at the level of specific hypothalamic circuits involved in body weight control have only begun to emerge.

Here, we will discuss recent evidence that sheds new light on the cellular and molecular mechanisms implicated in microglia–neuron communication in hypothalamic circuits that are relevant for body weight and food intake regulation. We will interrogate whether a defective dialogue between hypothalamic microglia and neurons may be causally linked with the etiology of DIO, while also highlighting the remaining outstanding questions on how this intercellular cross-talk operates to maintain energy homeostasis in the whole body.

2. The Emerging Neuroimmune Theory of Obesity

Bidirectional interactions between the immune system and the CNS enable immunological, physiological, and behavioral responses. Upon activation by infection or injury, for instance, peripheral cells of the innate immune system synthesize and release cytokines that serve as immune mediators. These peripheral signals travel to the brain and act through complex mechanisms of microglia–neuron communication, ultimately leading to non-specific symptoms of infection, including lethargy or listlessness, amongst others, which are commonly referred to as the "sickness response" [18].

In animal models of DIO, enhanced sickness behavior is observed after a single administration of the bacterial lipopolysaccharides (LPSs) [19]. This phenotype is associated with increased production of proinflammatory mediators in the hypothalamus, including tumor necrosis factor alpha (TNFα), interleukin 1 beta (IL-1β), interleukin 6 (IL-6), the inhibitor of nuclear factor kappa alpha (IκBα), and cyclooxygenase-2 (COX2) [20], thereby suggesting that whole-body neuroimmune axes coordinated by CNS microglia undergo defective regulation in obesity.

In addition to responding to injury or infections, CNS microglia can be also activated by dietary or hormonal factors that influence whole-body energy handling, including, for instance, lipids [21,22], carbohydrates [23], or the hormone leptin [20,24]. This suggests that microglial cell activity may play a critical role in maintaining systemic energy homeostasis. After long-term feeding with hypercaloric diets, an increased number of activated microglial cells has been consistently observed across different studies at the level of the mediobasal hypothalamus, where POMC and AgRP neurons reside [4,25–31]. These microglial changes are accompanied by the upregulation of cytokine-signaling-related pathways in hypothalamic neurons [4,25–31], and are not a rodent-specific phenomenon. Indeed, human obesity is also associated with radiological signs of gliosis [4,32,33]. Moreover, histological analyses of post-mortem brain tissues obtained from obese individuals has revealed the presence of microglial cells with aberrant morphologies at the level of hypothalamic areas [34]. Accordingly, body mass index (BMI) positively correlates with microglial soma size in subjects affected by type-2 diabetes (T2D), which is one of the major glucometabolic complications of obesity [35].

Based on these accumulated findings, one emerging theory in the field is that the exacerbated microglial activity induced by overfeeding may sustain and propagate a condition of low-grade chronic inflammation at both the central (hypothalamus) and systemic level, which contributes to the establishment and the progression of DIO and its associated glucometabolic complication. Several observations support this model: (i) inhibiting expansion of CNS microglia expansion in DIO mice hindered diet-induced weight gain and prevented central and peripheral inflammatory responses induced by overfeeding [15]; (ii) pharmacologically depleting microglia or selectively restraining microglial inflammatory signaling pathways in the hypothalamus sharply reduced hypothalamic microgliosis while also limiting diet-induced hyperphagia and weight gain in a murine model of DIO [36]; (iii) several molecules and pathways have been identified as candidate mediators of hypothalamic inflammatory reactions observed during chronic feeding with a hypercaloric diet (for a detailed review on this topic, see [37]). The targeted disruption of these pathways in a hypothalamus-specific manner, by pharmacologic or genetic means, limits the extent of DIO and its main systemic metabolic defects, e.g., impaired systemic glucose homeostasis [26–30,38].

Of note, emerging microglia-directed neuroimmune responses induced by diet and/or obesity share elements of similarity with other "classic" neuroimmune axes. For instance, a feature of the sickness behavior is the enhanced entry of peripheral immune cells into the CNS [18]. Likewise, during high-fat diet (HFD) consumption, changes in the peripheral and central immune systems are intimately interconnected.

In DIO mice, blood-borne monocyte-derived cells can extravasate the vasculature and enter the brain at the level of the hypothalamic arcuate nucleus (ARC) of the hypothalamus [36], which contains a leaky blood–brain barrier [39]. As they accumulate, these infiltrating cells resume a typical microglia morphology and contribute to propagating the diet-induced inflammatory response [36].

Additionally, diet-induced hypothalamic inflammation, and, similarly, other microglial neuroimmune responses, can have both positive and negative downstream neuronal effects. After brain injury or pathogens invasion, an initial microglial activation is typically observed and is necessary for neuroprotection, but increased or prolonged microglial activity can have neurotoxic effects [40]. Likewise, short-term microglial responses to diet-induced metabolic stress may initially be protective for neurons that are in strict communication with these glial cells, whereas in the long term, the communication process may become maladaptive, thus negatively influencing neurotransmission [41]. Supporting this view, ARC microglial activity and the resulting local increase in proinflammatory cytokine levels observed during DIO progression follow a dynamic pattern, as they occur rapidly during the first hours [42] and days [4] of overfeeding, are normalized for a few weeks during the dietary administration, and rise again thereafter [4]. Moreover, the establishment of a "low-grade" chronic microglial activation state in response to prolonged feeding with hypercaloric diets is often associated with hallmarks of hypothalamic neuronal dysfunction in animal models after long-term feeding of hypercaloric diets, including loss of synapses [5], impaired responsiveness to metabolic hormones [7,43,44], altered intracellular organelles function [4,29,45,46], and possibly cell death [47,48], although the latter was not observed in all of the studies [49].

Thus, an intimate and dynamic relationship links exacerbated hypothalamic microglial activity with hypothalamic neuronal dysfunction and DIO progression. Such a neuroimmune basis of obesity may either originate from or lead to a defective process of communication between neurons and surrounding microglial cells.

3. Role of Microglia in the Remodeling of Neuronal Networks in Obesity

Microglia can phagocytize neuronal spines, terminals, and cell bodies, and errors in this process can lead to impaired neuronal activity and possibly to cell death [13,50]. In response to pathogen invasion or neuronal damage, phagocytosis of microorganisms, dying cells, or neuronal debris from microglia occurs during the resolution phase of inflammation and involves the release of cytokines and lipid mediators that exert anti-inflammatory and prorepair properties.

Microglia can also remove synaptic elements from neuronal cell bodies, a process that allows the protection of neurons from excitotoxicity, e.g., the neuronal death caused by the overactivation of excitatory amino acid receptors [13]. Early in the course of a disease's process, it might be useful to strip away dying synapses so that healthy synaptic elements can take over. However, as the insult to the CNS persists, abnormal synapse removal can translate into neurodegenerative cellular states [13].

Consumption of dietary fat has been proposed to induce apoptosis of neurons in the arcuate nucleus, paraventricular nucleus, and lateral hypothalamus in mice [47,48], and such a neuronal loss is often paralleled by microgliosis [47,48], which makes one wonder if diet-induced neuronal death and exacerbated microglial activity in the hypothalamus are the two faces of the same phenomenon. However, the observation that DIO is associated with hypothalamic neuronal death has not been reproduced by all the studies [49]. This inconsistency may be due to differences in the diet-composition, caloric content, length of diet exposure, and technical approach used to assess neuronal counts across these studies [7]. Hence, only addressing whether and how these different contributing factors impact microglial phagocytic capacity and, eventually, hypothalamic neuronal death may provide a final answer.

Irrespective of this dispute, changes in microglial activity induced by overfeeding may have more "subtle" consequences on downstream neurons—for instance, on their synaptic transmission. LPS-mediated activation of microglial cells evokes excitatory responses in hypothalamic POMC neurons and inhibitory responses in AgRP neurons [51]. HFD-induced obesity is associated with the synaptic reorganization of hypothalamic POMC and AgRP neurons [5], which present fewer total synapses on their perikaryal [5]. These synaptic changes are intimately associated with the presence of increased hypothalamic microglial activity [5], which suggests that microglial-mediated effects may be directly implicated, although no direct causal evidence supporting this model is yet available.

Changes in microglial-mediated phagocytosis of synaptic elements (or microglial "pruning") seem to occur in response to overnutrition in extrahypothalamic areas—for instance, in the hippocampus. In a DIO animal model, low-grade chronic inflammation induced by a hypercaloric diet was associated with reduced cognitive function and with increased microglial pruning [52,53]. Memory impairment observed in these animals occurred along with reductions in hippocampal dendritic spines [53], and both the memory defects and the loss of neuronal spines could be rescued through pharmacological manipulation of microglial phagocytosis [53].

Whether or not microglia-mediated pruning contributes to the specific hypothalamic neuronal dysfunction observed during DIO has not yet been directly tested, but this hypothesis is supported by a series of indirect observations. For instance, in an animal model of obesity and type 2 diabetes (db/db mice), which was characterized by impaired hypothalamic neuronal synaptic plasticity [54,55], hypothalamic microglial cells presented a reduced expression of the phagocytic marker cluster of differentiation 68 (CD68) [56]. In this model, the deficit in hypothalamic neuronal function and the resulting obesity phenotype derived from a defective action of the metabolic hormone leptin.

Likewise, transgenic mice with a specific leptin receptor deficiency in myeloid cells (which include microglia) presented an obesity-like phenotype, a reduced number of hypothalamic ARC POMC-expressing neurons, and signs of impaired microglial phagocytic capacity in the hypothalamus [24]. Thus, microglia-mediated synaptic remodeling of hypothalamic neuronal cells may be under the influence of hormonal signals, such as leptin.

This implies that alterations in these microglial–neuroendocrine axes may contribute to DIO pathophysiology. Of note, leptin can potentiate the microglial response to LPS in an in vitro system, and this is linked to morphological changes that render the microglia more reactive [57], further suggesting that this hormone may have an important role in mediating microglial functions.

Notably, in addition to influencing synapse homeostasis through their phagocytic activity, microglial cells can also promote synapse formation [13], although no studies have thus far investigated whether changes in microglial-mediated synaptogenesis occur during diet-induced metabolic stress, which represents an interesting angle for future investigations.

Thus, hypercaloric diets may trigger microglial-mediated synaptic remodeling, which possibly disrupts the activity of AgRP/NPY and POMC neurons, thereby contributing to disease development. However, the defective synaptic plasticity observed in the ARC of DIO animal models likely involves synergy among not only neurons and microglia but also astrocytes [5], which are key components of the tripartite synapse. Thus, studying cell-type-specific mechanisms that link diet-induced inflammation with the impaired synaptic plasticity of ARC neurons is an interesting avenue for future research.

4. Impact of Fuel Substrates on Microglia–Neuron Communication in Obesity

Microglia are highly dynamic and continuously move around the brain parenchyma in response to chemotactic signals. Under activated states, these glial cells not only display increased phagocytic activity, but also augment the production of proinflammatory and neuromodulatory factors, which are all processes that require large amounts of energy. To meet these continuous and rapid changes in energy demands, microglia express transporters for glucose, fatty acids, and amino acids, the three main fuel substrates. Moreover, these glial cells are also capable of metabolizing alternative nutrients, such as glutamine [58].

The exact identities of the main extracellular influencing factors leading to reactive microglial states during overfeeding are under intense investigation and have not yet been clarified. However, saturated fat (SF) may be a key piece of the puzzle. In animals fed with a saturated fat (SF)-enriched diet, fatty acids, including palmitic acid, are conveyed to the brain after ingestion. They are then taken up by hypothalamic microglia, and ultimately contribute to inducing the hypothalamic inflammatory reaction [22].

Feeding mice with an SF-rich diet augments the expression of the heat-shock protein 72 (Hsp72), which is a chaperone protein that is implicated in cellular stress, in the mediobasal hypothalamus [22]. Such a molecular stress response is independent of the increased caloric intake associated with the SF-rich diet. Indeed, increased neuronal Hsp72 levels were observed by administering milk fat to control mice fed with a standard chow diet, but not following the administration of isocaloric olive oil [22]. Thus, increased SF intake—and not fat intake per se—accounts for the presence of signs of hypothalamic neuronal stress. Of note, depleting microglia via a genetic approach nearly abolishes the marked induction of neuronal Hsp72 that is otherwise seen in ARC neurons following SF-rich administration [22]. This suggests that changes in microglial SF sensing causally contribute to the neuronal dysfunction observed in DIO.

Saturated fat is likely not the only fuel substrate that mediates diet-induced microglial activation and the resulting negative downstream neuronal effects. Indeed, long-term feeding with a high-carbohydrate/high-fat diet (HCHF) also leads to increased hypothalamic microglial activity in mice, which is associated with the elevated production of advanced glycation end-products (AGEs) from hypothalamic POMC and AgRP neurons. AGEs are products of glycoxidation and lipoxidation reactions that originate from overexposure to sugars. These bioproducts are implicated in many degenerative conditions, including diabetes and Alzheimer's disease [59]. Certain types of AGEs—for instance, the product N(ε)-(carboxymethyl)-lysine (CML)—are mainly produced by neurons during diet-induced cellular stress, whereas their receptors are specifically localized in non-neuronal cells, including, but not limited to microglia [23]. Genetic deletion of CML receptors in mice fed with an HCHF diet resulted in less microglial reactivity in the hypothalamic ARC and also led to favorable antiobesity effects [23]. Thus, a putative model can be proposed whereby HCHF diets—and the consequent combined overload of both lipids and glucose in the mediobasal hypothalamus—trigger excessive production of neuronal metabolites (such as AGEs) that activate surrounding microglial cells. This results in a vicious feedback loop that impairs neuronal functions.

However, what are the main intracellular underpinnings of such a maladaptive crosstalk? A final answer is not yet available, but common intracellular nodes may integrate changes in fuel substrate handling between neurons and microglia. In both microglial and neuronal cells, maladaptive changes in mitochondrial function and cellular fuel utilization were observed after hypercaloric diet feeding, and in both cell types, these changes were, in part, mediated by the mitochondrial protein uncoupling protein 2 (UCP2) [60]. Of note, increased UCP2-dependent activity in either microglial or neuronal cells played a key role in mediating DIO susceptibility in murine models [61–63].

Thus, sensing of specific dietary components, such as fat and/or sugar, may lead to exacerbated cellular fuel metabolism in both microglial and neuronal cells located in the hypothalamus via common (UCP2-mediated) molecular pathways that are essential for proper mitochondrial activity. This implies that changes in fuel substrate uptake and/or metabolism induced by diet-induced caloric overload may directly impact the microglia–neuron crosstalk via mitochondria-mediated mechanisms and the subsequent production of yet-to-be identified energy metabolites that contribute to the intercellular communication.

Supporting this view, specific microglial genetic deletion of lipoprotein lipase (LPL), one of the key enzymes involved in cellular fuel uptake, caused mitochondrial dysmorphologies in both microglia and nearby ARC neurons in animals that were fed with a hypercaloric diet [64]. These transgenic mice that specifically lacked LPL activity in microglia also presented higher body weight gain relative to wild-type animals when fed with an obesogenic diet [64].

5. Role of Cytokines in Microglia–Neuron Communication in Obesity

Activated hypothalamic microglia produce a variety of proinflammatory cytokines that potentially influence metabolic or behavioral outputs by inducing changes in neuronal activity [65–67]. In vitro incubation of hypothalamic POMC neurons with the microglial-produced cytokine TNF-α led to an increased neuronal firing rate, increased intracellular ATP production, and mitochondrial fusion processes [68]. HCHF diet consumption is associated with elevated hypothalamic TNF-α levels [69] and with signs of POMC neuronal mitochondrial stress [68]. Hence, persistently elevated TNF-α signaling during DIO may cause functional impairment of hypothalamic POMC neurons as a result of alterations in neuronal mitochondrial activity [68]. This hypothesis is supported by multiple studies that showed that DIO leads to mitochondrial dysfunction in POMC neurons and that this specific pathogenic mechanism may promote dysregulated neuronal activity and obesity [7]. The molecular underpinnings of TNF-α-mediated neuronal dysfunction in obesity have yet to be fully elucidated, but cytokine-mediated changes in the c-Jun N-terminal kinase (JNK)/activator protein 1(AP-1) and NF-κB signal transduction pathways may have a central role. A detailed analysis of the neuronal molecular pathways implicated in hypothalamic inflammation is beyond the main scope of the present work, but more detailed information can be found elsewhere—for instance, in [37].

Although it is tempting to speculate that proinflammatory cytokines produced by hypothalamic microglial cells in response to overfeeding may globally contribute to the induction of neuronal dysfunction and obesity, this hypothesis appears to be too simplistic given that certain microglia-derived cytokines can have positive effects on metabolism and body weight regulation. For instance, activated hypothalamic microglia can produce the proinflammatory cytokines IL6 and IL-1β [70], which can have favorable effects on energy balance regulation. Administration or overexpression of these cytokines in the brain results in reduced food intake, increased energy expenditure, and decreased body weight gain [71–75]. This is likely due to changes in the activity of neurocircuitries located in the hypothalamus [76,77]. Accordingly, the central application of IL6 led to potent feeding suppression and to improvements in glucose tolerance in DIO mice via the modulation of neurons located in the paraventricular nucleus of the hypothalamus [78]. Thus, different proinflammatory cytokines can have either positive or negative effects on body weight control, which makes one wonder whether diet-induced hypothalamic inflammation is necessarily a pathophysiological mechanism that contributes to disease progression.

The fact that microglia-derived proinflammatory cytokines may induce dynamic downstream neuronal effects is probably part of the answer. For instance, among families of cytokines that orchestrate hypothalamic inflammatory reactions induced by overfeeding, chemokines act by attracting immune cells to the inflamed site [79]. When proinflammatory cytokines activity is enhanced in the brain of an animal, this generally leads to a condition of initial high-grade inflammation and to a transitory reduction in food intake, which is partly due to changes in the activity of hypothalamic neurons [80,81]. Nevertheless, such acute and robust inflammatory responses and the linked food-intake-reducing effect generally tends to diminish or, in some cases, to fade away a few hours after administration [80,81]. The metabolic or behavioral effects observed after short-term hypothalamic cytokines action during hypercaloric diets feeding may reflect a defense mechanism that limits the introduction of unhealthy diets. Nevertheless, this mechanism of "neuroprotection" may become maladaptive in the long term and have the opposite effects, thereby contributing to disease establishment and progression. Accordingly, certain chemokines, such as the C-X3-C motif chemokine ligand 1 (CX3CL1, or 'fractalkine'), promote obesity when chronically administered in the brain [82], whereas inhibition of CX3CL1 actions in the hypothalamus of DIO mice through an approach of small interfering RNA reduced local inflammation and led to weight loss and to improvements in systemic glucose metabolism [82]. Moreover, the expression of the C-X-C motif chemokine 12 (CXCL12) and its receptors was induced in the hypothalamus of mice fed with an HFD [83]. Such increased CXCL12 activity has been suggested to favor overfeeding by enhancing the synaptic activity of specific neurons located in this hypothalamic area that produce the hunger-promoting neuropeptide enkephalin [83].

From a broader perspective, cytokine-mediated changes in microglia–neuronal communication likely also involve other non-neuronal cells. As previously discussed, DIO leads to hypothalamic astrocyte activation, and following activation, these non-neuronal cells produce a variety of immune mediators, including TNFα, IL-1β, IL-6, or the monocyte chemoattractant protein-1 (MCP-1) [39,84], amongst others.

When cultured in vitro, astrocytes accumulate lipid droplets under a free fatty acid (FFA)-rich environment, while also producing a large amount of proinflammatory cytokines [84]. If cultured microglial cells are treated with a conditioned medium obtained from lipid-laden astrocytes, this leads to a markedly enhanced chemotactic activity of the microglial cells, which is due to MCP-1-mediated actions at the level of the microglial C-C chemokine receptor type 2 (CCR2) [84]. Thus, astrocytes-derived inflammatory signals may have an impact on neighboring CNS microglia, thereby alimenting the hypothalamic inflammatory response. Given that the chemokine MCP-1 disrupts the integrity of the blood–brain barrier (BBB) [85] and that the increased brain infiltration of bone-marrow-derived monocytes/macrophages contributes to the induction of hypothalamic inflammation [36] (see above), this vicious cycle may be also influenced by peripheral immune cells.

In conclusion, a multicellular model can be proposed whereby, in response to hypercaloric diets, astrocytes, local microglial cells, and peripheral monocytes/macrophages co-direct changes in the local production of cytokines and chemokines, which ultimately impact hypothalamic neuronal activity and energy homeostasis in a highly dynamic manner during the progression of DIO.

6. "ON/OFF" Communication between Microglia and Neurons in Obesity

While long-term changes in the activity of CNS microglia can impair neuronal functions, the opposite relationship can be also relevant, as discussed herein. However, can hypothalamic neurons be initiators of vicious interactions with surrounding non-neuronal cells in response to diet-induced metabolic stress? If one considers that neuronal cells in mammals can release ON signals that promote microglial recruitment and activity [82,86], this hypothesis seems plausible.

For instance, the energy substrates adenosine 5′- triphosphate (ATP) and adenosine 5′-diphosphate (ADP) can be released from the neuronal cell body via membrane channels or activity-dependent vesicle exocytosis [87,88]. Microglia contain specific purinergic receptors (such as P2Y12) that are expressed at the level of structural interaction sites located between neuronal cell bodies and microglial processes [88]. P2Y12-containing purinergic junctions allow microglia to constantly monitor the neuronal status. During traumatic neuronal events (e.g., brain stroke) these junctions initiate microglia-driven neuroprotective actions in a targeted manner on the neuronal cell body [88]. Moreover, purinergic receptors are implicated in controlling microglial motility and attraction to synapses [89].

Similar mechanisms of spatially controlled neuronal surveillance and protection may take place in hypothalamic neurons in response to diet-induced neuronal stress. DIO leads to POMC neuronal mitochondrial dysfunctions [7,68] and to a higher number of activated microglial cells that are in close contact with POMC neurons [68], e.g., potential sites of P2Y12-mediated communication. However, no studies have, thus far, directly addressed whether hypothalamic neuronal cells display a plastic ability to release purinergic signals that influence microglia recruitment and functions, which represent an interesting area for future investigation.

Of note, microglial cells in the healthy CNS are immunologically more quiescent than peripheral macrophages [90], which raises the question of whether brain neurons may release OFF signals that keep these glial cells in a non-active state. In hippocampal brain slices, the induction of the major histocompatibility complex class II (MHC-II) in microglia, which is indicative of increased microglial immune response [91], is restricted by neurotrophins released from electrically active neurons, for instance, by the brain-derived neurotrophic factor (BDNF [92]).

BDNF-producing hypothalamic neurons play a major role in energy balance regulation in response to the hormone leptin [93]. Accordingly, mutations in the BDNF gene lead to insatiable appetite and severe obesity [94]. Hence, BDNF might be a neuronal OFF signal that maintains a homeostatic process of neuron–microglia communication in the hypothalamus, while defective BDNF production may ultimately lead to exacerbated microglial activity and obesity.

In addition to neurons, microglia can also release BDNF, for instance, in response to the activation of ATP-sensitive purinergic receptors [95]. Moreover, BDNF-producing hematopoietic cells can migrate from circulation to the hypothalamus; they can make direct contact with neurons in response to the feeding status, thereby influencing whole-body metabolic and behavioral responses [96]. Mice with congenital BDNF deficiency specifically in these hematopoietic cells developed hyperphagia, obesity, and insulin resistance [96], suggesting that defective BDNF production from either neuronal and non-neuronal cells located in the hypothalamus may initiate a vicious cycle leading to defective neuronal functions and obesity.

Neurotransmitters may represent an additional layer of reciprocal ON/OFF regulation between microglia and neurons. Microglia can express receptors for the neurotransmitters glutamate, gamma-aminobutyric acid (GABA), noradrenaline, or dopamine (amongst others) in both in vitro and ex vivo systems [97], although evidence for microglial expression of these markers in vivo is still lacking. In vitro treatment of microglial cells with most of these neurotransmitters inhibits LPS-induced release of various proinflammatory factors from these glial cells [86]. Nevertheless, these neurotransmitter-mediated effects on microglia activity are unlikely to be only directly mediated. Instead, these influences could be indirectly influenced by extracellular ATP, which is released in response to changes in neurotransmission [98].

While several neurotransmitters may act as an OFF-signals that keep microglia in a quiescent state, one exception may be the excitatory neurotransmitter glutamate. Glutamate can trigger activation and release of the proinflammatory cytokine TNF-α from microglial cells [99,100]. Faced with proinflammatory stimuli, such as TNF-α, microglia can, in turn, release glutamate [101], and thus probably enter a vicious cycle that leads to the increased generation of both glutamate and TNF-α. The reciprocal control of glutamate production by microglial and neuronal cells may be part of a feedback loop that protects neurons from excitotoxicity. Alternatively, neurotransmitter-mediated cross-talks between neurons and microglia may serve as a defense mechanism that limits exacerbated cytokine action or excessive microglial phagocytic activity [86]. Whether or not DIO leads to defective neurotransmitters-mediated microglia–neurons communication has yet to be directly addressed. Given that animal models of DIO display impaired hypothalamic expression of glutamate transporters [102], understanding how and whether changes in the balance between both ON and OFF signals orchestrate the (dys) functional dialogue between microglia and neurons in obesity warrants further investigation.

7. Conclusions and Perspectives

A better understanding of the neurobiological basis of obesity is urgently needed given that this disease may have an origin in the brain [3] and that currently available antiobesity strategies have not yet achieved a transformative therapeutic impact [103].

New pharmacological tools that selectively and safely target hypothalamic neuronal circuits affected by obesity are on the horizon [38,104]. However, the cellular and molecular mechanisms leading to hypothalamic neurons' dysfunction in obesity have only begun to emerge, and additional research will be needed before functional optimization and clinical translation of these brain-directed agents.

These future efforts toward understanding the neurobiological basis of obesity should consider that hypothalamic neuronal cells are not isolated functional entities, and maladaptive reciprocal interactions between neurons and surrounding microglia cells likely contribute to the establishment and the progression of the disease (Figure 1).

Several outstanding questions related to the extracellular and intracellular mode(s) of action of this process of cell–cell communication remain. For instance, what are the main extrinsic factors that transform a melodious microglia–neuron interaction into a maladaptive process? The caloric content of the diet, diet composition, and duration of diet exposure, may influence hypothalamic neuroinflammatory responses through multiple mechanisms, including changes in gut microbiota composition [32]. Moreover, the overload of certain dietary components, such as fats or carbohydrates, may engage specific cellular metabolic pathways that interfere with mitochondrial function and with the extracellular release of chemical messengers (such as ATP or ADP) influencing the inter-cellular communication.

Of note, both neurons and microglia in the brain may secrete extracellular vesicles (EVs), which are known to contain a broad range of signals possibly involved in the maladaptive microglia–neuron communication, including chemokines, ATP, and inflammation-associated small, noncoding RNA (miRNAs) [105]. Whether or not mitochondrial or EV-mediated mechanisms of communication may contribute to diet-induced hypothalamic neuronal alterations, however, remains to be directly explored.

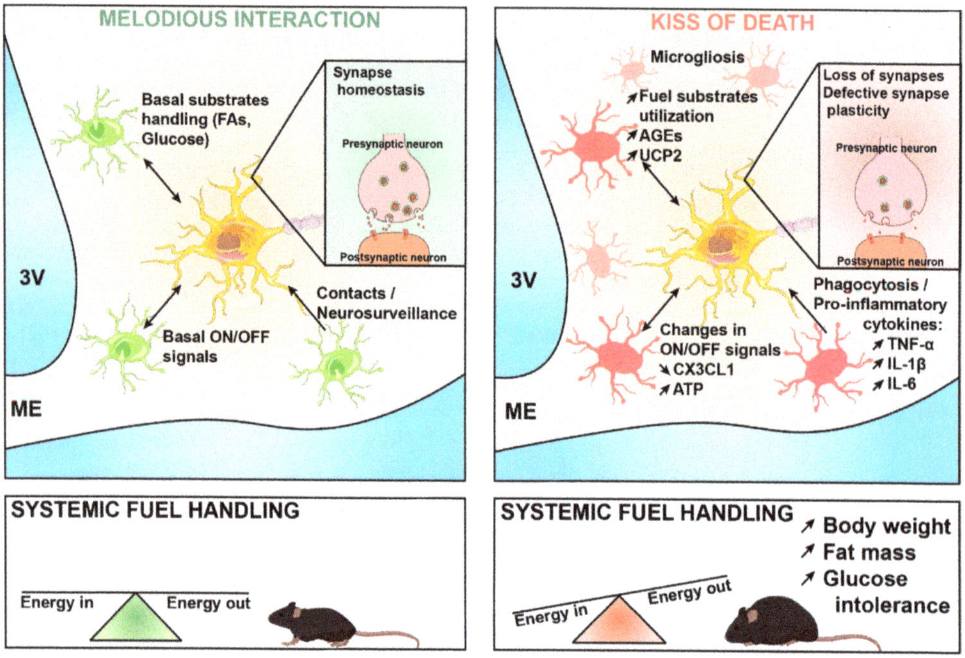

Figure 1. Under physiological conditions (**left**) microglia play a key role in fine-tuning neuronal synaptic function in response to changes in the energy status of the body. This involves different mechanisms including the release of ON-OFF signals, the intercellular production of metabolic messengers in response to the cellular metabolism of fuel substrates, and physical intercellular contacts that are necessary for neurosurveillance. This melodious interaction may become maladaptive during diet-induced obesity (**right**), thereby leading to impaired synaptic plasticity through multiple possible mechanisms, which are highlighted in the figure. The maladaptive cross-talk ultimately results in energy balance dysregulation, obesity, and systemic glucose intolerance. FA: fatty acids; AGEs: advanced glycation end-products; UCP2: uncoupling protein 2; CX3CL1: C-X3-C motif chemokine ligand 1; ATP: adenosine 5′- triphosphate; TNFα: tumor necrosis factor-alpha; IL-1β: interleukin 1 beta, IL6: interleukin 6.

In addition to being influenced by extrinsic factors, the maladaptive cross-talk between microglia and neurons observed in obesity may then originate from intrinsic variables, including genetic predisposition. Indeed, human subjects carrying common polymorphisms in the Jun N-terminal kinase (JNK) or the melanocortin 4 receptor (MC4R) gene are more susceptible to development of obesity and hypothalamic neuroinflammation [32]. Furthermore, gender may play a key role given that microglia isolated from adult brains present sex-specific molecular features [106–108] and that sex differences in microglial activity determine obesity susceptibility in mice [109,110].

Finally, the maladaptive microglia–neuron crosstalk that is observed in response to overnutrition likely involves synergy with other cell types—for instance, astrocytes [5,111], which are key components of the tripartite synapse. Moreover, peripheral monocytes/macrophages, which can extravasate from circulation to the brain during DIO progression, are known to contribute to neuroinflammation and microglia–neuron communication [36]. Tanycytes, which are specialized ependymocytes that form a blood–cerebrospinal fluid barrier in the hypothalamus, may have an additional influencing role [39] due to their ability to fine-tune the brain transport of nutrients and hormones implicated in metabolic control [112,113]. Additionally, endothelial cells located in the hypothalamus are emerging as key cellular mediators of obesity and related metabolic disorders [114] and may therefore contribute to the modulation of diet-induced inflammatory responses in hypothalamic neurocircuitries [37].

While trying to fill in this knowledge gap, cellular heterogeneity may represent a challenging obstacle, but also a relevant aspect. Indeed, microglia and neurons in the hypothalamus form highly diverse cellular clusters [115–118], and may therefore operate in response to diverse and cell-specific mechanisms of communication. Accordingly, different subpopulations of microglial cells have been found in the mediobasal hypothalamus of rodents and humans [119], and these subpopulations present a high degree of spatial localization and distinct molecular signatures in response to diet-induced inflammatory conditions [119].

Thus, addressing when, where, and how hypothalamic neurons and surrounding microglial cells communicate in response to diet-induced metabolic stress will be challenging, but this effort may ultimately favor progress into both the etiology and treatment of metabolic disorders.

Author Contributions: S.L.: writing original draft, A.N.: writing, C.Q.: conceptualization and writing. All authors have read and agreed to the published version of the manuscript.

Funding: C.Q. acknowledges support by INSERM, Agence Nationale de la Recherche (neuroIDobese, ANR-20-CE14-0046), Société Française d'Endocrinologie (Pfizer–SFE Prix de Recherche en Endocrinologie), Société Française de Nutrition and Société Francophone du Diabète.

Institutional Review Board Statement: Not applicable.

Informed Consent Statement: Not applicable.

Data Availability Statement: No new data were created or analyzed in this study. Data sharing is not applicable to this article.

Conflicts of Interest: Authors declare no conflict of interest.

References

1. Friedman, J. The Long Road to Leptin. *J. Clin. Investig.* **2016**, *126*, 4727–4734. [CrossRef]
2. Elmquist, J.K.; Maratos-Flier, E.; Saper, C.B.; Flier, J.S. Unraveling the Central Nervous System Pathways Underlying Responses to Leptin. *Nat. Neurosci.* **1998**, *1*, 445–450. [CrossRef]
3. Locke, A.E.; Kahali, B.; Berndt, S.I.; Justice, A.E.; Pers, T.H.; Day, F.R.; Powell, C.; Vedantam, S.; Buchkovich, M.L.; Yang, J.; et al. Genetic Studies of Body Mass Index Yield New Insights for Obesity Biology. *Nature* **2015**, *518*, 197–206. [CrossRef]
4. Thaler, J.P.; Yi, C.-X.; Schur, E.A.; Guyenet, S.J.; Hwang, B.H.; Dietrich, M.O.; Zhao, X.; Sarruf, D.A.; Izgur, V.; Maravilla, K.R.; et al. Obesity Is Associated with Hypothalamic Injury in Rodents and Humans. *J. Clin. Investig.* **2012**, *122*, 153–162. [CrossRef] [PubMed]
5. Horvath, T.L.; Sarman, B.; García-Cáceres, C.; Enriori, P.J.; Sotonyi, P.; Shanabrough, M.; Borok, E.; Argente, J.; Chowen, J.A.; Perez-Tilve, D.; et al. Synaptic Input Organization of the Melanocortin System Predicts Diet-Induced Hypothalamic Reactive Gliosis and Obesity. *Proc. Natl. Acad. Sci. USA* **2010**, *107*, 14875–14880. [CrossRef]
6. Paeger, L.; Pippow, A.; Hess, S.; Paehler, M.; Klein, A.C.; Husch, A.; Pouzat, C.; Brüning, J.C.; Kloppenburg, P. Energy Imbalance Alters Ca2+ Handling and Excitability of POMC Neurons. *eLife* **2017**, *6*, e25641. [CrossRef] [PubMed]
7. Quarta, C.; Fioramonti, X.; Cota, D. POMC Neurons Dysfunction in Diet-Induced Metabolic Disease: Hallmark or Mechanism of Disease? *Neuroscience* **2019**. [CrossRef] [PubMed]
8. Schwartz, M.W.; Woods, S.C.; Porte, D.; Seeley, R.J.; Baskin, D.G. Central Nervous System Control of Food Intake. *Nature* **2000**, *404*, 661–671. [CrossRef]
9. Cota, D.; Proulx, K.; Seeley, R.J. The Role of CNS Fuel Sensing in Energy and Glucose Regulation. *Gastroenterology* **2007**, *132*, 2158–2168. [CrossRef]
10. Koch, M.; Horvath, T.L. Molecular and Cellular Regulation of Hypothalamic Melanocortin Neurons Controlling Food Intake and Energy Metabolism. *Mol. Psychiatry* **2014**, *19*, 752–761. [CrossRef] [PubMed]
11. Dietrich, M.O.; Horvath, T.L. Hypothalamic Control of Energy Balance: Insights into the Role of Synaptic Plasticity. *Trends Neurosci.* **2013**, *36*, 65–73. [CrossRef]
12. Perea, G.; Navarrete, M.; Araque, A. Tripartite Synapses: Astrocytes Process and Control Synaptic Information. *Trends Neurosci.* **2009**, *32*, 421–431. [CrossRef]
13. Eroglu, C.; Barres, B.A. Regulation of Synaptic Connectivity by Glia. *Nature* **2010**, *468*, 223–231. [CrossRef] [PubMed]
14. Kettenmann, H.; Kirchhoff, F.; Verkhratsky, A. Microglia: New Roles for the Synaptic Stripper. *Neuron* **2013**, *77*, 10–18. [CrossRef]
15. André, C.; Guzman-Quevedo, O.; Rey, C.; Rémus-Borel, J.; Clark, S.; Castellanos-Jankiewicz, A.; Ladeveze, E.; Leste-Lasserre, T.; Nadjar, A.; Abrous, D.N.; et al. Inhibiting Microglia Expansion Prevents Diet-Induced Hypothalamic and Peripheral Inflammation. *Diabetes* **2017**, *66*, 908–919. [CrossRef]

16. Valdearcos, M.; Myers, M.G.; Koliwad, S.K. Hypothalamic Microglia as Potential Regulators of Metabolic Physiology. *Nat. Metab.* **2019**, *1*, 314–320. [CrossRef]
17. Szepesi, Z.; Manouchehrian, O.; Bachiller, S.; Deierborg, T. Bidirectional Microglia-Neuron Communication in Health and Disease. *Front. Cell. Neurosci.* **2018**, *12*, 323. [CrossRef]
18. Dantzer, R. Neuroimmune Interactions: From the Brain to the Immune System and Vice Versa. *Physiol. Rev.* **2018**, *98*, 477–504. [CrossRef] [PubMed]
19. Pohl, J.; Sheppard, M.; Luheshi, G.N.; Woodside, B. Diet-Induced Weight Gain Produces a Graded Increase in Behavioral Responses to an Acute Immune Challenge. *Brain Behav. Immun.* **2014**, *35*, 43–50. [CrossRef] [PubMed]
20. Pohl, J.; Woodside, B.; Luheshi, G.N. Leptin Modulates the Late Fever Response to LPS in Diet-Induced Obese Animals. *Brain Behav. Immun.* **2014**, *42*, 41–47. [CrossRef]
21. Leyrolle, Q.; Layé, S.; Nadjar, A. Direct and Indirect Effects of Lipids on Microglia Function. *Neurosci. Lett.* **2019**, *708*, 134348. [CrossRef]
22. Valdearcos, M.; Robblee, M.M.; Benjamin, D.I.; Nomura, D.K.; Xu, A.W.; Koliwad, S.K. Microglia Dictate the Impact of Saturated Fat Consumption on Hypothalamic Inflammation and Neuronal Function. *Cell Rep.* **2014**, *9*, 2124–2138. [CrossRef]
23. Gao, Y.; Bielohuby, M.; Fleming, T.; Grabner, G.F.; Foppen, E.; Bernhard, W.; Guzmán-Ruiz, M.; Layritz, C.; Legutko, B.; Zinser, E.; et al. Dietary Sugars, Not Lipids, Drive Hypothalamic Inflammation. *Mol. Metab.* **2017**, *6*, 897–908. [CrossRef] [PubMed]
24. Gao, Y.; Vidal-Itriago, A.; Milanova, I.; Korpel, N.L.; Kalsbeek, M.J.; Tom, R.Z.; Kalsbeek, A.; Hofmann, S.M.; Yi, C.-X. Deficiency of Leptin Receptor in Myeloid Cells Disrupts Hypothalamic Metabolic Circuits and Causes Body Weight Increase. *Mol. Metab.* **2018**, *7*, 155–160. [CrossRef] [PubMed]
25. Velloso, L.A.; Schwartz, M.W. Altered Hypothalamic Function in Diet-Induced Obesity. *Int. J. Obes.* **2011**, *35*, 1455–1465. [CrossRef] [PubMed]
26. Holland, W.L.; Bikman, B.T.; Wang, L.-P.; Yuguang, G.; Sargent, K.M.; Bulchand, S.; Knotts, T.A.; Shui, G.; Clegg, D.J.; Wenk, M.R.; et al. Lipid-Induced Insulin Resistance Mediated by the Proinflammatory Receptor TLR4 Requires Saturated Fatty Acid-Induced Ceramide Biosynthesis in Mice. *J. Clin. Investig.* **2011**, *121*, 1858–1870. [CrossRef]
27. Kleinridders, A.; Schenten, D.; Könner, A.C.; Belgardt, B.F.; Mauer, J.; Okamura, T.; Wunderlich, F.T.; Medzhitov, R.; Brüning, J.C. MyD88 Signaling in the CNS Is Required for Development of Fatty Acid-Induced Leptin Resistance and Diet-Induced Obesity. *Cell Metab.* **2009**, *10*, 249–259. [CrossRef]
28. Milanski, M.; Degasperi, G.; Coope, A.; Morari, J.; Denis, R.; Cintra, D.E.; Tsukumo, D.M.L.; Anhe, G.; Amaral, M.E.; Takahashi, H.K.; et al. Saturated Fatty Acids Produce an Inflammatory Response Predominantly through the Activation of TLR4 Signaling in Hypothalamus: Implications for the Pathogenesis of Obesity. *J. Neurosci.* **2009**, *29*, 359–370. [CrossRef]
29. Ozcan, L.; Ergin, A.S.; Lu, A.; Chung, J.; Sarkar, S.; Nie, D.; Myers, M.G.; Ozcan, U. Endoplasmic Reticulum Stress Plays a Central Role in Development of Leptin Resistance. *Cell Metab.* **2009**, *9*, 35–51. [CrossRef]
30. Zhang, X.; Zhang, G.; Zhang, H.; Karin, M.; Bai, H.; Cai, D. Hypothalamic IKKbeta/NF-KappaB and ER Stress Link Overnutrition to Energy Imbalance and Obesity. *Cell* **2008**, *135*, 61–73. [CrossRef] [PubMed]
31. De Souza, C.T.; Araujo, E.P.; Bordin, S.; Ashimine, R.; Zollner, R.L.; Boschero, A.C.; Saad, M.J.A.; Velloso, L.A. Consumption of a Fat-Rich Diet Activates a Proinflammatory Response and Induces Insulin Resistance in the Hypothalamus. *Endocrinology* **2005**, *146*, 4192–4199. [CrossRef]
32. Kreutzer, C.; Peters, S.; Schulte, D.M.; Fangmann, D.; Türk, K.; Wolff, S.; van Eimeren, T.; Ahrens, M.; Beckmann, J.; Schafmayer, C.; et al. Hypothalamic Inflammation in Human Obesity Is Mediated by Environmental and Genetic Factors. *Diabetes* **2017**, *66*, 2407–2415. [CrossRef]
33. Schur, E.A.; Melhorn, S.J.; Oh, S.-K.; Lacy, J.M.; Berkseth, K.E.; Guyenet, S.J.; Sonnen, J.A.; Tyagi, V.; Rosalynn, M.; De Leon, B.; et al. Radiologic Evidence That Hypothalamic Gliosis Is Associated with Obesity and Insulin Resistance in Humans. *Obesity* **2015**, *23*, 2142–2148. [CrossRef]
34. Baufeld, C.; Osterloh, A.; Prokop, S.; Miller, K.R.; Heppner, F.L. High-Fat Diet-Induced Brain Region-Specific Phenotypic Spectrum of CNS Resident Microglia. *Acta Neuropathol.* **2016**, *132*, 361–375. [CrossRef]
35. Kalsbeek, M.J.; Wolff, S.E.; Korpel, N.L.; la Fleur, S.E.; Romijn, J.A.; Fliers, E.; Kalsbeek, A.; Swaab, D.F.; Huitinga, I.; Hol, E.M.; et al. The Impact of Antidiabetic Treatment on Human Hypothalamic Infundibular Neurons and Microglia. *JCI Insight* **2020**, *5*. [CrossRef]
36. Valdearcos, M.; Douglass, J.D.; Robblee, M.M.; Dorfman, M.D.; Stifler, D.R.; Bennett, M.L.; Gerritse, I.; Fasnacht, R.; Barres, B.A.; Thaler, J.P.; et al. Microglial Inflammatory Signaling Orchestrates the Hypothalamic Immune Response to Dietary Excess and Mediates Obesity Susceptibility. *Cell Metab.* **2017**, *26*, 185–197.e3. [CrossRef]
37. Valdearcos, M.; Xu, A.W.; Koliwad, S.K. Hypothalamic Inflammation in the Control of Metabolic Function. *Annu. Rev. Physiol.* **2015**, *77*, 131–160. [CrossRef]
38. Quarta, C.; Clemmensen, C.; Zhu, Z.; Yang, B.; Joseph, S.S.; Lutter, D.; Yi, C.-X.; Graf, E.; García-Cáceres, C.; Legutko, B.; et al. Molecular Integration of Incretin and Glucocorticoid Action Reverses Immunometabolic Dysfunction and Obesity. *Cell Metab.* **2017**, *26*, 620–632.e6. [CrossRef]
39. Clasadonte, J.; Prevot, V. The Special Relationship: Glia-Neuron Interactions in the Neuroendocrine Hypothalamus. *Nat. Rev. Endocrinol.* **2018**, *14*, 25–44. [CrossRef]

40. Block, M.L.; Zecca, L.; Hong, J.-S. Microglia-Mediated Neurotoxicity: Uncovering the Molecular Mechanisms. *Nat. Rev. Neurosci.* **2007**, *8*, 57–69. [CrossRef]
41. Thaler, J.P.; Guyenet, S.J.; Dorfman, M.D.; Wisse, B.E.; Schwartz, M.W. Hypothalamic Inflammation: Marker or Mechanism of Obesity Pathogenesis? *Diabetes* **2013**, *62*, 2629–2634. [CrossRef]
42. Cansell, C.; Stobbe, K.; Sanchez, C.; Le Thuc, O.; Mosser, C.-A.; Ben-Fradj, S.; Leredde, J.; Lebeaupin, C.; Debayle, D.; Fleuriot, L.; et al. Dietary Fat Exacerbates Postprandial Hypothalamic Inflammation Involving Glial Fibrillary Acidic Protein-Positive Cells and Microglia in Male Mice. *Glia* **2021**, *69*, 42–60. [CrossRef] [PubMed]
43. Münzberg, H.; Flier, J.S.; Bjørbæk, C. Region-Specific Leptin Resistance within the Hypothalamus of Diet-Induced Obese Mice. *Endocrinology* **2004**, *145*, 4880–4889. [CrossRef] [PubMed]
44. Olofsson, L.E.; Unger, E.K.; Cheung, C.C.; Xu, A.W. Modulation of AgRP-Neuronal Function by SOCS3 as an Initiating Event in Diet-Induced Hypothalamic Leptin Resistance. *Proc. Natl. Acad. Sci. USA* **2013**, *110*, E697–E706. [CrossRef]
45. Cavadas, C.; Aveleira, C.A.; Souza, G.F.P.; Velloso, L.A. The Pathophysiology of Defective Proteostasis in the Hypothalamus—From Obesity to Ageing. *Nat. Rev. Endocrinol.* **2016**, *12*, 723–733. [CrossRef] [PubMed]
46. Ramírez, S.; Claret, M. Hypothalamic ER Stress: A Bridge between Leptin Resistance and Obesity. *FEBS Lett.* **2015**, *589*, 1678–1687. [CrossRef] [PubMed]
47. Moraes, J.C.; Coope, A.; Morari, J.; Cintra, D.E.; Roman, E.A.; Pauli, J.R.; Romanatto, T.; Carvalheira, J.B.; Oliveira, A.L.R.; Saad, M.J.; et al. High-Fat Diet Induces Apoptosis of Hypothalamic Neurons. *PLoS ONE* **2009**, *4*, e5045. [CrossRef]
48. Nyamugenda, E.; Trentzsch, M.; Russell, S.; Miles, T.; Boysen, G.; Phelan, K.D.; Baldini, G. Injury to Hypothalamic Sim1 Neurons Is a Common Feature of Obesity by Exposure to High-Fat Diet in Male and Female Mice. *J. Neurochem.* **2019**, *149*, 73–97. [CrossRef]
49. Lemus, M.B.; Bayliss, J.A.; Lockie, S.H.; Santos, V.V.; Reichenbach, A.; Stark, R.; Andrews, Z.B. A Stereological Analysis of NPY, POMC, Orexin, GFAP Astrocyte, and Iba1 Microglia Cell Number and Volume in Diet-Induced Obese Male Mice. *Endocrinology* **2015**, *156*, 1701–1713. [CrossRef]
50. Bessis, A.; Béchade, C.; Bernard, D.; Roumier, A. Microglial Control of Neuronal Death and Synaptic Properties. *Glia* **2007**, *55*, 233–238. [CrossRef]
51. Reis, W.L.; Yi, C.-X.; Gao, Y.; Tschöp, M.H.; Stern, J.E. Brain Innate Immunity Regulates Hypothalamic Arcuate Neuronal Activity and Feeding Behavior. *Endocrinology* **2015**, *156*, 1303–1315. [CrossRef]
52. Hao, S.; Dey, A.; Yu, X.; Stranahan, A.M. Dietary Obesity Reversibly Induces Synaptic Stripping by Microglia and Impairs Hippocampal Plasticity. *Brain Behav. Immun.* **2016**, *51*, 230–239. [CrossRef]
53. Cope, E.C.; LaMarca, E.A.; Monari, P.K.; Olson, L.B.; Martinez, S.; Zych, A.D.; Katchur, N.J.; Gould, E. Microglia Play an Active Role in Obesity-Associated Cognitive Decline. *J. Neurosci.* **2018**, *38*, 8889–8904. [CrossRef]
54. Mizuno, T.M.; Kleopoulos, S.P.; Bergen, H.T.; Roberts, J.L.; Priest, C.A.; Mobbs, C.V. Hypothalamic Pro-Opiomelanocortin MRNA Is Reduced by Fasting and [Corrected] in Ob/Ob and Db/Db Mice, but Is Stimulated by Leptin. *Diabetes* **1998**, *47*, 294–297. [CrossRef]
55. Horvath, T.L. Synaptic Plasticity in Energy Balance Regulation. *Obesity* **2006**, *14* (Suppl. S5), 228S–233S. [CrossRef]
56. Gao, Y.; Ottaway, N.; Schriever, S.C.; Legutko, B.; García-Cáceres, C.; de la Fuente, E.; Mergen, C.; Bour, S.; Thaler, J.P.; Seeley, R.J.; et al. Hormones and Diet, but Not Body Weight, Control Hypothalamic Microglial Activity. *Glia* **2014**, *62*, 17–25. [CrossRef]
57. Lafrance, V.; Inoue, W.; Kan, B.; Luheshi, G.N. Leptin Modulates Cell Morphology and Cytokine Release in Microglia. *Brain Behav. Immun.* **2010**, *24*, 358–365. [CrossRef]
58. Bernier, L.-P.; York, E.M.; Kamyabi, A.; Choi, H.B.; Weilinger, N.L.; MacVicar, B.A. Microglial Metabolic Flexibility Supports Immune Surveillance of the Brain Parenchyma. *Nat. Commun.* **2020**, *11*, 1559. [CrossRef]
59. Vistoli, G.; De Maddis, D.; Cipak, A.; Zarkovic, N.; Carini, M.; Aldini, G. Advanced Glycoxidation and Lipoxidation End Products (AGEs and ALEs): An Overview of Their Mechanisms of Formation. *Free Radic. Res.* **2013**, *47* (Suppl. S1), 3–27. [CrossRef]
60. Horvath, T.L.; Andrews, Z.B.; Diano, S. Fuel Utilization by Hypothalamic Neurons: Roles for ROS. *Trends Endocrinol. Metab.* **2009**, *20*, 78–87. [CrossRef]
61. Parton, L.E.; Ye, C.P.; Coppari, R.; Enriori, P.J.; Choi, B.; Zhang, C.-Y.; Xu, C.; Vianna, C.R.; Balthasar, N.; Lee, C.E.; et al. Glucose Sensing by POMC Neurons Regulates Glucose Homeostasis and Is Impaired in Obesity. *Nature* **2007**, *449*, 228–232. [CrossRef]
62. Andrews, Z.B.; Liu, Z.-W.; Walllingford, N.; Erion, D.M.; Borok, E.; Friedman, J.M.; Tschöp, M.H.; Shanabrough, M.; Cline, G.; Shulman, G.I.; et al. UCP2 Mediates Ghrelin's Action on NPY/AgRP Neurons by Lowering Free Radicals. *Nature* **2008**, *454*, 846–851. [CrossRef] [PubMed]
63. Kim, J.D.; Yoon, N.A.; Jin, S.; Diano, S. Microglial UCP2 Mediates Inflammation and Obesity Induced by High-Fat Feeding. *Cell Metab.* **2019**, *30*, 952–962.e5. [CrossRef]
64. Gao, Y.; Vidal-Itriago, A.; Kalsbeek, M.J.; Layritz, C.; García-Cáceres, C.; Tom, R.Z.; Eichmann, T.O.; Vaz, F.M.; Houtkooper, R.H.; van der Wel, N.; et al. Lipoprotein Lipase Maintains Microglial Innate Immunity in Obesity. *Cell Rep.* **2017**, *20*, 3034–3042. [CrossRef]
65. Nakanishi, M.; Niidome, T.; Matsuda, S.; Akaike, A.; Kihara, T.; Sugimoto, H. Microglia-Derived Interleukin-6 and Leukaemia Inhibitory Factor Promote Astrocytic Differentiation of Neural Stem/Progenitor Cells. *Eur. J. Neurosci.* **2007**, *25*, 649–658. [CrossRef]

66. Lambertsen, K.L.; Clausen, B.H.; Babcock, A.A.; Gregersen, R.; Fenger, C.; Nielsen, H.H.; Haugaard, L.S.; Wirenfeldt, M.; Nielsen, M.; Dagnaes-Hansen, F.; et al. Microglia Protect Neurons against Ischemia by Synthesis of Tumor Necrosis Factor. *J. Neurosci.* **2009**, *29*, 1319–1330. [CrossRef]
67. Ropelle, E.R.; Flores, M.B.; Cintra, D.E.; Rocha, G.Z.; Pauli, J.R.; Morari, J.; de Souza, C.T.; Moraes, J.C.; Prada, P.O.; Guadagnini, D.; et al. IL-6 and IL-10 Anti-Inflammatory Activity Links Exercise to Hypothalamic Insulin and Leptin Sensitivity through IKKbeta and ER Stress Inhibition. *PLoS Biol.* **2010**, *8*. [CrossRef] [PubMed]
68. Yi, C.X.; Walter, M.; Gao, Y.; Pitra, S.; Legutko, B.; Kälin, S.; Layritz, C.; García-Cáceres, C.; Bielohuby, M.; Bidlingmaier, M.; et al. TNFα Drives Mitochondrial Stress in POMC Neurons in Obesity. *Nat. Commun.* **2017**, *8*, 15143. [CrossRef]
69. Jais, A.; Brüning, J.C. Hypothalamic Inflammation in Obesity and Metabolic Disease. *J. Clin. Investig.* **2017**, *127*, 24–32. [CrossRef]
70. Bobbo, V.C.D.; Jara, C.P.; Mendes, N.F.; Morari, J.; Velloso, L.A.; Araújo, E.P. Interleukin-6 Expression by Hypothalamic Microglia in Multiple Inflammatory Contexts: A Systematic Review. Available online: https://www.hindawi.com/journals/bmri/2019/1365210/ (accessed on 23 February 2021).
71. Schöbitz, B.; Pezeshki, G.; Pohl, T.; Hemmann, U.; Heinrich, P.C.; Holsboer, F.; Reul, J.M. Soluble Interleukin-6 (IL-6) Receptor Augments Central Effects of IL-6 in Vivo. *FASEB J.* **1995**, *9*, 659–664. [CrossRef]
72. Wallenius, K.; Wallenius, V.; Sunter, D.; Dickson, S.L.; Jansson, J.-O. Intracerebroventricular Interleukin-6 Treatment Decreases Body Fat in Rats. *Biochem. Biophys. Res. Commun.* **2002**, *293*, 560–565. [CrossRef]
73. Wallenius, V.; Wallenius, K.; Ahrén, B.; Rudling, M.; Carlsten, H.; Dickson, S.L.; Ohlsson, C.; Jansson, J.-O. Interleukin-6-Deficient Mice Develop Mature-Onset Obesity. *Nat. Med.* **2002**, *8*, 75–79. [CrossRef]
74. Dascombe, M.J.; Rothwell, N.J.; Sagay, B.O.; Stock, M.J. Pyrogenic and Thermogenic Effects of Interleukin 1 Beta in the Rat. *Am. J. Physiol.* **1989**, *256*, E7–E11. [CrossRef] [PubMed]
75. Plata-Salamán, C.R. Cytokine-Induced Anorexia. Behavioral, Cellular, and Molecular Mechanisms. *Ann. N. Y. Acad. Sci.* **1998**, *856*, 160–170. [CrossRef] [PubMed]
76. Benrick, A.; Schéle, E.; Pinnock, S.B.; Wernstedt-Asterholm, I.; Dickson, S.L.; Karlsson-Lindahl, L.; Jansson, J.-O. Interleukin-6 Gene Knockout Influences Energy Balance Regulating Peptides in the Hypothalamic Paraventricular and Supraoptic Nuclei. *J. Neuroendocrinol.* **2009**, *21*, 620–628. [CrossRef]
77. Schéle, E.; Benrick, A.; Grahnemo, L.; Egecioglu, E.; Anesten, F.; Pálsdóttir, V.; Jansson, J.-O. Inter-Relation between Interleukin (IL)-1, IL-6 and Body Fat Regulating Circuits of the Hypothalamic Arcuate Nucleus. *J. Neuroendocrinol.* **2013**, *25*, 580–589. [CrossRef]
78. Timper, K.; Denson, J.L.; Steculorum, S.M.; Heilinger, C.; Engström-Ruud, L.; Wunderlich, C.M.; Rose-John, S.; Wunderlich, F.T.; Brüning, J.C. IL-6 Improves Energy and Glucose Homeostasis in Obesity via Enhanced Central IL-6 Trans-Signaling. *Cell Rep.* **2017**, *19*, 267–280. [CrossRef]
79. Le Thuc, O.; Stobbe, K.; Cansell, C.; Nahon, J.-L.; Blondeau, N.; Rovère, C. Hypothalamic Inflammation and Energy Balance Disruptions: Spotlight on Chemokines. *Front. Endocrinol.* **2017**, *8*. [CrossRef]
80. Plata-Salamán, C.R.; Borkoski, J.P. Chemokines/Intercrines and Central Regulation of Feeding. *Am. J. Physiol.* **1994**, *266*, R1711–R1715. [CrossRef]
81. Le Thuc, O.; Cansell, C.; Bourourou, M.; Denis, R.G.; Stobbe, K.; Devaux, N.; Guyon, A.; Cazareth, J.; Heurteaux, C.; Rostène, W.; et al. Central CCL2 Signaling onto MCH Neurons Mediates Metabolic and Behavioral Adaptation to Inflammation. *EMBO Rep.* **2016**, *17*, 1738–1752. [CrossRef]
82. Morari, J.; Anhe, G.F.; Nascimento, L.F.; de Moura, R.F.; Razolli, D.; Solon, C.; Guadagnini, D.; Souza, G.; Mattos, A.H.; Tobar, N.; et al. Fractalkine (CX3CL1) Is Involved in the Early Activation of Hypothalamic Inflammation in Experimental Obesity. *Diabetes* **2014**, *63*, 3770–3784. [CrossRef] [PubMed]
83. Poon, K.; Barson, J.R.; Ho, H.T.; Leibowitz, S.F. Relationship of the Chemokine, CXCL12, to Effects of Dietary Fat on Feeding-Related Behaviors and Hypothalamic Neuropeptide Systems. *Front. Behav. Neurosci.* **2016**, *10*, 51. [CrossRef] [PubMed]
84. Kwon, Y.-H.; Kim, J.; Kim, C.-S.; Tu, T.H.; Kim, M.-S.; Suk, K.; Kim, D.H.; Lee, B.J.; Choi, H.-S.; Park, T.; et al. Hypothalamic Lipid-Laden Astrocytes Induce Microglia Migration and Activation. *FEBS Lett.* **2017**, *591*, 1742–1751. [CrossRef]
85. Yao, Y.; Tsirka, S.E. Monocyte Chemoattractant Protein-1 and the Blood-Brain Barrier. *Cell. Mol. Life Sci.* **2014**, *71*, 683–697. [CrossRef]
86. Biber, K.; Neumann, H.; Inoue, K.; Boddeke, H.W.G.M. Neuronal "On" and "Off" Signals Control Microglia. *Trends Neurosci.* **2007**, *30*, 596–602. [CrossRef]
87. Haynes, S.E.; Hollopeter, G.; Yang, G.; Kurpius, D.; Dailey, M.E.; Gan, W.-B.; Julius, D. The P2Y12 Receptor Regulates Microglial Activation by Extracellular Nucleotides. *Nat. Neurosci.* **2006**, *9*, 1512–1519. [CrossRef]
88. Cserép, C.; Pósfai, B.; Lénárt, N.; Fekete, R.; László, Z.I.; Lele, Z.; Orsolits, B.; Molnár, G.; Heindl, S.; Schwarcz, A.D.; et al. Microglia Monitor and Protect Neuronal Function through Specialized Somatic Purinergic Junctions. *Science* **2020**, *367*, 528–537. [CrossRef]
89. Badimon, A.; Strasburger, H.J.; Ayata, P.; Chen, X.; Nair, A.; Ikegami, A.; Hwang, P.; Chan, A.T.; Graves, S.M.; Uweru, J.O.; et al. Negative Feedback Control of Neuronal Activity by Microglia. *Nature* **2020**, *586*, 417–423. [CrossRef]
90. Galea, I.; Bechmann, I.; Perry, V.H. What Is Immune Privilege (Not)? *Trends Immunol.* **2007**, *28*, 12–18. [CrossRef] [PubMed]

91. Mathys, H.; Adaikkan, C.; Gao, F.; Young, J.Z.; Manet, E.; Hemberg, M.; De Jager, P.L.; Ransohoff, R.M.; Regev, A.; Tsai, L.-H. Temporal Tracking of Microglia Activation in Neurodegeneration at Single-Cell Resolution. *Cell Rep.* **2017**, *21*, 366–380. [CrossRef] [PubMed]
92. Neumann, H.; Misgeld, T.; Matsumuro, K.; Wekerle, H. Neurotrophins Inhibit Major Histocompatibility Class II Inducibility of Microglia: Involvement of the P75 Neurotrophin Receptor. *Proc. Natl. Acad. Sci. USA* **1998**, *95*, 5779–5784. [CrossRef] [PubMed]
93. Wang, P.; Loh, K.H.; Wu, M.; Morgan, D.A.; Schneeberger, M.; Yu, X.; Chi, J.; Kosse, C.; Kim, D.; Rahmouni, K.; et al. A Leptin-BDNF Pathway Regulating Sympathetic Innervation of Adipose Tissue. *Nature* **2020**, *583*, 839–844. [CrossRef]
94. Xu, B.; Xie, X. Neurotrophic Factor Control of Satiety and Body Weight. *Nat. Rev. Neurosci.* **2016**, *17*, 282–292. [CrossRef] [PubMed]
95. Trang, T.; Beggs, S.; Wan, X.; Salter, M.W. P2X4-Receptor-Mediated Synthesis and Release of Brain-Derived Neurotrophic Factor in Microglia Is Dependent on Calcium and P38-Mitogen-Activated Protein Kinase Activation. *J. Neurosci.* **2009**, *29*, 3518–3528. [CrossRef]
96. Urabe, H.; Kojima, H.; Chan, L.; Terashima, T.; Ogawa, N.; Katagi, M.; Fujino, K.; Kumagai, A.; Kawai, H.; Asakawa, A.; et al. Haematopoietic Cells Produce BDNF and Regulate Appetite upon Migration to the Hypothalamus. *Nat. Commun.* **2013**, *4*, 1526. [CrossRef] [PubMed]
97. Pocock, J.M.; Kettenmann, H. Neurotransmitter Receptors on Microglia. *Trends Neurosci.* **2007**, *30*, 527–535. [CrossRef] [PubMed]
98. Fontainhas, A.M.; Wang, M.; Liang, K.J.; Chen, S.; Mettu, P.; Damani, M.; Fariss, R.N.; Li, W.; Wong, W.T. Microglial Morphology and Dynamic Behavior Is Regulated by Ionotropic Glutamatergic and GABAergic Neurotransmission. *PLoS ONE* **2011**, *6*, e15973. [CrossRef]
99. Shytle, R.D.; Mori, T.; Townsend, K.; Vendrame, M.; Sun, N.; Zeng, J.; Ehrhart, J.; Silver, A.A.; Sanberg, P.R.; Tan, J. Cholinergic Modulation of Microglial Activation by Alpha 7 Nicotinic Receptors. *J. Neurochem.* **2004**, *89*, 337–343. [CrossRef] [PubMed]
100. Noda, M.; Nakanishi, H.; Nabekura, J.; Akaike, N. AMPA-Kainate Subtypes of Glutamate Receptor in Rat Cerebral Microglia. *J. Neurosci.* **2000**, *20*, 251–258. [CrossRef] [PubMed]
101. Thomas, A.G.; O'Driscoll, C.M.; Bressler, J.; Kaufmann, W.; Rojas, C.J.; Slusher, B.S. Small Molecule Glutaminase Inhibitors Block Glutamate Release from Stimulated Microglia. *Biochem. Biophys. Res. Commun.* **2014**, *443*, 32–36. [CrossRef]
102. Fuente-Martín, E.; García-Cáceres, C.; Granado, M.; de Ceballos, M.L.; Sánchez-Garrido, M.Á.; Sarman, B.; Liu, Z.-W.; Dietrich, M.O.; Tena-Sempere, M.; Argente-Arizón, P.; et al. Leptin Regulates Glutamate and Glucose Transporters in Hypothalamic Astrocytes. *J. Clin. Investig.* **2012**, *122*, 3900–3913. [CrossRef] [PubMed]
103. The GBD 2015 Obesity Collaborators. Health Effects of Overweight and Obesity in 195 Countries over 25 Years. *N. Engl. J. Med.* **2017**, *377*, 13–27. [CrossRef] [PubMed]
104. Stemmer, K.; Müller, T.D.; DiMarchi, R.D.; Pfluger, P.T.; Tschöp, M.H. CNS-Targeting Pharmacological Interventions for the Metabolic Syndrome. *J. Clin. Investig.* **2019**, *129*, 4058–4071. [CrossRef] [PubMed]
105. Delpech, J.-C.; Herron, S.; Botros, M.B.; Ikezu, T. Neuroimmune Crosstalk through Extracellular Vesicles in Health and Disease. *Trends Neurosci.* **2019**, *42*, 361–372. [CrossRef] [PubMed]
106. Villa, A.; Gelosa, P.; Castiglioni, L.; Cimino, M.; Rizzi, N.; Pepe, G.; Lolli, F.; Marcello, E.; Sironi, L.; Vegeto, E.; et al. Sex-Specific Features of Microglia from Adult Mice. *Cell Rep.* **2018**, *23*, 3501–3511. [CrossRef]
107. Thion, M.S.; Low, D.; Silvin, A.; Chen, J.; Grisel, P.; Schulte-Schrepping, J.; Blecher, R.; Ulas, T.; Squarzoni, P.; Hoeffel, G.; et al. Microbiome Influences Prenatal and Adult Microglia in a Sex-Specific Manner. *Cell* **2018**, *172*, 500–516.e16. [CrossRef]
108. Guneykaya, D.; Ivanov, A.; Hernandez, D.P.; Haage, V.; Wojtas, B.; Meyer, N.; Maricos, M.; Jordan, P.; Buonfiglioli, A.; Gielniewski, B.; et al. Transcriptional and Translational Differences of Microglia from Male and Female Brains. *Cell Rep.* **2018**, *24*, 2773–2783.e6. [CrossRef]
109. Dorfman, M.D.; Krull, J.E.; Douglass, J.D.; Fasnacht, R.; Lara-Lince, F.; Meek, T.H.; Shi, X.; Damian, V.; Nguyen, H.T.; Matsen, M.E.; et al. Sex Differences in Microglial CX3CR1 Signalling Determine Obesity Susceptibility in Mice. *Nat. Commun.* **2017**, *8*, 14556. [CrossRef]
110. Argente-Arizón, P.; Díaz, F.; Ros, P.; Barrios, V.; Tena-Sempere, M.; García-Segura, L.M.; Argente, J.; Chowen, J.A. The Hypothalamic Inflammatory/Gliosis Response to Neonatal Overnutrition Is Sex and Age Dependent. *Endocrinology* **2018**, *159*, 368–387. [CrossRef]
111. García-Cáceres, C.; Balland, E.; Prevot, V.; Luquet, S.; Woods, S.C.; Koch, M.; Horvath, T.L.; Yi, C.-X.; Chowen, J.A.; Verkhratsky, A.; et al. Role of Astrocytes, Microglia, and Tanycytes in Brain Control of Systemic Metabolism. *Nat. Neurosci.* **2019**, *22*, 7–14. [CrossRef]
112. Yoo, S.; Cha, D.; Kim, S.; Jiang, L.; Cooke, P.; Adebesin, M.; Wolfe, A.; Riddle, R.; Aja, S.; Blackshaw, S. Tanycyte Ablation in the Arcuate Nucleus and Median Eminence Increases Obesity Susceptibility by Increasing Body Fat Content in Male Mice. *Glia* **2020**, *68*, 1987–2000. [CrossRef]
113. Balland, E.; Dam, J.; Langlet, F.; Caron, E.; Steculorum, S.; Messina, A.; Rasika, S.; Falluel-Morel, A.; Anouar, Y.; Dehouck, B.; et al. Hypothalamic Tanycytes Are an ERK-Gated Conduit for Leptin into the Brain. *Cell Metab.* **2014**, *19*, 293–301. [CrossRef] [PubMed]
114. Graupera, M.; Claret, M. Endothelial Cells: New Players in Obesity and Related Metabolic Disorders. *Trends Endocrinol. Metab.* **2018**, *29*, 781–794. [CrossRef] [PubMed]
115. Masuda, T.; Sankowski, R.; Staszewski, O.; Prinz, M. Microglia Heterogeneity in the Single-Cell Era. *Cell Rep.* **2020**, *30*, 1271–1281. [CrossRef]

116. Campbell, J.N.; Macosko, E.Z.; Fenselau, H.; Pers, T.H.; Lyubetskaya, A.; Tenen, D.; Goldman, M.; Verstegen, A.M.J.; Resch, J.M.; McCarroll, S.A.; et al. A Molecular Census of Arcuate Hypothalamus and Median Eminence Cell Types. *Nat. Neurosci.* **2017**, *20*, 484–496. [CrossRef]
117. Chen, R.; Wu, X.; Jiang, L.; Zhang, Y. Single-Cell RNA-Seq Reveals Hypothalamic Cell Diversity. *Cell Rep.* **2017**, *18*, 3227–3241. [CrossRef] [PubMed]
118. Quarta, C.; Claret, M.; Zeltser, L.M.; Williams, K.W.; Yeo, G.S.H.; Tschöp, M.H.; Diano, S.; Brüning, J.C.; Cota, D. POMC Neuronal Heterogeneity in Energy Balance and beyond: An Integrated View. *Nat. Metab.* **2021**, *3*, 299–308. [CrossRef]
119. Mendes, N.F.; Jara, C.P.; Zanesco, A.M.; de Araújo, E.P. Hypothalamic Microglial Heterogeneity and Signature under High Fat Diet–Induced Inflammation. *Int. J. Mol. Sci.* **2021**, *22*, 2256. [CrossRef] [PubMed]

International Journal of
Molecular Sciences

MDPI

Review

Hypothalamic Astrocytes as a Specialized and Responsive Cell Population in Obesity

Ismael González-García [1,2,*] and Cristina García-Cáceres [1,2,3]

1 Institute for Diabetes and Obesity, Helmholtz Diabetes Center at Helmholtz Zentrum München, 85764 Neuherberg, Germany; garcia-caceres@helmholtz-muenchen.de
2 German Center for Diabetes Research (DZD), 85764 Neuherberg, Germany
3 Medizinische Klinik und Poliklinik IV, Klinikum der Universität, Ludwig-Maximilians-Universität München, D-80336 Munich, Germany
* Correspondence: ismael.gonzalez@helmholtz-muenchen.de

Abstract: Astrocytes are a type of glial cell anatomically and functionally integrated into the neuronal regulatory circuits for the neuroendocrine control of metabolism. Being functional integral compounds of synapses, astrocytes are actively involved in the physiological regulatory aspects of metabolic control, but also in the pathological processes that link neuronal dysfunction and obesity. Between brain areas, the hypothalamus harbors specialized functional circuits that seem selectively vulnerable to metabolic damage, undergoing early cellular rearrangements which are thought to be at the core of the pathogenesis of diet-induced obesity. Such changes in the hypothalamic brain region consist of a rise in proinflammatory cytokines, the presence of a reactive phenotype in astrocytes and microglia, alterations in the cytoarchitecture and synaptology of hypothalamic circuits, and angiogenesis, a phenomenon that cannot be found elsewhere in the brain. Increasing evidence points to the direct involvement of hypothalamic astrocytes in such early metabolic disturbances, thus moving the study of these glial cells to the forefront of obesity research. Here we provide a comprehensive review of the most relevant findings of molecular and pathophysiological mechanisms by which hypothalamic astrocytes might be involved in the pathogenesis of obesity.

Keywords: astrocytes; neurons; hypothalamus; obesity

Citation: González-García, I.; García-Cáceres, C. Hypothalamic Astrocytes as a Specialized and Responsive Cell Population in Obesity. *Int. J. Mol. Sci.* **2021**, *22*, 6176. https://doi.org/10.3390/ijms22126176

Academic Editors: Rosalía Rodríguez-Rodríguez and Cristina Miralpeix

Received: 14 May 2021
Accepted: 4 June 2021
Published: 8 June 2021

1. The Hypothalamus: Integrator of Neuroendocrine and Systemic Metabolic Homeostasis

The organization and cellular components of mammalian neuroendocrine systems are highly evolutionarily conserved and require circuits precisely tuned throughout life. The central nervous system (CNS) senses and processes local and peripheral afferent metabolic signals emanating from circulation (hormones, nutrients), and translates information from sensory nerves into physiological responses regulating feeding behavior and energy expenditure [1–3]. The identification of the hypothalamus as a metabolic center arose from pioneered observations during the mid-20th century. These seminal studies demonstrated first that inflicting lesions or the presence of tumors in specific hypothalamic brain areas elicited either hyperphagic obesity or severe anorexia [4]. Subsequently, the advances in technologies and knowledge allowed for further confirmation that the key role of the hypothalamus is its ability to integrate and process metabolic cues of endocrine and autonomic nervous system origins; thus, positioning this brain area at center stage in the regulation of body weight and energy balance.

The hypothalamus is structured and organized into different nuclei with the arcuate nucleus (ARC) having been more deeply studied for containing specialized neuronal populations. These neurons are vastly interconnected between themselves and with other extra-hypothalamic areas, which allows for a high intercommunication and redundancy in the homeostatic mechanisms [5]. The ARC is located at the base of the hypothalamus in close proximity to the median eminence (ME) and it is there where two neuronal populations well-known for being fundamental for whole-body energy regulation reside: Neuropeptide Y and agouti-related peptide (NPY/AgRP) expressing neurons whose activation exert orexigenic effects and reduce energy expenditure, and proopiomelanocortin (POMC) expressing neurons with opposing effects on feeding behavior and energy expenditure compared to NPY/AgRP neurons [1]. Both sets of neurons express receptors for peripheral endocrine inputs (e.g., leptin, insulin, GLP1, free fatty acids, and ghrelin), but also for central-derived signals (e.g., NPY, GABA, serotonin, and melanocortin), and the overall result of the integration and processing of these signals is to influence the wiring and flexibility of neuronal circuits for the homeostatic maintenance of energy balance. Thus, the connectivity of these ARC neuronal circuits is flexible, varying depending on the energy storage and nutritional state of the entire organism. Such adaptive plasticity of neuronal connectivity requires the chemical and physical association and cooperation of adjacent glial cells, which are active units in the regulation of synaptic function.

In the last few years, hypothalamic astrocytes have started to draw some attention in the neuroendocrine field for acting as hypothalamic functioning regulators, particularly for their role in the central control of metabolism. Indeed, several findings emphasize the crucial involvement of astrocytes for their functional interactions with specific ARC-resident neuronal populations [6–10] together forming one unique functional circuit in metabolic regulation. The advances in neuroscience and molecular biological technologies designed for targeting non-neuronal cells are allowing us to advance current knowledge by providing novel concepts about the functional contribution of astrocytes in the physiological regulation of food intake and energy homeostasis.

2. Hypothalamic Astrocytes: A Specialized Population in Regulating Metabolism

The hypothalamus, in particular its ventral border where the ARC is located, is beneficially situated adjacent to the ME, a circumventricular organ with a particular angioarchitecture (i.e., fenestrated microvessels which lack blood brain barrier) positioning it as a more metabolically sensitive brain center over others in the CNS [11–13]. Such high and rapid responsiveness to metabolic disturbances is also due to being highly-enriched in a wide-array of metabolic receptors and transporters acting as primary sensors and responders to nutrients and endocrine cues emanating from the bloodstream [14–17]. Additionally, emerging insights highlight that hypothalamic astrocytes, tightly connected to ARC-resident neurons and blood vessels [18], are strategically situated to survey the organism's metabolic status and to, in turn, remodel local vascular beds thereby controlling the selective access of some circulating factors into the brain [19]. Earlier studies have also indicated that hypothalamic astrocytes are particularly and primarily affected in response to metabolic damage induced by hypercaloric diets [20], positioning these glial cells as the center of attention for obesity research. In a more physiological context though, current literature underlines the role of astrocytes in the control of metabolism by their role of nutrient and hormone sensing regulation, and via the release of gliotransmitters in crosstalk with neurons.

2.1. *Hypothalamic Astrocytes: In Bidirectional Tuning with Neurons for Hypothalamic Feeding Control*

In recent years, several studies based on Designer Receptor-Exclusively Activated by Designed Drug (DREADD) have revealed that changes in astrocyte activity can impact the feeding behavior response to leptin and ghrelin. In 2015, a study led by Yang and colleagues showed that activation of astrocytes within the mediobasal hypothalamus (MBH) leads to reduced feeding by inhibiting AgRP neurons via adenosine A1 receptors, both in basal conditions and after ghrelin-evoked feeding. Moreover, pharmacogenetic activation of MBH-astrocytes increases leptin's anorectic effect [6]. This study proposed a mechanism by which astrocytes participate in food intake regulation through the release of adenosine, a gliotransmitter known by its inhibitory action on both pre-and post- synaptic neurons [21], leading to the suppression of AgRP neuronal activity and its orexigenic action.

Contrary to this, a later study showed opposing results, observing an increase in AgRP neuronal stimulation and food intake after chemogenetic activation of astrocytes [7]. These discrepancies between studies could be explained by the heterogeneous nature of the glia–neuronal interconnections in the hypothalamic feeding circuits where their connectivity and plasticity can vary depending on health (also influenced by sexual hormones and age) and metabolic status of the animal. Therefore, small alterations in the experimental paradigm such as time, last meal, environment, the anatomical location, and number of hypothalamic astrocytes chemogenetically manipulated could shift the circuit specific manipulation for controlling feeding and thus explain the diverse result observed. A recent study has pointed out the essentiality of a regulatory role of AgRP neurons over adjacent astrocytes for dynamic fine-tuned adaptations of their activity as part of homeostatic control of feeding [10]. This study has shown that chemogenetic or ghrelin-evoked activation of AgRP neurons promotes the release of GABA to increase astrocytic endfeet coverage on the membrane of AgRP neurons and triggers prostaglandin E2 release to subsequently increase the excitability of AgRP neurons via EP2 receptors. These findings support a dynamic and bidirectional communication between AgRP neurons and astrocytes for a feed-forward autoactivation loop of hunger circuits. Other studies have also reported that a gliopeptide Acyl-CoA–binding protein–derived (ACBP-derived) and its product octadecaneuropeptide (ODN) secreted by astrocytes is important for hypothalamic astrocyte–neuron tuning in the regulation of energy balance. In particular, they showed that pan-brain deletion of the ACBP gene in astrocytes promotes diet-induced hyperphagia and obesity in mice. Mechanistically, they proposed a "gliogenic" endozepine released mechanism based on the ACBP–ODN signaling in ARC astrocytes to influence the activity of POMC neurons to reduce feeding and weight gain [9].

2.2. Hypothalamic Astrocytes: Nutrient and Hormone Sensing-Dependent Regulation

Lipid sensing. Astrocytes uptake lipids and store them in the form of droplets via several lipid transporters, including lipoprotein lipase (LPL), which is essential for the control of cellular lipid storage in the brain. When *lpl* is postnatally deleted in astrocytes, mice develop an aggravated obesogenic phenotype characterized by increased body weight gain and glucose intolerance in response to a high fat diet [22]. Mechanistically, it was found that these mice exhibited increased ceramide accumulation, specifically in hypothalamic neurons, a phenomenon observed in neurodegenerative processes induced by lipotoxicity [13,23]. In vitro studies have reported that hypothalamic astrocytes accumulate lipid droplets under saturated fatty acid-rich conditions, adopting a reactive inflammatory profile and eliciting chemotactic activity of microglia to produce inflammatory mediators [24]. Other in vitro work in human astrocytes has also pointed out a direct link between a disruption of lipid metabolism in astrocytes and insulin resistance [25]; however, the use in this case of cultured astrocytes would require additional confirmation for its relevance at hypothalamic level.

Ketone body sensing. Besides lipid homeostasis function, astrocytes play a key role in fatty acid oxidation by synthesizing the majority of ketone bodies (KB) in the CNS. Studies in mice have shown that short-term exposure to a hypercaloric diet (HFD) is associated with higher levels of astrocyte-derived KB in the hypothalamus and an acute decrease in food intake. Accordingly, pharmacological inhibition of ketone synthesis blunts HFD-induced anorexia, supporting a model where astrocytes regulate fatty acids and KB levels in the hypothalamus as a lipid sensing mechanism, which is disrupted in leptin resistant obese rats [26]. Further, in vitro studies with the ketone body β-hydroxybutyrate have shown the direct effect of KBs on lowering the consumption of glucose and higher ability of mitochondria to metabolize pyruvate in mouse astrocytes, and consequently affecting the delivery of glucose and lactate to neurons [27]. Overall, these in vitro and in vivo findings indicate that astrocytes play a key role in maintaining a tight regulation of hypothalamic lipid homeostasis necessary for correct metabolic control. Further studies will need to address in detail how the lipogenic/lipolytic state of hypothalamic astrocytes is able to determine its cellular bioenergetics and communication with the surrounding neurons for the regulation of metabolism.

Glucose sensing. The brain has high energetic requirements, with glucose being its preferred fuel. However, glucose or energy storage in the brain is rather limited and restricted to astrocytes that, unlike neurons, are able to store glucose in the form of glycogen deposits in order to energetically support themselves through its lactate-shuttle transport [28,29]. It must be noted though that other authors have raised some questions about whether lactate oxidation primarily occurs in neurons and therefore the lactate-shuttle transport model probably needs further validation [30]. Independent of this role of astrocytes in energy storing, they also behave as glucose sensors. One of the first pieces of evidence of astrocyte involvement in central glucose sensing was provided by studies based on glucose transporter 2 (GLUT2) interrogation, which demonstrated that its inhibition in the ARC impaired glucose sensing by inhibiting the insulin response in response to intra-carotid glucose injection [31]. Interestingly, it was described that GLUT2 is mainly expressed in astrocytes instead of neurons within the hypothalamus [32–34]. Later studies using *glut2* global knockout (KO) mice confirmed the relevance of astrocytes in central glucose sensing. The mice globally lacking GLUT2 exhibited elevated blood glucagon levels and a decreased glucagon secretion in response to fluctuations in systemic glucose, a phenotype which illustrates that GLUT2 is essential for the counterregulatory response to hypoglycemia. Further, transgenesis-induced GLUT2 rescue specifically in astrocytes was able to restore glucagon secretion [35]. However, these last effects were specifically observed in astrocytes located in the brainstem, a relevant brain region regarding astrocyte glucose sensing and metabolism [36–38]. Glucose sensing is also a mechanism involved in the control of food intake, and thus the effects derived from hypothalamic astrocytes affect the way energy balance is regulated. In this regard, a different study showed that transgenic mice with inhibited GLUT2-mediated glucose detection augmented daily food intake by a mechanism that increased the meal size, but not the number of meals [39]. GLUT2 is located not only in astrocytes [32,35,40], but also in tanycytes and neurons [41,42]; therefore, these findings cannot be attributed solely to the astrocytic glucose transporter. In parallel, astrocytes located within the hypothalamus were also associated with the glucose sensing mechanism via GLUT1 in streptozotocin (STZ)-treated rats. Virogenetic overexpression of GLUT1 in hypothalamic astrocytes of STZ-diabetic rats with reduced GLUT1 normalizes circulating glucose levels and restores hypothalamic glucose sensing in rats, namely it rescues glucose production measured in intrahypothalamic clamp studies [43]. These findings support hypothalamic astrocytes as being essential components of the glucose sensing mechanism.

Insulin sensing. Insulin receptor (IR) expressed in hypothalamic astrocytes has been proven critical for the entry of glucose into the brain and for the modulation of glucose metabolism within the brain. Indeed, mice with postnatal *ir* knockout in astrocytes develop a defective anorectic response to glucose and an altered glucose homeostasis [8,44]. Despite this, genetically ablated mice lack IR in the whole brain, the phenotype was reproduced under a virogenetic deletion approach, thus proving that astrocytes within the hypothalamus were sufficient to induce the changes in glucose homeostasis. Moreover, we observed that the glucose-evoked reduction in neuronal activation occurred specifically in the hypothalamus and particularly in POMC neurons, thus reinforcing the special relevance of astrocytes located in this brain region [8]. Subsequent in vitro studies confirmed that loss of IR in astrocytes downregulates the expression of GLUT1 [45], while insulin-like growth factor 1 receptor (IGF1R) cooperates together with insulin in order to stimulate glucose uptake by astrocytes [46]. Importantly, in vivo experiments have also shown that IR ablation in astrocytes induces a compensatory upregulation of IGF1 receptor [45] and therefore, the fact that these two receptors have high homology and convergent signaling pathways underlines the importance of understanding their dual functional role in astrocytes.

Leptin sensing. Initial studies in agouti viable yellow (Avy) mice, showed that leptin receptor (LepR) in astrocytes may play a role in leptin transport across the BBB and in the regulation of adult-onset obesity of these animals [47,48]. Molecular analysis of the brain in the same model, in addition to HFD-fed mice, showed that obesity rapidly increases LepR expression in astrocytes where it then influences the leptin-induced calcium signaling in astrocytes [49]. Later, additional studies showed that central and chronic leptin administration modifies hypothalamic astrocyte morphology and glutamate transporters [50,51]. The subsequent generation of specific astrocyte LepR ablation mouse models confirmed these initial findings. In this regard, constitutive astrocyte-specific *lepr* (KO) mice did not develop an altered phenotype in terms of energy balance, likely due to a compensatory mechanism [52]. However, inducible astrocyte-specific *lepr* KO mice exhibited a lack of proper anorectic response to leptin and showed an exacerbated refeeding after fasting. Interestingly, both physiological alterations were associated with changes in POMC neurons (i.e., lower activation after leptin treatment) and AgRP neurons (i.e., overactivation during the refeeding phase) [53]. Overall, these findings support the existence and functional relevance of LepR in astrocytes. In agreement with previous reports, our work has recently confirmed the specific presence of the truncated $LepR_a$ in fluorescence-activated cell sorting (FACS)-isolated hypothalamic astrocytes [19].

GLP1 sensing. Recent studies have highlighted the role of astrocytic glucagon-like peptide-1 (GLP-1) signaling in the regulation of brain glucose uptake [54]. When GLP-1R was postnatally and genetically ablated in hypothalamic astrocytes, mice displayed an increased brain glucose uptake, which was associated with an improvement in memory formation and systemic glucose response [54]. Further evidence suggested that these effects are mediated through fibroblast growth factor (FGF)-21 action, given that neutralization of central FGF21, as well as astrocyte-specific ablation of FGF21, abrogated the improvements in glucose metabolism observed in the astrocyte-specific *glp-1r* knockout mice [54].

3. Hypothalamic Astrocytes in the Physiopathological Development of Obesity

Based on human genome-wide association studies (GWAS) from obese patients, it has recently emerged that the brain, in particular the hypothalamus, controls most aspects of systemic metabolism over other classically described peripheral endocrine axes, suggesting that obesity is largely a brain disease [55]. Interestingly, clinical studies have reported that being obese is associated with brain atrophy; describing alterations in food reward-regulating circuits [56], nutrient and hormone sensing-dependent neuronal activity [57,58], hypothalamic gliosis [20,59–61], and angiopathy [62], all being signs of neuronal dysfunction and aging [63]. Most of these changes have been reported to originate from the hypothalamus, which is considered a selective vulnerable area in diet-induced obesity and has been suggested as a hallmark of neuronal dysfunction associated with the initiation and progression of this pathology [64,65]. In fact, it is thought that obesity-induced hypothalamic dysfunction emerges from aberrant glia responses (astrocyte and microglia) inducing impairments in the connectivity and activity (responsiveness) of ARC-resident neuronal circuits responsible for orchestrating metabolic control and contributing to the aggravation of this pathology [64–67]. Despite the relative contribution of each glial cell type [68] and their crosstalk [69] to obesity pathophysiology being a current topic of research in this section, we will focus in the astrocyte-related mechanisms.

3.1. *Inflammatory-Related Mechanisms*

One of the most established pathophysiological mechanisms to explain obesity development is the inflammatory and oxidative stress hypothesis, mediated through different pathways, occurring in different endocrine tissues, including in the hypothalamus. Early studies pointed out that rich caloric diets induce a low-grade inflammation within the hypothalamus characterized by a rapid rise in cytokines [70], gliosis, and metabolic stress leading to dysfunction of hypothalamic circuits and consequent obesity [20]. Moreover, hypothalamic inflammation has also been associated with the development of central resistance to endocrine signals before changes in body weight [20,70,71]. Indeed, hypothalamic inflammatory processes are tightly linked with hormone resistance and obesity. This is supported by studies reporting that amelioration of hypothalamic inflammatory processes i.e., a blunted tumor necrosis factor alpha (TNFα) signaling, an ablated suppressor of cytokine signaling-3 (Socs3), and the disruption of hypothalamic-specific interleukin (IL)-6) led to rescue of both leptin and insulin signaling and protection from the metabolic alterations induced by HFD [72–74]. Several pathways have been proposed as mediators of hypothalamic inflammatory responses in diet-induced obesity. Notably, some studies pointed out that the IkB kinase-β (IKKβ)/nuclear factor κB (NF-κB) pathway mediates the expression of several pro-inflammatory genes within the hypothalamus, including cytokines and chemokines [75], and leads to endoplasmic reticulum (ER) stress, a cellular process characterized by the accumulation of misfolded or unfolded proteins in the ER lumen [76,77]. Both signaling pathways are actually interconnected, and it was shown that HFD leads to increased ER stress and consequently activates IKKβ–NF-κB signaling in the hypothalamus, producing the altered response to hormones and ultimately leading to metabolic homeostasis dysregulation.

3.2. Reactive Astrogliosis in Obesity

This term, which recently reached a consensus agreement on its definition, together with "astrocyte reactivity" or "reactive astrocytes" [78], entails the morphological and/or functional changes seen in astrocytes responding to healing processes or pathophysiology [79]. Seminal studies led by Horvath and colleagues reported that diet-induced obesity in mice also induces reactive astrogliosis in the hypothalamus [18]. The ultrastructural analyses of these brains also showed that, concomitant with the findings, there were changes in the physical interactions of astrocytes with endothelial cells and neurons causing alterations of the cytoarchitecture and synaptology of hypothalamic circuits [18]. Surprisingly, these astrocytic morphological changes, as well as the increased cytokine production in the hypothalamus, occurred before any sign of systemic inflammation or body weight gain [20]. HFD-induced reactive astrogliosis is characterized by an upregulation of the structural protein glial fibrillary acidic protein (GFAP), promotion of a pro-inflammatory phenotype (producing and releasing cytokine markers), and the acquisition of a hypertrophic morphology, but without proliferation-unlike what was observed in other brain pathologies. Remarkably, a recent study has found that GFAP gene induction occurs very early, and even one hour of HFD exposure is sufficient to find a significant change. That short exposure interferes also in the neuropeptide and an inflammatory-like gene responses exerted by the chemogenetic activation of astrocytes [80].

Blocking inflammatory-induced astrocyte activation via expression of a dominant-negative form of the NFκB inhibitor, IκBα, in GFAP$^+$ astrocytes in a doxycycline-inducible manner showed an inhibition of HFD-induced reactive astrogliosis in the hypothalamus coupled with a promotion of hyperphagia, even 24 h following the dietary switch [81]. These findings suggest that the NFκB-IκBα pathway in astrocytes controls HFD-induced reactive astrogliosis illustrating acute morphological adaptations in astrocytes for precisely controlling food intake [81]. Of note, astrocyte changes described here were found in the MBH; the genetic mouse model was generated modifying all GFAP$^+$ astrocytes in the CNS and therefore an indirect effect via non-hypothalamic regions cannot be ruled out. Douglass and colleagues found that postnatal ablation of IKKβ specifically in astrocytes after HFD exposure has a protective effect on weight gain mediated through a reduction in food intake and an increase in energy expenditure, which was also accompanied by an improved glucose tolerance and insulin sensitivity. Importantly, both hypothalamic inflammation and reactive astrogliosis were diminished in HFD-fed IKKβKO mice [82]. These findings were similarly supported by other authors showing that astrocytic IKKβ/NF-κB upregulation is a key signaling pathway during overnutrition, which mediates some of the early-onset effects of obesity. Mechanistically, astrocytic activation via IKKβ/NF-κB seems to modify astrocytic morphology (i.e., decreased astrocytic process density), a downstream regulation of hypothalamic extracellular GABA levels, and BDNF expression within the hypothalamus [83]. That being said, hypothalamic inflammation and reactive astrogliosis have been used as generic terms to describe a variety of morphological and molecular changes occurring under the effects of high fat diet exposure, sometimes describing phenotypes under the same phenomena, which only partly fulfill the molecular and cellular characteristics of the other phenotypes reported. In this regard, there is a clear demand for neuroscientists to agree on a clear and meaningful nomenclature, which precisely describes the morphological and functional readouts to evaluate.

Some studies indicate that dietary factors such as fat are determinant triggers to the hypothalamic inflammatory response and the changes induced in glial cells. In vitro studies have pointed out that saturated fatty acids (i.e., palmitic, lauric, and stearic acid) induce the release of TNF-α and IL-6 from cultured astrocytes [84]. Mechanistically, saturated fatty acids require toll-like receptor 4 (TLR4) to induce cytokine release, while the pharmacological inhibition of p38 or p42/44 MAPK pathways precludes the pro-inflammatory actions of the saturated fatty acids [84]. In agreement with these findings, TLR2/4-deficient mice were protected from HFD-induced weight gain as well as glucose intolerance and insulin resistance. Moreover, C3H/HeJ/TLR2-deficient mice also showed a decreased expression of IL-6 in both basal and insulin-stimulated states. Following that, treatment with an IL-6-neutralizing antibody led to a rescue of insulin sensitivity in HFD-fed mice despite no changes being found in body weight [85]. Gao and colleagues found that while reactive astrogliosis was mainly a response to the fat content of the diet, it is the combination of sucrose and fat consumption that are responsible for stimulating some other aspects of hypothalamic inflammatory response, such as microgliosis and angiogenesis [86]. Lastly, other studies have highlighted that the astrocytes' response to develop an HFD-induced hypertrophic phenotype can vary depending on the age of the mice. Specifically, Lemus and colleagues observed that aged mice showed a higher GFAP expression in basal conditions than younger ones, which can affect the visualization of differences between mice fed with chow diet and HFD at a certain age [87]. Aside from diet-induced obesity, other studies have also reported the presence of reactive astrogliosis in monogenetic genetic obese mouse models [47,49,88], or by maternal dietary influence during the fetal and early stages of life of the offspring [51]. Therefore, a careful interpretation has to be taken about the context in which reactive astrogliosis is evaluated because the molecular pathways entailed and functional implications in obesity development could vary depending on the nature of the trigger (e.g., dietary factor, aberrant metabolic, or hormone inputs), the brain areas involved, the status of surrounding cellular compounds (microgliosis and neuronal connectivity), as well as other relevant parameters, such as health state and age of the animal that could interfere with the resulting outcomes and conclusions.

3.3. *Alterations in the Astroglia-Vascular Interface in Obesity*

HFD-induced reactive astrogliosis has been detected in specific hypothalamic regions, in particular, those ones close to the third ventricle, such as the MBH [18,64,88], which have a higher accessibility of circulating factors and are partly placed out of the BBB [88]. Indeed, HFD-fed mice show a significant increase in blood vessel length and density in the MBH, as well as a lack of BBB integrity [62]. Interestingly, human diabetic patients exhibited an increased number of arteries and arterioles in the hypothalamic infundibular nuclei, suggesting a potential link between hypothalamic vascular remodeling and the development of metabolic diseases in humans [62]. A recent study by our group has further dissected the pathogenic processes leading to such impairments, uncovering the role of hypothalamic astrocytes in promoting hypothalamic microangiopathy and the systemic arterial hypertension associated with obesity. By using virus-mediated gene transfer, it was found that astroglial vascular endothelial growth factor (VEGF) in the hypothalamus exerts significant effects on systemic blood pressure control via tuning sympathetic outflow. These new findings supporting a novel mechanism with hypothalamic angiopathies at its center are important as they could explain how obesity-associated hyperleptinemia rather than adiposity causes hypertension [19] (Figure 1).

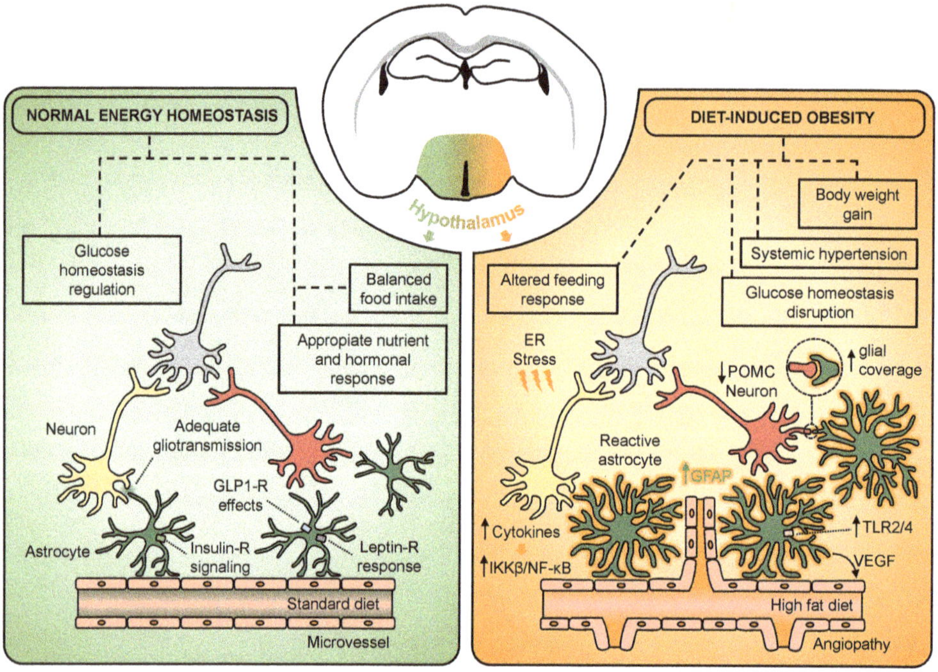

Figure 1. Hypothalamic astrocytes participate in the regulation of metabolic homeostasis and act as key players in the development of diet-induced obesity. Astrocytes located in the hypothalamus express several endocrine receptors such as, insulin, leptin, and GLP-1. The action of hormones and pharmacological analogues on hypothalamic astrocytes has been described to influence several aspects of normal energy homeostasis. Under energy-dense diet exposure, the hypothalamus experiences a low-grade inflammation characterized by a rapid rise in cytokines, which trigger downstream pathways such as the IKKβ/NF-κB and TLR2/4. This also leads to increased hypothalamic ER stress, which seems to interfere with the adequate response to hormones. Together, there is an HFD-reactive astrogliosis characterized by an overexpression of GFAP, the acquisition of a hypertrophic morphology, and a synaptic input reorganization of the melanocortin system, which may suggest a direct involvement of hypothalamic astrocytes in the pathogenesis of diet-induced obesity. At vascular level, obesogenic diets trigger astroglial VEGF-driving angiogenesis in the hypothalamus to promote the development of systemic hypertension. ER, endoplasmic reticulum; IKKβ, IkB kinase-β; GFAP, glial fibrillary acidic protein; GLP-1, glucagon-like peptide-1; NF-κB, nuclear factor κB; POMC, proopiomelanocortin; TLR2/4, toll-like receptor 2/4; VEGF, vascular endothelial growth factor.

4. Outlook

The hypothalamus, with its different nuclei, is a good example of how different neuronal types and connections translate to functional specialized roles. Given the complex nature of hypothalamic cytoarchitecture where circuits that play opposing roles in the control of metabolism are tightly intermingled, it will not be surprising that the study of astrocytes in the control of metabolism should be approached to specifically address the precise anatomical location of the circuits and the interaction among them. To date, several studies have contributed to understanding the functional relevance of astrocytes in the development of obesity; however, a big gap is still present on how these alterations affect the neural networks in which astrocytes are embedded. Thus, the idea of a disruption in the astrocyte-neuron communication within the hypothalamus seems a plausible, yet poorly investigated, hypothesis that could contribute to the pathophysiology of obesity. Research on how astrocyte–neuron communication is altered by exposure to obesogenic insults will allow a better understanding of the hypothalamic alterations that occur in this context.

Another important factor linked to the regional specialization is astrocyte heterogeneity. Regardless of the classical morphological taxonomy (i.e., fibrous and protoplasmic astrocytes), the specific molecular markers label a wide range of astrocyte subpopulations, which could also indicate a complex array of functional actions. In the upcoming years, an important topic to address will be the relative contribution of each subpopulation to the development of obesity. Taking advantage of the novel technological advances in the study of cell- and synapse-specific functionality by using combined two-photon imaging, patch-clamp recordings, cellular mapping of transcriptomic profiles in the tissue context using single-cell RNA sequencing, and new in situ hybridization technologies. In combination, researchers will be able to understand the tissue structure and function on a single astrocyte level, helping to identify specific hypothalamic astrocyte subpopulations, which might be critical for the pathophysiology and development of obesity.

Author Contributions: Conceptualization, I.G.-G.; writing—original draft and figure preparation, I.G.-G.; writing—review and editing, I.G.-G. and C.G.-C. All authors have read and agreed to the published version of the manuscript.

Funding: This research was funded by European Research Council ERC (StG AstroNeuroCrosstalk no. 757393), the German Research Foundation DFG under Germany's Excellence Strategy within the framework of the Munich Cluster for Systems Neurology (EXC 2145 SyNergy-ID 390857198) and Helmholtz Excellence Network, the German Research Foundation DFG Nutripathos Project ANR-15-CE14-0030, the DZD, and the Initiative and Networking Fund of the Helmholtz Association. IG-G is a recipient of a fellowship from European Union's Horizon 2020 research and innovation program under the Marie Sklodowska-Curie actions (842080-H2020-MSCA-IF-2018).

Institutional Review Board Statement: Not applicable.

Informed Consent Statement: Not applicable.

Data Availability Statement: Not applicable.

Acknowledgments: The authors thank Cassie Holleman (Helmholtz Zentrum München; Germany) for her comments.

Conflicts of Interest: The authors declare no conflict of interest. The funders had no role in study design, data collection, or analyses, decision to publish, or preparation of the manuscript.

References

1. Schneeberger, M.; Gomis, R.; Claret, M. Hypothalamic and brainstem neuronal circuits controlling homeostatic energy balance. *J. Endocrinol.* **2014**, *220*, T25–T46. [CrossRef]
2. Timper, K.; Bruning, J.C. Hypothalamic circuits regulating appetite and energy homeostasis: Pathways to obesity. *Dis. Model. Mech.* **2017**, *10*, 679–689. [CrossRef] [PubMed]
3. Kim, K.S.; Seeley, R.J.; Sandoval, D.A. Signalling from the periphery to the brain that regulates energy homeostasis. *Nat. Rev. Neurosci.* **2018**, *19*, 185–196. [CrossRef]
4. Hetherington, A.W.; Ranson, S.W. Hypothalamic lesions and adiposity in the rat. *Anat. Rec.* **1940**, *78*, 149–172. [CrossRef]
5. Waterson, M.J.; Horvath, T.L. Neuronal Regulation of Energy Homeostasis: Beyond the Hypothalamus and Feeding. *Cell Metab.* **2015**, *22*, 962–970. [CrossRef]
6. Yang, L.; Qi, Y.; Yang, Y. Astrocytes control food intake by inhibiting AGRP neuron activity via adenosine A1 receptors. *Cell Rep.* **2015**, *11*, 798–807. [CrossRef]
7. Chen, N.; Sugihara, H.; Kim, J.; Fu, Z.; Barak, B.; Sur, M.; Feng, G.; Han, W. Direct modulation of GFAP-expressing glia in the arcuate nucleus bi-directionally regulates feeding. *Elife* **2016**, *5*, e18716. [CrossRef] [PubMed]
8. Garcia-Caceres, C.; Quarta, C.; Varela, L.; Gao, Y.; Gruber, T.; Legutko, B.; Jastroch, M.; Johansson, P.; Ninkovic, J.; Yi, C.X.; et al. Astrocytic Insulin Signaling Couples Brain Glucose Uptake with Nutrient Availability. *Cell* **2016**, *166*, 867–880. [CrossRef]
9. Bouyakdan, K.; Martin, H.; Lienard, F.; Budry, L.; Taib, B.; Rodaros, D.; Chretien, C.; Biron, E.; Husson, Z.; Cota, D.; et al. The gliotransmitter ACBP controls feeding and energy homeostasis via the melanocortin system. *J. Clin. Investig.* **2019**, *129*, 2417–2430. [CrossRef] [PubMed]
10. Varela, L.; Stutz, B.; Song, J.E.; Kim, J.G.; Liu, Z.W.; Gao, X.B.; Horvath, T.L. Hunger-promoting AgRP neurons trigger an astrocyte-mediated feed-forward auto-activation loop in mice. *J. Clin. Investig.* **2021**. [CrossRef]
11. Miyata, S. New aspects in fenestrated capillary and tissue dynamics in the sensory circumventricular organs of adult brains. *Front. Neurosci.* **2015**, *9*, 390. [CrossRef] [PubMed]

12. Moulle, V.S.; Picard, A.; Le Foll, C.; Levin, B.E.; Magnan, C. Lipid sensing in the brain and regulation of energy balance. *Diabetes Metab.* **2014**, *40*, 29–33. [CrossRef] [PubMed]
13. Gonzalez-Garcia, I.; Ferno, J.; Dieguez, C.; Nogueiras, R.; Lopez, M. Hypothalamic Lipids: Key Regulators of Whole Body Energy Balance. *Neuroendocrinology* **2017**, *104*, 398–411. [CrossRef] [PubMed]
14. Morton, G.J.; Cummings, D.E.; Baskin, D.G.; Barsh, G.S.; Schwartz, M.W. Central nervous system control of food intake and body weight. *Nature* **2006**, *443*, 289–295. [CrossRef] [PubMed]
15. Myers, M.G., Jr.; Munzberg, H.; Leinninger, G.M.; Leshan, R.L. The geometry of leptin action in the brain: More complicated than a simple ARC. *Cell Metab.* **2009**, *9*, 117–123. [CrossRef]
16. Rodriguez, E.M.; Blazquez, J.L.; Guerra, M. The design of barriers in the hypothalamus allows the median eminence and the arcuate nucleus to enjoy private milieus: The former opens to the portal blood and the latter to the cerebrospinal fluid. *Peptides* **2010**, *31*, 757–776. [CrossRef] [PubMed]
17. Langlet, F. Tanycytes: A gateway to the metabolic hypothalamus. *J. Neuroendocrinol.* **2014**, *26*, 753–760. [CrossRef] [PubMed]
18. Horvath, T.L.; Sarman, B.; Garcia-Caceres, C.; Enriori, P.J.; Sotonyi, P.; Shanabrough, M.; Borok, E.; Argente, J.; Chowen, J.A.; Perez-Tilve, D.; et al. Synaptic input organization of the melanocortin system predicts diet-induced hypothalamic reactive gliosis and obesity. *Proc. Natl. Acad. Sci. USA* **2010**, *107*, 14875–14880. [CrossRef]
19. Gruber, T.; Pan, C.; Contreras, R.E.; Wiedemann, T.; Morgan, D.A.; Skowronski, A.A.; Lefort, S.; de Bernardis Murat, C.; Le Thuc, O.; Legutko, B.; et al. Obesity-associated hyperleptinemia alters the gliovascular interface of the hypothalamus to promote hypertension. *Cell Metab.* **2021**, *33*, 1155–1170.e10. [CrossRef]
20. Thaler, J.P.; Yi, C.X.; Schur, E.A.; Guyenet, S.J.; Hwang, B.H.; Dietrich, M.O.; Zhao, X.; Sarruf, D.A.; Izgur, V.; Maravilla, K.R.; et al. Obesity is associated with hypothalamic injury in rodents and humans. *J. Clin. Investig.* **2012**, *122*, 153–162. [CrossRef]
21. Thompson, S.M.; Haas, H.L.; Gahwiler, B.H. Comparison of the actions of adenosine at pre- and postsynaptic receptors in the rat hippocampus in vitro. *J. Physiol.* **1992**, *451*, 347–363. [CrossRef]
22. Gao, Y.; Layritz, C.; Legutko, B.; Eichmann, T.O.; Laperrousaz, E.; Moulle, V.S.; Cruciani-Guglielmacci, C.; Magnan, C.; Luquet, S.; Woods, S.C.; et al. Disruption of Lipid Uptake in Astroglia Exacerbates Diet-Induced Obesity. *Diabetes* **2017**, *66*, 2555–2563. [CrossRef]
23. Chaurasia, B.; Summers, S.A. Ceramides—Lipotoxic Inducers of Metabolic Disorders. *Trends Endocrinol. Metab.* **2015**, *26*, 538–550. [CrossRef]
24. Kwon, Y.H.; Kim, J.; Kim, C.S.; Tu, T.H.; Kim, M.S.; Suk, K.; Kim, D.H.; Lee, B.J.; Choi, H.S.; Park, T.; et al. Hypothalamic lipid-laden astrocytes induce microglia migration and activation. *FEBS Lett.* **2017**, *591*, 1742–1751. [CrossRef] [PubMed]
25. Heni, M.; Eckstein, S.S.; Schittenhelm, J.; Bohm, A.; Hogrefe, N.; Irmler, M.; Beckers, J.; Hrabe de Angelis, M.; Haring, H.U.; Fritsche, A.; et al. Ectopic fat accumulation in human astrocytes impairs insulin action. *R. Soc. Open Sci.* **2020**, *7*, 200701. [CrossRef] [PubMed]
26. Le Foll, C.; Dunn-Meynell, A.A.; Miziorko, H.M.; Levin, B.E. Regulation of hypothalamic neuronal sensing and food intake by ketone bodies and fatty acids. *Diabetes* **2014**, *63*, 1259–1269. [CrossRef]
27. Valdebenito, R.; Ruminot, I.; Garrido-Gerter, P.; Fernandez-Moncada, I.; Forero-Quintero, L.; Alegria, K.; Becker, H.M.; Deitmer, J.W.; Barros, L.F. Targeting of astrocytic glucose metabolism by beta-hydroxybutyrate. *J. Cereb. Blood Flow Metab.* **2016**, *36*, 1813–1822. [CrossRef] [PubMed]
28. Pellerin, L.; Pellegri, G.; Bittar, P.G.; Charnay, Y.; Bouras, C.; Martin, J.L.; Stella, N.; Magistretti, P.J. Evidence supporting the existence of an activity-dependent astrocyte-neuron lactate shuttle. *Dev. Neurosci.* **1998**, *20*, 291–299. [CrossRef] [PubMed]
29. Chih, C.P.; Roberts, E.L., Jr. Energy substrates for neurons during neural activity: A critical review of the astrocyte-neuron lactate shuttle hypothesis. *J. Cereb. Blood Flow Metab.* **2003**, *23*, 1263–1281. [CrossRef] [PubMed]
30. Hertz, L. The astrocyte-neuron lactate shuttle: A challenge of a challenge. *J. Cereb. Blood Flow Metab.* **2004**, *24*, 1241–1248. [CrossRef]
31. Leloup, C.; Orosco, M.; Serradas, P.; Nicolaidis, S.; Penicaud, L. Specific inhibition of GLUT2 in arcuate nucleus by antisense oligonucleotides suppresses nervous control of insulin secretion. *Mol. Brain Res.* **1998**, *57*, 275–280. [CrossRef]
32. Leloup, C.; Arluison, M.; Lepetit, N.; Cartier, N.; Marfaing-Jallat, P.; Ferre, P.; Penicaud, L. Glucose transporter 2 (GLUT 2): Expression in specific brain nuclei. *Brain Res.* **1994**, *638*, 221–226. [CrossRef]
33. Arluison, M.; Quignon, M.; Nguyen, P.; Thorens, B.; Leloup, C.; Penicaud, L. Distribution and anatomical localization of the glucose transporter 2 (GLUT2) in the adult rat brain—an immunohistochemical study. *J. Chem. Neuroanat.* **2004**, *28*, 117–136. [CrossRef]
34. Arluison, M.; Quignon, M.; Thorens, B.; Leloup, C.; Penicaud, L. Immunocytochemical localization of the glucose transporter 2 (GLUT2) in the adult rat brain. II. Electron microscopic study. *J. Chem. Neuroanat.* **2004**, *28*, 137–146. [CrossRef]
35. Marty, N.; Dallaporta, M.; Foretz, M.; Emery, M.; Tarussio, D.; Bady, I.; Binnert, C.; Beermann, F.; Thorens, B. Regulation of glucagon secretion by glucose transporter type 2 (glut2) and astrocyte-dependent glucose sensors. *J. Clin. Investig.* **2005**, *115*, 3545–3553. [CrossRef]
36. McDougal, D.H.; Hermann, G.E.; Rogers, R.C. Astrocytes in the nucleus of the solitary tract are activated by low glucose or glucoprivation: Evidence for glial involvement in glucose homeostasis. *Front. Neurosci.* **2013**, *7*, 249. [CrossRef]
37. McDougal, D.H.; Viard, E.; Hermann, G.E.; Rogers, R.C. Astrocytes in the hindbrain detect glucoprivation and regulate gastric motility. *Auton. Neurosci.* **2013**, *175*, 61–69. [CrossRef]

38. Rogers, R.C.; McDougal, D.H.; Hermann, G.E. Hindbrain Astrocyte Glucodetectors and Counterregulation. In *Appetite and Food Intake: Central Control*, 2nd ed.; Harris, R.B.S., Ed.; CRC Press: Boca Raton, FL, USA, 2017; pp. 205–228. [CrossRef]
39. Stolarczyk, E.; Guissard, C.; Michau, A.; Even, P.C.; Grosfeld, A.; Serradas, P.; Lorsignol, A.; Penicaud, L.; Brot-Laroche, E.; Leturque, A.; et al. Detection of extracellular glucose by GLUT2 contributes to hypothalamic control of food intake. *Am. J. Physiol. Endocrinol. Metab.* **2010**, *298*, E1078–E1087. [CrossRef] [PubMed]
40. Young, J.K.; McKenzie, J.C. GLUT2 immunoreactivity in Gomori-positive astrocytes of the hypothalamus. *J. Histochem. Cytochem.* **2004**, *52*, 1519–1524. [CrossRef]
41. Kang, L.; Routh, V.H.; Kuzhikandathil, E.V.; Gaspers, L.D.; Levin, B.E. Physiological and molecular characteristics of rat hypothalamic ventromedial nucleus glucosensing neurons. *Diabetes* **2004**, *53*, 549–559. [CrossRef] [PubMed]
42. Garcia, M.; Millan, C.; Balmaceda-Aguilera, C.; Castro, T.; Pastor, P.; Montecinos, H.; Reinicke, K.; Zuniga, F.; Vera, J.C.; Onate, S.A.; et al. Hypothalamic ependymal-glial cells express the glucose transporter GLUT2, a protein involved in glucose sensing. *J. Neurochem.* **2003**, *86*, 709–724. [CrossRef]
43. Chari, M.; Yang, C.S.; Lam, C.K.; Lee, K.; Mighiu, P.; Kokorovic, A.; Cheung, G.W.; Lai, T.Y.; Wang, P.Y.; Lam, T.K. Glucose transporter-1 in the hypothalamic glial cells mediates glucose sensing to regulate glucose production in vivo. *Diabetes* **2011**, *60*, 1901–1906. [CrossRef]
44. Gonzalez-Garcia, I.; Gruber, T.; Garcia-Caceres, C. Insulin action on astrocytes: From energy homeostasis to behaviour. *J. Neuroendocrinol.* **2021**, *33*, e12953. [CrossRef]
45. Hernandez-Garzon, E.; Fernandez, A.M.; Perez-Alvarez, A.; Genis, L.; Bascunana, P.; Fernandez de la Rosa, R.; Delgado, M.; Angel Pozo, M.; Moreno, E.; McCormick, P.J.; et al. The insulin-like growth factor I receptor regulates glucose transport by astrocytes. *Glia* **2016**, *64*, 1962–1971. [CrossRef]
46. Fernandez, A.M.; Hernandez-Garzon, E.; Perez-Domper, P.; Perez-Alvarez, A.; Mederos, S.; Matsui, T.; Santi, A.; Trueba-Saiz, A.; Garcia-Guerra, L.; Pose-Utrilla, J.; et al. Insulin Regulates Astrocytic Glucose Handling Through Cooperation with IGF-I. *Diabetes* **2017**, *66*, 64–74. [CrossRef] [PubMed]
47. Pan, W.; Hsuchou, H.; He, Y.; Sakharkar, A.; Cain, C.; Yu, C.; Kastin, A.J. Astrocyte leptin receptor (ObR) and leptin transport in adult-onset obese mice. *Endocrinology* **2008**, *149*, 2798–2806. [CrossRef] [PubMed]
48. Pan, W.; Hsuchou, H.; Xu, C.; Wu, X.; Bouret, S.G.; Kastin, A.J. Astrocytes modulate distribution and neuronal signaling of leptin in the hypothalamus of obese A vy mice. *J. Mol. Neurosci.* **2011**, *43*, 478–484. [CrossRef]
49. Hsuchou, H.; He, Y.; Kastin, A.J.; Tu, H.; Markadakis, E.N.; Rogers, R.C.; Fossier, P.B.; Pan, W. Obesity induces functional astrocytic leptin receptors in hypothalamus. *Brain* **2009**, *132*, 889–902. [CrossRef]
50. Garcia-Caceres, C.; Fuente-Martin, E.; Burgos-Ramos, E.; Granado, M.; Frago, L.M.; Barrios, V.; Horvath, T.; Argente, J.; Chowen, J.A. Differential acute and chronic effects of leptin on hypothalamic astrocyte morphology and synaptic protein levels. *Endocrinology* **2011**, *152*, 1809–1818. [CrossRef]
51. Fuente-Martin, E.; Garcia-Caceres, C.; Granado, M.; de Ceballos, M.L.; Sanchez-Garrido, M.A.; Sarman, B.; Liu, Z.W.; Dietrich, M.O.; Tena-Sempere, M.; Argente-Arizon, P.; et al. Leptin regulates glutamate and glucose transporters in hypothalamic astrocytes. *J. Clin. Investig.* **2012**, *122*, 3900–3913. [CrossRef] [PubMed]
52. Jayaram, B.; Pan, W.; Wang, Y.; Hsuchou, H.; Mace, A.; Cornelissen-Guillaume, G.G.; Mishra, P.K.; Koza, R.A.; Kastin, A.J. Astrocytic leptin-receptor knockout mice show partial rescue of leptin resistance in diet-induced obesity. *J. Appl. Physiol.* **2013**, *114*, 734–741. [CrossRef] [PubMed]
53. Kim, J.G.; Suyama, S.; Koch, M.; Jin, S.; Argente-Arizon, P.; Argente, J.; Liu, Z.W.; Zimmer, M.R.; Jeong, J.K.; Szigeti-Buck, K.; et al. Leptin signaling in astrocytes regulates hypothalamic neuronal circuits and feeding. *Nat. Neurosci.* **2014**, *17*, 908–910. [CrossRef]
54. Timper, K.; Del Rio-Martin, A.; Cremer, A.L.; Bremser, S.; Alber, J.; Giavalisco, P.; Varela, L.; Heilinger, C.; Nolte, H.; Trifunovic, A.; et al. GLP-1 Receptor Signaling in Astrocytes Regulates Fatty Acid Oxidation, Mitochondrial Integrity, and Function. *Cell Metab.* **2020**, *31*, 1189–1205.e13. [CrossRef]
55. Locke, A.E.; Kahali, B.; Berndt, S.I.; Justice, A.E.; Pers, T.H.; Day, F.R.; Powell, C.; Vedantam, S.; Buchkovich, M.L.; Yang, J.; et al. Genetic studies of body mass index yield new insights for obesity biology. *Nature* **2015**, *518*, 197–206. [CrossRef]
56. Shin, A.C.; Filatova, N.; Lindtner, C.; Chi, T.; Degann, S.; Oberlin, D.; Buettner, C. Insulin Receptor Signaling in POMC, but Not AgRP, Neurons Controls Adipose Tissue Insulin Action. *Diabetes* **2017**, *66*, 1560–1571. [CrossRef] [PubMed]
57. Kullmann, S.; Kleinridders, A.; Small, D.M.; Fritsche, A.; Haring, H.U.; Preissl, H.; Heni, M. Central nervous pathways of insulin action in the control of metabolism and food intake. *Lancet Diabetes Endocrinol.* **2020**, *8*, 524–534. [CrossRef]
58. Kullmann, S.; Valenta, V.; Wagner, R.; Tschritter, O.; Machann, J.; Haring, H.U.; Preissl, H.; Fritsche, A.; Heni, M. Brain insulin sensitivity is linked to adiposity and body fat distribution. *Nat. Commun.* **2020**, *11*, 1–6. [CrossRef]
59. Lee, D.; Thaler, J.P.; Berkseth, K.E.; Melhorn, S.J.; Schwartz, M.W.; Schur, E.A. Longer T(2) relaxation time is a marker of hypothalamic gliosis in mice with diet-induced obesity. *Am. J. Physiol. Endocrinol. Metab.* **2013**, *304*, E1245–E1250. [CrossRef] [PubMed]
60. Berkseth, K.E.; Guyenet, S.J.; Melhorn, S.J.; Lee, D.; Thaler, J.P.; Schur, E.A.; Schwartz, M.W. Hypothalamic gliosis associated with high-fat diet feeding is reversible in mice: A combined immunohistochemical and magnetic resonance imaging study. *Endocrinology* **2014**, *155*, 2858–2867. [CrossRef] [PubMed]

61. Schur, E.A.; Melhorn, S.J.; Oh, S.K.; Lacy, J.M.; Berkseth, K.E.; Guyenet, S.J.; Sonnen, J.A.; Tyagi, V.; Rosalynn, M.; de Leon, B.; et al. Radiologic evidence that hypothalamic gliosis is associated with obesity and insulin resistance in humans. *Obesity* **2015**, *23*, 2142–2148. [CrossRef]
62. Yi, C.X.; Gericke, M.; Kruger, M.; Alkemade, A.; Kabra, D.G.; Hanske, S.; Filosa, J.; Pfluger, P.; Bingham, N.; Woods, S.C.; et al. High calorie diet triggers hypothalamic angiopathy. *Mol. Metab.* **2012**, *1*, 95–100. [CrossRef] [PubMed]
63. O'Brien, P.D.; Hinder, L.M.; Callaghan, B.C.; Feldman, E.L. Neurological consequences of obesity. *Lancet Neurol.* **2017**, *16*, 465–477. [CrossRef]
64. Garcia-Caceres, C.; Yi, C.X.; Tschop, M.H. Hypothalamic astrocytes in obesity. *Endocrinol. Metab. Clin.* **2013**, *42*, 57–66. [CrossRef]
65. Lee, C.H.; Suk, K.; Yu, R.; Kim, M.S. Cellular Contributors to Hypothalamic Inflammation in Obesity. *Mol. Cells* **2020**, *43*, 431–437. [CrossRef]
66. Kalin, S.; Heppner, F.L.; Bechmann, I.; Prinz, M.; Tschop, M.H.; Yi, C.X. Hypothalamic innate immune reaction in obesity. *Nat. Rev. Endocrinol.* **2015**, *11*, 339–351. [CrossRef] [PubMed]
67. Douglass, J.D.; Dorfman, M.D.; Thaler, J.P. Glia: Silent partners in energy homeostasis and obesity pathogenesis. *Diabetologia* **2017**, *60*, 226–236. [CrossRef]
68. Robb, J.L.; Morrissey, N.A.; Potter, P.G.W.; Smithers, H.E.; Beall, C.; Ellacott, K.L.J. Immunometabolic Changes in Glia—A Potential Role in the Pathophysiology of Obesity and Diabetes. *Neuroscience* **2020**, *447*, 167–181. [CrossRef]
69. Rahman, M.H.; Kim, M.S.; Lee, I.K.; Yu, R.; Suk, K. Interglial Crosstalk in Obesity-Induced Hypothalamic Inflammation. *Front. Neurosci.* **2018**, *13*, 939. [CrossRef]
70. De Souza, C.T.; Araujo, E.P.; Bordin, S.; Ashimine, R.; Zollner, R.L.; Boschero, A.C.; Saad, M.J.; Velloso, L.A. Consumption of a fat-rich diet activates a proinflammatory response and induces insulin resistance in the hypothalamus. *Endocrinology* **2005**, *146*, 4192–4199. [CrossRef]
71. Milanski, M.; Degasperi, G.; Coope, A.; Morari, J.; Denis, R.; Cintra, D.E.; Tsukumo, D.M.; Anhe, G.; Amaral, M.E.; Takahashi, H.K.; et al. Saturated fatty acids produce an inflammatory response predominantly through the activation of TLR4 signaling in hypothalamus: Implications for the pathogenesis of obesity. *J. Neurosci.* **2009**, *29*, 359–370. [CrossRef]
72. Uysal, K.T.; Wiesbrock, S.M.; Marino, M.W.; Hotamisligil, G.S. Protection from obesity-induced insulin resistance in mice lacking TNF-alpha function. *Nature* **1997**, *389*, 610–614. [CrossRef]
73. Howard, J.K.; Cave, B.J.; Oksanen, L.J.; Tzameli, I.; Bjorbaek, C.; Flier, J.S. Enhanced leptin sensitivity and attenuation of diet-induced obesity in mice with haploinsufficiency of Socs3. *Nat. Med.* **2004**, *10*, 734–738. [CrossRef] [PubMed]
74. Ropelle, E.R.; Flores, M.B.; Cintra, D.E.; Rocha, G.Z.; Pauli, J.R.; Morari, J.; de Souza, C.T.; Moraes, J.C.; Prada, P.O.; Guadagnini, D.; et al. IL-6 and IL-10 anti-inflammatory activity links exercise to hypothalamic insulin and leptin sensitivity through IKKbeta and ER stress inhibition. *PLoS Biol.* **2010**, *8*, e1000465. [CrossRef]
75. Zhang, X.; Zhang, G.; Zhang, H.; Karin, M.; Bai, H.; Cai, D. Hypothalamic IKKbeta/NF-kappaB and ER stress link overnutrition to energy imbalance and obesity. *Cell* **2008**, *135*, 61–73. [CrossRef] [PubMed]
76. Walter, P.; Ron, D. The unfolded protein response: From stress pathway to homeostatic regulation. *Science* **2011**, *334*, 1081–1086. [CrossRef]
77. Hetz, C. The unfolded protein response: Controlling cell fate decisions under ER stress and beyond. *Nat. Rev. Mol. Cell Biol.* **2012**, *13*, 89–102. [CrossRef]
78. Escartin, C.; Galea, E.; Lakatos, A.; O'Callaghan, J.P.; Petzold, G.C.; Serrano-Pozo, A.; Steinhauser, C.; Volterra, A.; Carmignoto, G.; Agarwal, A.; et al. Reactive astrocyte nomenclature, definitions, and future directions. *Nat. Neurosci.* **2021**, *24*, 312–325. [CrossRef] [PubMed]
79. Hamby, M.E.; Sofroniew, M.V. Reactive astrocytes as therapeutic targets for CNS disorders. *Neurotherapeutics* **2010**, *7*, 494–506. [CrossRef]
80. Cansell, C.; Stobbe, K.; Sanchez, C.; Le Thuc, O.; Mosser, C.A.; Ben-Fradj, S.; Leredde, J.; Lebeaupin, C.; Debayle, D.; Fleuriot, L.; et al. Dietary fat exacerbates postprandial hypothalamic inflammation involving glial fibrillary acidic protein-positive cells and microglia in male mice. *Glia* **2021**, *69*, 42–60. [CrossRef]
81. Buckman, L.B.; Thompson, M.M.; Lippert, R.N.; Blackwell, T.S.; Yull, F.E.; Ellacott, K.L. Evidence for a novel functional role of astrocytes in the acute homeostatic response to high-fat diet intake in mice. *Mol. Metab.* **2015**, *4*, 58–63. [CrossRef]
82. Douglass, J.D.; Dorfman, M.D.; Fasnacht, R.; Shaffer, L.D.; Thaler, J.P. Astrocyte IKKbeta/NF-kappaB signaling is required for diet-induced obesity and hypothalamic inflammation. *Mol. Metab.* **2017**, *6*, 366–373. [CrossRef] [PubMed]
83. Zhang, Y.; Reichel, J.M.; Han, C.; Zuniga-Hertz, J.P.; Cai, D. Astrocytic Process Plasticity and IKKbeta/NF-kappaB in Central Control of Blood Glucose, Blood Pressure, and Body Weight. *Cell Metab.* **2017**, *25*, 1091–1102.e4. [CrossRef] [PubMed]
84. Gupta, S.; Knight, A.G.; Gupta, S.; Keller, J.N.; Bruce-Keller, A.J. Saturated long-chain fatty acids activate inflammatory signaling in astrocytes. *J. Neurochem.* **2012**, *120*, 1060–1071. [CrossRef]
85. Sartorius, T.; Lutz, S.Z.; Hoene, M.; Waak, J.; Peter, A.; Weigert, C.; Rammensee, H.G.; Kahle, P.J.; Haring, H.U.; Hennige, A.M. Toll-like receptors 2 and 4 impair insulin-mediated brain activity by interleukin-6 and osteopontin and alter sleep architecture. *FASEB J.* **2012**, *26*, 1799–1809. [CrossRef] [PubMed]
86. Gao, Y.; Bielohuby, M.; Fleming, T.; Grabner, G.F.; Foppen, E.; Bernhard, W.; Guzman-Ruiz, M.; Layritz, C.; Legutko, B.; Zinser, E.; et al. Dietary sugars, not lipids, drive hypothalamic inflammation. *Mol. Metab.* **2017**, *6*, 897–908. [CrossRef] [PubMed]

87. Lemus, M.B.; Bayliss, J.A.; Lockie, S.H.; Santos, V.V.; Reichenbach, A.; Stark, R.; Andrews, Z.B. A stereological analysis of NPY, POMC, Orexin, GFAP astrocyte, and Iba1 microglia cell number and volume in diet-induced obese male mice. *Endocrinology* **2015**, *156*, 1701–1713. [CrossRef]
88. Buckman, L.B.; Thompson, M.M.; Moreno, H.N.; Ellacott, K.L. Regional astrogliosis in the mouse hypothalamus in response to obesity. *J. Comp. Neurol.* **2013**, *521*, 1322–1333. [CrossRef]

International Journal of *Molecular Sciences*

MDPI

Review

Hypothalamic Microglial Heterogeneity and Signature under High Fat Diet–Induced Inflammation

Natália Ferreira Mendes [1,2], Carlos Poblete Jara [1,2], Ariane Maria Zanesco [2,3] and Eliana Pereira de Araújo [1,2,*]

[1] Faculty of Nursing, University of Campinas, Campinas 13083-887, Brazil; natalia.mendesss@gmail.com (N.F.M.); blinkeado@gmail.com (C.P.J.)
[2] Laboratory of Cell Signaling, Obesity and Comorbidities Research Center, University of Campinas, Campinas 13083-864, Brazil; arianemariazanesco@gmail.com
[3] School of Medical Sciences, University of Campinas, Campinas 13083-887, Brazil
* Correspondence: earaujo@unicamp.br

Abstract: Under high-fat feeding, the hypothalamus atypically undergoes pro-inflammatory signaling activation. Recent data from transcriptomic analysis of microglia from rodents and humans has allowed the identification of several microglial subpopulations throughout the brain. Numerous studies have clarified the roles of these cells in hypothalamic inflammation, but how each microglial subset plays its functions upon inflammatory stimuli remains unexplored. Fortunately, these data unveiling microglial heterogeneity have triggered the development of novel experimental models for studying the roles and characteristics of each microglial subtype. In this review, we explore microglial heterogeneity in the hypothalamus and their crosstalk with astrocytes under high fat diet–induced inflammation. We present novel currently available ex vivo and in vivo experimental models that can be useful when designing a new research project in this field of study. Last, we examine the transcriptomic data already published to identify how the hypothalamic microglial signature changes upon short-term and prolonged high-fat feeding.

Keywords: microglia; gliosis; hypothalamus; high-fat diet; obesity; transcriptomic

Citation: Mendes, N.F.; Jara, C.P.; Zanesco, A.M.; de Araújo, E.P. Hypothalamic Microglial Heterogeneity and Signature under High Fat Diet–Induced Inflammation. *Int. J. Mol. Sci.* **2021**, *22*, 2256. https://doi.org/10.3390/ijms22052256

Academic Editor: Rosalía Rodríguez-Rodríguez

Received: 15 January 2021
Accepted: 19 February 2021
Published: 24 February 2021

1. Introduction

Hypothalamic inflammation is a condition frequently observed in experimental models of diet-induced obesity (DIO) [1–3] and obese humans [4–6]. This inflammatory response is mainly triggered by excessive saturated fatty acids (SFAs) from the diet [7–9], which reach the neural tissue mainly through the median eminence (ME), where fenestrated vascular endothelium lacks a blood–brain barrier (BBB) [10,11]. Brain perivascular macrophages (PVMs) also react to excessive free fatty acids (FFAs) circulating in the blood vessels, with a consequent increase in BBB permeability [12,13]. Glial cells, such as astrocytes and microglia, quickly sense and react to the presence of those SFAs in the hypothalamic parenchyma, releasing pro-inflammatory cytokines, chemokines, and reactive oxygen species (ROS) [14,15]. If the stimulus persists, the hypothalamic neuronal network may be damaged, resulting in neuro-inflammation, which eventually leads to energy balance disruption [16,17], and finally, to neuronal dysfunction/apoptosis [18].

Researchers have shown that hypothalamic inflammation initiates just a few hours/days upon high-fat feeding [5,15,19]. After the onset of the inflammatory response, bone marrow–derived cells (BMDC) and other peripheral immune cells, such as neutrophils, lymphocytes, and regulatory T (Treg) cells can arise into the hypothalamic parenchyma in a time-dependent manner, directly affecting glial functions [12,20,21]. To avoid further metabolic complications, hypothalamic neuronal and non-neuronal cells, along with peripheral immune cells, should act together in an orchestrated mode beginning with the earliest phase of the inflammatory response.

The molecular mechanisms underlying microglial immune and metabolic interactions with other cell types under high-fat diet (HFD)-induced hypothalamic inflammation still require a more detailed exploration [22]. Recent data unveiling microglial diversity and signatures throughout the brain have contributed to the development of novel state-of-the-art approaches in experimental studies, which can be valuable for this field in the coming years.

In this review, we explore novel findings of hypothalamic microglial diversity from rodents and humans. We examine the subtypes of microglia that may be involved in HFD-induced hypothalamic inflammation and investigate how these cells interact with astrocytes upon high-fat feeding. Beyond that, we discuss which models can be useful to get the most reliable data when studying distinct subsets of microglia, myeloid cells, and border-associated macrophages (BAMs). Last, we investigate novel transcriptomic data already published to clarify how the hypothalamic microglia signature changes under saturated fat consumption.

2. Microglial Heterogeneity in the Hypothalamus

Microglia were identified by Pío del Río-Hortega in 1919, but only in the last two decades has the interest in these cells grown exponentially, with the discovery of their unique origin in the yolk sac and motile capacity [23,24]. Since their identification, microglia have been studied as a unique macrophage-like cell type in the central nervous system (CNS), able to quickly react to a wide range of stimuli by switching their phenotype activation between M1 and M2 subtypes [25,26]. According to this past view, microglia represent a naïve cell type that could equally react to any stimuli by acquiring a predetermined phenotype.

Most recently, advances in genetic tools have been developed and extensively employed in experimental research, enabling a deeper understanding of microglial diversity. Stratoulias et al. [27] recently proposed a new classification, in which microglia constitute a heterogeneous cell group, and each subtype has distinct properties and physiological functions, reacting differently to stimuli. This new view is based on the regional steady-state heterogeneity of microglia and their broad gene marker diversity. Curiously, from six putative microglial subtypes with unique specializations presented by these authors, the only one found in the hypothalamic area is dark microglia (DM). Despite their classification in subtypes, it is important to highlight the existence of many other types of microglia showing distinct features and functions in the hypothalamic area, which were not considered, or at least did not show up, in their categorization.

DM are also found in the hippocampus, cerebral cortex, and amygdala [28]. These cells can only be visualized through high-spatial-resolution transmission electron microscopy and display markers of oxidative stress, such as a condensed, electron-dense cytoplasm and nucleoplasm, dilatation of the Golgi apparatus and endoplasmic reticulum, mitochondrial alteration, and a partial to complete loss of the heterochromatin pattern. Functionally, DM are very active and show extremely thin processes, which allow them to make contact with synaptic elements [28]. Despite being rarely observed in healthy young adult mice, DM are widespread upon chronic stress, ageing, in CX3C chemokine receptor 1 (CX3CR1)-knockout (KO) mice, and in Alzheimer's disease pathology (APP/PS1 model) [27].

It remains unknown through which mechanisms DM are involved in the central inflammatory response, but it is reasonable to speculate that DM functions, similarly to other microglial subtypes, are influenced by peripheral signals. Savage et al. [29] showed that 24 h after an acute systemic injection of lipopolysaccharide (LPS), there are alterations in the inflammatory profile and in the microglial ultrastructure in the hippocampus, with no direct impact on DM. Curiously, DM are largely found in C-C chemokine receptor type 2 (CCR2-KO) mice, which present impaired recruitment of peripheral monocytes to the brain [30]. This finding is particularly interesting because upon prolonged HFD consumption, these peripheral cells are recruited to the hypothalamic parenchyma, as previously mentioned, similarly to what happens in many types of acute CNS injury [21].

Since their first description in 2016 [28], studies have described several DM features in rodents [31–36] and humans [37]. However, none of these studies have focused on DM roles specifically in the hypothalamus. Advances in this subject remain necessary to clarify how DM crosstalk with other microglia under hypothalamic inflammation, whether DM are involved in this recruitment of immune cells to the brain, and which peripheral stimuli can affect hypothalamic DM functions.

Microglia simultaneously express several hallmarks, such as *Iba1, Cx3cr1, P2ry12, Tmem119, Trem2, Cd11b, Hexb, Csf1r, Itgam,* and *Siglec*, among others [38–43]. However, there is no unique expression pattern of these transcripts in these cells, that is, it varies according to the pathological condition, age, sex, species, and brain area [27]. Particularly in mice, Valdearcos et al. [20] recently showed through immunofluorescence assays that the microglial signature varies according to hypothalamic nuclei and the dietary fat content. Thus, at least in mice, some CX3CR1+ cells in the arcuate nucleus (ARC) overlap with Tmem119+ or P2ry12+ cells, while other CX3CR1+ cells do not. When gliosis occurs upon consuming a HFD, ionized calcium binding adaptor molecule 1 (Iba1+) cells become widespread in the ME, ARC, and ventromedial nucleus (VMH), while Tmem119+ and P2ry12+ cells remain more restricted to the VMH when compared with the brain slices obtained from mice fed on chow diet. Some studies involving DM also show this heterogeneity, because these cells barely express *Iba1*, *Cx3cr1*, and *P2ry12*, but robustly express cluster of differentiation molecule 11B (*Cd11b*), which is involved in their synaptic pruning role [44,45].

These data indicate that some hypothalamic microglial subsets may have anti-inflammatory functions, depending on the transcriptomic profile of each subtype and their status (steady-state or reactive). It is plausible to consider that it may also happen with other hallmarks, which explains why using only one surface marker for microglial staining, choosing a single cell line for ex vivo experiments, or even examining a single transgenic mouse model for manipulating microglia implicates a methodology bias.

Disease-associated microglia (DAM) are another subset of CNS resident macrophages that have been recently identified in experimental models of neurodegeneration and demyelination [46]. These cells are characterized by the expression of several genes, such as *Apoe*, *Clec7a*, *Cst7*, and *Spp1*. They are mainly found at sites of neurodegeneration and might play a protective role [47]. Interestingly, DAM hallmarks are also observed in human Alzheimer's disease post-mortem brains [46,48]. Triggering receptor expressed on myeloid cells-2 (TREM2) is also highly expressed by DAM; the activation of its intracellular signaling is essential for the transition of homeostatic microglia to the DAM state [49]. Both Toll-like receptor 4 (TLR4) and TREM2 can recognize different pathogen-associated molecular patterns [50,51] and other ligands, such as gram-negative bacteria [52], lipids [53], apolipoproteins (ApoE, ApoJ, and ApoA) [54–56], and nucleic acids released by apoptotic cells [57]. Because TREM2 and TLR4 share some ligands, it is not easy to identify which signaling pathway is activated through each receptor by these stimuli. However, experiments regarding TREM2 inhibition in combination with TLR4 stimulation by LPS have started to clarify the involvement of each receptor in neuro-inflammation [58].

An important characteristic observed in HFD-induced hypothalamic inflammation is the activation of nuclear factor kappa-light-chain-enhancer of activated B cells (NF-κB) [14,59–62]. Conversely, inhibition of IKKβ/NF-κB signaling in microglia expressing CX3CR1 ameliorates DIO and hypothalamic inflammation [20]. In the CNS, TREM2 is widely expressed in microglia [63,64], where it acts by negatively regulating the activation of NF-κB [65]. In BV2 cell culture, TREM2 overexpression inhibits PI3K/AKT and NF-κB signaling pathways [66,67]. Interestingly, LPS reduces the expression of TREM2 in these cells through the activation of JNK and NF-κB, resulting in a vicious cycle [66]. Recently, Zhang et al. [68] treated BV2 cells with curcumin, a bioactive compound isolated from *Curcuma longa* with anti-inflammatory and antioxidant activities, and LPS; they observed reduced microglial activation via TLR4/NF-κB when compared with microglial cells treated only with LPS, which occurred in parallel to the increased TREM2 expression, reinforcing the potential anti-inflammatory role of this receptor.

Microglial dynamics and density vary between brain areas and LPS doses during infection-induced inflammation [69]. In addition, considering that both LPS and SFAs can activate TLR4 and TREM2 intracellular signaling pathways, it is reasonable to speculate that a HFD could also have some effect on these microglial characteristics, especially in the hypothalamus. Yet, whether continued activation of TREM2+ cells under prolonged HFD-induced inflammation is implicated in the development of neurodegenerative diseases remains unexplored. The broad diversity of hypothalamic microglia at steady state and during the HFD-induced inflammatory response is shown in Figure 1.

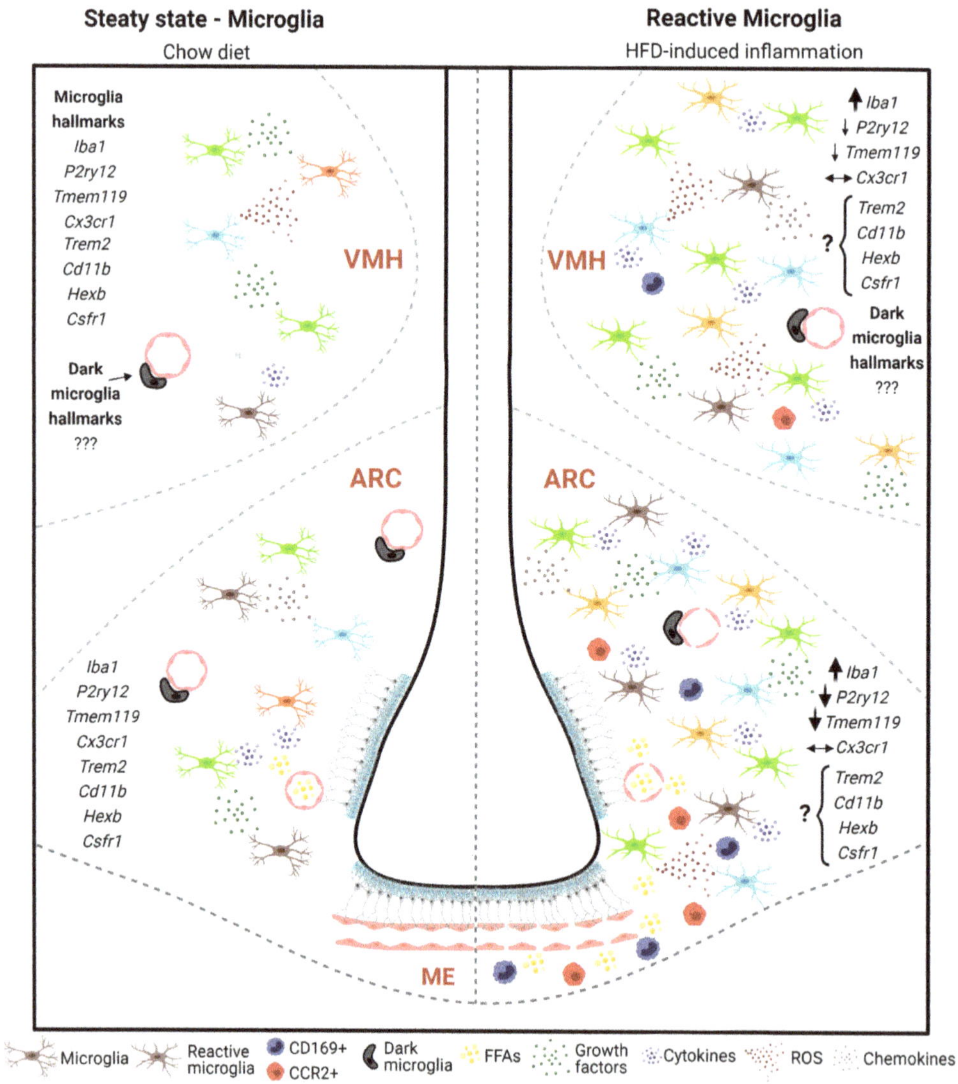

Figure 1. Microglial heterogeneity in the hypothalamus of rodents fed on chow or a high-fat diet (HFD). At steady state, microglia can be targeted by several hallmarks and are widely distributed in the hypothalamic parenchyma. Dark microglia (DM)

are only visualized by transmission electron microscopy and are close to blood vessels and neuronal synapses. Hallmarks of DM are well known in hippocampus, but remains barely explored in the hypothalamus. In HFD-induced hypothalamic inflammation, reactive gliosis is observed and microglia change their spatial distribution and molecular signature. Microglia cells react by increasing the release of pro-inflammatory cytokines, reactive oxygen species (ROS), and growth factors. In the ARC, ME, VMH, there is a huge increase in Iba1+ cells (wide arrow) and a decrease in P2ry12+ and Tmem119+ cells, which become more restricted to the VMH (thinner arrows when compared to their arrows in the ARC). How other microglial hallmarks (e.g., *Trem2, Cd11b, Hexb, Csfr1*, among others) behave under HFD have not been studied yet in this inflammatory process. The role of DM and changes in their hallmarks in this specific inflammatory response have still not been explored. A leaky blood–brain barrier (BBB) allows free fatty acids (FFAs) accumulation in the hypothalamic parenchyma, boosting inflammation. If HFD persists for some weeks, peripheral myeloid cells, such as CD169+ and CCR2+ cells, are chemoattracted and infiltrate the hypothalamic parenchyma, but their functions in the HFD-induced inflammation need to be further studied. Abbreviations: ARC, arcuate nucleus of the hypothalamus; VMH, ventromedial nucleus of the hypothalamus; ME, median eminence.

Most transcriptomic data currently available about microglia heterogeneity is from studies with rodents [39,70–72]. By contrast, human microglial diversity has only begun to be comprehended in the last few years. Masuda et al. [39] recently identified several clusters of microglia in healthy human brains and in the brains of patients with multiple sclerosis, but these authors evaluated only the cortex and temporal lobe. Similarly, Sankowski et al. [73] identified several microglial subsets in the human brain during homeostasis and some diseases through the combination of two high-dimensional technologies, single-cell RNA sequencing (scRNA-seq) and multiplexed mass cytometry (CyTOF), although the hypothalamus was not included in the analysis. Likewise, Böttcher et al. [74] applied multiplexed CyTOF to detect microglia regional heterogeneity in human postmortem samples from the subventricular zone, thalamus, cerebellum, temporal lobe, and frontal lobe.

Unfortunately, there is still a lack of data from hypothalamic human microglia in the hypothalamus because there are several methodological difficulties in collecting human brain samples and keeping them cryopreserved without damage. These issues have limited large-scale studies. In addition, while recent advances in transcriptomics, multiplex protein expression analysis, and methods to detect chromatin structure have revealed many facets and details about microglia in rodents and humans, the correlation between the genome-related information and their features remains to be explored.

3. Hypothalamic Microglia–Astrocyte Crosstalk

CNS homeostasis, development, injury, and repair are precisely controlled by appropriate cell–cell communication. Nevertheless, studies investigating microglial functions have been focused on microglia isolation, neglecting that physiological actions of these cells are part of a complex network involving other cell types. Therefore, new tools have been widely applied in experimental studies, allowing the determination of an infinite number of intercellular interactions by connecting ligands to target genes across cell types and tissues [75–77]. Unsurprisingly, the inflammatory process in the hypothalamus, observed in obesity, drives multiple harmful outcomes that can affect the interactions between microglia and other cell types [78,79]. As astrocytes and microglia play active roles from the onset of the hypothalamic inflammatory response, here we focus on their crosstalk under this process.

While microglia comprise only 5% to 10% of the total number of CNS cells in humans and rodents [80], astrocytes are the largest cell component of the brain, comprising at least 50% of all CNS cells [81]. Astrocytes make intimate contacts with synapses, blood vessels, and other glial cells, thus controlling synaptic transmission, BBB structure and function, and sensing nutrients and hormones in the blood [82,83]. They are also involved in the regulation of lipid metabolism and storage [84] and some studies have been indicating that astrocytes can also express some hormone receptors, such as leptin, insulin, ghrelin, and glucocorticoid receptors [85–88].

Coordinated crosstalk between glial cells is a determinant in homeostatic and pathological conditions [89]. Microglia and astrocytes interact via contact-dependent and secreted factors, such as growth factors, neurotransmitters, cytokines, chemokines, innate-immunity mediators, mitogenic factors, ROS, and metabolic mediators such as glutamate, which can be used for cell metabolism and may also mediate tissue changes [90]. These glial cells play a crucial role in synapse development and function in the healthy CNS, forming the 'quad-partite' synapse [91], which is essential for neuro-immune communication [90], critically contributing to brain homeostasis [92].

Both microglia and astrocytes quickly become activated upon an injury or any inflammatory stimulus [93]. Reactive microglia release several pro-inflammatory cytokines that induce astrocytic activation [75,93]. Likewise, HFD-induced microglial activation results in astrocytic proliferation, morphological changes, and increased production of cytokines and growth factors [61,84,94–96]. Classically, these morphologic and functional adaptations by glial cells under inflammation are known as gliosis [5,97] and are usually described in rodents fed on HFD for weeks [20,98], or even after very short periods of high-fat feeding [15].

In the hypothalamus, all these astrocytic characteristics are essential for controlling energy homeostasis. NF-κB signaling inhibition by IKKβ deletion in astrocytes (GFAP-Cre mice) reduces HFD-induced hypothalamic inflammation and reactive astrogliosis and attenuates DIO and glucose intolerance [14].

Similarly, knockdown of TGF-β1—which is predominantly synthesized by astrocytes, specifically in the ARC of HFD-fed mice—reduces TGF-β/SMAD and NF-κB signaling pathways and, consequently, also attenuates the inflammation [99]. Interestingly, this approach of hypothalamic TGF-β1 knockdown presents many metabolic benefits, preventing obesity development even under HFD. Countless microglial functions depend on TGF-β1, including their maturation and activation, both in homeostatic conditions and in response to any inflammatory stimuli [100,101]. In addition, the accurate functioning of the TGF-β1 signaling pathway is crucial to prevent excessive microglia activation [102,103].

In astrocyte cell culture, treatment with palmitic acid, an SFA largely found in HFD, evokes lipid droplet accumulation [95], culminating in an inflammatory response, characterized by an increased production of chemokines, such as chemokine C-C motif ligand-2 (CCL2), also called monocyte chemoattractant protein-1 (MCP-1). Curiously, the main chemotactic mechanism described in the hypothalamic inflammation is mediated by fractalkine (CX3CL1), a chemokine produced by hypothalamic neurons that acts through the fractalkine receptor (CX3CR1) [104]. However, another possible chemokine involved in the peripheral recruitment of immune cells to the hypothalamus is MCP-1 [105]. Peripheral immune CCR2+ and CD169+ cells arise in the hypothalamic parenchyma upon chronic periods of HFD [12,20]. CCR2+ cells of obese mice can enter into their white adipose tissue, and this chemo-attraction is mediated by the MCP-1/CCR2 axis [106].

In the CNS, the MCP-1/CCR2 axis has also been presented as an important chemotactic mechanism involved in the recruitment of peripheral immune cells to the paraventricular nucleus of the hypothalamus (PVH) upon inflammatory stimuli [107]. These authors showed that blockage of this recruitment by the peripheral administration of a CCR2 antagonist results in a reduction in local inflammation, suggesting that the MCP-1/CCR2 axis may also be involved in the chemotaxis of peripheral myeloid cells seen in HFD-induced hypothalamic inflammation. Curiously, according to the scRNA-seq data from ARC and ME published by Campbell et al. [108], MCP-1 expression in these areas is mainly observed in microglia. Hence, upon high-fat feeding, neurons could be the main cells involved in fractalkine production, while astrocytes and microglia are responsible for CCL2 (MCP-1) release. Based on these data, activated astrocytes not only directly modulate microglial activity, through the increased synthesis of pro-inflammatory cytokines and growth factors, but also mediate and stimulate the recruitment of peripheral macrophages to the hypothalamus, thus contributing to sustained microglial activation.

A leaky BBB in the hypothalamic area has been observed in the HFD-induced inflammatory response [109]. The mechanism underlying this BBB disruption may be due to various factors, such as altered function/structure of non-fenestrated brain endothelial cells, tanycytes, pericytes, neurons, or glia [11]. These cells constitute the neurovascular unit and confer integrity to the BBB under physiological conditions [110]. However, in response to some inflammatory stimuli, astrocytes and tanycytes increase the secretion of vascular endothelial growth factor (VEGF), increasing BBB permeability [110,111]. It is not well established whether this mechanism is crucial for the entrance of FFAs and peripheral cells into the hypothalamic parenchyma upon chronic HFD intake. Therefore, Lee et al. [12] showed that PVMs in the hypothalamic area produce inducible nitric oxide synthase (iNOS) abundantly under high-fat feeding, resulting in disruption of BBB integrity and in the spread of monocyte-derived macrophages in the ARC. As expected, specific hypothalamic macrophage iNOS inhibition completely abrogates astrocytic lipid droplets and macrophage accumulation and activation in the ARC of obese mice.

Another possible mechanism by which microglia and astrocytes contribute to the leaky BBB – observed in several inflammatory diseases – is the downregulation of tight-junction proteins, namely claudin-5 (CLDN5), occludin, and zonula occludens-1, in activated glial cells [112,113]. Interestingly, under systemic inflammation, vessel-associated microglia are able to phagocytize astrocytic end-feet, an action that damages BBB function [114]. However, how hypothalamic microglial and astrocytic crosstalk impairs BBB integrity under HFD-induced inflammation has been minimally explored.

Unsurprising, upon hypothalamic inflammation, microglial activity can also modulate astrocytic functions. Studies using co-culture of microglia and astrocytes have shown that microglial production of pro-inflammatory cytokines increases astrocytic glucose uptake, thus reducing intercellular glucose trafficking [115] and inhibiting astrocytic gap junctions [116,117]. Phenotypic alterations observed both in microglia and astrocytes at the onset of the inflammatory response are mediated by metabolic changes, switching from mitochondrial oxidative phosphorylation to glycolysis [118]. An important mechanism involved in HFD-induced hypothalamic inflammation in rodents comprises the increased levels of microglial UCP2 levels after HFD consumption, which drives a disruption in microglial mitochondrial dynamics [119]. These authors found that UCP2 deletion specifically in microglia reduced the number of GFAP+ cells in the ARC of mice fed a HFD, suggesting that mitochondrial dynamics in microglia regulate astrogliosis in HFD-induced hypothalamic inflammation.

The schematic representation of the crosstalk between microglia and astrocytes under homeostasis and HFD-induced hypothalamic inflammation is shown in Figure 2.

Over the past decade, accumulating evidence has demonstrated that astrocytes also display high heterogeneity. These cells present many subpopulations spread throughout the CNS, which can vary depending on age, species, and sex, and astrocytic state (surveillance or reactive) [120–122]. Given their complexity, a consensus statement about reactive astrocytes [123] recently recommended that astrocyte phenotypes should be defined by a combination of molecular markers and functional readouts. There is evidence that heterogeneity of astrocytes and microglial hallmarks expression can be bi-directionally controlled [124]. Although understanding how different glial subpopulations regulate their local niche under HFD-induced hypothalamic inflammation still need to be investigated. Future studies regarding single-cell transcriptomic heterogeneity of hypothalamic astrocytes and microglia will elucidate which of their sub-populations are paired, their features, and how their hallmarks change upon HFD-induced reactivity.

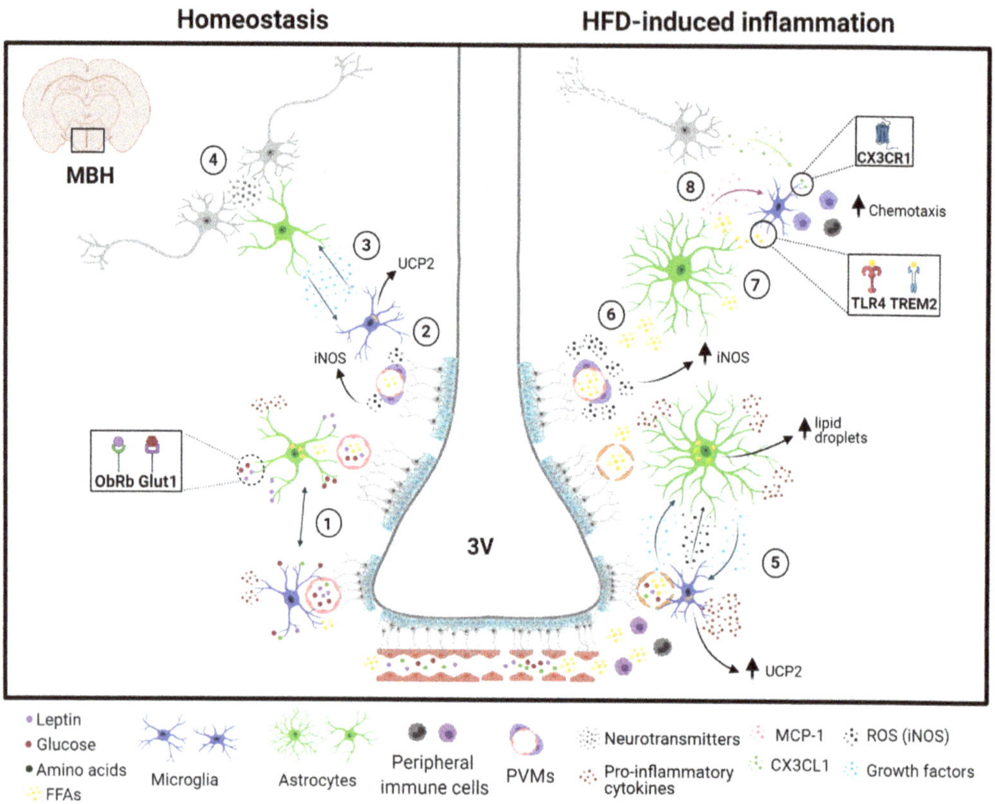

Figure 2. Microglia and astrocyte crosstalk in the hypothalamus. In homeostatic conditions: (**1**) microglia and astrocytes sense nutrients (amino acids, glucose, and free fatty acids [FFAs], among others) and hormones (e.g., leptin) that enter the hypothalamic parenchyma and react to those environmental changes. Astrocytes regulate lipid metabolism and storage and trigger metabolic changes due to hormone binding to their surface receptors (e.g., leptin and glucose receptors). (**2**) UCP2 expression in microglia is tightly controlled preserving their mitochondrial function while PVMs release normal levels of iNOS in response to FFAs concentration in blood vessels. (**3**) Microglia and astrocytes interact via contact-dependent and secreted factors and (**4**) control synapse development and function. Under high-fat feeding: (**5**) Glial cells quickly sense environmental changes and become reactive, releasing several pro-inflammatory cytokines and growth factors (e.g., VEGF and TGF-β1), and increasing reactive oxygen species (ROS) production. Astrocytes store the excessive FFAs in lipid droplets while microglial UCP2 increases, promoting astrogliosis and mitochondrial dysfunction in microglia. (**6**) PVMs surrounding blood vessels increase the release of iNOS promoting leaking BBB and facilitating the entrance of FFAs into the hypothalamic parenchyma, (**7**) which can stimulate IKKβ/NF-κB signaling by activating TLR4 and TREM2 in microglial cells. (**8**) Glial cells and neurons increase the synthesis of chemokines (e.g., MCP-1 and fractalkine), leading to the recruitment of peripheral myeloid cells to the central nervous system. If the inflammatory stimulus persists, the astrocytic control of synaptic functions become impaired, resulting in neuronal dysfunction. Abbreviations: 3V, third ventricle; MBH, mediobasal hypothalamus.

4. Advances in Experimental Manipulation of Microglia

The identification of the microglial transcriptomic signature and heterogeneity has given rise to the development of new experimental models for manipulating or labelling specific subtypes of microglia. The hypothalamus has many microglial subpopulations that distinctly react to hormones and nutrients from the diet. Thus, the availability of these new models can provide a significant advance to comprehend the involvement of each microglial subtype in the HFD-induced hypothalamic inflammation and other inflammatory or neurodegenerative diseases. To achieve that, researchers should be able to choose the best model to answer their specific questions. In this topic, we discuss the newest approaches that have been developed.

4.1. Ex Vivo Models

To understand how microglia react to different stimuli, most studies conducted in the past decades have employed ex vivo strategies. A large variety of mouse microglia cell lines have been generated over the years, but the most used has been the murine microglial BV2 cell line. These cells were originally obtained and immortalized from the cerebral cortex of neonatal mice [125]. Other similar cell lines have been obtained from the whole brain, cerebellum, or cortex of adult/embryo rodents and have been less applied in basic research and barely cited in the literature compared with BV2 [126].

Most studies investigating microglial hypothalamic features using ex vivo methods have been conducted with BV2 cells [127–129]. It is well known that there are many phenotypic and genomic differences between microglia from the cortex and from the hypothalamus or microglia from a neonatal mouse compared to an adult or embryo – not to mention the fact that these cells were immortalized 30 years ago, and have very likely already suffered several transcriptomic modifications since that time. Another concern that must be considered when interpreting data is whether these cells are at the surveillance or reactive state when cultivated with no other neuronal or glial cells. Many questions and worries emerged when microglia stopped being seen as a simple macrophage of the CNS. Hence, to reduce any possible bias, published data and ongoing experiments with BV2 cells should be carefully planned, analyzed, and always accompanied by other in vivo strategies.

The development of new transgenic mouse models that express a fluorescent protein driven by microglial hallmarks promises an appealing alternative for researchers interested in cultured microglia. Magnetic-activated cell sorting (MACS) [130] and fluorescence-activated cell sorting (FACS) [131,132] using these fluorescent reporter mice or even microglia-specific surface markers allow the physical isolation of various subsets of microglia, which can be obtained from primary cultures or co-cultures with other cells. Using these approaches, the brain area of interest can be precisely harvested after a specific dietary or treatment protocol, and the researcher can freely choose the age and the sex of the rodents used in the experiments. These strategies render the collected data much more reliable when compared to studies using only BV2 cells. Although, when choosing cell sorter techniques researchers should keep in mind that any steps of protocols can trigger some reaction in microglia morphology or gene expression and even impair the cell viability (Figure 3).

Grassivaro et al. [133] recently applied FACS to isolate CNS resident microglia and peripheral myeloid cells from some models of neuroinflammation, such as experimental autoimmune encephalomyelitis (EAE). These authors collected samples during embryonic and postnatal periods and subsequently targeted cells from the brain and liver through labelling microglia and peripheral myeloid cells with fluorescent antibodies. By this approach, they observed extraordinary differences and transcriptomic details between those cells. Unfortunately, there is still a lack of data obtained through FACS or MACS for isolating microglia from the hypothalamus of experimental models of obesity.

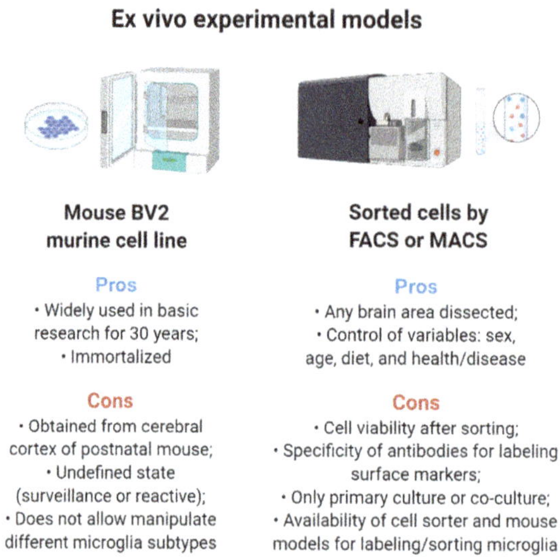

Figure 3. Comparisons between ex vivo experimental models for studying cultured microglia. Pros (in blue) and Cons (in red) of each approach are depicted in figure. Abbreviations: FACS, fluorescence-associated cell sorting; MACS, magnetic-activated cell sorting.

4.2. *In Vivo Models*

Most studies need to evaluate microglial functions in vivo. Conveniently, at least in rodents, functional contributions of specific cell populations can be explored using Cre recombinase–mediated mutagenesis. The identification and cloning of CX3CR1 triggered the development of new tools for studying microglia in vivo [134–136]. Studies regarding the mechanisms through microglia-mediated HFD-induced hypothalamic inflammation have been mostly conducted with a transgenic mouse known as CX3CR1-Cre, which expresses Cre recombinase under the direction of the *Cx3cr1* promoter in the mononuclear phagocyte system, and CX3CR1^{EGFP} knock-in/knock-out reporter mice, which expresses a green fluorescent protein (EGFP) in monocytes, dendritic cells, natural killer cells, and brain microglia under the control of the same locus [20,137]. Consequently, all mechanisms regarding microglial reactivity and gliosis in those research articles can only be interpreted as adaptations and features of CX3CR1+ microglia. Thus, when researchers activate/inhibit microglia in CX3CR1-Cre mice through DREADD, optogenetic, or other Cre recombinase-targeted manipulation, they can only interpret the obtained data in the context of the manipulation of CX3CR1+ microglial, and not to other cells.

CX3XR1-Cre^{ERT2} is another *Cx3xr1*-driven model widely used in experimental research since it expresses Cre recombinase in a tamoxifen-inducible manner [138–140]. This tamoxifen-inducible model allows the temporal manipulation of microglia. Therefore, Van Hove et al. [141] have recently warned that the use of this strain is not indicated for all types of microglia-related investigations, such as fate-mapping studies, once it exhibits considerable leakiness in the absence of tamoxifen.

For many years, *Cx3cr1*-driven lines were largely employed in basic research – and they still are. However, they are haploinsufficient for *Cx3cr1*, which means that EGFP or Cre coding regions have been designed to replace the endogenous *Cx3cr1* locus. Some studies suggest that this haploinsufficiency could affect microglial functions, such as synaptic plasticity [142–144]. Given this background, all Cx3cr1-drive lines should be used with this limitation in mind. It is fortunate that new transgenic mouse models for genetic labelling and manipulating other microglial subtypes, myeloid cells and border-associated macrophages (BAMs) found in the hypothalamus and in other brain areas have been generated in recent years, allowing a deep investigation of new microglial phenotypes and functions (Table 1).

Table 1. Novel in vivo experimental models for studying various subtypes of microglia, myeloid cells, and BAMS.

Manipulating Microglia	Labeling Microglia	Depleting Microglia	Targeting Myeloid Cells	Targeting Microglia and PVMs
CX3CR1-Cre CX3CR1-Cre^{ERT2} P2ry12-Cre^{ERT2} [148] Tmem119-Cre^{ERT2} [147] Hexb-Cre^{ERT2} [150]	CX3CR1^{EGFP} Tmem119^{EGFP} [147] Tmem119^{TdTomato} [149] Hexb^{TdTomato} [150]	Csf1r-FIRE$^{\Delta/\Delta}$ [145]	LysM^{EGFP} LysM-Cre	Sall1^{ncre}:Cx3cr1^{ccre} [146] Lyve1^{ncre}: Cx3cr1^{ccre} [146]

Masuda et al. [150] recently developed another of these new microglia gene-targeting models. For this endeavor, the authors initially analyzed microglia and CNS-associated macrophage (CAM) signatures obtained from scRNA-seq, during homeostasis and disease, and identified beta-hexosaminidase subunit beta (*Hexb*) as a stably expressed microglia core gene. Based on that, they generated Hexb^{TdTomato} and Hexb-Cre^{ERT2} mouse strains, which express a red fluorescent protein (TdTomato) or Cre recombinase protein under the control of the *Hexb* locus upon tamoxifen administration, respectively. Although this research article was published a few months ago, there is no data available in the literature about the participation of microglial cells Hexb+ in hypothalamic inflammation. Curiously, *Hexb* is highly expressed in the ARC and ME [108], but in these areas, it is not tightly restricted to microglial when compared with other markers that can also be manipulated through transgenic mouse strains (*Cx3cr1*, *P2ry12*, or *Tmem119*) (Figure 4). Nevertheless, future studies should evaluate whether these Hexb+ microglia are important in some stage of HFD-induced hypothalamic inflammation and characterize their interactions with other glial and non-glial cells in the hypothalamus.

Transmembrane protein 119 (*Tmem119*) and purinergic receptor P2Y12 (*P2ry12*) loci have also been targets for the development of new mouse models (Table 1). Kaiser and Feng [147] generated Tmem119^{EGFP} mice and Tmem119-Cre^{ERT2} mice, which express EGFP or Cre^{ERT2} under the control of the Tmem119 coding region, respectively. More recently, Ruan et al. [149] generated another reporter mouse, Tmem119^{TdTomato}, which expresses TdTomato rather than EGFP in microglia. All these Tmem119-driven lines have been validated and are now valuable tools to study specifically the role of Tmem119+ microglia in health and disease. McKinsey et al. [148] recently published the new P2ry12-Cre^{ERT2} mouse line. As stated by the authors, they chose P2ry12 because it appeared to be the most restricted to brain myeloid cells compared with other markers. Following the pattern, this model expresses Cre protein recombinase under the control of the P2ry12 locus. They also suggested that this model could be useful for studies about middle cerebral artery occlusion–induced ischemic stroke as well as EAE.

As previously mentioned, Valdearcos et al. [20] identified that both Tmem119+ and P2ry12+ cells react differently to Cx3cr1+ or Iba1+ cells in the ARC and the VMH under prolonged HFD intake. Thus, these novel *Tmem119*- and *P2ry12*-driven strains can also be promising to comprehend the mechanisms through which these specific microglial subtypes react to SFAs from the diet in this specific inflammatory condition.

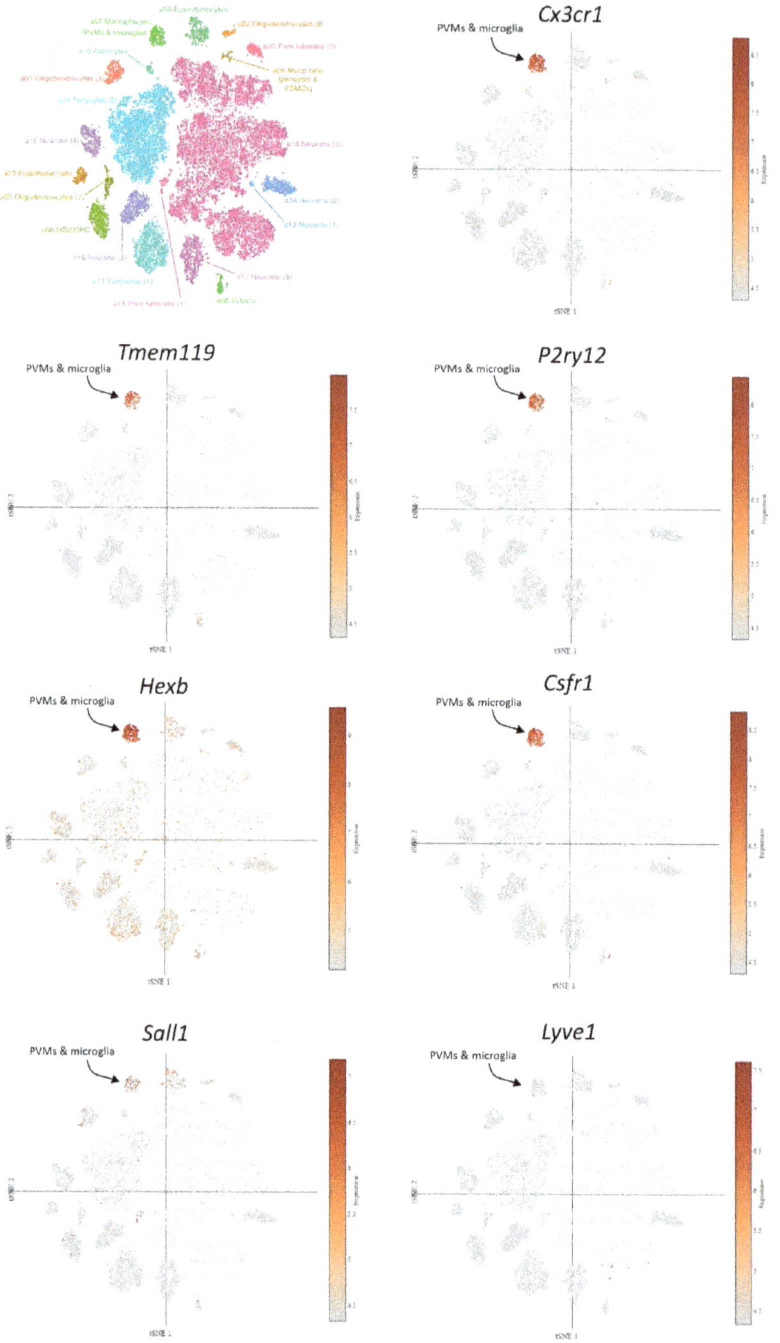

Figure 4. Distribution of genes in the arcuate nucleus (ARC) and median eminence (ME) chosen to generate mouse models that target microglia and perivascular macrophages (PVMs). Scatter plot images were obtained from https://singlecell.broadinstitute.org/ accessed on 20 November 2020, ARC-ME DROPSEQ data is from Campbell et al. [108].

Beyond Cre-driven and fluorescent reporter mouse lines, many researchers have chosen to deplete microglia to evaluate their role in some pathologic condition. There are many microglial depletion strategies, which were recently remarkably reviewed by Miron and Priller [151], including the use of liposomes, global knockout of genes required for microglial development or survival, or even transgenic or pharmacological induction of microglial death. Unfortunately, all these methods have some bias and do not allow the depletion of only microglial cells, because BAMs and monocytes may also be targeted. The use of some knockout mice, such as the knockout for colony stimulating factor 1 receptor (*Csfr1*-KO), is also not a good choice, because these mice lack all CNS macrophages; they show serious brain abnormalities and usually die within weeks after birth [152].

Due this dilemma, a novel model has been developed in which microglia are depleted, but BAMs are widely preserved after the genomic deletion of a super enhancer in the colony stimulating factor 1 receptor (*Csf1r*) coding region (fms-intronic regulatory element; Csf1r-FIRE$^{\Delta/\Delta}$) [145]. A few studies have employed microglial depletion to study microglial immune and metabolic functions in the hypothalamus [137,153], because depletion disrupts immune balance and energy homeostasis [154]. Fortunately, Csf1r is also expressed by microglia from the ARC and ME [108] (Figure 4). CSF1R inhibition through a pharmacologic inhibitor (PLX5622) can improve metabolic outcomes in middle-aged female mice [155]. Thus, investigating the role of Csfr1+ microglial cells in the hypothalamic inflammation, through this Csf1r-FIRE$^{\Delta/\Delta}$ mouse model, could also provide new, important information regarding this subtype of microglia.

As previously described, PVMs are also involved in HFD-induced hypothalamic inflammation [12]. This study was conducted with transgenic mouse models targeting lysozyme M (LysM)-expressing myeloid cells. Although many experimental models have been developed in the past years, there are still some methodological challenges with regard to manipulating specifically microglia or other CAMs/BAMs, such as PVMs. Even the above-mentioned transgenic lines used to study several microglial subsets (Table 1) have the mutation driven by a locus common to microglia and PVMs, at least in the ARC and ME [108] (Figure 4).

Some studies using scRNA-seq have already analyzed these cells separately [71], through dissection of leptomeninges, the perivascular space and parenchyma, and the choroid plexus; however, experimental manipulation of microglia and other CAMs/BAMs independently remains a challenge. Fortunately, Kim et al. [146] have just developed a binary transgenic system involving complementation-competent NCre and CCre fragments. Their expression is driven by two distinct promoters: Sall1^{ncre}:Cx3cr1^{ccre}, which specifically targets parenchymal microglia, and Lyve1^{ncre}: Cx3cr1^{ccre}, which allows the ability to target various CAMs throughout the brain. These new models are very promising and were developed using CRISPR/Cas9 technology. According to the authors, a CCre cassette was inserted after the *Cx3cr1* gene, and Ncre partner transgenes were inserted into *Sall1* and *Lyve1* loci, respectively. In fact, *Lyve1* is scarcely expressed by microglia, at least in the ARC and ME [108] (Figure 4). Hence, both models may be useful to study separately the functions of microglia and CAMs in HFD-induced inflammation, because the knowledge about the role of each glial cell type in this specific condition still need to be better clarified.

5. Microglial Signature Changes Upon HFD-Induced Hypothalamic Inflammation

There are sufficient clues indicating that microglia present distinct activated signatures under inflammatory conditions. Sousa et al. [72] performed scRNA-seq to investigate the microglial profile in the brain of LPS-injected mice. They found that microglia isolated from these mice exhibit a downregulation of their homeostatic signature together with an upregulation of inflammatory genes. They obtained this data by excluding other CNS-resident immune cells and peripheral cells in the analysis.

To explore deeply how HFD-induced inflammation affects microglial transcriptomics, we have searched for transcriptomic data published from HFD-fed rodents at a single-cell resolution. As already mentioned, Campbell et al. [108] performed Drop-seq profiling on the ARC and ME from mice across different feeding conditions, including 1-week HFD (60% calories from fat). They found that microglia present high *P2ry12* and low *Mrc1* while PVMs show low *P2ry12* and high *Mcr1*. However, when comparing low fat diet–fed mice with HFD-fed mice, they found the transcripts from these clusters downregulated, but they did not describe which HFD-sensitive genes were modulated.

On the other hand, in an experimental model of prolonged high-fat feeding, C57BL/6J male mice were fed a HFD (45% calories from fat) for 10 weeks, and their ARC was collected for single-nucleus RNA-seq (snRNA-seq) [156]. After consuming a HFD for 10-weeks, some genes were upregulated in the cluster of microglia (*Nudt5*, *Gsk3a*, *Oxt*, *Lars2*, and *Il17ra*) while other genes (*Sun1*, *Tmem173* and *Anxa3*) were downregulated (Figure 5).

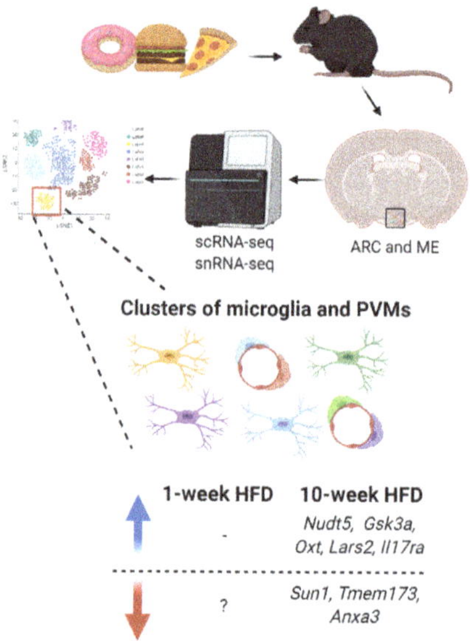

Figure 5. Transcriptomic signature changes in microglia (ramified cells) and perivascular macrophages (PVMs) (elongated cells around blood vessels) from the arcuate nucleus (ARC) and median eminence (ME) under short-term and prolonged HFD-induced hypothalamic inflammation. Blue arrow indicates upregulation while red arrow indicates downregulation.

Some of them have also been described in obesity-related studies. Methylation in *Lars2*, which is a mitochondrial gene, has been reported in an epigenome-wide association study as a gene associated with increased body mass index and waist circumference [157]. The receptor for interleukin-17 (IL-17), known as *Il17ra*, is found in pro-opiomelanocortin (POMC) and agouti-related peptide (AgRP) neurons in the hypothalamus and modulates food intake after IL-17 binding, without affecting whole-body energy expenditure [158].

Although the ARC is the main nucleus involved in energy metabolism and food intake control, functional impairments in neurons and glial cells from other hypothalamic nuclei are also involved in DIO establishment [159,160]. Recently, Rossi et al. [161] identified in the lateral hypothalamic area (LHA) thousands of genes altered upon prolonged HFD intake. This research article highlighted only dynamic transcriptomic details in neurons; thus, the authors did not deeply explore the transcription differences in microglia and other myeloid cell lineages upon. However, the authors distinctively represent these cells by using *Cx3cr1* hallmark for microglia and *Lys2* for myeloid cells, which was not considered in data from ARC published by Deng et al. [156].

Interestingly, from all transcriptomic data obtained from the ARC–ME or LHA, researchers can apply a bioinformatic analysis to better investigate how microglia and other myeloid cells changes upon prolonged HFD. In the face of the recent and continuous development of novel transcriptomic tools, such as scRNA-seq, snRNA-seq, and CyTOF, among others, future studies should be conducted to better clarify the main changes in microglial signature throughout the hypothalamus under different stages of high-fat feeding. For accurate data interpretation, microglia and other CNS-associated macrophages should be clustered and analyzed separately. The detailed identification of those cells will be valuable to answer more precisely several outstanding questions.

6. Conclusions

Microglia were first recognized as macrophage-like cells from the CNS a century ago, but for a long time their complexity was unknown. Luckily, the development of assorted transcriptomic tools has boosted the knowledge about these cells in recent years. Currently, it is well known that the hypothalamus presents several microglial subsets that can be identified by their hallmarks: *Iba1*, *Cx3Cr1*, *Tmem119*, *P2ry12*, *Trem2*, *Hexb*, and *Csfr1*, among many other. DM are also found in the hypothalamic area, but their roles in HFD-induced inflammation has been poorly investigated. In fact, microglia play a pivotal role in different stages of the hypothalamic inflammatory process, but how each microglial subtype reacts to SFAs from the diet, communicates with other cells, or even leads to the recruitment of peripheral myeloid cells remains to be explored. Beyond microglia, astrocytes also display high heterogeneity throughout the CNS, and in the hypothalamus, its diversity and its crosstalk with microglia require to be better elucidated. Novel experimental models for manipulating or labelling microglia have been developed and will be useful to answer that question in forthcoming research. On the other hand, the hypothalamic microglial signature under HFD-induced hypothalamic inflammation should still be further studied using different transcriptomic approaches. Together, these advances will allow researchers to take full advantage of crucial insights we have gained about microglial heterogeneity (provided by transcriptomic data) and to exploit this knowledge to determine the mechanisms by which microglia are involved in the inflammatory process observed in obesity.

Author Contributions: Conceptualization, N.F.M. and E.P.d.A.; investigation, N.F.M., C.P.J., and A.M.Z.; writing—original draft preparation, N.F.M., C.P.J., and A.M.Z.; writing—review and editing, N.F.M.; supervision, E.P.d.A.; funding acquisition, N.F.M., C.P.J., A.M.Z., and E.P.d.A. All authors have read and agreed to the published version of the manuscript.

Funding: This research was funded by the Sao Paulo Research Foundation (FAPESP: 2013/07607-8, 2016/17810-3, 2017/22511-8, and 2019/00330-7) and by the Coordination for the Improvement of Higher Education Personnel (CNPq).

Institutional Review Board Statement: Not applicable.

Informed Consent Statement: Not applicable.

Data Availability Statement: Not applicable.

Acknowledgments: Figures 1–3 and 5 were created with BioRender.com and Figure 4 was created with Corel Draw Graphics Suite 2020.

Conflicts of Interest: The authors declare no conflict of interest.

Abbreviations

ARC	Arcuate nucleus of the hypothalamus
BAMs	Border-associated macrophages
BBB	Blood–brain barrier
BMDC	Bone-marrow-derived cells

CAMs	CNS-associated macrophages
CCL2	Chemokine C-C motif ligand-2
CCR2	C-C chemokine receptor type 2
CD11b	Cluster of differentiation molecule 11B
CD169	Sialoadhesin
CNS	Central nervous system
CSF1R	Colony stimulating factor 1 receptor
CX3CR1	CX3C chemokine receptor 1
CyTOF	Mass cytometry
DAM	Disease-associated microglia
DIO	Diet-induced obesity
DM	Dark microglia
EAE	Experimental autoimmune encephalomyelitis
FACS	Fluorescence-activated cell sorting
FFAs	Free fatty acids
GFAP	Glial fibrillary acidic protein
Hexb	Beta-hexosaminidase subunit beta
HFD	High-fat diet
IBA1	Ionized calcium binding adaptor molecule 1
iNOS	Inducible nitric oxide synthase
ITGAM	Integrin subunit alpha M
LHA	Lateral hypothalamic area
LPS	Lipopolysaccharide
MACS	Magnetic-activated cell sorting
MCP-1	Monocyte chemoattractant protein-1
ME	Median eminence
P2ry12	Purinergic receptor P2Y12
PVH	Paraventricular nucleus of the hypothalamus
PVMs	Perivascular macrophages
ROS	Reactive oxygen species
scRNA-seq	Single-cell RNA sequencing
SFAs	Saturated fatty acids
Siglec	Sialic acid-binding immunoglobulin super-family lectin
snRNA-seq	Single-nucleus RNA sequencing
TGF-β	Transforming growth factor beta
TLR4	Toll-like receptor 4
Tmem119	Transmembrane protein 119
TREM2	Triggering receptor expressed on myeloid cells-2
VEGF	Vascular endothelial growth factor
VMH	Ventromedial nucleus of the hypothalamus

References

1. De Souza, C.T.; Araujo, E.P.; Bordin, S.; Ashimine, R.; Zollner, R.L.; Boschero, A.C.; Saad, M.J.A.; Velloso, L.A. Consumption of a Fat-Rich Diet Activates a Proinflammatory Response and Induces Insulin Resistance in the Hypothalamus. *Endocrinology* **2005**, *146*, 4192–4199. [CrossRef] [PubMed]
2. Posey, K.A.; Clegg, D.J.; Printz, R.L.; Byun, J.; Morton, G.J.; Vivekanandan-Giri, A.; Pennathur, S.; Baskin, D.G.; Heinecke, J.W.; Woods, S.C.; et al. Hypothalamic Proinflammatory Lipid Accumulation, Inflammation, and Insulin Resistance in Rats Fed a High-Fat Diet. *Am. J. Physiol. Endocrinol. Metab.* **2009**, *296*. [CrossRef] [PubMed]
3. Klöckener, T.; Hess, S.; Belgardt, B.F.; Paeger, L.; Verhagen, L.A.W.; Husch, A.; Sohn, J.W.; Hampel, B.; Dhillon, H.; Zigman, J.M.; et al. High-Fat Feeding Promotes Obesity via Insulin Receptor/PI3K-Dependent Inhibition of SF-1 VMH Neurons. *Nat. Neurosci.* **2011**, *14*, 911–918. [CrossRef] [PubMed]
4. Van De Sande-Lee, S.; Pereira, F.R.S.; Cintra, D.E.; Fernandes, P.T.; Cardoso, A.R.; Garlipp, C.R.; Chaim, E.A.; Pareja, J.C.; Geloneze, B.; Li, M.; et al. Partial Reversibility of Hypothalamic Dysfunction and Changes in Brain Activity after Body Mass Reduction in Obese Subjects. *Diabetes* **2011**, *60*, 1699–1704. [CrossRef] [PubMed]
5. Thaler, J.P.; Yi, C.X.; Schur, E.A.; Guyenet, S.J.; Hwang, B.H.; Dietrich, M.O.; Zhao, X.; Sarruf, D.A.; Izgur, V.; Maravilla, K.R.; et al. Obesity Is Associated with Hypothalamic Injury in Rodents and Humans. *J. Clin. Investig.* **2012**, *122*, 153–162. [CrossRef]

6. Kreutzer, C.; Peters, S.; Schulte, D.M.; Fangmann, D.; Türk, K.; Wolff, S.; Van Eimeren, T.; Ahrens, M.; Beckmann, J.; Schafmayer, C.; et al. Hypothalamic Inflammation in Human Obesity Is Mediated by Environmental and Genetic Factors. *Diabetes* **2017**, *66*, 2407–2415. [CrossRef] [PubMed]
7. Milanski, M.; Degasperi, G.; Coope, A.; Morari, J.; Denis, R.; Cintra, D.E.; Tsukumo, D.M.L.; Anhe, G.; Amaral, M.E.; Takahashi, H.K.; et al. Saturated Fatty Acids Produce an Inflammatory Response Predominantly through the Activation of TLR4 Signaling in Hypothalamus: Implications for the Pathogenesis of Obesity. *J. Neurosci.* **2009**, *29*, 359–370. [CrossRef]
8. Rebelos, E.; Hirvonen, J.; Bucci, M.; Pekkarinen, L.; Nyman, M.; Hannukainen, J.C.; Iozzo, P.; Salminen, P.; Nummenmaa, L.; Ferrannini, E.; et al. Brain Free Fatty Acid Uptake Is Elevated in Morbid Obesity, and Is Irreversible 6 Months after Bariatric Surgery: A Positron Emission Tomography Study. *Diabetes Obes. Metab.* **2020**, *22*, 1074–1082. [CrossRef]
9. Sergi, D.; Williams, L.M. Potential Relationship between Dietary Long-Chain Saturated Fatty Acids and Hypothalamic Dysfunction in Obesity. *Nutr. Rev.* **2020**, *78*, 261–277. [CrossRef]
10. Rodríguez, E.M.; Blázquez, J.L.; Guerra, M. The Design of Barriers in the Hypothalamus Allows the Median Eminence and the Arcuate Nucleus to Enjoy Private Milieus: The Former Opens to the Portal Blood and the Latter to the Cerebrospinal Fluid. *Peptides* **2010**, *31*, 757–776. [CrossRef]
11. Haddad-Tóvolli, R.; Dragano, N.R.V.; Ramalho, A.F.S.; Velloso, L.A. Development and Function of the Blood-Brain Barrier in the Context of Metabolic Control. *Front. Neurosci.* **2017**, *11*, 224. [CrossRef]
12. Lee, C.H.; Kim, H.J.; Lee, Y.S.; Kang, G.M.; Lim, H.S.; Lee, S.H.; Song, D.K.; Kwon, O.; Hwang, I.; Son, M.; et al. Hypothalamic Macrophage Inducible Nitric Oxide Synthase Mediates Obesity-Associated Hypothalamic Inflammation. *Cell Rep.* **2018**, *25*, 934–946.e5. [CrossRef] [PubMed]
13. Lee, C.H.; Shin, S.H.; Kang, G.M.; Kim, S.; Kim, J.; Yu, R.; Kim, M.S. Cellular Source of Hypothalamic Macrophage Accumulation in Diet-Induced Obesity. *J. Neuroinflamm.* **2019**, *16*. [CrossRef] [PubMed]
14. Douglass, J.D.; Dorfman, M.D.; Fasnacht, R.; Shaffer, L.D.; Thaler, J.P. Astrocyte IKKβ/NF-KB Signaling Is Required for Diet-Induced Obesity and Hypothalamic Inflammation. *Mol. Metab.* **2017**, *6*, 366–373. [CrossRef]
15. Cansell, C.; Stobbe, K.; Sanchez, C.; Le Thuc, O.; Mosser, C.A.; Ben-Fradj, S.; Leredde, J.; Lebeaupin, C.; Debayle, D.; Fleuriot, L.; et al. Dietary Fat Exacerbates Postprandial Hypothalamic Inflammation Involving Glial Fibrillary Acidic Protein-Positive Cells and Microglia in Male Mice. *Glia* **2020**. [CrossRef]
16. Enriori, P.J.; Evans, A.E.; Sinnayah, P.; Jobst, E.E.; Tonelli-Lemos, L.; Billes, S.K.; Glavas, M.M.; Grayson, B.E.; Perello, M.; Nillni, E.A.; et al. Diet-Induced Obesity Causes Severe but Reversible Leptin Resistance in Arcuate Melanocortin Neurons. *Cell Metab.* **2007**, *5*, 181–194. [CrossRef] [PubMed]
17. Dalvi, P.S.; Chalmers, J.A.; Luo, V.; Han, D.Y.; Wellhauser, L.; Liu, Y.; Tran, D.Q.; Castel, J.; Luquet, S.; Wheeler, M.B.; et al. High Fat Induces Acute and Chronic Inflammation in the Hypothalamus: Effect of High-Fat Diet, Palmitate and TNF-α on Appetite-Regulating NPY Neurons. *Int. J. Obes.* **2017**, *41*, 149–158. [CrossRef]
18. Moraes, J.C.; Coope, A.; Morari, J.; Cintra, D.E.; Roman, E.A.; Pauli, J.R.; Romanatto, T.; Carvalheira, J.B.; Oliveira, A.L.R.; Saad, M.J.; et al. High-Fat Diet Induces Apoptosis of Hypothalamic Neurons. *PLoS ONE* **2009**, *4*, e5045. [CrossRef]
19. Carraro, R.S.; Souza, G.F.; Solon, C.; Razolli, D.S.; Chausse, B.; Barbizan, R.; Victorio, S.C.; Velloso, L.A. Hypothalamic Mitochondrial Abnormalities Occur Downstream of Inflammation in Diet-Induced Obesity. *Mol. Cell. Endocrinol.* **2018**, *460*, 238–245. [CrossRef] [PubMed]
20. Valdearcos, M.; Douglass, J.D.; Robblee, M.M.; Dorfman, M.D.; Stifler, D.R.; Bennett, M.L.; Gerritse, I.; Fasnacht, R.; Barres, B.A.; Thaler, J.P.; et al. Microglial Inflammatory Signaling Orchestrates the Hypothalamic Immune Response to Dietary Excess and Mediates Obesity Susceptibility. *Cell Metab.* **2017**, *26*, 185–197.e3. [CrossRef] [PubMed]
21. Greenhalgh, A.D.; David, S.; Bennett, F.C. Immune Cell Regulation of Glia during CNS Injury and Disease. *Nat. Rev. Neurosci.* **2020**, *21*, 139–152. [CrossRef] [PubMed]
22. Mendes, N.F.; Kim, Y.-B.; Velloso, L.A.; Araújo, E.P. Hypothalamic Microglial Activation in Obesity: A Mini-Review. *Front. Neurosci.* **2018**. [CrossRef] [PubMed]
23. Nimmerjahn, A.; Kirchhoff, F.; Helmchen, F. Neuroscience: Resting Microglial Cells Are Highly Dynamic Surveillants of Brain Parenchyma in Vivo. *Science* **2005**, *308*, 1314–1318. [CrossRef]
24. Ginhoux, F.; Greter, M.; Leboeuf, M.; Nandi, S.; See, P.; Gokhan, S.; Mehler, M.F.; Conway, S.J.; Ng, L.G.; Stanley, E.R.; et al. Fate Mapping Analysis Reveals That Adult Microglia Derive from Primitive Macrophages. *Science* **2010**, *330*, 841–845. [CrossRef] [PubMed]
25. Martinez, F.O.; Gordon, S. The M1 and M2 Paradigm of Macrophage Activation: Time for Reassessment. *F1000Prime Rep.* **2014**, *6*. [CrossRef] [PubMed]
26. Ransohoff, R.M. A Polarizing Question: Do M1 and M2 Microglia Exist. *Nat. Neurosci.* **2016**, *19*, 987–991. [CrossRef] [PubMed]
27. Stratoulias, V.; Venero, J.L.; Tremblay, M.; Joseph, B. Microglial Subtypes: Diversity within the Microglial Community. *EMBO J.* **2019**, *38*. [CrossRef]
28. Bisht, K.; Sharma, K.P.; Lecours, C.; Gabriela Sánchez, M.; El Hajj, H.; Milior, G.; Olmos-Alonso, A.; Gómez-Nicola, D.; Luheshi, G.; Vallières, L.; et al. Dark Microglia: A New Phenotype Predominantly Associated with Pathological States. *Glia* **2016**, *64*, 826–839. [CrossRef]
29. Savage, J.C.; St-Pierre, M.K.; Hui, C.W.; Tremblay, M.E. Microglial Ultrastructure in the Hippocampus of a Lipopolysaccharide-Induced Sickness Mouse Model. *Front. Neurosci.* **2019**, *13*. [CrossRef]

30. Mildner, A.; Schmidt, H.; Nitsche, M.; Merkler, D.; Hanisch, U.K.; Mack, M.; Heikenwalder, M.; Brück, W.; Priller, J.; Prinz, M. Microglia in the Adult Brain Arise from Ly-6ChiCCR2+ Monocytes Only under Defined Host Conditions. *Nat. Neurosci.* **2007**, *10*, 1544–1553. [CrossRef]
31. Hui, C.W.; St-Pierre, A.; El Hajj, H.; Remy, Y.; Hébert, S.S.; Luheshi, G.N.; Srivastava, L.K.; Tremblay, M.È. Prenatal Immune Challenge in Mice Leads to Partly Sex-Dependent Behavioral, Microglial, and Molecular Abnormalities Associated with Schizophrenia. *Front. Mol. Neurosci.* **2018**, *11*. [CrossRef]
32. Elgayar, S.A.M.; Abdel-hafez, A.A.M.; Gomaa, A.M.S.; Elsherif, R. Vulnerability of Glia and Vessels of Rat Substantia Nigra in Rotenone Parkinson Model. *Ultrastruct. Pathol.* **2018**, *42*, 181–192. [CrossRef] [PubMed]
33. El Hajj, H.; Savage, J.C.; Bisht, K.; Parent, M.; Vallières, L.; Rivest, S.; Tremblay, M.È. Ultrastructural Evidence of Microglial Heterogeneity in Alzheimer's Disease Amyloid Pathology. *J. Neuroinflamm.* **2019**, *16*. [CrossRef]
34. Kavetsky, L.; Green, K.K.; Boyle, B.R.; Yousufzai, F.A.K.; Padron, Z.M.; Melli, S.E.; Kuhnel, V.L.; Jackson, H.M.; Blanco, R.E.; Howell, G.R.; et al. Increased Interactions and Engulfment of Dendrites by Microglia Precede Purkinje Cell Degeneration in a Mouse Model of Niemann Pick Type-C. *Sci. Rep.* **2019**, *9*. [CrossRef] [PubMed]
35. Calì, C.; Agus, M.; Kare, K.; Boges, D.J.; Lehväslaiho, H.; Hadwiger, M.; Magistretti, P.J. 3D Cellular Reconstruction of Cortical Glia and Parenchymal Morphometric Analysis from Serial Block-Face Electron Microscopy of Juvenile Rat. *Prog. Neurobiol.* **2019**, *183*. [CrossRef]
36. Pedicone, C.; Fernandes, S.; Dungan, O.M.; Dormann, S.M.; Viernes, D.R.; Adhikari, A.A.; Choi, L.B.; De Jong, E.P.; Chisholm, J.D.; Kerr, W.G. Pan-SHIP1/2 Inhibitors Promote Microglia Effector Functions Essential for CNS Homeostasis. *J. Cell Sci.* **2020**, *133*. [CrossRef]
37. Uranova, N.A.; Vikhreva, O.V.; Rakhmanova, V.I.; Orlovskaya, D.D. Ultrastructural Pathology of Oligodendrocytes Adjacent to Microglia in Prefrontal White Matter in Schizophrenia. *Npj Schizophr.* **2018**, *4*. [CrossRef]
38. Whitelaw, B.S.; Matei, E.K.; Majewska, A.K. Phosphoinositide 3-Kinase γ Is Not a Predominant Regulator of ATP-Dependent Directed Microglial Process Motility or Experience-Dependent Ocular Dominance Plasticity. *Eneuro* **2020**. [CrossRef]
39. Masuda, T.; Sankowski, R.; Staszewski, O.; Böttcher, C.; Amann, L.; Sagar; Scheiwe, C.; Nessler, S.; Kunz, P.; van Loo, G.; et al. Spatial and Temporal Heterogeneity of Mouse and Human Microglia at Single-Cell Resolution. *Nature* **2019**, *566*, 388–392. [CrossRef] [PubMed]
40. Gratuze, M.; Leyns, C.E.G.; Sauerbeck, A.D.; St-Pierre, M.K.; Xiong, M.; Kim, N.; Serrano, J.R.; Tremblay, M.È.; Kummer, T.T.; Colonna, M.; et al. Impact of TREM2R47Hvariant on Tau Pathology-Induced Gliosis and Neurodegeneration. *J. Clin. Investig.* **2020**, *130*, 4954–4968. [CrossRef]
41. Liu, W.; Taso, O.; Wang, R.; Bayram, S.; Graham, A.C.; Garcia-Reitboeck, P.; Mallach, A.; Andrews, W.D.; Piers, T.M.; Botia, J.A.; et al. Trem2 Promotes Anti-Inflammatory Responses in Microglia and Is Suppressed under pro-Inflammatory Conditions. *Hum. Mol. Genet.* **2020**, *29*, 3224–3248. [CrossRef]
42. Li, F.; Xu, D.; Hou, K.; Gou, X.; Li, Y. The Role of P2Y12 Receptor Inhibition in Ischemic Stroke on Microglia, Platelets and Vascular Smooth Muscle Cells. *J. Thromb. Thrombolysis* **2020**, *50*, 874–885. [CrossRef] [PubMed]
43. Bohnert, S.; Seiffert, A.; Trella, S.; Bohnert, M.; Distel, L.; Ondruschka, B.; Monoranu, C.M. TMEM119 as a Specific Marker of Microglia Reaction in Traumatic Brain Injury in Postmortem Examination. *Int. J. Legal Med.* **2020**, *134*. [CrossRef] [PubMed]
44. Stevens, B.; Allen, N.J.; Vazquez, L.E.; Howell, G.R.; Christopherson, K.S.; Nouri, N.; Micheva, K.D.; Mehalow, A.K.; Huberman, A.D.; Stafford, B.; et al. The Classical Complement Cascade Mediates CNS Synapse Elimination. *Cell* **2007**, *131*, 1164–1178. [CrossRef] [PubMed]
45. Schafer, D.P.; Lehrman, E.K.; Kautzman, A.G.; Koyama, R.; Mardinly, A.R.; Yamasaki, R.; Ransohoff, R.M.; Greenberg, M.E.; Barres, B.A.; Stevens, B. Microglia Sculpt Postnatal Neural Circuits in an Activity and Complement-Dependent Manner. *Neuron* **2012**, *74*, 691–705. [CrossRef] [PubMed]
46. Keren-Shaul, H.; Spinrad, A.; Weiner, A.; Matcovitch-Natan, O.; Dvir-Szternfeld, R.; Ulland, T.K.; David, E.; Baruch, K.; Lara-Astaiso, D.; Toth, B.; et al. A Unique Microglia Type Associated with Restricting Development of Alzheimer's Disease. *Cell* **2017**, *169*, 1276–1290.e17. [CrossRef] [PubMed]
47. Deczkowska, A.; Keren-Shaul, H.; Weiner, A.; Colonna, M.; Schwartz, M.; Amit, I. Disease-Associated Microglia: A Universal Immune Sensor of Neurodegeneration. *Cell* **2018**, *173*, 1073–1081. [CrossRef] [PubMed]
48. Friedman, B.A.; Srinivasan, K.; Ayalon, G.; Meilandt, W.J.; Lin, H.; Huntley, M.A.; Cao, Y.; Lee, S.H.; Haddick, P.C.G.; Ngu, H.; et al. Diverse Brain Myeloid Expression Profiles Reveal Distinct Microglial Activation States and Aspects of Alzheimer's Disease Not Evident in Mouse Models. *Cell Rep.* **2018**, *22*, 832–847. [CrossRef]
49. Schlepckow, K.; Monroe, K.M.; Kleinberger, G.; Cantuti-Castelvetri, L.; Parhizkar, S.; Xia, D.; Willem, M.; Werner, G.; Pettkus, N.; Brunner, B.; et al. Enhancing Protective Microglial Activities with a Dual Function TREM 2 Antibody to the Stalk Region. *EMBO Mol. Med.* **2020**, *12*. [CrossRef]
50. Kigerl, K.A.; de Rivero Vaccari, J.P.; Dietrich, W.D.; Popovich, P.G.; Keane, R.W. Pattern Recognition Receptors and Central Nervous System Repair. *Exp. Neurol.* **2014**, *258*, 5–16. [CrossRef]
51. Painter, M.M.; Atagi, Y.; Liu, C.C.; Rademakers, R.; Xu, H.; Fryer, J.D.; Bu, G. TREM2 in CNS Homeostasis and Neurodegenerative Disease. *Mol. Neurodegener.* **2015**, *10*, 1–10. [CrossRef]
52. N'Diaye, E.N.; Branda, C.S.; Branda, S.S.; Nevarez, L.; Colonna, M.; Lowell, C.; Hamerman, J.A.; Seaman, W.E. TREM-2 (Triggering Receptor Expressed on Myeloid Cells 2) Is a Phagocytic Receptor for Bacteria. *J. Cell Biol.* **2009**, *184*, 215–223. [CrossRef]

53. Wang, Y.; Cella, M.; Mallinson, K.; Ulrich, J.D.; Young, K.L.; Robinette, M.L.; Gilfillan, S.; Krishnan, G.M.; Sudhakar, S.; Zinselmeyer, B.H.; et al. TREM2 Lipid Sensing Sustains the Microglial Response in an Alzheimer's Disease Model. *Cell* **2015**, *160*, 1061–1071. [CrossRef] [PubMed]
54. Atagi, Y.; Liu, C.-C.; Painter, M.M.; Chen, X.-F.; Verbeeck, C.; Zheng, H.; Li, X.; Rademakers, R.; Kang, S.S.; Xu, H.; et al. Apolipoprotein E Is a Ligand for Triggering Receptor Expressed on Myeloid Cells 2 (TREM2). *J. Biol. Chem.* **2015**, *290*, 26043–26050. [CrossRef] [PubMed]
55. Bailey, C.C.; Devaux, L.B.; Farzan, M. The Triggering Receptor Expressed on Myeloid Cells 2 Binds Apolipoprotein E. *J. Biol. Chem.* **2015**, *290*, 26033–26042. [CrossRef] [PubMed]
56. Yeh, F.L.; Wang, Y.; Tom, I.; Gonzalez, L.C.; Sheng, M. TREM2 Binds to Apolipoproteins, Including APOE and CLU/APOJ, and Thereby Facilitates Uptake of Amyloid-Beta by Microglia. *Neuron* **2016**, *91*, 328–340. [CrossRef] [PubMed]
57. Kawabori, M.; Kacimi, R.; Kauppinen, T.; Calosing, C.; Kim, J.Y.; Hsieh, C.L.; Nakamura, M.C.; Yenari, M.A. Triggering Receptor Expressed on Myeloid Cells 2 (TREM2) Deficiency Attenuates Phagocytic Activities of Microglia and Exacerbates Ischemic Damage in Experimental Stroke. *J. Neurosci.* **2015**, *35*, 3384–3396. [CrossRef] [PubMed]
58. Turnbull, I.R.; Gilfillan, S.; Cella, M.; Aoshi, T.; Miller, M.; Piccio, L.; Hernandez, M.; Colonna, M. Cutting Edge: TREM-2 Attenuates Macrophage Activation. *J. Immunol.* **2006**, *177*, 3520–3524. [CrossRef]
59. Zhang, X.; Zhang, G.; Zhang, H.; Karin, M.; Bai, H.; Cai, D. Hypothalamic IKKβ/NF-KB and ER Stress Link Overnutrition to Energy Imbalance and Obesity. *Cell* **2008**, *135*, 61–73. [CrossRef]
60. Tam, A.B.; Mercado, E.L.; Hoffmann, A.; Niwa, M. ER Stress Activates NF-KB by Integrating Functions of Basal IKK Activity, IRE1 and PERK. *PLoS ONE* **2012**, *7*. [CrossRef]
61. Zhang, Y.; Reichel, J.M.; Han, C.; Pablo Zuniga-Hertz, J.; Correspondence, D.C.; Zuniga-Hertz, J.P.; Cai, D. Astrocytic Process Plasticity and IKKb/NF-KB in Central Control of Blood Glucose, Blood Pressure, and Body Weight Cell Metabolism Article Astrocytic Process Plasticity and IKKb/NF-KB in Central Control of Blood Glucose, Blood Pressure, and Body Weight. *Cell Metab.* **2017**, *25*. [CrossRef]
62. Cai, D.; Khor, S. "Hypothalamic Microinflammation" Paradigm in Aging and Metabolic Diseases. *Cell Metab.* **2019**, *30*, 19–35. [CrossRef]
63. Schmid, C.D.; Sautkulis, L.N.; Danielson, P.E.; Cooper, J.; Hasel, K.W.; Hilbush, B.S.; Sutcliffe, J.G.; Carson, M.J. Heterogeneous Expression of the Triggering Receptor Expressed on Myeloid Cells-2 on Adult Murine Microglia. *J. Neurochem.* **2002**, *83*, 1309–1320. [CrossRef]
64. Hickman, S.E.; El Khoury, J. TREM2 and the Neuroimmunology of Alzheimer's Disease. *Biochem. Pharmacol.* **2014**, *88*, 495–498. [CrossRef] [PubMed]
65. Colonna, M. Trems in the Immune System and Beyond. *Nat. Rev. Immunol.* **2003**, *3*, 445–453. [CrossRef]
66. Zhong, L.; Zhang, Z.L.; Li, X.; Liao, C.; Mou, P.; Wang, T.; Wang, Z.; Wang, Z.; Wei, M.; Xu, H.; et al. TREM2/DAP12 Complex Regulates Inflammatory Responses in Microglia via the JNK Signaling Pathway. *Front. Aging Neurosci.* **2017**, *9*. [CrossRef] [PubMed]
67. Li, J.T.; Zhang, Y. TREM2 Regulates Innate Immunity in Alzheimer's Disease. *J. Neuroinflamm.* **2018**, *15*, 1–7. [CrossRef]
68. Zhang, J.; Zheng, Y.; Luo, Y.; Du, Y.; Zhang, X.; Fu, J. Curcumin Inhibits LPS-Induced Neuroinflammation by Promoting Microglial M2 Polarization via TREM2/ TLR4/ NF-KB Pathways in BV2 Cells. *Mol. Immunol.* **2019**, *116*, 29–37. [CrossRef] [PubMed]
69. Furube, E.; Kawai, S.; Inagaki, H.; Takagi, S.; Miyata, S. Brain Region-Dependent Heterogeneity and Dose-Dependent Difference in Transient Microglia Population Increase during Lipopolysaccharide-Induced Inflammation. *Sci. Rep.* **2018**, *8*. [CrossRef] [PubMed]
70. Hammond, T.R.; Dufort, C.; Dissing-Olesen, L.; Giera, S.; Young, A.; Wysoker, A.; Walker, A.J.; Gergits, F.; Segel, M.; Nemesh, J.; et al. Single-Cell RNA Sequencing of Microglia throughout the Mouse Lifespan and in the Injured Brain Reveals Complex Cell-State Changes. *Immunity* **2019**, *50*, 253–271.e6. [CrossRef] [PubMed]
71. Jordão, M.J.C.; Sankowski, R.; Brendecke, S.M.; Sagar; Locatelli, G.; Tai, Y.H.; Tay, T.L.; Schramm, E.; Armbruster, S.; Hage-meyer, N.; et al. Neuroimmunology: Single-Cell Profiling Identifies Myeloid Cell Subsets with Distinct Fates during Neuroinflammation. *Science* **2019**, *363*. [CrossRef] [PubMed]
72. Sousa, C.; Golebiewska, A.; Poovathingal, S.K.; Kaoma, T.; Pires-Afonso, Y.; Martina, S.; Coowar, D.; Azuaje, F.; Skupin, A.; Balling, R.; et al. Single-cell Transcriptomics Reveals Distinct Inflammation-induced Microglia Signatures. *EMBO Rep.* **2018**, *19*. [CrossRef] [PubMed]
73. Sankowski, R.; Böttcher, C.; Masuda, T.; Geirsdottir, L.; Sagar; Sindram, E.; Seredenina, T.; Muhs, A.; Scheiwe, C.; Shah, M.J.; et al. Mapping Microglia States in the Human Brain through the Integration of High-Dimensional Techniques. *Nat. Neurosci.* **2019**, *22*, 2098–2110. [CrossRef]
74. Böttcher, C.; Schlickeiser, S.; Sneeboer, M.A.M.; Kunkel, D.; Knop, A.; Paza, E.; Fidzinski, P.; Kraus, L.; Snijders, G.J.L.; Kahn, R.S.; et al. Human Microglia Regional Heterogeneity and Phenotypes Determined by Multiplexed Single-Cell Mass Cytometry. *Nat. Neurosci.* **2019**, *22*, 78–90. [CrossRef]
75. Liddelow, S.A.; Marsh, S.E.; Stevens, B. Microglia and Astrocytes in Disease: Dynamic Duo or Partners in Crime? *Trends Immunol.* **2020**, *41*, 820–835. [CrossRef] [PubMed]

76. Esaulova, E.; Cantoni, C.; Shchukina, I.; Zaitsev, K.; Bucelli, R.C.; Wu, G.F.; Artyomov, M.N.; Cross, A.H.; Edelson, B.T. Single-Cell RNA-Seq Analysis of Human CSF Microglia and Myeloid Cells in Neuroinflammation. *Neurol. Neuroimmunol. Neuroinflamm.* **2020**, *7*, 732. [CrossRef] [PubMed]
77. Geirsdottir, L.; David, E.; Keren-Shaul, H.; Weiner, A.; Bohlen, S.C.; Neuber, J.; Balic, A.; Giladi, A.; Sheban, F.; Dutertre, C.A.; et al. Cross-Species Single-Cell Analysis Reveals Divergence of the Primate Microglia Program. *Cell* **2019**, *179*, 1609–1622.e16. [CrossRef]
78. Rahman, M.H.; Kim, M.-S.; Lee, I.-K.; Yu, R.; Suk, K. Interglial Crosstalk in Obesity-Induced Hypothalamic Inflammation. *Front. Neurosci.* **2018**, *12*. [CrossRef] [PubMed]
79. Lee, D.A.; Bedont, J.L.; Pak, T.; Wang, H.; Song, J.; Miranda-Angulo, A.; Takiar, V.; Charubhumi, V.; Balordi, F.; Takebayashi, H.; et al. Tanycytes of the Hypothalamic Median Eminence Form a Diet-Responsive Neurogenic Niche. *Nat. Neurosci.* **2012**, *15*, 700–702. [CrossRef]
80. Li, Q.; Barres, B.A. Microglia and Macrophages in Brain Homeostasis and Disease. *Nat. Rev. Immunol.* **2018**, *18*, 225–242. [CrossRef] [PubMed]
81. Giovannoni, F.; Quintana, F.J. The Role of Astrocytes in CNS Inflammation. *Trends Immunol.* **2020**, *41*, 805–819. [CrossRef]
82. Tsai, H.H.; Li, H.; Fuentealba, L.C.; Molofsky, A.V.; Taveira-Marques, R.; Zhuang, H.; Tenney, A.; Murnen, A.T.; Fancy, S.P.J.; Merkle, F.; et al. Regional Astrocyte Allocation Regulates CNS Synaptogenesis and Repair. *Science* **2012**, *337*, 358–362. [CrossRef]
83. Ortiz-Rodriguez, A.; Arevalo, M.A. The Contribution of Astrocyte Autophagy to Systemic Metabolism. *Int. J. Mol. Sci.* **2020**, *21*, 2479. [CrossRef]
84. Lee, C.H.; Suk, K.; Yu, R.; Kim, M.S. Cellular Contributors to Hypothalamic Inflammation in Obesity. *Mol. Cells* **2020**, *43*, 431–437.
85. García-Cáceres, C.; Quarta, C.; Varela, L.; Gao, Y.; Gruber, T.; Legutko, B.; Jastroch, M.; Johansson, P.; Ninkovic, J.; Yi, C.X.; et al. Astrocytic Insulin Signaling Couples Brain Glucose Uptake with Nutrient Availability. *Cell* **2016**, *166*, 867–880. [CrossRef] [PubMed]
86. Hsuchou, H.; Pan, W.; Barnes, M.J.; Kastin, A.J. Leptin Receptor MRNA in Rat Brain Astrocytes. *Peptides* **2009**, *30*, 2275–2280. [CrossRef] [PubMed]
87. Wiktorowska, L.; Bilecki, W.; Tertil, M.; Kudla, L.; Szumiec, L.; Mackowiak, M.; Przewlocki, R. Knockdown of the Astrocytic Glucocorticoid Receptor in the Central Nucleus of the Amygdala Diminishes Conditioned Fear Expression and Anxiety. *Behav. Brain Res.* **2021**, *402*. [CrossRef]
88. Fuente-Martín, E.; Garciá-Cáceres, C.; Argente-Arizón, P.; Diáz, F.; Granado, M.; Freire-Regatillo, A.; Castro-González, D.; Ceballos, M.L.; Frago, L.M.; Dickson, S.L.; et al. Ghrelin Regulates Glucose and Glutamate Transporters in Hypothalamic Astrocytes. *Sci. Rep.* **2016**, *6*, 1–15. [CrossRef]
89. Jha, M.K.; Jo, M.; Kim, J.H.; Suk, K. Microglia-Astrocyte Crosstalk: An Intimate Molecular Conversation. *Neuroscientist* **2019**, *25*, 227–240. [CrossRef]
90. Matejuk, A.; Ransohoff, R.M. Crosstalk Between Astrocytes and Microglia: An Overview. *Front. Immunol.* **2020**, *11*, 1416. [CrossRef] [PubMed]
91. Schafer, D.P.; Lehrman, E.K.; Stevens, B. The "Quad-Partite" Synapse: Microglia-Synapse Interactions in the Developing and Mature CNS. *Glia* **2013**, *61*, 24–36. [CrossRef] [PubMed]
92. Cohen-Salmon, M.; Slaoui, L.; Mazaré, N.; Gilbert, A.; Oudart, M.; Alvear-Perez, R.; Elorza-Vidal, X.; Chever, O.; Boulay, A. Astrocytes in the Regulation of Cerebrovascular Functions. *Glia* **2020**, glia.23924. [CrossRef]
93. Liu, L.R.; Liu, J.C.; Bao, J.S.; Bai, Q.Q.; Wang, G.Q. Interaction of Microglia and Astrocytes in the Neurovascular Unit. *Front. Immunol.* **2020**, *11*, 1024. [CrossRef] [PubMed]
94. Yan, J.; Zhang, H.; Yin, Y.; Li, J.; Tang, Y.; Purkayastha, S.; Li, L.; Cai, D. Obesity-and Aging-Induced Excess of Central Transforming Growth Factor-β 2 Potentiates Diabetic Development via an RNA Stress Response. *Nat. Med.* **2014**, *20*, 1001–1008. [CrossRef]
95. Kwon, Y.H.; Kim, J.; Kim, C.S.; Tu, T.H.; Kim, M.S.; Suk, K.; Kim, D.H.; Lee, B.J.; Choi, H.S.; Park, T.; et al. Hypothalamic Lipid-Laden Astrocytes Induce Microglia Migration and Activation. *FEBS Lett.* **2017**, *591*, 1742–1751. [CrossRef]
96. Douglass, J.D.; Dorfman, M.D.; Thaler, J.P. Glia: Silent Partners in Energy Homeostasis and Obesity Pathogenesis. *Diabetologia* **2017**, *60*, 226–236. [CrossRef] [PubMed]
97. Horvath, T.L.; Sarman, B.; García-Cáceres, C.; Enriori, P.J.; Sotonyi, P.; Shanabrough, M.; Borok, E.; Argente, J.; Chowen, J.A.; Perez-Tilve, D.; et al. Synaptic Input Organization of the Melanocortin System Predicts Diet-Induced Hypothalamic Reactive Gliosis and Obesity. *Proc. Natl. Acad. Sci. USA* **2010**, *107*, 14875–14880. [CrossRef]
98. Baufeld, C.; Osterloh, A.; Prokop, S.; Miller, K.R.; Heppner, F.L. High-Fat Diet-Induced Brain Region-Specific Phenotypic Spectrum of CNS Resident Microglia. *Acta Neuropathol.* **2016**, *132*, 361–375. [CrossRef]
99. Mendes, N.F.; Gaspar, J.M.; Lima-Júnior, J.C.; Donato, J.; Velloso, L.A.; Araújo, E.P. TGF-B1 down-Regulation in the Mediobasal Hypothalamus Attenuates Hypothalamic Inflammation and Protects against Diet-Induced Obesity. *Metabolism* **2018**, *85*. [CrossRef]
100. Tichauer, J.E.; Flores, B.; Soler, B.; Eugenín-von Bernhardi, L.; Ramírez, G.; Von Bernhardi, R. Age-Dependent Changes on TGFβ1 Smad3 Pathway Modify the Pattern of Microglial Cell Activation. *Brain Behav. Immun.* **2014**, *37*, 187–196. [CrossRef]
101. Dudiki, T.; Meller, J.; Mahajan, G.; Liu, H.; Zhevlakova, I.; Stefl, S.; Witherow, C.; Podrez, E.; Kothapalli, C.R.; Byzova, T.V. Microglia Control Vascular Architecture via a TGFβ1 Dependent Paracrine Mechanism Linked to Tissue Mechanics. *Nat. Commun.* **2020**, *11*. [CrossRef] [PubMed]

102. Spittau, B.; Dokalis, N.; Prinz, M. The Role of TGFβ Signaling in Microglia Maturation and Activation. *Trends Immunol.* **2020**, *41*, 836–848. [CrossRef] [PubMed]
103. Attaai, A.; Neidert, N.; von Ehr, A.; Potru, P.S.; Zöller, T.; Spittau, B. Postnatal Maturation of Microglia Is Associated with Alternative Activation and Activated TGFβ Signaling. *Glia* **2018**, *66*, 1695–1708. [CrossRef]
104. Morari, J.; Anhe, G.F.; Nascimento, L.F.; De Moura, R.F.; Razolli, D.; Solon, C.; Guadagnini, D.; Souza, G.; Mattos, A.H.; Tobar, N.; et al. Fractalkine (CX3CL1) Is Involved in the Early Activation of Hypothalamic Inflammation in Experimental Obesity. *Diabetes* **2014**, *63*, 3770–3784. [CrossRef]
105. Deshmane, S.L.; Kremlev, S.; Amini, S.; Sawaya, B.E. Monocyte Chemoattractant Protein-1 (MCP-1): An Overview. *J. Interf. Cytokine Res.* **2009**, *29*, 313–325. [CrossRef]
106. Ito, A.; Suganami, T.; Yamauchi, A.; Degawa-Yamauchi, M.; Tanaka, M.; Kouyama, R.; Kobayashi, Y.; Nitta, N.; Yasuda, K.; Hirata, Y.; et al. Role of CC Chemokine Receptor 2 in Bone Marrow Cells in the Recruitment of Macrophages into Obese Adipose Tissue. *J. Biol. Chem.* **2008**, *283*, 35715–35723. [CrossRef]
107. Ataka, K.; Asakawa, A.; Nagaishi, K.; Kaimoto, K.; Sawada, A.; Hayakawa, Y.; Tatezawa, R.; Inui, A.; Fujimiya, M. Bone Marrow-Derived Microglia Infiltrate into the Paraventricular Nucleus of Chronic Psychological Stress-Loaded Mice. *PLoS ONE* **2013**, *8*. [CrossRef]
108. Campbell, J.N.; Macosko, E.Z.; Fenselau, H.; Pers, T.H.; Lyubetskaya, A.; Tenen, D.; Goldman, M.; Verstegen, A.M.J.; Resch, J.M.; McCarroll, S.A.; et al. A Molecular Census of Arcuate Hypothalamus and Median Eminence Cell Types. *Nat. Neurosci.* **2017**, *20*, 484–496. [CrossRef] [PubMed]
109. Ramalho, A.F.; Bombassaro, B.; Dragano, N.R.; Solon, C.; Morari, J.; Fioravante, M.; Barbizan, R.; Velloso, L.A.; Araujo, E.P. Dietary Fats Promote Functional and Structural Changes in the Median Eminence Blood/Spinal Fluid Interface-the Protective Role for BDNF. *J. Neuroinflamm.* **2018**, *15*. [CrossRef] [PubMed]
110. Langlet, F. Tanycytes: A Gateway to the Metabolic Hypothalamus. *J. Neuroendocrinol.* **2014**, *26*, 753–760. [CrossRef]
111. Argaw, A.T.; Asp, L.; Zhang, J.; Navrazhina, K.; Pham, T.; Mariani, J.N.; Mahase, S.; Dutta, D.J.; Seto, J.; Kramer, E.G.; et al. Astrocyte-Derived VEGF-A Drives Blood-Brain Barrier Disruption in CNS Inflammatory Disease. *J. Clin. Investig.* **2012**, *122*, 2454–2468. [CrossRef]
112. Obermeier, B.; Daneman, R.; Ransohoff, R.M. Development, Maintenance and Disruption of the Blood-Brain Barrier. *Nat. Med.* **2013**, *19*, 1584–1596. [CrossRef] [PubMed]
113. Keaney, J.; Campbell, M. The Dynamic Blood-Brain Barrier. *FEBS J.* **2015**, *282*, 4067–4079. [CrossRef] [PubMed]
114. Haruwaka, K.; Ikegami, A.; Tachibana, Y.; Ohno, N.; Konishi, H.; Hashimoto, A.; Matsumoto, M.; Kato, D.; Ono, R.; Kiyama, H.; et al. Dual Microglia Effects on Blood Brain Barrier Permeability Induced by Systemic Inflammation. *Nat. Commun.* **2019**, *10*. [CrossRef]
115. Retamal, M.A.; Froger, N.; Palacios-Prado, N.; Ezan, P.; Sáez, P.J.; Sáez, J.C.; Giaume, C. Cx43 Hemichannels and Gap Junction Channels in Astrocytes Are Regulated Oppositely by Proinflammatory Cytokines Released from Activated Microglia. *J. Neurosci.* **2007**, *27*, 13781–13792. [CrossRef] [PubMed]
116. Même, W.; Calvo, C.; Froger, N.; Ezan, P.; Amigou, E.; Koulakoff, A.; Giaume, C. Proinflammatory Cytokines Released from Microglia Inhibit Gap Junctions in Astrocytes: Potentiation by Β-amyloid. *FASEB J.* **2006**, *20*, 494–496. [CrossRef] [PubMed]
117. Watanabe, M.; Masaki, K.; Yamasaki, R.; Kawanokuchi, J.; Takeuchi, H.; Matsushita, T.; Suzumura, A.; Kira, J.I. Th1 Cells Downregulate Connexin 43 Gap Junctions in Astrocytes via Microglial Activation. *Sci. Rep.* **2016**, *6*. [CrossRef]
118. Nair, S.; Sobotka, K.S.; Joshi, P.; Gressens, P.; Fleiss, B.; Thornton, C.; Mallard, C.; Hagberg, H. Lipopolysaccharide-Induced Alteration of Mitochondrial Morphology Induces a Metabolic Shift in Microglia Modulating the Inflammatory Response in Vitro and in Vivo. *Glia* **2019**, *67*, 1047–1061. [CrossRef] [PubMed]
119. Kim, J.D.; Yoon, N.A.; Jin, S.; Diano, S. Microglial UCP2 Mediates Inflammation and Obesity Induced by High-Fat Feeding. *Cell Metab.* **2019**, *30*, 952–962.e5. [CrossRef] [PubMed]
120. John Lin, C.C.; Yu, K.; Hatcher, A.; Huang, T.W.; Lee, H.K.; Carlson, J.; Weston, M.C.; Chen, F.; Zhang, Y.; Zhu, W.; et al. Identification of Diverse Astrocyte Populations and Their Malignant Analogs. *Nat. Neurosci.* **2017**, *20*, 396–405. [CrossRef] [PubMed]
121. Liddelow, S.A.; Guttenplan, K.A.; Clarke, L.E.; Bennett, F.C.; Bohlen, C.J.; Schirmer, L.; Bennett, M.L.; Münch, A.E.; Chung, W.S.; Peterson, T.C.; et al. Neurotoxic Reactive Astrocytes Are Induced by Activated Microglia. *Nature* **2017**, *541*, 481–487. [CrossRef]
122. Miller, S.J. Astrocyte Heterogeneity in the Adult Central Nervous System. *Front. Cell. Neurosci.* **2018**, *12*, 15. [CrossRef]
123. Escartin, C.; Galea, E.; Oamp, J.P.; Petzold, G.C.; Serrano-Pozo, A.; Steinhamp, C.; Volterra, A.; Carmignoto, G.; Agarwal, A.; Allen, N.J.; et al. Reactive Astrocyte Nomenclature, Definitions, and Future Directions. *Nat. Neurosci.* **2021**, 1–14. [CrossRef]
124. Han, R.T.; Kim, R.D.; Molofsky, A.V.; Liddelow, S.A. Astrocyte-Immune Cell Interactions in Physiology and Pathology. *Immunity* **2021**, *54*, 211–224. [CrossRef]
125. Blasi, E.; Barluzzi, R.; Bocchini, V.; Mazzolla, R.; Bistoni, F. Immortalization of Murine Microglial Cells by a V-Raf/v-Myc Carrying Retrovirus. *J. Neuroimmunol.* **1990**, *27*, 229–237. [CrossRef]
126. Timmerman, R.; Burm, S.M.; Bajramovic, J.J. An Overview of in Vitro Methods to Study Microglia. *Front. Cell. Neurosci.* **2018**, *12*, 242. [CrossRef] [PubMed]
127. Liu, J.J.; Mustafa, S.; Barratt, D.T.; Hutchinson, M.R. Corticosterone Preexposure Increases NF-KB Translocation and Sensitizes IL-1β Responses in BV2 Microglia-like Cells. *Front. Immunol.* **2018**, *9*. [CrossRef]

128. Duffy, C.M.; Xu, H.; Nixon, J.P.; Bernlohr, D.A.; Butterick, T.A. Identification of a Fatty Acid Binding Protein4-UCP2 Axis Regulating Microglial Mediated Neuroinflammation. *Mol. Cell. Neurosci.* **2017**, *80*, 52–57. [CrossRef]
129. Zhu, L.; Liu, X.; Nemeth, D.P.; DiSabato, D.J.; Witcher, K.G.; Mckim, D.B.; Oliver, B.; Le, X.; Gorantla, G.; Berdysz, O.; et al. Interleukin-1 Causes CNS Inflammatory Cytokine Expression via Endothelia-Microglia Bi-Cellular Signaling. *Brain Behav. Immun.* **2019**, *81*, 292–304. [CrossRef] [PubMed]
130. He, Y.; Yao, X.; Taylor, N.; Bai, Y.; Lovenberg, T.; Bhattacharya, A. RNA Sequencing Analysis Reveals Quiescent Microglia Isolation Methods from Postnatal Mouse Brains and Limitations of BV2 Cells. *J. Neuroinflamm.* **2018**, *15*. [CrossRef] [PubMed]
131. Bennett, M.L.; Bennett, F.C.; Liddelow, S.A.; Ajami, B.; Zamanian, J.L.; Fernhoff, N.B.; Mulinyawe, S.B.; Bohlen, C.J.; Adil, A.; Tucker, A.; et al. New Tools for Studying Microglia in the Mouse and Human CNS. *Proc. Natl. Acad. Sci. USA* **2016**, *113*, E1738–E1746. [CrossRef] [PubMed]
132. Bohlen, C.J.; Bennett, F.C.; Bennett, M.L. Isolation and Culture of Microglia. *Curr. Protoc. Immunol.* **2019**, *125*, e70. [CrossRef] [PubMed]
133. Grassivaro, F.; Menon, R.; Acquaviva, M.; Ottoboni, L.; Ruffini, F.; Bergamaschi, A.; Muzio, L.; Farina, C.; Martino, G. Convergence between Microglia and Peripheral Macrophages Phenotype during Development and Neuroinflammation. *J. Neurosci.* **2019**, *40*, 784–795. [CrossRef] [PubMed]
134. Imai, T.; Hieshima, K.; Haskell, C.; Baba, M.; Nagira, M.; Nishimura, M.; Kakizaki, M.; Takagi, S.; Nomiyama, H.; Schall, T.J.; et al. Identification and Molecular Characterization of Fractalkine Receptor CX3CR1, Which Mediates Both Leukocyte Migration and Adhesion. *Cell* **1997**, *91*, 521–530. [CrossRef]
135. Combadiere, C.; Gao, J.; Tiffany, H.L.; Murphy, P.M. Gene Cloning, RNA Distribution, and Functional Expression of MCX3CR1, a Mouse Chemotactic Receptor for the CX3C Chemokine Fractalkine. *Biochem. Biophys. Res. Commun.* **1998**, *253*, 728–732. [CrossRef]
136. Nishiyori, A.; Minami, M.; Ohtani, Y.; Takami, S.; Yamamoto, J.; Kawaguchi, N.; Kume, T.; Akaike, A.; Satoh, M. Localization of Fractalkine and CX3CR1 MRNAs in Rat Brain: Does Fractalkine Play a Role in Signaling from Neuron to Microglia? *FEBS Lett.* **1998**, *429*, 167–172. [CrossRef]
137. Valdearcos, M.; Robblee, M.M.; Benjamin, D.I.; Nomura, D.K.; Xu, A.W.; Koliwad, S.K. Microglia Dictate the Impact of Saturated Fat Consumption on Hypothalamic Inflammation and Neuronal Function. *Cell Rep.* **2014**, *9*, 2124–2138. [CrossRef] [PubMed]
138. Joost, E.; Jordão, M.J.C.; Mages, B.; Prinz, M.; Bechmann, I.; Krueger, M. Microglia Contribute to the Glia Limitans around Arteries, Capillaries and Veins under Physiological Conditions, in a Model of Neuroinflammation and in Human Brain Tissue. *Brain Struct. Funct.* **2019**, *224*, 1301–1314. [CrossRef] [PubMed]
139. Von Ehr, A.; Attaai, A.; Neidert, N.; Potru, P.S.; Ruß, T.; Zöller, T.; Spittau, B. Inhibition of Microglial TGFβ Signaling Increases Expression of Mrc1. *Front. Cell. Neurosci.* **2020**, *14*. [CrossRef] [PubMed]
140. Lee, Y.; Lee, J.W.; Nam, H.; Yu, S.W.; Yu, S.W. Cx3cr1 CreERT2-Driven Atg7 Deletion in Adult Mice Induces Intestinal Adhesion. *Mol. Brain* **2020**, *13*. [CrossRef]
141. Van Hove, H.; Antunes, A.R.P.; De Vlaminck, K.; Scheyltjens, I.; Van Ginderachter, J.A.; Movahedi, K. Identifying the Variables That Drive Tamoxifen-Independent CreERT2 Recombination: Implications for Microglial Fate Mapping and Gene Deletions. *Eur. J. Immunol.* **2020**, *50*, 459–463. [CrossRef] [PubMed]
142. Hickman, S.E.; Allison, E.K.; Coleman, U.; Kingery-Gallagher, N.D.; El Khoury, J. Heterozygous CX3CR1 Deficiency in Microglia Restores Neuronal β-Amyloid Clearance Pathways and Slows Progression of Alzheimer's Like-Disease in PS1-APP Mice. *Front. Immunol.* **2019**, *10*. [CrossRef] [PubMed]
143. Lee, S.; Varvel, N.H.; Konerth, M.E.; Xu, G.; Cardona, A.E.; Ransohoff, R.M.; Lamb, B.T. CX3CR1 Deficiency Alters Microglial Activation and Reduces Beta-Amyloid Deposition in Two Alzheimer's Disease Mouse Models. *Am. J. Pathol.* **2010**, *177*, 2549–2562. [CrossRef]
144. Rogers, J.T.; Morganti, J.M.; Bachstetter, A.D.; Hudson, C.E.; Peters, M.M.; Grimmig, B.A.; Weeber, E.J.; Bickford, P.C.; Gemma, C. CX3CR1 Deficiency Leads to Impairment of Hippocampal Cognitive Function and Synaptic Plasticity. *J. Neurosci.* **2011**, *31*, 16241–16250. [CrossRef]
145. Rojo, R.; Raper, A.; Ozdemir, D.D.; Lefevre, L.; Grabert, K.; Wollscheid-Lengeling, E.; Bradford, B.; Caruso, M.; Gazova, I.; Sánchez, A.; et al. Deletion of a Csf1r Enhancer Selectively Impacts CSF1R Expression and Development of Tissue Macrophage Populations. *Nat. Commun.* **2019**, *10*. [CrossRef] [PubMed]
146. Kim, J.-S.; Kolesnikov, M.; Peled-Hajaj, S.; Scheyltjens, I.; Xia, Y.; Trzebanski, S.; Haimon, Z.; Shemer, A.; Lubart, A.; Van Hove, H.; et al. A Binary Cre Transgenic Approach Dissects Microglia and CNS Border-Associated Macrophages. *Immunity* **2020**. [CrossRef]
147. Kaiser, T.; Feng, G. Tmem119-EGFP and Tmem119-CreERT2 Transgenic Mice for Labeling and Manipulating Microglia. *Eneuro* **2019**, *6*. [CrossRef]
148. McKinsey, G.L.; Lizama, C.O.; Keown-Lang, A.E.; Niu, A.; Santander, N.; Larpthaveesarp, A.; Chee, E.; Gonzalez, F.F.; Arnold, T.D. A New Genetic Strategy for Targeting Microglia in Development and Disease. *eLife* **2020**, *9*, 1–34. [CrossRef]
149. Ruan, C.; Sun, L.; Kroshilina, A.; Beckers, L.; De Jager, P.; Bradshaw, E.M.; Hasson, S.A.; Yang, G.; Elyaman, W. A Novel Tmem119-TdTomato Reporter Mouse Model for Studying Microglia in the Central Nervous System. *Brain Behav. Immun.* **2020**, *83*, 180–191. [CrossRef]
150. Masuda, T.; Amann, L.; Sankowski, R.; Staszewski, O.; Lenz, M.; d´Errico, P.; Snaidero, N.; Costa Jordão, M.J.; Böttcher, C.; Kierdorf, K.; et al. Novel Hexb-Based Tools for Studying Microglia in the CNS. *Nat. Immunol.* **2020**, *21*, 802–815. [CrossRef]

151. Miron, V.E.; Priller, J. Investigating Microglia in Health and Disease: Challenges and Opportunities. *Trends Immunol.* **2020**, *41*, 785–793. [CrossRef] [PubMed]
152. Erblich, B.; Zhu, L.; Etgen, A.M.; Dobrenis, K.; Pollard, J.W. Absence of Colony Stimulation Factor-1 Receptor Results in Loss of Microglia, Disrupted Brain Development and Olfactory Deficits. *PLoS ONE* **2011**, *6*. [CrossRef]
153. De Luca, S.N.; Sominsky, L.; Soch, A.; Wang, H.; Ziko, I.; Rank, M.M.; Spencer, S.J. Conditional Microglial Depletion in Rats Leads to Reversible Anorexia and Weight Loss by Disrupting Gustatory Circuitry. *Brain Behav. Immun.* **2019**, *77*, 77–91. [CrossRef]
154. Lee, B.; Qiao, L.; Kinney, B.; Feng, G.S.; Shao, J. Macrophage Depletion Disrupts Immune Balance and Energy Homeostasis. *PLoS ONE* **2014**, *9*. [CrossRef] [PubMed]
155. Ali, S.; Mansour, A.G.; Huang, W.; Queen, N.J.; Mo, X.; Anderson, J.M.; Hassan, Q.N.; Patel, R.S.; Wilkins, R.K.; Caligiuri, M.A.; et al. CSF1R Inhibitor PLX5622 and Environmental Enrichment Additively Improve Metabolic Outcomes in Middle-Aged Female Mice. *Aging (Albany N.Y.)* **2020**, *12*, 2101–2122. [CrossRef]
156. Deng, G.; Morselli, L.L.; Wagner, V.A.; Balapattabi, K.; Sapouckey, S.A.; Knudtson, K.L.; Rahmouni, K.; Cui, H.; Sigmund, C.D.; Kwitek, A.E.; et al. Single-Nucleus RNA Sequencing of the Hypothalamic Arcuate Nucleus of C57BL/6J Mice after Prolonged Diet-Induced Obesity. *Hypertension* **2020**, *76*, 589–597. [CrossRef] [PubMed]
157. Dhana, K.; Braun, K.V.E.; Nano, J.; Voortman, T.; Demerath, E.W.; Guan, W.; Fornage, M.; Van Meurs, J.B.J.; Uitterlinden, A.G.; Hofman, A.; et al. An Epigenome-Wide Association Study of Obesity-Related Traits. *Am. J. Epidemiol.* **2018**, *187*, 1662–1669. [CrossRef]
158. Nogueira, G.; Solon, C.; Carraro, R.S.; Engel, D.F.; Ramalho, A.F.; Sidarta-Oliveira, D.; Gaspar, R.S.; Bombassaro, B.; Vasques, A.C.; Geloneze, B.; et al. Interleukin-17 Acts in the Hypothalamus Reducing Food Intake. *Brain Behav. Immun.* **2020**, *87*, 272–285. [CrossRef] [PubMed]
159. Gumbs, M.C.R.; Eggels, L.; Kool, T.; Unmehopa, U.A.; van den Heuvel, J.K.; Lamuadni, K.; Mul, J.D.; la Fleur, S.E. Neuropeptide Y Signaling in the Lateral Hypothalamus Modulates Diet Component Selection and Is Dysregulated in a Model of Diet-Induced Obesity. *Neuroscience* **2020**, *447*, 28–40. [CrossRef]
160. Li, M.M.; Madara, J.C.; Steger, J.S.; Krashes, M.J.; Balthasar, N.; Campbell, J.N.; Resch, J.M.; Conley, N.J.; Garfield, A.S.; Lowell, B.B. The Paraventricular Hypothalamus Regulates Satiety and Prevents Obesity via Two Genetically Distinct Circuits. *Neuron* **2019**, *102*, 653–667.e6. [CrossRef]
161. Rossi, M.A.; Basiri, M.L.; McHenry, J.A.; Kosyk, O.; Otis, J.M.; Van Den Munkhof, H.E.; Bryois, J.; Hübel, C.; Breen, G.; Guo, W.; et al. Obesity Remodels Activity and Transcriptional State of a Lateral Hypothalamic Brake on Feeding. *Science* **2019**, *364*, 1271–1274. [CrossRef] [PubMed]

International Journal of
Molecular Sciences

MDPI

Review

Adaptive Changes in the Central Control of Energy Homeostasis Occur in Response to Variations in Energy Status

Cassandra Gastelum [1,†], Lynnea Perez [1,†], Jennifer Hernandez [1], Nikki Le [1], Isabella Vahrson [1], Sarah Sayers [1] and Edward J. Wagner [1,2,*]

1 Graduate College of Biomedical Sciences, Western University of Health Sciences, Pomona, CA 91766, USA; cassandra.gastelum@westernu.edu (C.G.); lynnea.perez@westernu.edu (L.P.); jenniferhernandez@westernu.edu (J.H.); nikki.le@westernu.edu (N.L.); isabella.vahrson@westernu.edu (I.V.); sarah.sayers@westernu.edu (S.S.)
2 College of Osteopathic Medicine of the Pacific, Western University of Health Sciences, Pomona, CA 91766, USA
* Correspondence: ewagner@westernu.edu; Tel.: +1-909-469-8348; Fax: +1-909-469-5698
† These authors contribute equally to this work.

Abstract: Energy homeostasis is regulated in coordinate fashion by the brain-gut axis, the homeostatic energy balance circuitry in the hypothalamus and the hedonic energy balance circuitry comprising the mesolimbcortical A_{10} dopamine pathway. Collectively, these systems convey and integrate information regarding nutrient status and the rewarding properties of ingested food, and formulate it into a behavioral response that attempts to balance fluctuations in consumption and food-seeking behavior. In this review we start with a functional overview of the homeostatic and hedonic energy balance circuitries; identifying the salient neural, hormonal and humoral components involved. We then delve into how the function of these circuits differs in males and females. Finally, we turn our attention to the ever-emerging roles of nociceptin/orphanin FQ (N/OFQ) and pituitary adenylate cyclase-activating polypeptide (PACAP)—two neuropeptides that have garnered increased recognition for their regulatory impact in energy homeostasis—to further probe how the imposed regulation of energy balance circuitry by these peptides is affected by sex and altered under positive (e.g., obesity) and negative (e.g., fasting) energy balance states. It is hoped that this work will impart a newfound appreciation for the intricate regulatory processes that govern energy homeostasis, as well as how recent insights into the N/OFQ and PACAP systems can be leveraged in the treatment of conditions ranging from obesity to anorexia.

Keywords: sex difference; estradiol; nociceptin/orphanin FQ; pituitary adenylate cyclase-activating polypeptide; obesity; fasting

Citation: Gastelum, C.; Perez, L.; Hernandez, J.; Le, N.; Vahrson, I.; Sayers, S.; Wagner, E.J. Adaptive Changes in the Central Control of Energy Homeostasis Occur in Response to Variations in Energy Status. *Int. J. Mol. Sci.* **2021**, 22, 2728. https://doi.org/10.3390/ijms22052728

Academic Editor: Rosalía Rodríguez-Rodríguez

Received: 5 February 2021
Accepted: 4 March 2021
Published: 8 March 2021

Publisher's Note: MDPI stays neutral with regard to jurisdictional claims in published maps and institutional affiliations.

1. The Hypothalamic Energy Balance Circuit in Homeostatic Feeding

Energy homeostasis, the intricate balance between energy intake and expenditure, is regulated in coordinate fashion by homeostatic and hedonic neural circuits [1]. Aberrations in these circuits are implicated in the pathophysiology of conditions such as obesity, type-II diabetes, and food addiction [2–5]. Homeostatic control of energy balance is attributed to the hypothalamic energy balance neural circuitry, which integrates information relayed from the brainstem regarding the energy/nutritional status of the organism, based on chemical and mechanical cues from the gastrointestinal (GI) tract [6,7]. These communications are ultimately encoded by various hypothalamic nuclei and their associated neuronal populations, including those in the arcuate nucleus (ARC), ventromedial nucleus (VMN), lateral hypothalamus (LH), dorsomedial nucleus (DMN), and the paraventricular nucleus (PVN); producing orexigenic and anorexigenic signals to, respectively, stimulate and suppress energy intake as well as altering energy expenditure [1,8].

Excitatory input from steroidogenic factor (SF-1) neurons, located in the VMN, impinging on proopiomelanocortin (POMC) neurons in the ARC represent a critical anorexigenic synapse in homeostatic energy balance that, when activated, suppresses energy intake and enhances energy expenditure [5,8–12]. SF-1 is a transcription factor encoded by the NR5A1 gene, and activation of these neurons in the VMN leads to glutamatergic stimulation of POMC neurons [2,5,8–11]. Activation of POMC expressing neurons within the hypothalamic melanocortin system ultimately leads to formation of POMC posttranslational cleavage products such as α-melanocortin stimulating hormone (α-MSH), which following release from axon terminals go on to bind downstream effectors like melanocortin 4 receptors (MC4R) expressed on corticotropin-releasing hormone (CRH) neurons in the PVN [13–17]. In line with α-MSH functioning as a satiety mediator, the percentage of α-MSH neurons colocalizing with c-Fos in the ARC is greatest at the end of a meal, compared to the beginning of or hours after consumption [18]. In addition to α-MSH, POMC neurons also release β-endorphin (following posttranslational modification), and co-express cocaine- and amphetamine- regulated transcript (CART) [19,20]. Overall, POMC signaling reduces food intake, increases energy expenditure, and regulates glucose metabolism [16,17,19,21]. Prevailing glucose concentrations play a key role in relaying nutrition/energy state cues to the homeostatic energy balance circuity, where anorexigenic and orexigenic ARC neurons have glucose sensing abilities. ARC POMC/CART neurons are categorized as glucose-responsive neurons and take up glucose via a glucose transporter (GLUT2) where it then metabolizes, producing ATP and thereby promoting the closure of ATP-dependent potassium (K_{ATP}) channels to reduce the outflow of K^+, ultimately leading to the depolarization of the cell [22,23]. This aligns POMC cellular excitability and firing rate in direct proportion to glucose concentrations, with satiety signaling accentuated as glucose levels rise (e.g., during or shortly following a meal) [22,23]. It should therefore not be surprising that perturbations in signaling at this VMN SF-1/ARC POMC synapse can pose detrimental consequences for energy balance, where null mutations in POMC, its cleavage enzymes or downstream receptors, as well as lesioning in the VMN, ultimately leads to hyperphagia and obese phenotypes in rodents and humans [16,17,24].

At the opposing end of hypothalamic energy balance control are neuronal populations that promote orexigenic or appetite-stimulating effects. Alongside POMC neurons, the ARC houses neuropeptide Y (NPY) and agouti-related peptide (AgRP) co-expressing neurons, as well as ghrelin-containing somata [25–27]. Activation of this subset of neurons pleiotropically dampens the aforementioned anorexigenic signaling. For example, following NYP/AgRP neuronal activation, direct inhibition of neighboring POMC neurons is achieved via synaptic release of the inhibitory amino acid neurotransmitter γ-aminobutyric acid (GABA) on POMC soma [28]. Further downstream modulation of POMC signaling is mediated by AgRP antagonism on MC4R or via NPY acting on various receptor subtypes, ultimately impeding the anorexigenic signaling of α-MSH to induce feeding and reduce energy expenditure [15,23]. Additionally, NPY/AgRP neurons in the ARC are glucose-sensitive neurons and their activity/firing rate is inversely proportional to ambient glucose levels [23]. Neurons found within the LH also relay orexigenic signals, which are mediated by neurotransmission of melanin-concentrating hormone (MCH) neurons and orexin neurons [29,30]. Indeed, orexin has been shown to electrically silence POMC neurons by enhancing GABAergic and diminishing glutamatergic inputs onto these cells [31]. In addition, endocannabinoids elicit their orexigenic effects in part through retrograde inhibition of GABAergic inputs onto MCH neurons [32].

In addition to the ascending inputs from the brainstem, the neurons comprising the hypothalamic energy balance circuit are also susceptible to the influence of circulating peripheral hormones—leptin, ghrelin, insulin and sex hormones. Sensitivity to these transient peripheral hormones (and other periphery signaling molecules) is especially true for ARC neuronal populations, as this region lies in close proximity to the third ventricle (3V) and median eminence; a circumventricular organ with an incomplete blood–brain barrier [33,34]. Therefore, the ARC is in direct contact with systemic circulating hormones, permitting critical communication about the energy/nutritional status of the body to ARC neurons that promotes energy homeostasis. Leptin is synthesized by white adipose tissue (WAT), with levels fluctuating in proportion to fat mass. Leptin acts as a potent suppressor of food intake, while also stimulating metabolism and reducing excessive stored energy [35]. POMC and SF-1 neurons are depolarized by leptin receptor (LEPR) activation via JAK/signal-transducer-and-activator-of-transcription (STAT) and phosphatidylinositol-4,5-bisphosphate 3-kinase (PI3K) pathway signaling [36–39], while LEPR activity on AgRP neurons activates ATP-gated K^+ (K_{ATP}) channels leading to potassium outflow, hyperpolarization and decreases in firing [40–42]. The anorexigenic effect of leptin also involves a reduction in hypothalamic endocannabinoid levels [43]. Conversely, hypothalamic levels of endocannabinoids are increased in leptin receptor-deficient *fa/fa* rats and leptin-deficient *ob/ob* mice [43]. Similar to leptin, insulin signaling in the hypothalamus promotes an anorexigenic tone and elicits transient receptor potential (TRP)C5 channel-induced excitation of POMC neurons following receptor activation [44]. While for insulin, several prior studies reported inhibitory responses in POMC neurons due to activation of K_{ATP} channels, it is now known that the proportion of excitatory vs. inhibitory insulin-induced responses is dependent on ambient levels of protein tyrosine phosphatase 1B (PTP1B) and T-cell protein tyrosine phosphatase (TCPTP) activity [45–47]. Both enzymes are expressed in ARC neurons integral to the regulation of energy balance. PTP1B and TCPTP are key regulators of cell metabolism, as PTP1B decreases leptin activity and TCPTP attenuates insulin signaling via dephosphorylation of JAK2 tyrosine kinase and the insulin receptor, respectively, in POMC neurons, whereas in NPY/AgRP neurons TCPTP attenuates insulin but not leptin signaling [48–50]. Deletion of both enzymes in POMC neurons, or of TCPTP in NPY/AgRP neurons, from obese mice results in promoted weight loss due to decreased food consumption and increased WAT browning along with elevated uncoupling protein (UCP)-1 expression due to enhanced leptin and insulin signaling [48,49].

Conversely, levels of the orexigenic gut-derived peptide ghrelin increase in response to negative energy balance. As a result, the peptide binds to receptors on ARC NPY/AgRP neurons to stimulate these neurons, and thereby promote feeding behavior and energy storage [35]. The orexigenic effect of ghrelin also depends on enhanced hypothalamic production of endocannabinoids and activation of the energy-sensing signaling molecule AMP-dependent protein kinase (AMPK) that, in turn, elicits retrograde inhibition of excitatory input impinging on parvocellular neurons in the PVN [51]. These actions require functionally intact cannabinoid CB1 and ghrelin receptor systems [51,52]. AMPK functions as an important signaling molecule within the hypothalamic energy balance circuitry, and exists as a heterotrimeric complex comprising α-, β-, and γ-subunits [53–56]. The α subunit is the catalytic subunit, while the β and γ subunits are involved in glycogen and AMP/ATP binding, respectively [53–56]. AMPK activity is stimulated by phosphorylation via two upstream kinases, human tumor suppressor LKB1 or Ca^{2+}/calmodulin-dependent protein kinase kinase-β (CaMKKβ) that is triggered by cellular stress, cytokines, hormones like those mentioned above, as well as by increases in AMP/ATP ratio [53–56]. This in turn inhibits anabolic pathways and activates catabolic pathways to generate ATP [53–56]. Leptin, insulin, α-MSH, high plasma glucose concentrations and refeeding all inhibit AMPK whereas AgRP increases it [57]. On the other hand, AMPK activation is necessary for leptin to inhibit fatty acid synthesis and thereby promote fatty acid oxidation in skeletal muscle via phosphorylation of acetyl coenzyme A carboxylase [58]. This is consistent with other examples of opposing cannabinoid- and ghrelin-induced effects on AMPK activity in the central nervous system (CNS) versus the periphery [59]. When the effects of AMPK are constitutively manifest in transgenic mice bearing a gain-of-function mutation in the $\gamma2$ subunit, this brings about ghrelin-dependent hyperphagia that leads to obesity, glucose intolerance and hyperinsulinemia [60]. The dynamic interplay between the aforementioned peripheral hormones and the homeostatic energy balance circuitry is graphically depicted in Figure 1.

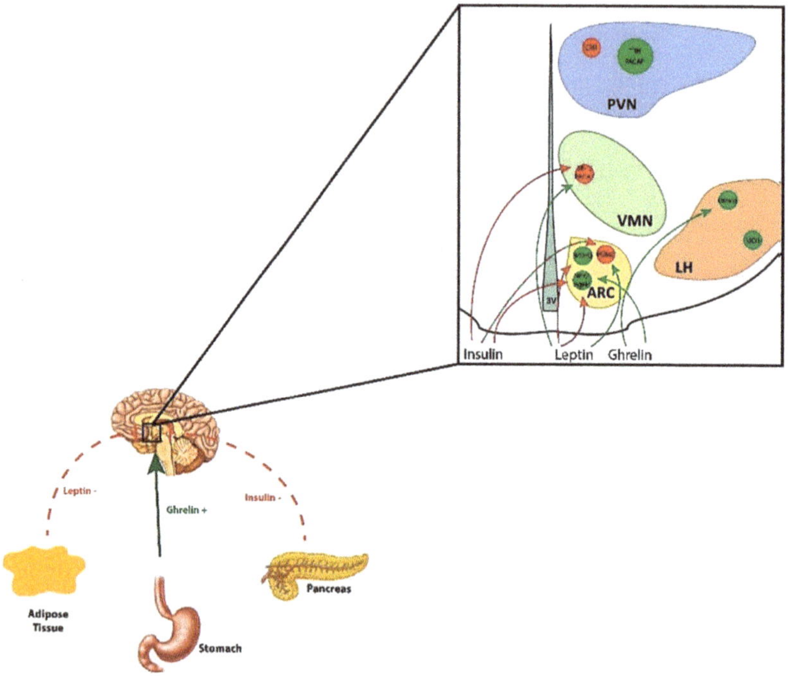

Figure 1. Schematic diagram illustrating the interplay between peripheral hormones and the homeostatic energy balance circuitry. Hormones like leptin and insulin released from adipose tissue and the endocrine pancreas, respectively, exert anorexigenic effects, whereas ghrelin released from the gastric mucosa exerts orexigenic effects. Leptin's appetite-suppressing actions are mediated via excitatory effects on anorexigenic POMC and SF-1/PACAP neurons, as well as inhibitory effects on orexigenic NPY/AgRP and N/OFQ neurons. Insulin also inhibits NPY/AgRP neurons and, paradoxically, SF-1/PACAP neurons as well. Insulin's effects on POMC neurons are dependent upon prevailing levels of tyrosine protein phosphatases. On the other hand, ghrelin's appetite-promoting effects are due to its excitatory effects on NPY/AgRP and orexin neurons.

2. The Mesolimbic Dopamine Network and Hedonic Feeding Behavior

Alongside the homeostatic-hypothalamic circuitry, energy balance is also modulated by hedonic aspects of feeding behavior pertaining to reward-based food intake, or eating for pleasure. Amounting evidence in rodents and humans now support the theory that both drugs of abuse and the consumption of highly palatable foods converge on a shared pathway within the limbic system to mediate motivated behaviors [61–63]. Therefore, the hedonic consumption of palatable foods involves the mesolimbic dopamine (A_{10}) neurons that emanate from the ventral tegmental area (VTA) and project onto structures such as the nucleus accumbens (NAc), prefrontal cortex (PFC), hippocampus, and amygdala [64–66]. The VTA is comprised of three main neuronal phenotypes including dopaminergic, GABAergic, and glutamatergic neurons. The rate limiting enzyme for catecholamine synthesis is tyrosine hydroxylase (TH), which catalyzes the hydroxylation of tyrosine to 3,4-dihydroxyphenylalanine that is then rapidly decarboxylated to produce dopamine and, in some neuronal populations, norepinephrine and epinephrine [67]. Optogenetic studies have shown that local activation of A_{10} dopamine neurons or their terminals within the NAc promotes responses of increased reward-seeking behavior [68–71]. On the other hand, activation of VTA GABAergic neurons promotes aversive responses, while their inhibition promotes reward, by dampening or promoting A_{10} dopamine signaling, respectively [72,73]. Photoactivation studies, following an intracranial self-stimulation operant paradigm protocol [74], further support the role of VTA A_{10} dopamine neurons in reward-seeking. This reward-seeking role is exemplified through data showing that photoactivation of VTA glutamatergic neurons expressing cation channel rhodopsin-2 (ChR2) under a vesicular glutamate transporter-2 (VGlut2) promoter caused conditioned place preference to a photostimulation-paired chamber and motivated operant responding to earn optical intracranial self-stimulation in mice [75]. These rewarding effects are attributed to local excitatory input onto the VTA A_{10} dopamine neurons [75]. A_{10} dopamine neurons thereby encode reward processing for natural and drug-induced rewards and are implicated in increasing incentive salience for palatable foods, food-seeking behavior and impulsivity that could, under the right circumstances, lead to binge-feeding behavior [4,64,66,76–78].

To fully grasp the distinct role A_{10} dopamine neurons play in hedonic feeding patterns, it is imperative to delineate the multifaceted aspects of global reward processing. The influential theories proposed by Berridge and colleagues make the case that when examining the role of food reward in feeding behavior, distinctions must be made between what he coined as reward 'liking' and reward 'wanting' [79,80]. 'Liking' is associated with the hedonic impact, or the brain reaction underlying sensory pleasure triggered by a rewarding stimulus, such as a highly palatable food [79,80]. Hedonic pleasure ('liking') has been reliably measured through observation of facial affective reactions prompted by exposure to a natural taste stimulus, where sweet tastes elicit positive 'liking' patterns of distinct orofacial expressions (e.g., rhythmic or lateral tongue protrusions) and bitter tastes alternatively evoke "disliking' expressions (e.g., gapes) [81]. These varying patterns of orofacial reactions are homologous to those observed in human infants, orangutan, chimpanzees, monkeys, rats, and mice, insinuating evolutionary conservation of the underlying brain circuits involved [80–83]. Data collected from taste reactivity studies allowed further insights into the neural underpinning of hedonic impact reactions, illuminating involvement of hedonic hotspots in the rostrodorsal quadrant of the medial shell of the NAc, ventral pallidum, and brainstem regions such as the parabrachial nucleus in the pons [80,84,85].

Conversely, hedonic impact ('liking') is distinguishable from 'wanting', which is related to incentive salience or motivation in reward-seeking [79,80]. Further, incentive salience (in the form of cue-triggered 'wanting') is mediated by separate neural networks, like those originating from A_{10} dopamine neurons in the VTA, though integration of signals from both components is mechanistically crucial to produce the full phenomenon we typically think of as reward [79,80]. As previously mentioned, VTA dopamine signaling is indicated in incentive salience, a notion vastly supported by evidence wherein rewarding stimuli leads to enhanced dopamine transmission, while suppression of dopamine signaling lessens the motivation for rewards including food, sex, and drugs [65,76,86–88]. However, taste reactivity studies utilizing mutant mice with genetically abolished neural dopamine, or impaired ventral striatal dopamine caused by neurochemical 6-OHDA lesions, have reiterated this point, showing no detectable effects on ability to register the pleasurability/hedonic impact of tastes in the absence of dopamine [80,89]. Recognition of these separate neuronal networks encompassing overall reward is key to be able to discern the contributing neural pathways, and perturbations of these pathways, that lead to pathological reward-seeking behaviors (e.g., in food and drug addiction). In regard our recent study described below in Section 4, we mainly focus on 'wanting' behaviors in hedonic feeding and modulations of A_{10} dopamine signaling in the context of food reward and binge feeding.

It should be noted that A_{10} dopamine neurons also play a critical role in aversion processing in ways that are related to their firing pattern as well as the topographical pattern of dopaminergic innervation of the NAc. In vivo photometry studies have revealed that responses to aversive stimuli are mediated by A_{10} dopamine neurons terminating in the ventromedial shell of the NAc [90]. Aversive responses are induced by excitation of these neurons that occurs, at least in part, via glutamatergic input from the lateral hypothalamus [91]. Optogenetic studies indicate that prolonged stimulation of these A_{10} dopamine neurons activates dopamine D1 and D2 receptor-bearing medium spiny neurons that, in turn, increases the firing rate of GABAergic neurons in the ventral pallidum, decreases the excitability of A_{10} dopamine neurons and reduces cocaine reward [92].

As with the hypothalamic energy balance circuit, the neurons comprising the hedonic reward circuitry, namely the A_{10} dopamine neurons, are also sensitive to circulating leptin, insulin, and ghrelin levels, relaying signals to regulate food-seeking behavior and ultimately changes in body mass [93]. Leptin binding to and activating LEPRs expressed on VTA A_{10} dopamine neurons leads to phosphorylation of STAT3, membrane hyperpolarization and reduced firing in these neurons [94,95]. Moreover, site-specific ablation of LEPRs in the VTA heightens the sensitivity of mice to the rewarding aspect of highly palatable foods (e.g., sucrose), while microinjection of leptin into the VTA reduces food intake [95]. Secondary regulation of VTA dopamine neurons by leptin is mediated via direct leptin action on LH LEPR expressing neurons which in turn relay signals onto the A_{10} dopamine population, via synaptic contact, to promote decreases in food intake and concomitant decreases in body weight [96]. Interestingly, LH LEPR neurons have been shown to represent a unique neuronal population distinct from the previously mentioned orexin or MCH neurons also found in the LH [96]. Anterograde tract- and retrograde tracing further confirmed that LH LEPR neurons project caudally to densely innervate the VTA, with few to no projections seen in hypothalamic regions (including the ARC) or in the striatum (including the NAc). Therefore, LH LEPR neurons may play a unique role in A_{10} dopamine regulation compared to other LEPR expressing neurons in the hypothalamus [96]. This interaction of LH signaling to control VTA neurons importantly highlights how homeostatic and hedonic neural circuits may dynamically and coordinately interact with one another to promote global energy balance. Insulin receptors are also present on VTA (and substantia nigra) dopamine neurons and can induce expression of the dopamine transporter (DAT) in these neuronal populations, as was demonstrated following intracerebroventricular (i.c.v.) insulin treatment [93]. Increased DAT results in quicker dopamine reuptake from the synaptic cleft back into presynaptic neurons, thereby halting stimulation of postsynaptic neurons and ultimately working to decrease the rewarding effect of food [93]. Furthermore, ghrelin binding to and activating ghrelin receptors expressed on A_{10} dopamine neurons has been shown to elevate the frequency of action potentials in these neurons, as well as induce increased dopamine turnover in the NAc, promoting appetite [97]. Interestingly, the ghrelin-induced increase in locomotor activity and dopamine release in the NAc is negated by CB1 receptor antagonism with rimonabant [98]. Thus, the abundant presence of communication between the brains reward circuitry and fluctuating hormones and neuromodulators, that typically relay nutritional/energy state cues, underscores the important role of A_{10} dopamine signaling in energy balance and potential links to feeding pathologies.

The intrinsic separation of the neural systems encoding hedonic impact from pleasure and the incentive salience or motivation for a reward, can give way to possible explanations for aberrant feeding behaviors that may underlie feeding pathologies [80]. Individuals with substance abuse disorders seem to take drugs compulsively even when they no longer derive pleasure from them ('liking'), and their motivation to take the drug ('wanting') may persist due to long-lasting sensitization of their brain mesolimbic systems, brought on by repeated binges [76,80,88]. Food is a natural reward with reinforcing properties, similar to rewards such as sex and drugs, and activates the dopaminergic mesolimbic system by elevating extracellular dopamine concentrations in the NAc [99–101]. In particular, highly palatable, calorie rich foods can critically effect A_{10} dopamine neurotransmission [102,103], analogous to modulations caused by other potent reinforcing stimuli, e.g., cocaine, amphetamines, opiate-like drugs, cannabinoids, alcohol, and nicotine [104–108]. Some evidence suggests that similar sensitization-like changes can be induced by exposure to certain regimens of food and restriction, modelling oscillations between dieting and binging on palatable foods [109–113]. Wherein, rats exposed to brief, intermittent bouts of sucrose access (sucrose binges), express sensitization-like changes, especially when binges are cycled with food-restriction. Observed changes include: increasing propensity to over-consume when allowed, an enduring enhanced neural response to the presentation of food reward and cues, and an over-response to the psychostimulant effects of drugs such as amphetamine (a typical behavioral marker of drug-induced neural sensitization, which suggests a common underlying mechanism) [110,113–115]. Indeed, dopamine release into the NAc and c-fos expression in A_{10} dopamine neurons is increased during binge-feeding episodes [77,116], lending further credence to this notion. Additionally, obesity is associated with dysfunction of dopaminergic systems. Obese patients present with reduced striatal dopamine D2 receptor, as measured by positron emission tomography [117]. This similarity between food reward and drug reward gives rise to the notion that feeding disorders and drug abuse and/or dependence share common mechanisms, as neuropsychological diseases involving negative alterations of the neural networks associated with the reinforcing properties of rewards [118–120].

3. Influences of Sex and Diet on Central Energy Balance Circuits

Sex differences are abundantly present in the context of energy homeostasis and although the prevalence of obesity is similar between men and women, women seem to have a greater risk of developing eating disorders and extreme obesity [121,122]. Indeed, sex is thought to represent one of the main risk factors for food-related disorders, including binge eating disorder [123]. Other lines of evidence illustrating sex differences in food-based reward processing include women having a reduced ability to control food desire, higher cortical and limbic activation when presented with visual, gustatory and olfactory cues, as well as increased susceptibility to episodes of food-craving and lack of control for sugary foods, compared to men [124–128]. The role of sex hormones, particularly the fluctuating estrogen levels in females throughout the estrous cycle, is of interest in these disparities seen between males and females for reward-based consumption.

Estrogens elicit inhibitory effects on food intake, attributable to activation of estrogen receptors in key brain regions responsible for food intake control and body weight, such as the hypothalamus and nucleus tractus solitarius [5,129–132]. In regard to sex hormones, studies on the cannabinoid system have elucidated activation effects of gonadal steroid hormones on hypothalamic energy balance neural circuitry. Estradiol (E_2) in females has been shown to attenuate cannabinoid-induced hyperphagia and hypothermia, as well the decrease in glutamatergic input onto POMC neurons [133]. These estrogenic actions occur through activation of estrogen receptor (ER) and the G_q-coupled membrane ER (G_q-mER), which triggers a signaling pathway involving PI3K, protein kinase C (PKC), protein kinase A (PKA) and neuronal nitric oxide synthase (nNOS) [131,132,134]. This, in turn, diminishes endocannabinoid tone at VMN SF-1/ARC POMC synapses, thereby relieving the retrograde inhibition of the glutamatergic input [10,11]. Regarding nNOS, while multiple isoforms with differing efficacies in eliciting downstream nitric oxide signaling have been characterized in brain [135], it is currently unknown which isoform mediates the estrogenic diminution in endocannabinoid tone at these synapses. Therefore, it will be necessary for future studies to elucidate the isoform responsible.

Interestingly, functional glutamatergic synapses at these VMN SF-1/ARC POMC synapses are largely silenced in obese females [11]. Indeed, studies of both POMC and NPY/AgRP neurons from *ob/ob* mice reveal that they undergo extensive synaptic plasticity under conditions of obesity; with the former receiving significantly more inhibitory inputs concomitant with a reduction in excitatory inputs, and the latter having appreciably more excitatory inputs and fewer inhibitory ones impinging upon them [136]. Obesity also correlates with chronic inflammation and resistance to leptin and insulin not only in the CNS but also peripherally [137–140]. However, the inflammation, reactive gliosis and subsequent neuronal injury observed in the mediobasal hypothalamus develops more rapidly than for similar maladaptations occurring in peripheral organs [141].

By contrast, testosterone in males rapidly increases energy intake and is reversed by the CB1 receptor antagonist AM251 and the diacylglycerol lipase (DAGL) inhibitor

Orlistat, and potentiates the cannabinoid-induced decrease in glutamatergic input onto POMC neurons [10,142]. This androgen-induced increase in endocannabinoid tone is due to activation of AMPK, which augments retrograde inhibition of glutamate release at VMN SF-1/ARC POMC synapses [10,142]. These effects are further magnified in obese males due to reduced PI3K signaling in the ARC [11,143]. Lastly, the development of central insulin resistance brought on by diet-induced obesity is sexually differentiated. Males are more susceptible to the attenuated activation of both TRPC5 channels in POMC neurons and K_{ATP} channels in NPY/AgRP neurons than are females under the protection of E_2, which prevents the respective increase in suppressor of cytokine signaling-3 and decrease in PI3K signaling that drives central insulin resistance [143,144].

Estrogen receptors are also expressed within the VTA. Evidence indicates that estrogens increase self-administration of rewards like psychomotor stimulants and alcohol [145–147]. In animal studies, female rats have been more motivated to work for cocaine during the estrus phase, compared to other phases of the estrous cycle [148,149], and E_2-treated ovariectomized (OVX) female rats exhibited increased motivation to self-administer cocaine [150]. The effects of estrogen on reward neural circuitry are also evident in motivation for food rewards. In opposition of the findings from self-administration of drug rewards, intra-VTA injections of E_2 significantly reduced the motivation to work for sucrose rewards in a progressive ratio operant conditioning task within 1 h after injection, while overall food intake was not altered by this treatment [151]. Additionally, a study comparing the self-administration of chocolate-flavored beverage (CFB) and concomitant changes in extracellular dopamine in the dialysate obtained from the NAc, between male as well as intact and OVX female rats, showed that female rats in the proestrus and estrus phases of the cycle had reduced lever responding for, and amount of self-administered CFB, paired with lowered extracellular dopamine in the dialysate from the NAc shell [152]. These variable findings between food versus drug rewards raise questions about the role estrogens play in food reward processing and how they may potentially explain the disparate prevalence rates between males and females in feeding behavior pathologies.

4. Nociceptin/Orphanin FQ Regulation of Homeostatic and Hedonic Energy Balance Circuits

The neuropeptide nociceptin/Orphanin FQ (N/OFQ) is an endogenous opioid hepadecapeptide that is encoded by the prepronociceptin gene [153–155]. N/OFQ binds with high affinity to its cognate $G_{i/o}$-coupled nociceptin opioid peptide (NOP) receptor, and despite high structure homology, has minimal affinity for classic opioid receptors (mu, kappa, or delta opioid receptors), nor do classical opioid receptor ligands (e.g., naloxone, endorphin, dynorphin) have high affinity binding for NOP [155,156]. Initial studies on N/OFQ indicated the peptide attenuates locomotor activity, increases sensitivity to pain, while blocking the antinociceptive activity of mu, delta, and kappa analgesics following i.c.v. injections; thus garnering the name nociceptin [154,155,157]. The NOP receptor is expressed extensively throughout the central nervous system (CNS) [153,156,158,159], subserving a role in an array of central processes including pain, learning and memory, emotional states, neuroendocrine control, food intake, and motor control (see [160] for thorough review).

The actions of N/OFQ in these disparate neuronal systems is accredited to peptide binding and subsequent activation of its NOP receptor. In line with agonist activation in all GPCRs, following NOP activation by N/OFQ the $G\alpha$ and $G\beta\gamma$ subunits dissociate to then promote the various effector pathways [155,158]. Therein, NOP receptor activation inhibits adenylate cyclase (AC) activity and couples to pertussis-toxin-sensitive G-proteins resulting in decreased cyclic adenosine monophosphate (cAMP) production [154,155,160, 161]. Further, NOP receptors canonically couple to G protein-gated inwardly rectifying potassium (GIRK)3 and both N-type as well as P/Q type voltage-gated Ca2+ channels. Upon NOP receptor activation, Ca2+ currents are reduced and GIRK channels activated, causing K^+ efflux, cellular hyperpolarization and attenuated neural activity [162–164].

Additionally, NOP receptor signaling has been shown to promote activation of PKC as well as phospholipase A2 and C [165–167].

As mentioned, the N/OFQ/NOP receptor system is expressed widely throughout the brain in mice, rats, and humans, and importantly is expressed densely in the ARC and VTA [153,156,159,168,169]. The abundant expression of N/OFQ and NOP in these key regions supports a role in both homeostatic and hedonic energy balance regulation. Indeed, i.c.v. injections into the lateral ventricle (LV), 3V, and intranuclear injections into the ARC of N/OFQ or NOP agonists have been shown to produce a dose-dependent increase in feeding behavior even in satiated animals; intra-ARC injections proving to be the most efficacious to induce hyperphagia [170–174]. Additionally, chronic i.c.v. infusions of the neuropeptide have been associated with increased body weight via modifications in feeding and metabolism in mice [175]. In addition, we have found that the rebound hyperphagia seen upon refeeding in fasted NOP receptor null mice is significantly blunted compared to their wildtype littermate controls [176]. The appetite-stimulating properties of N/OFQ can be attributed, at least in part, to the NOP receptor-mediated inhibition of POMC neurons via activation of GIRK channels, as seen with either exogenous bath application of the neuropeptide or high-frequency optogenetic stimulation of ARC N/OFQ neurons [5,176–179]. The observed effects of N/OFQ are not limited to anorexigenic pathways, however, and have been observed to also influence orexigenic circuitry. NOP receptors are expressed in NPY/AgRP neurons, and N/OFQ increases AgRP release from mediobasal hypothalamic explants [180]. Interestingly, these ARC N/OFQ neurons co-release GABA upon low-frequency optogenetic stimulation, which can provide additional inhibitory input onto POMC neurons [181]. They are also regulated by ambient levels of extracellular glucose, and inhibited by leptin [181]. On the other hand, N/OFQ administered into the perifornical/lateral hypothalamic area exerts a hypophagic effect attributed to a NOP receptor-mediated inhibition of orexin neurons via activation of K_{ATP} channels [182]. Nevertheless, the prevailing sentiment based on the totality of the available evidence clearly indicates that N/OFQ is profoundly hyperphagic via its collective actions within the homeostatic energy balance circuitry.

Moreover, several lines of in vivo and in vitro evidence illustrate the effects of N/OFQ on the A_{10} dopamine system, providing credence to the notion that N/OFQ may be inherently linked to reward system processing, and thus to hedonic feeding behavior. Data obtained from in vivo studies have shown that N/OFQ decreases the outflow of dopamine in the NAc following intracerebral injections, dampens the morphine-induced dopamine release in the NAc, and blocks the acquisition of morphine-dependent place preference [183–186]. Further, bath application of N/OFQ during intracellular electrophysiology recordings in rat midbrain slices caused membrane hyperpolarization and reduced firing under current-clamp, which was associated with an outward current under voltage clamp [187]. These inhibitory effects of N/OFQ in the VTA were reduced by the NOP receptor antagonist [Phe11jCH2-NH)Gly2]NC(1 $\pm$ 13)NH2 (1 μM) but were unaltered by presence of tetrodotoxin or the opioid receptor antagonist naloxone [187]. Additionally, activation of NOP receptors expressed in the NAc and dorsal striatum work to pre-synaptically inhibit dopamine synthesis and tyrosine hydroxylase (TH, the rate-limiting enzyme in dopamine production) phosphorylation and act post-synaptically on dopamine responsive neurons by decreasing dopamine D1 receptor signaling via suppression of cAMP/PKA activity [188]. Lastly, microdialysis studies have revealed that N/OFQ significantly reduces extracellular DA levels in both the VTA and NAc [184].

The preceding section on sex differences in energy balance control is particularly relevant when it comes to NOP receptor-mediated regulation. In homeostatic energy balance, males are significantly more sensitive to the inhibitory effects of exogenous N/OFQ on excitatory neurotransmission at VMN SF-1/ARC POMC synapses than are females during E_2-dominated phases of the estrous cycle in two ways: (1) the direct hyperpolarization/cessation of firing of both VMN SF-1 and ARC POMC neurons, as well as the underlying outward current, due to the activation of GIRK channels is greater in males than in females and (2) the presynaptic inhibition of glutamate released of VMN SF-1 neurons at these synapses is more substantive in males than in females [5,189,190]. Additionally, E_2 exerts powerful activational effects by diminishing the inhibitory effects of exogenous N/OFQ at VMN SF-1/ARC POMC synapses, as well as the decreased excitability of POMC neurons caused by optogenetic stimulation of ARC N/OFQ neurons, and protecting against the aberrant hyperphagia and reduction in energy expenditure caused by exogenous N/OFQ administered directly into the ARC of obese OVX females [5,178,179,191]. E_2 also attenuates the pleiotropic actions of N/OFQ on POMC neurons by binding to either ERα or the G_q-mER, which leads to a signaling cascade that includes phospholipase C (PLC), PI3K, PKC, PKA and nNOS [191]. Furthermore, progesterone administered to OVX, estrogen-primed females restores the sensitivity of POMC neurons to these pre- and postsynaptic actions of N/OFQ [192]. The precise actions of N/OFQ and NOP receptor activation in the aforementioned neural circuitry, its effects on feeding-behavior, and potential role in feeding-related pathologies will be discussed in further detail below.

In addition to sex differences, diet modifications resulting lean or obese phenotypes alter the signaling effects of N/OFQ in homeostatic and hedonic neural circuits, and increases the risk for aberrant feeding-behavioral patterns to form. Diet-induced obesity increases the intrinsic excitability of ARC N/OFQ neurons, which augments the inhibitory GABAergic tone received by POMC neurons [181]. Conversely, ablation of ARC N/OFQ neurons hinders the development of obesity caused by a four-week exposure to a HFD [181]. Moreover, diet-induced obesity renders VMN SF-1/ARC POMC synapses more susceptible to the inhibitory effects of exogenous N/OFQ in males and hypoestrogenic OVX females [5]. This greatly curtails anorexigenic signaling at these synapses; causing exaggerated N/OFQ-induced increases in energy intake and decreases in energy expenditure, and is entirely in keeping with our recent demonstration that N/OFQ administered directly into the ARC significantly enhances binge-feeding behavior caused by short-term intermittent exposure to a HFD [179].

Endogenous N/OFQ signaling may also be intrinsically associated with or escalate aberrant feeding patterns associated with diet-induced obesity and/or binge-feeding. In one instance, Sprague Dawley rats that had previously been determined as "fat-preferring", were particularly susceptible to N/OFQ-induced acute hyperphagia following i.c.v. injection [173]. Additionally, NOP receptor knockout mice displayed significantly reduced levels of HFD food consumption, compared to their wildtype littermate controls [193]. Additionally, administration of the novel NOP antagonist LY2940094, effectively increases lipid utilization metabolism and reduces fasting-induced hyperphagia of chow in wildtype 129S6 mice, but not in those with genetic deletion of the NOP receptor [194,195]. Further, LY2940094 reduces HFD consumption measured over a 5-h exposure period and also hindered weight gain over 3 days of HFD exposure [195]. Providing further support for the intrinsic contribution of endogenous N/OFQ/NOP signaling in the pathogenesis of obesity and eating disorders, systemic administration of LY2940094 reduced intake of HFD in diet-induced obese rats and mice, and also improved metabolic parameters by reducing the respiratory quotient in mice with access to HFD in their metabolic feeding chamber [195]. In relation to N/OFQ and binge-feeding, systemic treatment with the selective NOP antagonist SB 612111 produced a dose-dependent decrease in intermittent HFD binge eating, but not a change in the total 24-h food intake of mice who were either on an intermittent-HFD or continuous-HFD feeding regimen [196].

While the last few studies mentioned regarding altered feeding behavior certainly provide evidence that the N/OFQ-NOP system is involved in the neuropathology of obesity and related eating disorders, they were not designed to offer insight into the specific neuronal populations acted upon that leads to such pathogeneses. In addition, the exact mechanisms through which sex and diet interact to modulate NOP receptor-mediated inhibition of reward encoding A_{10} dopamine neurons and hedonic feeding remained a mystery until very recently. We have discovered that the endogenous release of N/OFQ caused by high-frequency optogenetic stimulation of VTA neurons in mesencephalic slices from

N/OFQ-cre mice powerfully inhibits neighboring VTA neurons; an effect that is faithfully recapitulated by exogenous bath application of N/OFQ during recordings of A_{10} dopamine neurons in slices from TH-cre mice [179]. The membrane hyperpolarizations and underlying outward currents were attenuated by E_2 in females, and accentuated by diet-induced obesity in males [179]. These N/OFQ-induced inhibitory effects on A_{10} dopamine neurons functionally translated into sex- and diet-dependent changes in binge-eating behavior, as N/OFQ delivered into the VTA decreased the rampant consumption seen during the binge episodes in obese but not lean males, and in both lean and obese females [179]. Just as we saw with the inhibitory effects of N/OFQ in A_{10} dopamine neurons from female mice, E_2 counteracted the inhibitory effect of intra-VTA N/OFQ on binge feeding [179]. Thus, it is clear that N/OFQ exerts multifaceted effects on energy balance via NOP receptor-mediated regulation of homeostatic and hedonic circuits that are site-specific as well as sex hormone- and diet-dependent. The site specificity underscores the fact that despite the inhibitory effect of N/OFQ on A_{10} dopamine neurons and the associated dampening of binge-feeding behavior, there is clearly a net hyperphagic response caused by the peptide that is due largely, if not exclusively, to its effects within the homeostatic energy balance circuitry.

5. Pituitary Adenylate Cyclase-Activating Polypeptide (PACAP) Regulation of Homeostatic and Hedonic Energy Balance Circuits

Pituitary adenylate cyclase-activating polypeptide (PACAP) is a hypophysiotropic neurohormone belonging to the largest family of developmental and regulatory peptides that includes vasoactive intestinal polypeptide (VIP)-secretin-growth hormone-releasing hormone (GHRH)-glucagon superfamily, along with secretin, glucagon, human growth hormone-releasing factor (hGRF), and VIP [197]. Containing the most conserved sequence throughout evolution in its family, PACAP is encoded by the ADCYAP1 gene as a 38-amino acid C-terminally α-amidated polypeptide [197]. It is transcribed in hypothalamic neurons and peripheral organs, such as the GI tract, pancreas, and gonads; exerting pleiotropic physiological effects such as the regulation of neurotransmitter release and secretion, vasodilation, energy balance, as well as stimulation and inhibition of cell proliferation and/or differentiation [198–200]. The PACAP precursor yields two different forms of PACAP, $PACAP_{1-38}$ and $PACAP_{1-27}$, as well as PACAP-related peptide (PRP) [200]. Given its widespread distribution in the CNS and periphery, PACAP is well-equipped to act as a hormone, a neurotransmitter, and a trophic factor in various tissue types [200].

The effects of PACAP are mediated through two classes of PACAP receptors: the pituitary adenylate cyclase-activating polypeptide-specific 1 receptor (PAC1) and two subtypes of VIP/PACAP-receptors termed VPAC1 and VPAC2 [197,200]. PAC1 receptor exhibits almost a twofold higher affinity for PACAP than for VIP, while the VPAC1 and VPAC2 receptors recognize both PACAP and VIP with equally high affinity [197]. PAC1, a metabotropic receptor, is found in various hypothalamic structures including the supraoptic nucleus (SON), PVN, ARC, LH, VMN, as well as extrahypothalamic regions of the brain like the cerebral cortex, Broca's area, the hippocampus, among others [197,200]. VPAC1 and VPAC2 receptors are appreciably expressed in peripheral organs including the lung, duodenum, and thymus, although with less abundance than PAC1 receptors [201,202]. There is considerable evidence demonstrating that the PAC1 receptor signals through G_q and G_s. For example, physiological studies have determined that PACAP acts on PAC1 receptor in mouse neural stem cells, and upon receptor activation the signal generated is carried via a G_q-mediated PLC/diacylglycerol/inositol 1,4,5-trisphosphate (IP3)-dependent signaling pathway [203]. In the hypothalamus, PACAP binds to PAC1 receptors that induce G_q-coupled stimulation of PLC, PI3K, and PKC to ultimately activate TRPC5 channels [12]. In the neurohypophysis, activation of PAC1 receptors by PACAP can lead to signaling via the G_s pathway [204] that increases firing and depolarizes the membrane potential of magnocellular neurons in rat brain slices via the activation of adenylate cyclase/cAMP/PKA signaling [204–206]. In doing so, PACAP stimulates the release of oxytocin and vasopressin from the posterior pituitary [204,207]. Activation of PAC1 receptors has also been shown to elevate intracellular Ca^{2+} concentrations in dissociated magnocellular neurons from rat SON, and stimulate a quinine- sensitive K^+ outward current in murine microglia [207,208].

In the immune system, PACAP decreases chemotaxis of thymocytes and lymphocytes both via the activation of the PKA pathway [209,210]. Additionally, PACAP inhibits tumor necrosis factor-α and both interleukin-6 and interleukin-12 release, while enhancing the production of the cytokine interleukin-10 in lipopolysaccharide-activated macrophages; suggesting it acts as a protective agent that regulates the release of proinflammatory and anti-inflammatory cytokines [211–214]. The effects of PACAP on cell proliferation/survival are dependent on the downstream channel targeted as well as transcriptional cues. In vitro experiments have demonstrated that the effect of PACAP on cell survival is regulated via the activation of the G_s pathway, contributing to the phosphorylation of the extracellular signal-regulated (ERK)-type of mitogen-activated protein (MAP) kinase and enhanced c-fos gene expression [215–217]. On the other hand, PACAP dose-dependently inhibits concanavalin A-induced cell proliferation in murine splenocytes [218]. Moreover, PACAP stimulates Ca^{2+} mobilization and blocks K^+ currents in a variety of neuronal cell types (e.g., magnocellular neurons, cerebellar granule cells), two processes intimately linked to PACAP-induced enhancements in cell excitability [206,219–221]. Furthermore, PACAP stimulates postprandial glucagon-like peptide, leptin and insulin secretion, and also has been shown to promote insulin secretion from pancreatic β-cells via Ca^{2+} influx through L-type Ca^{2+} channels [222–226]. This latter effect may be dependent on ambient glucose concentration and its ability to enhance ATP production and thereby negatively gate ATP-sensitive potassium ($K^+{}_{ATP}$) channels [227,228]. This indicates that endogenous PACAP acts as a physiological regulator of pancreatic β-cell activity linked to $K^+{}_{ATP}$ channels in a manner similar to that described for the vasodilatory and neuroprotective effects of the peptide. [229,230]. Thus, it is clear that PACAP regulates a wide array of bodily functions including hypophysiotropic neurosecretion, glial function, immunomodulation, cell proliferation/survival, glucose homeostasis, vasodilation, neuroprotection and energy homeostasis via G_s- and G_q-mediated signaling. Moreover, given the inexorable link between obesity and inflammation [137–139,141], it is entirely conceivable that the overall anti-inflammatory effect of PACAP contributes significantly to lean phenotypes promoted by this peptide [226,231].

PACAP exerts myriad effects on energy balance at all levels of the brain-gut axis. For example, intravenous injection of PACAP on rats causes secretion of saliva from the submandibular and parotid glands, whereas in the stomach, PACAP decreases histamine- and pentagastrin-activated gastric acid secretion; the latter suggesting that PACAP acts indirectly to regulate gastric acid release [232–234]. In addition, intravenous injection of PACAP increases bicarbonate secretion and chloride secretion in the duodenum and in the distal colon, respectively [235,236]. Moreover, while PACAP administered centrally increases gastric motility, peripherally it evokes a dose-dependent relaxation of the gastric smooth muscles, decreases gastric motility and therefore delays stomach clearing [237–239]. Likewise, PACAP stimulates intestinal smooth muscles to relax in rats and other species [240–243].

Concerning peripheral glucose and lipid homeostasis, PACAP exerts a more potent action in stimulating glucose output from a perfused rat liver as compared to VIP [244]. While PACAP can clearly act centrally to regulate glucose homeostasis, the hyperglycemic role of PACAP in vivo can also be attributed to both an indirect action via increase in plasma glucagon and/or catecholamines, which increase glycogenolysis and gluconeogenesis [245]. Finally, in regard to energy metabolism, PACAP is known to accelerate lipolysis via the sympathetic nervous system (SNS). This suggests that hypothalamic PACAP signaling may promote the use of catabolized lipids as a viable energy source [197,231].

The hypothalamic regions that play a role in the regulation of thermogenesis, energy expenditure, and energy intake such as the PVN, VMN, and ARC abundantly express PACAP and the PAC1 receptor, suggesting that PACAP plays a vital role in the control of these processes [200,201,246]. There are two major populations of hypothalamic PACAP neurons—one with cell bodies residing in the PVN and the other with somata in the

VMN [9,12,247]. The PVN PACAP neurons are thought to promote appetite through synaptic connections with, and excitation of, NPY/AgRP neurons [9]. It is the loss of signaling via this population of PACAP neurons that may explain reports of hypoinsulinemia, decreased adiposity, lower body weight and increased insulin sensitivity seen in transgenic PACAP-null mice [248]. The VMN PACAP neurons exhibit extensive colocalization with SF-1, which along with leptin drives PACAP expression [12,247,249]. These neurons are reported to be glucose inhibited, and their selective activation reduces circulating insulin concentrations and glucose tolerance [250,251]. VMN PACAP neurons make synaptic contact with POMC neurons, and like POMC neurons, are excited by leptin [9,12,36,37,252]. As such, this VMN population of PACAP neurons is poised to suppress appetite and enhance energy expenditure. In accordance, studies show that PACAP injected into the VMN causes an increase in body temperature via adaptive thermogenesis and increased levels of UCP-1 [197,247,253,254]. The adaptive thermogenesis brought on in part by increasing WAT browning in rodents increases energy expenditure and suppresses diet-induced obesity and glucose intolerance [49,50]. PACAP also controls activity, as an injection of PACAP given i.c.v. or into the VMN increases locomotion in rodents concomitant with an increase in O_2 consumption [247,253,254]. A systemic injection of PACAP into wild type (WT) mice dose-dependently lowers cumulative energy intake and decreases various indices of meal pattern like meal-size and rate of consumption, which correlates with reduced ghrelin secretion [226]. PACAP delivered directly into the VMN or PVN also reduces energy intake; however, these effects are coupled with somewhat disparate effects on meal pattern. PACAP injected directly into the PVN decreases meal size, rate of consumption, duration, total time spent eating and increased latency to meal initiation, whereas PACAP injected into the VMN only evokes an increase in the latency to meal initiation and a decrease in the rate of eating [254]. Thus, PACAP signaling throughout the homeostatic energy balance circuitry exerts far-reaching effects on energy intake, meal pattern and energy expenditure.

Additional studies investigating the homeostatic energy balance circuitry have demonstrated that PAC1 receptors are expressed in POMC neurons, and that PACAP administered to ad libitum-fed animals elevates POMC expression, c-Fos expression in POMC neurons, and MC4R receptor mRNA expression when injected i.c.v. or into the VMN, as well as enhancing α-MSH release from hypothalamic explants [250,253,255]. In contrast, PACAP had no effect on AgRP, CART or NPY mRNA levels [250]. In addition, PAC1 receptor blockade with $PACAP_{6-38}$ or PACAP deficiency seen in Adcyap1$^{-/-}$mice significantly decreases the leptin-induced hypophagia, hyperthermia, and increase in WAT sympathetic nerve activity in vivo [231,247]. We have recently shown that the profound influences PACAP exerts in the homeostatic control of energy balance in ad libitum-fed mice are diet- and sex-dependent [12]. We observed that PACAP evokes an inward current associated with an increase in firing in ARC POMC neurons that was abolished by PAC1 receptor antagonism and TRPC5 channel blockade, and augmented by E_2 [12]. The inward current was significantly attenuated upon inhibition of PLC, PI3K and PKC, but not PKA; suggesting that the PACAP-induced activation of POMC neurons was mediated via, PI3K and G_q-mediated signaling pathways [12]. The stimulation of ARC POMC neurons caused by PACAP administered directly into the ARC suppresses energy intake and enhances energy metabolism, and these effects were markedly attenuated under conditions of diet-induced obesity [12]. These effects of exogenously applied PACAP were effectively mirrored by chemogenetic and optogenetic stimulation of VMN PACAP neurons [12]. Collectively, these data suggest that under ad libitum-fed conditions PACAP functions through a PI3K/PLC/PKC pathway to activate POMC neurons via TRPC5 channels upon PAC1 receptor stimulation via G_q-mediated signaling [12]. Thus, VMN PACAP/ARC POMC synapses constitute a critical anorexigenic component of the homeostatic energy balance circuitry, one that is accentuated by E_2 in females and attenuated by obesity in males. These findings are consistent with other examples of positive estrogenic modulation of the PACAP/PAC1 receptor system that occur during the response to stress. E_2 stimulates PACAP and PAC1 receptor expression in the bed nucleus of the stria terminalis (BTNS) as well as in the medial basal hypothalamus (MBH) compared to levels seen in the oil-treated OVX controls [256]. Likewise, PACAP expression in the PVN and anterior pituitary is heightened during the elevated E_2 levels seen during proestrus [256]. Concerning the hedonic energy balance circuitry and the consumption of palatable food, prior studies have shown that when PACAP is injected into the NAc it mimics the inhibitory effect of GABA receptor agonists on binge-like feeding behavior and decreases firing in NAc neurons [257]. This effectively reduces hedonic drive for palatable food as gauged by decreases in appetitive orofacial responses to sucrose, as well as by increases in aversive responses when PACAP is administered into the caudal NAc [258].

Whereas obesity exemplifies a state of positive energy balance, fasting represents a state of negative energy balance. Food-restricted mice show low levels of POMC and PACAP mRNA expression coupled with an increase in NPY mRNA expression, and i.c.v. injections of PACAP decrease energy intake after 30 min of refeeding [247,250]. Surprisingly, in recent experiments we found that during voltage clamp recordings in POMC neurons from eGFP-POMC mice subjected to an 18-h fast for five consecutive days, the PACAP response reversed polarity from a predominantly excitatory inward current seen under ad libitum conditions to an inhibitory outward one (Figure 2A,B,E–H); see Supplementary Materials for full material and methods The inset I/V plots corresponding to the representative current traces in Figure 2A,B revealed that the reversal potential shifted from ~−20 mV under ad libitum conditions (indicative of a mixed cation current) to ~−90 mV under fasted conditions (indicative of a K^+ current). The PACAP-induced inhibitory outward current and the corresponding increase in conductance seen under fasting conditions were virtually abolished by the K_{ATP} channel blocker tolbutamide and the PAC1 receptor antagonist $PACAP_{6–38}$ (Figure 2C,D,G,H; 2G: one-way ANOVA/LSD, $F = 25.71$, $df = 3$, $p < 0.0001$; 2H: one-way ANOVA/LSD, $F = 9.78$, $df = 3$, $p < 0.0001$).

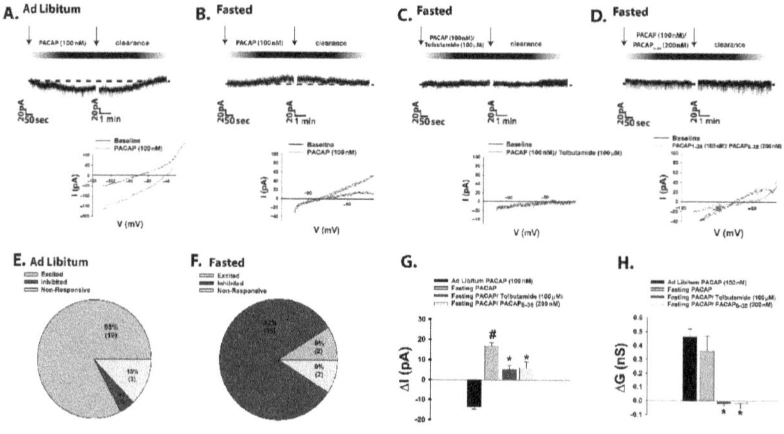

Figure 2. Fasting reverses the polarity of postsynaptic PACAP response in POMC neurons by switching the coupling of PAC1 receptors from TRPC5 to K_{ATP} channels. (**A–D**), Representative current traces from voltage clamp recordings of POMC neurons that depict the response elicited by 100 nM PACAP under ad libitum and fasting conditions, the latter of which is abrogated by blockade of K_{ATP} channels with tolbutamide (100 μM) and PAC1 receptors with $PACAP_{6–38}$ (200 nM). The inset I/V plots illustrate that the fasting-induced change in polarity is due to a switch from a mixed cation conductance to a PAC1 receptor-mediated K^+ conductance via K_{ATP} channels. (**E,F**), Pie charts that show the proportion of POMC neurons that are excited by, inhibited by, or unresponsive to PACAP under ad libitum (to the left) and fasting (to the right) conditions. (**G,H**) highlight the PACAP-induced changes in membrane current (ΔI) and conductance (ΔG); alone ($n = 12$) and in conjunction with tolbutamide ($n = 12$) and $PACAP_{6–38}$ ($n = 8$). Bars represent means and lines 1 SEM. #, $p < 0.05$ relative to PACAP under ad libitum conditions. *, $p < 0.05$ relative to PACAP under fasting conditions, one-way ANOVA/LSD. Figure adapted from [12].

A nearly identical switch in polarity of the PACAP response in POMC neurons was observed during voltage clamp recordings in EtOH vehicle-treated slices from fasted OVX female eGFP-POMC mice (Figure 3A,C–E). Bath application of E_2 in slices from fasted OVX female eGFP-POMC mice reduced the magnitude of the PACAP-induced outward current by ~50%, and markedly attenuated the increase in K^+ conductance (Figure 3B,D: Student's t-test, $t = 1.51$, $p < 0.15$; 3E: Student's t-test, $t = 2.664$, $p < 0.02$).

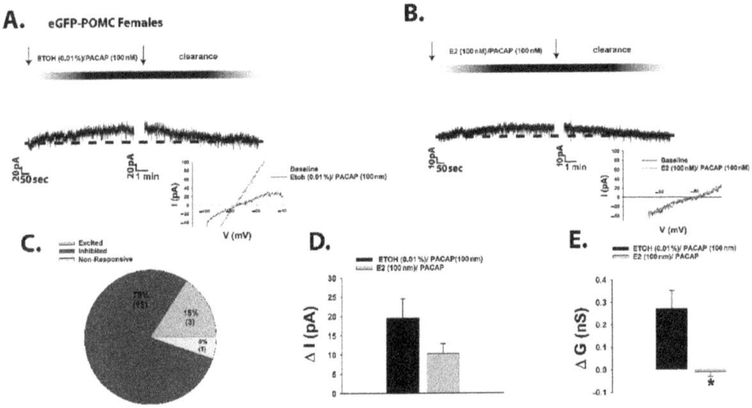

Figure 3. E_2 attenuates the PACAP-induced outward current in POMC neurons observed under fasting conditions. (**A**,**B**) are representative membrane current traces during voltage clamp recordings in EtOH vehicle- (0.01% (v:v); $n = 11$) and E_2-treated (100 nM; $n = 8$) slices from OVX females that illustrate the estrogenic diminution of the robust and reversible PACAP-induced outward current and change in K^+ conductance (as seen from the inset I/V plots). (**C**), Pie chart that indicates the percentage of POMC neurons from OVX females that are excited by, inhibited by, or unresponsive to PACAP under fasting conditions. (**D**,**E**) show the composite data underscoring the ability of E_2 to negatively modulate the PACAP response. Bars represent means and lines 1 SEM of the PACAP-induced ΔI and ΔG in POMC neurons from OVX females under fasting conditions. *, $p < 0.05$ relative to EtOH vehicle, Student's t-test.

Current clamp recordings in POMC neurons from fasted male eGFP-POMC mice uncovered a more hyperpolarized resting membrane potential (RMP) than those from their ad libitum-fed counterparts (Figure 4A: Student's t-test, $t = 2.237$, $p < 0.04$). Once again, the switch in polarity of the PACAP response was evident, such that the PACAP-induced depolarization of POMC neurons and the associated increase in firing seen in the representative trace and composite data from the ad libitum-fed state transformed into a hyperpolarization and a suppression of firing in the fasted state (Figure 4B–E; 4D: Student's t-test, $t = 9.501$, $p < 0.0001$; 4E: Kruskal–Wallis/median-notched box-and-whisker analysis, test statistic = 15.6201, $p < 0.002$).

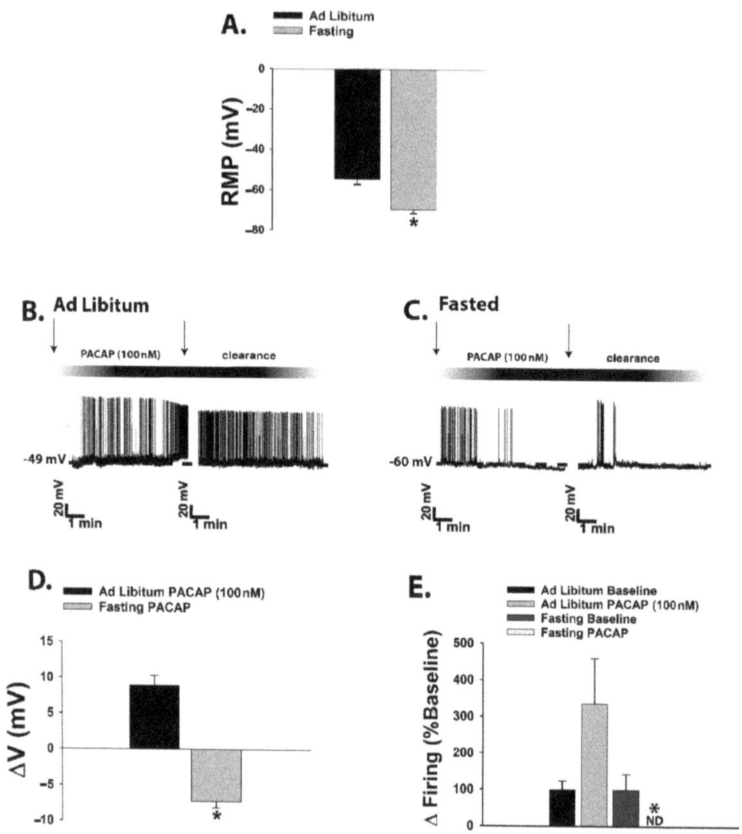

Figure 4. The PACAP-induced outward current observed during fasting conditions is associated with a hyperpolarization and a decrease in firing. (**A**), Composite bar graph that demonstrates the more hyperpolarized RMP of POMC neurons under fasting conditions. (**B**,**C**), Representative current clamp traces from POMC neurons showing the reversible PACAP-induced depolarization and increase in firing under ad libitum-fed conditions ($n = 10$) and the reversible hyperpolarization and suppression of firing seen under fasting conditions ($n = 10$). Comparable effects are seen during recordings in vehicle pre-treated slices from OVX females. (**D**,**E**), Composite data illustrating the PACAP-induced changes in membrane potential (ΔV) and firing rate under ad libitum-fed and fasting conditions. Bars represent means and lines 1 SEM. * $p < 0.05$, relative to ad libitum-fed conditions, Student's *t*-test (**D**); relative to baseline, Kruskal–Wallis/median-notched box-and-whisker analysis (**E**). Figure adapted from [12].

We corroborated these findings during optogenetic recordings in POMC neurons (Figure 5A,C,D) from PACAP-cre/eGFP POMC mice. Selective activation of VMN PACAP neurons (Figure 5B,E) elicited inward currents that were associated with depolarizations and increases in firing in the ad libitum-fed state (Figure 5F,H,J,L–N), and outward currents that were associated with hyperpolarizations and cessation of firing in the fasting state (Figure 5G,I,K–N; 5L: Student's *t*-test, $t = 6.444$, $p < 0.0001$; 5M: Student's *t*-test, $t = 6.97$, $p < 0.0001$; 5N: Kruskal–Wallis/median-notched box-and-whisker analysis, test statistic = 13.8222, $p < 0.004$).

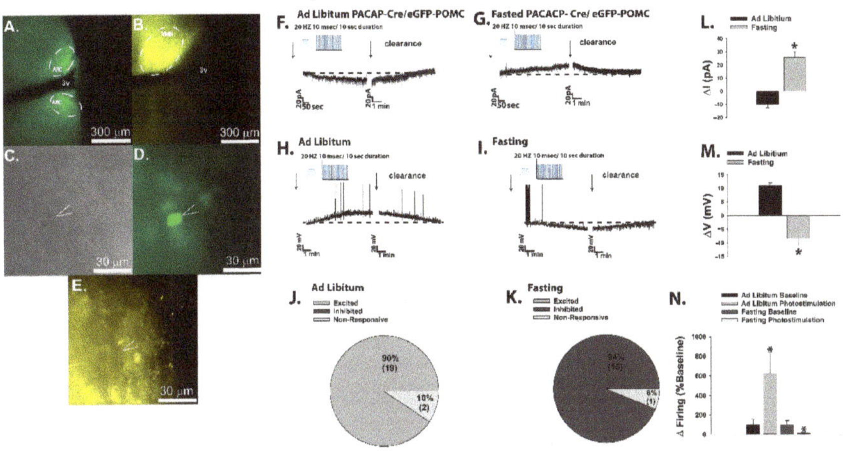

Figure 5. Optogenetic stimulation of VMN PACAP neurons depolarizes POMC neurons and increases their firing under ad libitum-fed conditions, effects which are flipped under fasting conditions. (**A**), Low power (4×) image of ARC POMC neurons taken from a PACAP-Cre/eGFP POMC mouse. (**B**), Photomicrograph (4×) showing the channel rhodopsin-2 expression in VMN PACAP neurons two weeks after AAV injection as visualized by eYFP. (C), Differential interference contrast image (40×) of a recorded POMC neuron and the corresponding eGFP fluorescence signal from the same neuron (**D**). (**E**), 40X image showing the eYFP-labeled fibers in the immediate vicinity of the recorded neuron. Photostimulation (10-ms pulses delivered at 20 Hz for 10 s) of male VMN PACAP neurons produces a reversible inward current linked to membrane depolarization and increase in the firing of ARC POMC under ad libitum-fed conditions neurons (**F,H,J,L–N**); n = 7–11), and the exact opposite is seen under fasting conditions (**G,I,K–N**); n = 11–13). Bars represent means, and lines 1 SEM of the light-induced change in ΔI (**L**), ΔV (**M**) and firing rate (**N**). * $p < 0.05$, relative to ad libitum-fed conditions, Student's t-test (**L,M**); relative to baseline, Kruskal–Wallis/median-notched box-and-whisker analysis (**N**). Figure adapted from [12].

Given that the tyrosine phosphatases PTP1B and TCPTP as well as AMPK figure prominently in orchestrating the cellular response to a state of negative energy balance, we then evaluated the role these signaling molecules play in the reversed polarity of the PACAP response in POMC neurons seen during fasting. In slices from fasted male eGFP-POMC mice we found that pretreatment with the PTP1B/TCPTP inhibitor CX08005 dramatically switched the PACAP response in POMC neurons from robust and reversible outward currents and hyperpolarizations (Figure 6A,C,E–I) to robust and reversible inward currents and depolarizations typically seen in the ad libitum-fed state (Figure 6B,D–I; 6E: Student's t-test, t = 8.194, $p < 0.0001$; 6F: Student's t-test, t = 6.878, $p < 0.0001$; 6G: Kruskal–Wallis/median-notched box-and-whisker analysis, test statistic = 13.5957, $p < 0.004$).

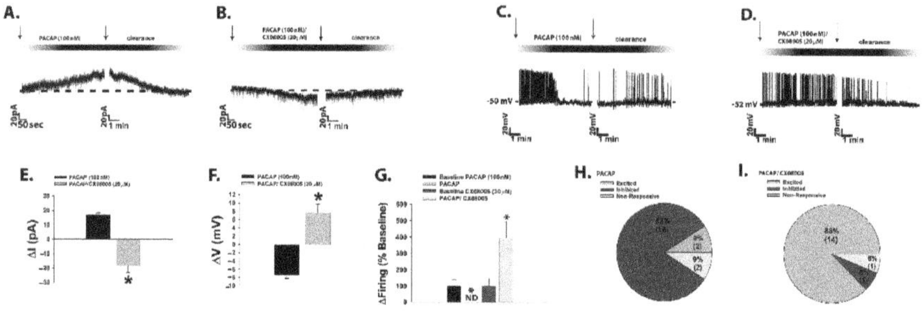

Figure 6. The fasting-induced switch in the polarity of the PACAP response in POMC neurons involves the activation of protein tyrosine phosphatases. Representative traces show that the PACAP-induced outward current (**A**; $n = 12$) and hyperpolarization (**C**; $n = 10$) seen under fasting conditions reverts back to an inward current (**B**; $n = 9$) and depolarization (**D**; $n = 7$) in the presence of the PTP1B/TCPTP inhibitor CX08005 (20 μM; **B**). This is further substantiated by the composite bar graphs in (**E–G**) as well as the pie charts in (**H**,**I**). Bars represent means, and lines 1 SEM of the PACAP-induced change ΔI, ΔV or normalized firing rate under fasted conditions, alone and in combination with CX08005, Compound C or metformin. * $p < 0.05$, relative to PACAP alone, Student's *t*-test (**E**,**F**); relative to baseline, Kruskal–Wallis/median-notched box-and-whisker analysis (**G**).

An equally dramatic switch in the polarity of the PACAP response in POMC neurons was observed during voltage clamp recordings in slices from both fasted eGFP-POMC mice that were pre-treated with AMPK inhibitor Compound C (Figure 7A,B,D–F), and from ad libitum-fed animals that were pre-treated with the AMPK activator metformin (Figure 7C,D,G; 7D: one-way ANOVA/LSD, $F = 41.35$, $df = 2$, $p < 0.0001$).

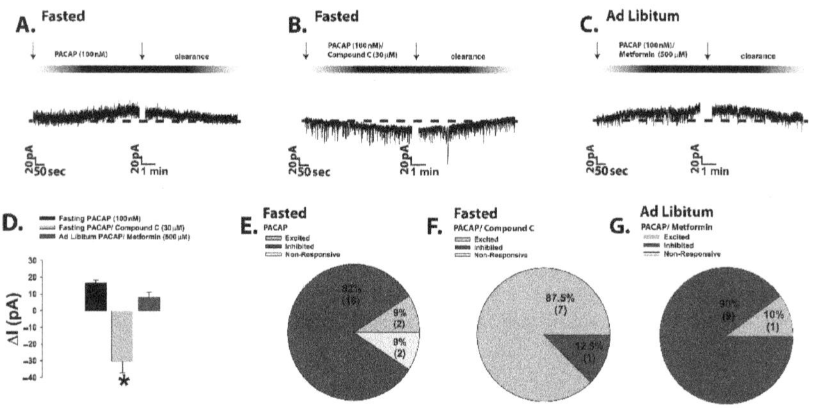

Figure 7. The fasting-induced reversal of the PACAP response in POMC neurons is also dependent upon activation of AMPK. The outward current caused by PACAP under fasting conditions in POMC neurons (**A**; $n = 12$) is transformed into an inward current in the presence of the AMPK inhibitor Compound C (30 μM; **B**; $n = 8$) The PACAP-induced outward current in (**A**) was reproduced under ad libitum-fed conditions in the presence of the AMPK activator metformin (500 μM; **C**; $n = 10$). The data from these representative traces is summarized in composite form by the bar graph in (**D**) and the pie charts in (**E–G**). Bars represent means and lines 1 SEM. *, $p < 0.05$ relative to PACAP alone, one-way ANOVA/LSD.

Consistent with these changes in cellular signaling, we also observed through immunohistofluorescent staining of coronal hypothalamic slices that fasting significantly increased the number of PTP1B- and pAMPK-positive ARC neurons as compared to that seen in the ad libitum-fed state (Figure 8; 8C: Student's *t*-test, $t = 7.748$, $p < 0.002$; 8F: Student's *t*-test, $t = 5.598$, $p < 0.006$).

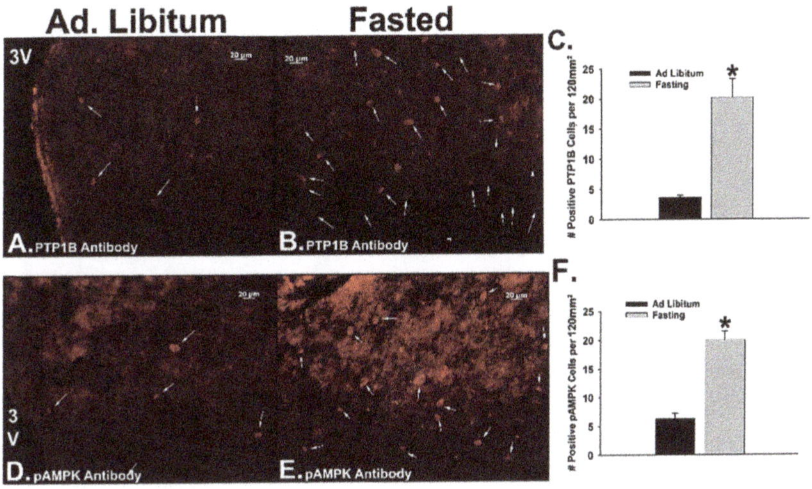

Figure 8. Fasting augments the activity and expression of PTP1B and AMPK in the ARC. The four 20× images depict the PTP1B (**A**,**B**; 1:100) and pAMPK (**D**,**E**; 1:100) immunoreactivity under ad libitum-fed (**A**,**D**) and fasting (**B**,**E**) as visualized with AF546 (1:600). The composite data in the bar graphs summarize the fasting-induced increase in the number of PTP1B- (**C**) and pAMPK-immunoreactive (**F**) cells per capita in the ARC. Bars represent means and lines 1 SEM. * $p < 0.05$, relative to ad libitum-fed conditions, Student's t-test.

Moreover, the flipped PACAP-induced change in the excitability of POMC neurons yielded parallel alterations in energy intake. Indeed, the pronounced anorexigenic response generated by PACAP administered directly into the ARC of ad libitum-fed wildtype male mice (Figure 9A; repeated measures multifactorial ANOVA/LSD, $F_{time} = 45.21$, df = 1, $p < 0.0001$; $F_{PACACP} = 71.82$, df = 1, $p < 0.0001$; $F_{interaction} = 0.72$, df = 1, $p < 0.40$) was completely reversed in fasted animals, such that there was no change in cumulative energy intake at three hours post-injection and a significant increase six hours after administration (Figure 9B; repeated measures multifactorial ANOVA/LSD, $F_{time} = 28.18$, df = 1, $p < 0.0001$; $F_{PACAP} = 4.32$, df = 1, $p < 0.04$; $F_{interaction} = 4.79$, df = 1, $p < 0.04$; one-way ANOVA/LSD, F = 12.88, df = 3, $p < 0.0001$). A similar profile was observed in OVX wildtype female mice; with PACAP decreasing cumulative energy intake in ad libitum-fed, sesame oil vehicle-treated OVX females, which was significantly potentiated in estradiol benzoate (EB)-treated OVX females (Figure 9C; repeated measures multi-factorial ANOVA/LSD: $F_{time} = 417.17$, df = 1, $p < 0.0001$), $F_{PACAP} = 9.00$, df = 1, $p < 0.004$, $F_{EB} = 20.59$, df = 1, $p < 0.0001$, $F_{interaction} = 0.91$, df = 1, $p < 0.35$). Conversely, the decrease in cumulative energy intake caused by PACAP in ad libitum-fed, vehicle-treated OVX females was once again completely reversed under fasting conditions. Surprisingly, EB per se failed to exert its prototypical anorexigenic effect in fasted OVX females, but it did abrogate the PACAP-induced increase in cumulative energy intake seen at six hours post-administration (Figure 9D; repeated measures multi-factorial ANOVA/LSD: $F_{time} = 74.67$, df = 1, $p < 0.0001$), $F_{PACAP} = 8.64$, df = 1, $p < 0.004$, $F_{EB} = 3.18$, df = 1, $p < 0.08$, $F_{interaction} = 14.88$, df = 1, $p < 0.0003$; one-way ANOVA/LSD, F = 15.53, df = 7, $p < 0.0001$).

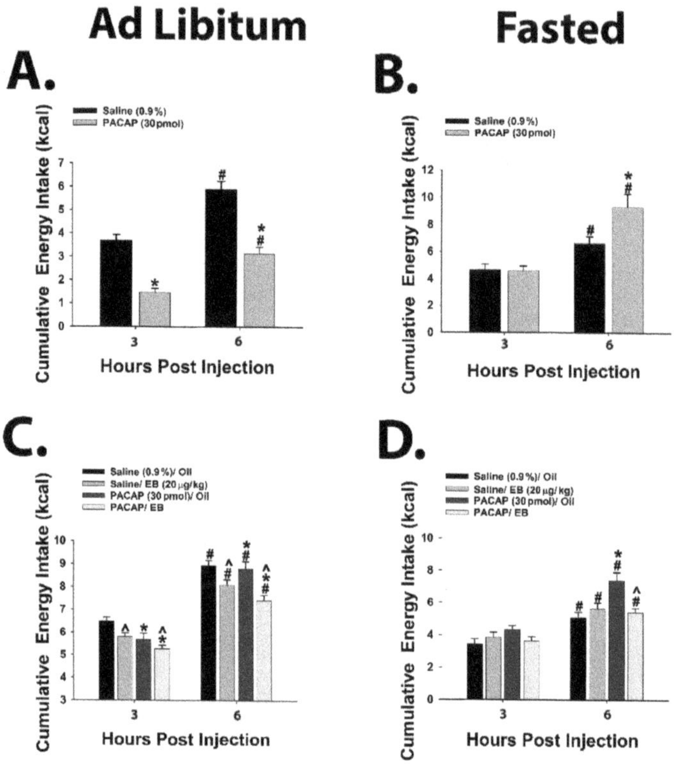

Figure 9. Fasting completely reverses the effect of a direct injection of PACAP into the ARC on energy intake, which is attenuated by EB in OVX females. Under ad libitum-fed conditions, PACAP (30pmo; n = 61) significantly decreases cumulative energy intake in wildtype males compared to saline-treated controls (0.2 μL; **A**; n = 6). This PACAP-induced decrease in energy intake is no longer apparent in fasted males, and PACAP actually causes an increase in cumulative consumption which is evident at six hours post-administration (**B**; n = 7–9). PACAP also decreases energy intake in ad libitum-fed OVX wildtype females, and this effect is potentiated by EB (20 μg/kg; s.c.; **C**; n = 6). Again, the effect of PACAP on consummatory behavior in fasted OVX females is exactly the opposite of that seen under ad libitum conditions, as is the modulatory effect of EB (**D**; n = 6). Bars represent means and lines 1 S.E.M. of the cumulative energy intake seen in ad libitum-fed or fasted mice injected with either PACAP or its saline vehicle. #, $p < 0.05$ relative to cumulative energy intake seen at three hours after PACAP injection, repeated-measures, multi-factorial ANOVA/LSD; *, $p < 0.05$ relative to saline vehicle, repeated measures, multi-factorial ANOVA/LSD; ^, $p < 0.05$ relative to sesame oil vehicle, repeated measures, multi-factorial ANOVA/LSD. Figure adapted from [12].

Taken together, this demonstrates that under conditions of negative energy balance, PAC1 receptor/effector coupling reverts from TRPC5 channel-induced excitation to K_{ATP} channel-induced inhibition, which completely reverses the effect of PACAP on energy intake. Thus, it is clear that postsynaptic PAC1 receptors at VMN PACAP/ARC POMC synapses effectively serve as metabolic switches that provide flexibility in the face of dynamic changes in energy status.

6. Concluding Remarks

To summarize, the N/OFQ/NOP and PACAP/PAC1 systems exert pleiotropic actions in the homeostatic and hedonic regulation of energy balance. In short, the N/OFQ/NOP system elicits a net orexigenic effect via the homeostatic energy balance circuitry, and dampens the consumption of palatable food via inhibitory actions within the hedonic energy balance circuitry. These effects are sexually differentiated, accentuated by diet-induced obesity in males and hypoestrogenic females, and attenuated by E2 in OVX females. On the other hand, the PACAP/PAC1 system contributes a net anorexigenic effect via the homeostatic and hedonic energy balance circuitries. A thorough investigation of the anorexigenic VMN PACAP/ARC POMC synapse reveals that the neurophysiological and accompanying behavioral effects described above are diminished by diet-induced obesity in males, potentiated by E2 in OVX females, and completely reversed under fasting conditions. The manner in which alterations in energy balance status can influence NOP and PAC1 receptor-mediated signaling in POMC neurons is summarized schematically in Figure 10.

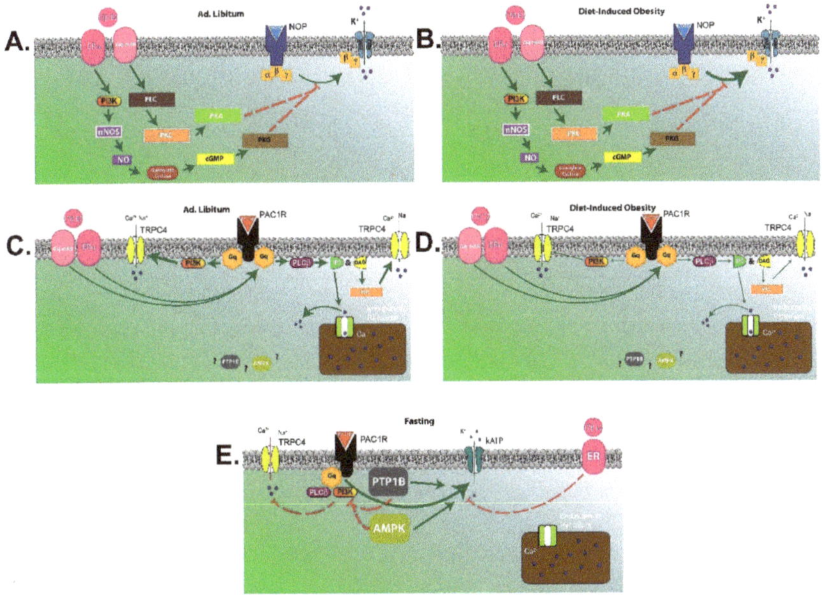

Figure 10. Schematic diagrams illustrating how PAC1 and NOP receptor signaling in POMC neurons is altered under various energy balance states. (**A**), In POMC neurons from ad libitum-fed animals, N/OFQ activates the NOP receptor that initiates $G_{i/o}$-mediated signaling and subsequent activation of GIRK channels through positive allosteric modification by the $\beta\gamma$ complex. This in turn promotes K^+ efflux and inhibition of POMC neurons, effects which are dampened by E_2 acting through ERα and G_q-mER to stimulate PI3K and nNOS as well as PLC, PKC and PKA signaling pathways, respectively. (**B**), In POMC neurons from obese animals, NOP receptor/effector coupling is enhanced; leading to a greater inhibitory effect of N/OFQ on POMC neurons. This N/OFQ-induced inhibition of POMC neurons is once again abrogated by E_2 in POMC neurons from obese females. (**C**), Under ad libitum conditions, PACAP activates its cognate PAC1 receptor to elicit G_q-mediated signaling; working through PI3K as well as PLC, IP3, DAG and PKC to promote Ca^{2+} mobilization from intracellular stores and the coupling of PAC1 receptors to TRCP5 channels. This leads to cation flux through the channel pore that depolarizes and thereby excites POMC neurons. In females, E_2 can act via ERα and Gq-mER to potentiate PAC1 receptor/TRPC5 channel coupling and PACAP-induced excitation of POMC neurons. (**D**), Under conditions of diet-induced obesity, the PAC1 receptor-mediated activation of TRPC5 channels in male POMC neurons is attenuated. However, in obese females the PACAP-induced excitation of POMC neurons is maintained due to the potentiating effect of E_2. (**E**), Under conditions of fasting, the expression and activity of AMPK and protein tyrosine phosphatases like PTP1B is elevated in POMC neurons. This triggers a switch in the coupling of PAC1 receptors, such that they now are no longer linked with TRPC5 channels and instead inhibit rather than excite POMC neurons via activation of K_{ATP} channels. This inhibitory effect of PACAP in POMC neurons from fasted animals is diminished by E_2 in females.

Over the past few decades, the neuroscience and neuroendocrine communities have made great strides in advancing our understanding of how the brain coordinates energy balance regulation via the homeostatic and hedonic energy balance circuitries. With the plethora of methodological tools currently at our disposal (e.g., optogenetics, chemogenetics, proteomics and transcriptomics, to name but a few), it is more than reasonable to expect that further advances will be readily and rapidly achieved. Just like the case made for PACAP in this very piece, it will be imperative to systematically evaluate all of the major players implicated in regulating energy homeostasis (including, but certainly not limited to, endocannabinoids, N/OFQ, NPY/AgRP neurons, POMC neurons, A10 dopamine neurons) not only under normophysiologic conditions but also under negative (e.g., fasting) and positive (e.g., obesity) energy balance states. Only in this way will we develop a comprehensive picture of how all of these functioning components are altered under these adaptive (e.g., fasting) and maladaptive (e.g., obesity) scenarios. In addition, it will be equally important to thoroughly dissect the impact of sex on the functioning of the central energy balance circuitries, and the cross-talk between gonadal hormones, the neurotransmitter/modulator systems and the signaling molecules involved. One final critical crowning achievement will be realized once we gain a firmer grasp on how the output from the homeostatic and hedonic energy balance circuitries functionally translates into clearly defined changes in the brain-gut axis and autonomic tone, as well as changes in gustatory and motivated behavior. We currently stand on the shoulders of pioneering giants who have paved the way by elucidating the mechanisms that provide the basis of our current understanding involved in the central control of energy homeostasis. In looking to the future, we welcome the next generation of innovative scientists to carry the torch and further advance our understanding.

Supplementary Materials: The following are available online at https://www.mdpi.com/1422-0067/22/5/2728/s13.S1: Methods and Materials.

Author Contributions: C.G. and J.H. performed all stereotaxic and survival surgeries. C.G. and J.H. performed all electrophysiological recordings. C.G., L.P., J.H., N.L., I.V. and S.S. performed all metabolic studies. C.G., J.H. and E.J.W. performed data analysis for all electrophysiology and metabolic studies. C.G., J.H., and E.J.W. created all figures and performed all statistical analyses. E.J.W. generated the manuscript, while all authors edited the final manuscript. E.J.W. and C.G. designed the experiments. All authors have read and agreed to the published version of the manuscript.

Funding: This study was supported by PHS Grant DA024314 and intramural funding from Western University of Health Sciences.

Institutional Review Board Statement: All care of animals and procedures were in accordance with the NIH Guide for the Care and Use of Laboratory Animals, and approved by the Western University of Health Sciences' IACUC (protocol code R18/IACUC/015,08/06/2020; R19/IACUC/027, 06/19/2020)."

Informed Consent Statement: Not applicable.

Data Availability Statement: The data presented in this study are available in Gastelum, C.; Perez, L.; Hernandez, J.; Le, N.; Vahrson, I.; Sayers, S.; Wagner, E.J. Adaptive Changes in the Central Control of Energy Homeostasis Occur in Response to Variations in Energy Status. Int. J. Mol. Sci. 2021, 22, 2728. doi.org/10.3390/ijms22052728.

Conflicts of Interest: The authors declare no conflict of interest.

References

1. Berthoud, H.-R. The neurobiology of food intake in an obesogenic environment. *Proc. Nutr. Soc.* **2012**, *71*, 478–487. [CrossRef] [PubMed]
2. Majdic, G.; Young, M.; Gomez-Sanchez, E.; Anderson, P.; Szczepaniak, L.S.; Dobbins, R.L.; McGarry, J.D.; Parker, K.L. Knockout mice lacking steroidogenic factor 1 are a novel genetic model of hypothalamic obesity. *Endocrinology* **2002**, *143*, 607–614. [CrossRef] [PubMed]

3. Fordahl, S.C.; Jones, S.R. High-fat-diet induced deficits in dopamine terminal function are reversed by restoring insulin signaling. *ACS Chem. Neurosci.* **2017**, *8*, 290–299. [CrossRef] [PubMed]
4. Volkow, N.D.; Wise, R.A.; Baler, R. The dopamine motive system: Implications for drug and food addiction. *Nat. Rev. Neurosci.* **2017**, *18*, 741–752. [CrossRef]
5. Hernandez, J.; Fabelo, C.; Perez, L.; Moore, C.; Chang, R.; Wagner, E.J. Nociceptin/orphanin FQ modulates energy homeostasis through inhibition of neurotransmission at VMN SF-1/ARC POMC synapses in a sex- and diet-dependent manner. *Biol. Sex Diff.* **2019**, *10*. [CrossRef]
6. Browning, K.N.; Kalyuzhny, A.E.; Travagli, R.A. Opioid peptides inhibit excitatory but not inhibitory synaptic transmission in the rat dorsal motor nucleus of the vagus. *J. Neurosci.* **2002**, *22*, 2998–3004. [CrossRef]
7. Sobrino Crespo, C.; Perianes Cachero, A.; Puebla Jimenez, L.; Barrios, V.; Arilla Ferreiro, E. Peptides and food intake. *Front. Endocrinol.* **2014**, *5*. [CrossRef]
8. Sternson, S.M.; Shepherd, G.M.G.; Friedman, J.M. Topographic mapping of VMH-arcuate nucleus microcirsuits and their reoganization by fasting. *Nat. Neurosci.* **2005**, *8*, 1356–1363. [CrossRef] [PubMed]
9. Krashes, M.J.; Shah, B.P.; Madara, J.C.; Olson, D.P.; Strochlic, D.E.; Garfield, A.S.; Vong, L.; Pei, H.; Watabe-Uchida, M.; Uchida, N.; et al. An excitatory paraventricular nucleus to AgRP neuron circuit that drives hunger. *Nature* **2014**, *507*, 238–242. [CrossRef] [PubMed]
10. Conde, K.; Fabelo, C.; Krause, W.C.; Propst, R.; Goethel, J.; Fischer, D.; Hur, J.; Meza, C.; Ingraham, H.A.; Wagner, E.J. Testosterone rapidly augments retrograde endocannabinoid signaling in proopiomelanocortin neurons to suppress glutamatergic input from steroidogenic factor 1 neurons via upregulation of diacylglycerol lipase-α. *Neuroendocrinology* **2017**, *105*, 341–356. [CrossRef]
11. Fabelo, C.; Hernandez, J.; Chang, R.S.S.; Alicea, N.; Tian, S.; Conde, K.; Wagner, E.J. Endocannabinoid signaling at hypothalamic steroidogenic factor-1/proopiomelanocortin synapses is sex- and diet-sensitive. *Front. Mol. Neurosci.* **2018**, *11*. [CrossRef] [PubMed]
12. Chang, R.; Hernandez, J.; Gastelum, C.; Guadagno, K.; Perez, L.; Wagner, E.J. Pituitary adenylate cyclase-activating polypeptide excites proopiomelanocortin neurons: Implications for the regulation of energy homeostasis. *Neuroendocrinology* **2020**. [CrossRef] [PubMed]
13. Jacobowitz, D.M.; O'Donohue, T.L. α-Melanocyte stimulating hormone: Immunolohistochemical identification and mapping in neurons of rat brain. *Proc. Natl. Acad. Sci. USA* **1978**, *75*, 6300–6304. [CrossRef]
14. Kiss, J.Z.; Cassell, M.D.; Palkovits, M. Analysis of the ACTH/β-end/α-MSH-immunoreactive afferent input to the hypothalamic paraventricular nucleus of rat. *Brain Res.* **1984**, *324*, 91–99. [CrossRef]
15. Fan, W.; Boston, B.A.; Kesterson, R.A.; Hruby, V.J.; Cone, R.D. Role of melanocortinergic neurons in feeding and the agouti obesity syndrome. *Nature* **1997**, *385*, 165–168. [CrossRef] [PubMed]
16. Krude, H.; Biebermann, H.; Luck, W.; Horn, R.; Brabant, G.; Gruters, A. Severe early-onset obesity, adrenal insufficiency and red hair pigmentation caused by POMC mutations in humans. *Nat. Genet.* **1998**, *19*, 155–157. [CrossRef] [PubMed]
17. Yaswen, L.; Diehl, N.; Brennan, M.B.; Hochgeschwender, U. Obesity in the mouse model of pro-opiomelanocortin deficiency responds to peripheral melanocortin. *Nat. Med.* **1999**, *5*, 1066–1070. [CrossRef]
18. Bomberg, E.M.; Grace, M.K.; Levine, A.S.; Olszewski, P.K. Functional interaction between nociceptin/orphanin FQ and α-melanocyte-stimulating hormone in the regulation of feeding. *Peptides* **2006**, *27*, 1827–1834. [CrossRef]
19. Appleyard, S.M.; Hayward, M.; Young, J.I.; Butler, A.A.; Cone, R.D.; Rubinstein, M.; Low, M.J. A role for endogenous β-endorphin in energy homeostasis. *Endocrinology* **2003**, *144*, 1753–1760. [CrossRef]
20. Horvath, T.L. The hardship of obesity: A soft-wired hypothalamus. *Nat. Neurosci.* **2005**, *8*, 561–565. [CrossRef]
21. Parton, L.E.; Ye, C.P.; Coppari, R.; Enriori, P.; Choi, B.; Zhang, C.-Y.; Xu, C.; Vianna, C.R.; Balthasar, N.; Lee, C.E.; et al. Glucose sensing by POMC neurons regulates glucose homestasis and is impaired in obesity. *Nature* **2007**, *449*, 228–232. [CrossRef] [PubMed]
22. Burdakov, D.; Luckman, S.M.; Verkhratsky, A. Glucose-sensing neurons of the hypothalamus. *Phil. Trans. R. Soc. B* **2005**, *360*, 2227–2235. [CrossRef] [PubMed]
23. Belgardt, B.F.; Okamura, T.; Brüning, J.C. Hormone and glucose signalling in POMC and AgRP neurons. *J. Physiol.* **2009**, *587*, 5305–5314. [CrossRef] [PubMed]
24. Stricker, E.M. Hyperphagia. *N. Engl. J. Med.* **1978**, *298*, 1010–1013. [CrossRef]
25. Chronwall, B.M. Anatomy and physiology of the neuroendocrine arcuate nucleus. *Peptides* **1985**, *6*, 1–11. [CrossRef]
26. Schwartz, M.W.; Woods, S.C.; Porte, D.; Seeley, R.J.; Baskin, D.G. Central nervous system control of food intake. *Nature* **2000**, *404*, 661–671. [CrossRef]
27. Krashes, M.J.; Koda, S.; Ye, C.; Rogan, S.C.; Adams, A.C.; Cusher, D.S.; Maratos-Flier, E.; Roth, B.L.; Lowell, B.B. Rapid, reversible activation of AgRP neurons drives feeding behavior in mice. *J. Clin. Investig.* **2011**, *121*, 1424–1428. [CrossRef] [PubMed]
28. Tong, Q.; Ye, C.-P.; Jones, J.E.; Elmquist, J.K.; Lowell, B.B. Synaptic release of GABA by AgRP neurons is required for normal regulation of energy balance. *Nat. Neurosci.* **2008**, *11*, 998–1000. [CrossRef]
29. Edwards, C.M.B.; Abusnana, S.; Sunter, D.; Murphy, K.G.; Ghatei, M.A.; Bloom, S.R. The effect of the orexins on food intake: Comparison with neuropeptide Y, melanin-concentrating hormone and galanin. *J. Endocrinol.* **1999**, *160*, R7–R12. [CrossRef]
30. Wagner, E.J. Sex differences in cannabinoid-regulated biology: A focus on energy homeostasis. *Front. Neuroendocrinol.* **2016**, *40*, 101–109. [CrossRef] [PubMed]

31. Ma, X.; Zubcevic, L.; Brüning, J.C.; Ashcroft, F.M.; Burdakov, D. Electrical inhibition of identified anorexigenic POMC neurons by orexin/hypocretin. *J. Neurosci.* **2007**, *27*, 1529–1533. [CrossRef] [PubMed]
32. Jo, Y.-H.; Chen, Y.-J.L.; Chua, S.C.; Talmage, D.A.; Role, L.W. Integration of endocannabinoid and leptin signaling in an appetite-related neural circuit. *Neuron* **2005**, *48*, 1055–1066. [CrossRef] [PubMed]
33. Mullier, A.; Bouret, S.G.; Prevot, V.; Dehouck, B. Differential distribution of tight junction proteins suggests a role for tanycytes in blood-hypothalamus barrier regulation in the adult mouse brain. *J. Comp. Neurol.* **2010**, *518*, 943–962. [CrossRef] [PubMed]
34. Rodriguez, E.M.; Blazquez, J.L.; Guerra, M. The design of barriers in the hypothalamus allows the median eminence and the arcuate nucleus to enjoy private milieus: The former opens to the portal blood and the latter to the cerebrospinal fluid. *Peptides* **2010**, *31*, 757–776. [CrossRef]
35. Zigman, J.M.; Elmquist, J.K. Minireview: From anorexia to obesity—The yin and yang of body weight control. *Endocrinology* **2003**, *144*, 3749–3756. [CrossRef]
36. Cowley, M.A.; Cone, R.D.; Enriori, P.; Louiselle, I.; Williams, S.M.; Evans, A.E. Electrophysiological actions of peripheral hormones on melanocortin neurons. *Ann. N. Y. Acad. Sci.* **2003**, *994*, 175–186. [CrossRef] [PubMed]
37. Dhillon, H.; Zigman, J.M.; Ye, C.; Lee, C.E.; McGovern, R.A.; Tang, V.; Kenny, C.D.; Christiansen, L.M.; White, R.D.; Edelstein, E.A.; et al. Leptin directly activates SF1 neurons in the VMH, and this action by leptin is required for normal body-weight homeostasis. *Neuron* **2006**, *49*, 191–203. [CrossRef] [PubMed]
38. Hill, J.W.; Williams, K.W.; Ye, C.; Luo, J.; Balthasar, N.; Coppari, R.; Cowley, M.A.; Cantley, L.C.; Lowell, B.B.; Elmquist, J.K. Acute effects of leptin require PI3K signaling in hypothalamic proopiomelanocortin neurons in mice. *J. Clin. Investig.* **2008**, *118*, 1796–1805. [CrossRef]
39. Qiu, J.; Fang, Y.; Rønnekleiv, O.K.; Kelly, M.J. Leptin excites proopiomelanocortin neurons via activation of TRPC channels. *J. Neurosci.* **2010**, *30*, 1560–1565. [CrossRef]
40. Van den Top, M.; Lee, K.; Whyment, A.D.; Blanks, A.M.; Spanswick, D. Orexigen-sensitive NPY/AgRP pacemaker neurons in the hypothalamic arcuate nucleus. *Nat. Neurosci.* **2004**, *5*, 493–494. [CrossRef]
41. Wei, W.; Pham, K.; Gammons, J.W.; Sutherland, D.; Liu, Y.; Smith, A.; Kaczorowski, C.C.; O'Connell, K.M. Diet composition, not calorie intake, rapidly alters intrinsic excitability of hypothalamic AgRP/NPY neurons in mice. *Sci. Rep.* **2015**, *5*. [CrossRef]
42. Qiu, J.; Wagner, E.J.; Ronnekleiv, O.K.; Kelly, M.J. Insulin and leptin excite anorexigenic pro-opiomelanocortin neurones via activation of TRPC5 channels. *J. Neuroendocrinol.* **2017**, *30*, e12501. [CrossRef]
43. Di Marzo, V.; Goparahu, S.K.; Wang, L.; Liu, J.; Bátkai, S.; Járai, Z.; Fezza, F.; Miura, G.I.; Palmiter, R.D.; Sugiura, T.; et al. Leptin-regulated endocannabinoids are involved in maintaining food intake. *Nature* **2001**, *410*, 822–825. [CrossRef] [PubMed]
44. Qiu, J.; Zhang, C.; Borgquist, A.; Nestor, C.C.; Smith, A.W.; Bosch, M.A.; Ku, S.; Wagner, E.J.; Rønnekleiv, O.K.; Kelly, M.J. Insulin excites anorexigenic proopiomelanocortin neurons via activation of canonical transient receptor potential channels. *Cell Metab.* **2014**, *19*, 682–693. [CrossRef] [PubMed]
45. Hill, J.W.; Elias, C.F.; Fukuda, M.; Williams, K.W.; Berglund, E.D.; Holland, W.L.; Cho, Y.-R.; Chuang, J.-C.; Xu, Y.; Choi, M.; et al. Direct insulin and leptin action on pro-opiomelanocortin neurons is required for normal glucose homeostasis and fertility. *Cell Metab.* **2010**, *11*, 286–297. [CrossRef] [PubMed]
46. Williams, K.W.; Margatho, L.O.; Lee, C.E.; Choi, M.; Lee, S.; Scott, M.M.; Elias, C.F.; Elmquist, J.K. Segregation of acute leptin and insulin effects in distinct populations of arcuate proopiomelanocortin neurons. *J. Neurosci.* **2010**, *30*, 2472–2479. [CrossRef] [PubMed]
47. Dodd, G.T.; Michael, N.J.; Lee-Young, R.S.; Mangiafico, S.P.; Pryor, J.T.; Munder, A.C.; Simonds, S.E.; Bruning, J.C.; Zhang, Z.-Y.; Cowley, M.A.; et al. Insulin regulates POMC neuronal plasticity to control glucose metabolism. *eLife* **2018**, *7*. [CrossRef] [PubMed]
48. Dodd, G.T.; Decherf, S.; Loh, K.; Simonds, S.E.; Wiede, F.; Balland, E.; Merry, T.L.; Munzberg, H.; Zhang, Z.-Y.; Kahn, B.B.; et al. Leptin and insulin act on POMC neurons to promote the browning of white fat. *Cell* **2015**, *160*, 88–104. [CrossRef] [PubMed]
49. Dodd, G.T.; Andrews, Z.B.; Simonds, S.E.; Michael, N.J.; DeVeer, M.; Bruning, J.C.; Spanswick, D.; Cowley, M.A.; Tiganis, T. A Hypothalamic Phosphatase Switch Coordinates Energy Expenditure with Feeding. *Cell Metab.* **2017**, *26*, 375–393. [CrossRef]
50. Dodd, G.T.; Xirouchaki, C.E.; Eramo, M.; Mitchell, C.A.; Andrews, Z.B.; Henry, B.A.; Cowley, M.A.; Tiganis, T. Intranasal targeting of hypothalamic PTP1B and TCPTP reinstates leptin and insulin sensitivity and promotes weight loss in obesity. *Cell Rep.* **2019**, *28*, 2905–2922. [CrossRef] [PubMed]
51. Kola, B.; Farkas, I.; Christ-Crain, M.; Wittmann, G.; Lolli, F.; Amin, F.; Harvey-White, J.; Liposits, Z.; Kunos, G.; Grossman, A.B.; et al. The orexigenic effect of ghrelin is mediated through central activation of the endogenous cannabinoid system. *PLoS ONE* **2008**, *3*, e1797. [CrossRef] [PubMed]
52. Lim, C.T.; Kola, B.; Feltrin, D.; Perez-Tilve, D.; Tschöp, M.H.; Grossman, A.B.; Korbonits, M. Ghrelin and cannabinoids require the ghrelin receptor to affect cellular energy metabolism. *Mol. Cell. Endocrinol.* **2013**, *365*, 303–308. [CrossRef] [PubMed]
53. Hardie, D.G.; Scott, J.W.; Pan, D.A.; Hudson, E.R. Management of cellular energy by the AMP-activated protein kinase system. *FEBS Lett.* **2003**, *546*, 113–120. [CrossRef]
54. Hardie, D.G. The AMP-activated protein kinase pathway—New players upstream and downstream. *J. Cell Sci.* **2004**, *117*, 5479–5487. [CrossRef]
55. Towler, M.C.; Hardie, D.G. AMP-activated protein kinase in metabolic control and insulin signaling. *Circ. Res.* **2007**, *100*, 328–341. [CrossRef]

56. Hardie, D.G.; Ross, F.A.; Hawley, S.A. AMPK: A nutrient and energy sensor that maintains energy homeostasis. *Nat. Rev.* **2012**, *13*, 251–262. [CrossRef]
57. Varela, L.; Horvath, T.L. Leptin and insulin pathways in POMC and AgRP neurons that modulate energy balance and glucose homeostasis. *EMBO Rep.* **2012**, *13*, 1079–1086. [CrossRef]
58. Minokoshi, Y.; Kim, Y.-B.; Peroni, O.D.; Fryer, L.G.D.; Muller, C.; Carling, D.; Kahn, B.B. Leptin stimulates fatty-acid oxidation by activating AMP-activated protein kinase. *Nature* **2002**, *415*, 339–343. [CrossRef]
59. Kola, B.; Hubina, E.; Tucci, S.A.; Kirkham, T.C.; Garcia, E.A.; Mitchell, S.E.; Williams, L.M.; Hawley, S.A.; Hardie, D.G.; Grossman, A.B.; et al. Cannabinoids and ghrelin have both central and peripheral metabolic and cardiac effects via AMP-activated protein kinase. *J. Biol. Chem.* **2005**, *280*, 25196–25201. [CrossRef]
60. Yavari, A.; Stocker, C.J.; Ghaffari, S.; Wargent, E.T.; Steeples, V.; Czibik, G.; Pinter, K.; Bellahcene, M.; Woods, A.; de Morentin, P.B.M.; et al. Chronic Activation of γ2 AMPK Induces Obesity and Reduces β Cell Function. *Cell Metab.* **2016**, *23*, 821–836. [CrossRef]
61. DiLeone, R.J.; Georgescu, D.; Nestler, E.J. Lateral hypothalamic neuropeptides in reward and drug addiction. *Life Sci.* **2003**, *73*, 759–768. [CrossRef]
62. Georgescu, D.; Sears, R.M.; Hommel, J.D.; Barrot, M.; Bolaños, C.A.; Marsh, D.J.; Bednarek, M.A.; Bibb, J.A.; Maratos-Flier, E.; Nestler, E.J.; et al. The hypothalamic neuropeptide melanin-concentrating hormone acts in the nucleus accumbens to modulate feeding behavior and forced-swim performance. *J. Neurosci.* **2005**, *25*, 2933–2940. [CrossRef] [PubMed]
63. McClung, C.A.; Nestler, E.J.; Zachariou, V. Regulation of gene expression by chronic morphine and morphine withdrawal in the locus ceruleus and ventral tegmental area. *J. Neurosci.* **2005**, *25*, 6005–6015. [CrossRef] [PubMed]
64. Kalivas, P.W.; Duffy, P. Sensitization to repeated morphine injection in the rat: Possible involvement of A_{10} dopamine neurons. *J. Pharmacol. Exp. Ther.* **1987**, *241*, 204–212.
65. Wise, R.A. Role of brain dopamine in food reward and reinforcement. *Phil. Trans. R. Soc. B* **2006**, *361*, 1149–1158. [CrossRef] [PubMed]
66. Durst, M.; Konczol, K.; Balazsa, T.; Eyre, M.D.; Toth, Z.E. Reward-representing D1-type neurons in the medial shell of the accumbens nucleus regulate palatable food intake. *Int. J. Obes.* **2019**, *43*, 917–927. [CrossRef] [PubMed]
67. Moore, K.E. Hypothalamic dopaminergic neuronal systems. In *Psychopharmacology: The Third Generation of Progress*; Meltzer, H.Y., Ed.; Raven Press: New York, NY, USA, 1987; pp. 127–139.
68. Tsai, H.-C.; Zhang, F.; Adamantidis, A.; Stuber, G.D.; Bonci, A.; De Lecea, L.; Deisseroth, K. Phasic firing in dopaminergic neurons is sufficient for behavioral conditioning. *Science* **2009**, *324*, 1080–1084. [CrossRef]
69. Witten, I.B.; Steinberg, E.E.; Lee, S.Y.; Davidson, T.J.; Zalocusky, K.A.; Brodsky, M.; Yizhar, O.; Cho, S.L.; Gong, S.; Ramakrishnan, C.; et al. Recombinase-driver rat lines: Tools, techniques, and optogenetic application to dopamine-mediated reinforcement. *Neuron* **2011**, *72*, 721–733. [CrossRef]
70. Ilango, A.; Kesner, A.J.; Keller, K.L.; Stuber, G.D.; Bonci, A.; Ikemoto, S. Similar roles of substantia nigra and ventral tegmental dopamine neurons in reward and aversion. *J. Neurosci.* **2014**, *34*, 817–822. [CrossRef]
71. Steinberg, E.E.; Boivin, J.R.; Saunders, B.T.; Witten, I.B.; Deisseroth, K.; Janak, P.H. Positive reinforcement mediated by midbrain dopamine neurons requires D1 and D2 receptor activation in the nucleus accumbens. *PLoS ONE* **2014**, *9*. [CrossRef]
72. Tan, K.R.; Yvon, C.; Turiault, M.; Mirzabekov, J.J.; Doehner, J.; Labouebe, G.; Deisseroth, K.; Tye, K.M.; Luscher, C. GABA neurons of the VTA drive conditioned place aversion. *Neuron* **2012**, *73*, 1173–1183. [CrossRef]
73. Jennings, J.H.; Sparta, D.R.; Stamatakis, A.M.; Ung, R.L.; Pleil, K.E.; Kash, T.L.; Stuber, G.D. Distinct extended amygdala circuits for divergent motivational states. *Nature* **2013**, *496*, 224–228. [CrossRef]
74. Carlezon, W.A., Jr.; Chartoff, E.H. Intracranial self-stimulation (ICSS) in rodents to study the neurobiology of motivation. *Nat. Protoc.* **2007**, *2*, 2987–2995. [CrossRef]
75. Wang, H.-L.; Qi, J.; Zhang, S.; Wang, H.; Morales, M. Rewarding Effects of Optical Stimulation of Ventral Tegmental Area Glutamatergic Neurons. *J. Neurosci.* **2015**, *35*, 15948–15954. [CrossRef] [PubMed]
76. Robinson, T.E.; Berridge, K.C. The incentive sensitization theory of addiction: Some current issues. *Phil. Trans. R. Soc. B* **2008**, *363*, 3137–3146. [CrossRef]
77. Avena, N.M.; Bocarsly, M.E. Dysregulation of brain reward systems in eating disorders: Neurochemical information from animal models of binge eating, bulimia nervosa, and anorexia nervosa. *Neuropharmacology* **2012**, *63*, 87–96. [CrossRef] [PubMed]
78. Carlin, J.L.; McKee, S.E.; Hill-Smith, T.; Grissom, N.M.; George, R.; Lucki, I.; Reyes, T.M. Removal of high-fat diet after chronic exposure drives binge behavior and dopaminergic dysregulation in female mice. *Neuroscience* **2016**, *326*, 170–179. [CrossRef]
79. Berridge, K.C. Food reward: Brain substrates of wanting and liking. *Neurosci. Biobehav. Rev.* **1996**, *20*, 1–25. [CrossRef]
80. Berridge, K.C. 'Liking' and 'wanting' food rewards: Brain substrates and roles in eating disorders. *Physiol. Behav.* **2009**, *97*, 537–550. [CrossRef] [PubMed]
81. Berridge, K.C. Measuring hedonic impact in animals and infants: Microstructure of affective taste reactivity patterns. *Neurosci. Biobehav. Rev.* **2000**, *24*, 173–198. [CrossRef]
82. Grill, H.J.; Norgren, R. The taste reactivity test. I. Mimetic responses to gustatory stimuli in neurologically normal rats. *Brain Res.* **1978**, *143*, 263–279. [CrossRef]
83. Steiner, J.E.; Glaser, D.; Hawilo, M.E.; Berridge, K.C. Comparative expression of hedonic impact: Affective reactions to taste by human infants and other primates. *Neurosci. Biobehav. Rev.* **2001**, *25*, 53–74. [CrossRef]

84. Smith, K.S.; Berridge, K.C. The ventral pallidum and hedonic reward: Neurochemical maps of sucrose "liking" and food intake. *J. Neurosci.* **2005**, *25*, 8637–8649. [CrossRef]
85. Mahler, S.V.; Smith, K.S.; Berridge, K.C. Endocannabinoid hedonic hotspot for sensory pleasure: Anandamide in nucleus accumbens shell enhances 'liking' of a sweet reward. *Neuropsychopharmacology* **2007**, *32*, 2267–2278. [CrossRef] [PubMed]
86. Koob, G.F.; Le Moal, M. Drug abuse: Hedonic homeostatic dysregulation. *Science* **1997**, *278*, 52–58. [CrossRef] [PubMed]
87. Dickinson, A.; Smith, J.; Mirenowicz, J. Dissociation of Pavlovian and instrumental incentive learning under dopamine antagonists. *Behav. Neurosci.* **2000**, *114*, 468–483. [CrossRef]
88. Everitt, B.J.; Robbins, T.W. Neural systems of reinforcement for drug addiction: From actions to habits to compulsion. *Nat. Neurosci.* **2005**, *8*, 1481–1489. [CrossRef] [PubMed]
89. Cannon, C.M.; Palmiter, R.D. Reward without dopamine. *J. Neurosci.* **2003**, *23*, 10827–10831. [CrossRef]
90. Yuan, L.; Dou, Y.-N.; Sun, Y.-G. Topography of Reward and Aversion Encoding in the Mesolimbic Dopaminergic System. *J. Neurosci.* **2019**, *39*, 6472–6481. [CrossRef]
91. De Jong, J.W.; Afjei, S.A.; Dorocic, I.P.; Peck, J.R.; Liu, C.; Kim, C.K.; Tian, L.; Deisseroth, K.; Lammel, S. A Neural Circuit Mechanism for Encoding Aversive Stimuli in the Mesolimbic Dopamine System. *Neuron* **2019**, *101*, 133–151. [CrossRef]
92. Soares-Cunha, C.; de Vasconcelos, N.A.P.; Coimbra, B.; Domingues, A.V.; Silva, J.M.; Loureiro-Campos, E.; Gaspar, R.; Sotiropoulos, I.; Sousa, N.; Rodrigues, A.J. Nucleus accumbens medium spiny neurons subtypes signal both reward and aversion. *Mol. Psychiatry* **2020**, *25*, 3241–3255. [CrossRef] [PubMed]
93. Figlewicz, D.P.; Benoit, S.C. Insulin, leptin, and food reward: Update 2008. *Am. J. Physiol. Regul. Integr. Comp. Physiol.* **2009**, *296*, R9–R19. [CrossRef] [PubMed]
94. Fulton, S.; Pissios, P.; Manchon, R.P.; Stiles, L.; Frank, L.; Pothos, E.N.; Maratos-Flier, E.; Flier, J.S. Leptin regulation of the mesoaccumbens dopamine pathway. *Neuron* **2006**, *51*, 811–822. [CrossRef]
95. Hommel, J.D.; Trinko, R.; Sears, R.M.; Georgescu, D.; Liu, Z.-W.; Gao, X.-B.; Thurmon, J.J.; Marinelli, M.; DiLeone, R.J. Leptin receptor signaling in midbrain dopamine neurons regulates feeding. *Neuron* **2006**, *51*, 801–810. [CrossRef]
96. Leinninger, G.M.; Jo, Y.-H.; Leshan, R.L.; Louis, G.W.; Yang, H.; Barrera, J.G.; Wilson, H.; Opland, D.M.; Faouzi, M.A.; Gong, Y.; et al. Leptin acts via leptin receptor-expressing lateral hypothalamic neurons to modulate the mesolimbic dopamine system and suppress feeding. *Neuron* **2009**, *10*, 89–98. [CrossRef]
97. Abizaid, A.; Liu, Z.-W.; Andrews, Z.B.; Shanabrough, M.; Borok, E.; Elsworth, J.D.; Roth, R.H.; Sleeman, M.W.; Picciotto, M.R.; Tschöp, M.H.; et al. Ghrelin modulates the activity and synaptic input organization of midbrain dopamine neurons while promoting appetite. *J. Clin. Investig.* **2006**, *116*, 3229–3239. [CrossRef] [PubMed]
98. Kalafateli, A.L.; Vallöf, D.; Jörnulf, J.W.; Heilig, M.; Jerlhag, E. A cannabinoid receptor antagonist attenuates ghrelin-induced activation of the mesolimbic dopamine system in mice. *Physiol. Behav.* **2018**, *184*, 211–219. [CrossRef] [PubMed]
99. Hernandez, L.; Hoebel, B.G. Food reward and cocaine increase extracellular dopamine in the nucleus accumbens as measured by microdialysis. *Life Sci.* **1988**, *42*, 1705–1712. [CrossRef]
100. Martel, P.; Fantino, M. Mesolimbic dopaminergic system activity as a function of food reward: A microdialysis study. *Pharmacol. Biochem. Behav.* **1996**, *53*, 221–226. [CrossRef]
101. Bassareo, V.; Di Chiara, G. Modulation of feeding-induced activation of mesolimbic dopamine transmission by appetitive stimuli and its relation to motivational state. *Eur. J. Neurosci.* **1999**, *11*, 4389–4397. [CrossRef] [PubMed]
102. Rada, P.; Avena, N.M.; Hoebel, B.G. Daily bingeing on sugar repeatedly releases dopamine in the accumbens shell. *Neuroscience* **2005**, *134*, 737–744. [CrossRef] [PubMed]
103. Rada, P.; Avena, N.M.; Barson, J.R.; Hoebel, B.G.; Leibowitz, S.F. A high-fat meal, or intraperitoneal administration of a fat emulsion, increases extracellular dopamine in the nucleus accumbens. *Brain Sci.* **2012**, *2*, 242–253. [CrossRef] [PubMed]
104. Di Ciano, P.; Coury, A.; Depoortere, R.Y.; Egilmez, Y.; Lane, J.D.; Emmett-Oglesby, M.W.; Lepiane, F.G.; Phillips, A.G.; Blaha, C.D. Comparison of changes in extracellular dopamine concentrations in the nucleus accumbens during intravenous self-administration of cocaine or *d*-amphetamine. *Behav. Pharmacol.* **1995**, *6*, 311–322. [CrossRef] [PubMed]
105. Kiyatkin, E.A. Functional significance of mesolimbic dopamine. *Neurosci. Biobehav. Rev.* **1995**, *19*, 573–598. [CrossRef]
106. Di Chiara, G.; Acquas, E.; Tanda, G. Ethanol as a neurochemical surrogate of conventional reinforcers: The dopamine-opioid link. *Alcohol* **1996**, *13*, 13–17. [CrossRef]
107. Tanda, G.; Pontieri, F.E.; Di Chiara, G. Cannabinoid and heroin activation of mesolimbic dopamine transmission by a μ_1 opioid receptor mechanism. *Science* **1997**, *276*, 2048–2050. [CrossRef] [PubMed]
108. Cheer, J.F.; Wassum, K.M.; Sombers, L.A.; Heien, M.L.A.V.; Ariansen, J.L.; Aragona, B.J.; Phillips, P.E.M.; Wightman, R.M. Phasic dopamine release evoked by abused substances requires cannabinoid receptor activation. *J. Neurosci.* **2007**, *27*, 791–795. [CrossRef] [PubMed]
109. Bell, S.M.; Stewart, R.B.; Thompson, S.C.; Meisch, R.A. Food-deprivation increases cocaine-induced conditioned place preference and locomotor activity in rats. *Psychopharmacology* **1997**, *131*, 1–8. [CrossRef]
110. Colantuoni, C.; Schwenker, J.; McCarthy, J.; Rada, P.; Ladenheim, B.; Cadet, J.L.; Schwartz, G.J.; Moran, T.H.; Hoebel, B.G. Excessive sugar intake alters binding to dopamine and mu-opioid receptors in the brain. *Neuroreport* **2001**, *12*, 49–52. [CrossRef] [PubMed]
111. Avena, N.M.; Hoebel, B.G. A diet promoting sugar dependency causes behavioral cross-sensitization to a low dose of amphetamine. *Neuroscience* **2003**, *122*, 17–20. [CrossRef]

112. Avena, N.M.; Hoebel, B.G. Amphetamine-sensitized rats show sugar-induced hyperactivity (cross-sensitization) and sugar hyperphagia. *Pharmacol. Biochem. Behav.* **2003**, *74*, 635–639. [CrossRef]
113. Bello, N.T.; Sweigart, K.L.; Lakoski, J.M.; Norgren, R.; Hajnal, A. Restricted feeding with scheduled sucrose access results in an upregulation of the rat dopamine transporter. *Am. J. Physiol. Regul. Integr. Comp. Physiol.* **2003**, *284*, R1260–R1268. [CrossRef] [PubMed]
114. Bakshi, V.P.; Kelley, A.E. Sensitization and conditioning of feeding following multiple morphine microinjections into the nucleus accumbens. *Brain Res.* **1994**, *648*, 342–346. [CrossRef]
115. Nocjar, C.; Panksepp, J. Chronic intermittent amphetamine pretreatment enhances future appetitive behavior for drug- and natural-reward: Interaction with environmental variables. *Behav. Brain Res.* **2002**, *128*, 189–203. [CrossRef]
116. Valdivia, S.; Cornejo, M.P.; Reynaldo, M.; De Francesco, P.N.; Perello, M. Escalation in high fat intake in a binge eating model differentially engages dopamine neurons of the ventral tegmental area and requires ghrelin signaling. *Psychoneuroendocrinology* **2015**, *60*, 206–216. [CrossRef] [PubMed]
117. Wang, G.J.; Volkow, N.D.; Logan, J.; Pappas, N.R.; Wong, C.T.; Zhu, W.; Netusil, N.; Fowler, J.S. Brain dopamine and obesity. *Lancet* **2001**, *357*, 354–357. [CrossRef]
118. Di Segni, M.; Patrono, E.; Patella, L.; Puglisi-Allegra, S.; Ventura, R. Animal models of compulsive eating behavior. *Nutrients* **2014**, *6*, 4591–4609. [CrossRef]
119. Jauch-Chara, K.; Oltmanns, K.M. Obesity—A neuropsychological disease? Systematic review and neuropsychological model. *Prog. Neurobiol.* **2014**, *114*, 84–101. [CrossRef]
120. Kreisler, A.D.; Garcia, M.G.; Spierling, S.R.; Hui, B.E.; Zorilla, E.P. Extended vs. brief intermittent access to palatable food differently promote binge-like intake, rejection of less preferred food, and weight cycling in female rats. *Physiol. Behav.* **2017**, *177*, 305–316. [CrossRef]
121. Hoek, H.W. Incidence, prevalence and mortality of anorexia nervosa and other eating disorders. *Curr. Opin. Psychiatry* **2006**, *19*, 389–394. [CrossRef] [PubMed]
122. Yang, L.; Colditz, G.A. Prevalence of Overweight and Obesity in the United States, 2007–2012. *JAMA Intern. Med.* **2015**, *175*, 1412–1413. [CrossRef]
123. Hallam, J.; Boswell, R.G.; DeVito, E.E.; Kober, H. Gender-related Differences in Food Craving and Obesity. *Yale J. Biol. Med.* **2016**, *89*, 161–173. [PubMed]
124. Weingarten, H.P.; Elston, D. Food cravings in a college population. *Appetite* **1991**, *17*, 167–175. [CrossRef]
125. Zellner, D.A.; Garriga-Trillo, A.; Rohm, E.; Centeno, S.; Parker, S. Food liking and craving: A cross-cultural approach. *Appetite* **1999**, *33*, 61–70. [CrossRef] [PubMed]
126. Wang, G.J.; Volkow, N.D.; Telang, F.; Jayne, M.; Ma, J.; Rao, M.; Zhu, W.; Wong, C.T.; Pappas, N.R.; Geliebter, A.; et al. Exposure to appetitive food stimuli markedly activates the human brain. *Neuroimage* **2004**, *21*, 1790–1797. [CrossRef]
127. Uher, R.; Treasure, J.; Heining, M.; Brammer, M.J.; Campbell, I.C. Cerebral processing of food-related stimuli: Effects of fasting and gender. *Behav. Brain Res.* **2006**, *169*, 111–119. [CrossRef]
128. Wang, G.J.; Volkow, N.D.; Telang, F.; Jayne, M.; Ma, Y.; Pradhan, K.; Zhu, W.; Wong, C.T.; Thanos, P.K.; Geliebter, A.; et al. Evidence of gender differences in the ability to inhibit brain activation elicited by food stimulation. *Proc. Natl. Acad. Sci. USA* **2009**, *106*, 1249–1254. [CrossRef]
129. Asarian, L.; Geary, N. Estradiol enhances cholecystokinin-dependent lipid-induced satiation and activates estrogen receptor-α-expressing cells in the nucleus tractus solitarius of ovariectomized rats. *Endocrinology* **2007**, *148*, 5656–5666. [CrossRef] [PubMed]
130. Thammacharoen, S.; Lutz, T.A.; Geary, N.; Asarian, L. Hindbrain administration of estradiol inhibits feeding and activates estrogen receptor-α-expressing cells in the nucleus tractus solitarius of ovariectomized rats. *Endocrinology* **2008**, *149*, 1609–1617. [CrossRef]
131. Borgquist, A.; Meza, C.; Wagner, E.J. Role of neuronal nitric oxide synthase in the estrogenic attenuation of cannabinoid-induced changes in energy homeostasis. *J. Neurophysiol.* **2015**, *113*, 904–914. [CrossRef] [PubMed]
132. Mela, V.; Vargas, A.; Meza, C.; Kachani, M.; Wagner, E.J. Modulatory influences of estradiol and other anorexigenic hormones on metabotropic, Gi/o-coupled receptor function in the hypothalamic control of energy homeostasis. *J. Steroid Biochem. Mol. Biol.* **2016**, *160*, 15–26. [CrossRef] [PubMed]
133. Kellert, B.A.; Nguyen, M.C.; Nguyen, C.; Nguyen, Q.H.; Wagner, E.J. Estrogen rapidly attenuates cannabinoid-induced changes in energy homeostasis. *Eur. J. Pharmacol.* **2009**, *622*, 15–24. [CrossRef] [PubMed]
134. Washburn, N.; Borgquist, A.; Wang, K.; Jeffery, G.S.; Kelly, M.J.; Wagner, E.J. Receptor subtypes and signal transduction mechanisms contributing to the estrogenic attenuation of cannabinoid-induced changes in energy homeostasis. *Neuroendocrinology* **2013**, *97*, 160–175. [CrossRef]
135. Ihara, H.; Kitamura, A.; Kasamatsu, S.; Ida, T.; Kakihana, Y.; Tsutsuki, H.; Sawa, T.; Watanabe, Y.; Akaike, T. Superoxide generation from nNOS splice variants and its potential involvement in redox signal regulation. *Biochem. J.* **2017**, *474*, 1149–1162. [CrossRef] [PubMed]
136. Pinto, S.; Roseberry, A.G.; Liu, H.; Diano, S.; Shanabrough, M.; Cai, X.; Friedman, J.M.; Horvath, T.L. Rapid rewiring of arcuate nucleus feeding circuits by leptin. *Science* **2004**, *304*, 110–115. [CrossRef]

137. Hotamisligil, G.S.; Shargill, N.S.; Spiegelman, B.M. Adipose expression of tumor necrosis factor-alpha: Direct role in obesity-linked insulin resistance. *Science* **1993**, *259*, 87–91. [CrossRef]
138. Deveaux, V.; Cadoudal, T.; Ichigotani, Y.; Teixeira-Clerc, F.; Louvet, A.; Manin, S.; Nhieu, J.T.-V.; Belot, M.P.; Zimmer, A.; Even, P.; et al. Cannabinoid CB2 receptor potentiates obesity-associated inflammation, insulin resistance and hepatic steatosis. *PLoS ONE* **2009**, *4*, e5844. [CrossRef]
139. Muccioli, G.G.; Naslain, D.; Bäckhed, F.; Reigstad, C.S.; Lambert, D.M.; Delzenne, N.M.; Cani, P.D. The endocannabinoid system links gut microbiota to adipogenesis. *Mol. Syst. Biol.* **2010**, *6*, 392. [CrossRef]
140. Konner, A.C.; Bruning, J.C. Selective insulin and leptin resistance in metabolic disorders. *Cell Metab.* **2012**, *16*, 144–152. [CrossRef]
141. Thaler, J.P.; Yi, C.-X.; Schur, E.A.; Guyenet, S.J.; Hwang, B.H.; Dietrich, M.O.; Zhao, X.; Sarruf, D.A.; Izgur, V.; Maravilla, K.R.; et al. Obesity is associated with hypothalamic injury in rodents and humans. *J. Clin. Investig.* **2012**, *122*, 153–162. [CrossRef] [PubMed]
142. Borgquist, A.; Meza, C.; Wagner, E.J. The role of AMP-activated protein kinase in the androgenic potentiation of cannabinoid-induced changes in energy homeostasis. *Am. J. Physiol. Endocrinol. Metab.* **2015**, *308*, E482–E495. [CrossRef]
143. Qiu, J.; Bosch, M.A.; Meza, C.; Navarro, U.-V.; Nestor, C.C.; Wagner, E.J.; Rønnekleiv, O.K.; Kelly, M.J. Estradiol protects proopiomelanocortin neurons against insulin resistance. *Endocrinology* **2018**, *159*, 647–664. [CrossRef] [PubMed]
144. Qiu, J.; Bosch, M.A.; Zhang, C.; Ronnekleiv, O.K.; Kelly, M.J. Estradiol protects neuropeptide Y/agouti-related peptide neurons against insulin resistance in females. *Neuroendocrinology* **2020**, *110*, 105–118. [CrossRef]
145. Hu, M.; Becker, J.B. Effects of sex and estrogen on behavioral sensitization to cocaine in rats. *J. Neurosci.* **2003**, *23*, 693–699. [CrossRef]
146. Evans, S.M.; Foltin, R.W. Does the response to cocaine differ as a function of sex or hormonal status in human and non-human primates? *Horm. Behav.* **2009**, *58*, 13–21. [CrossRef] [PubMed]
147. Vandegrift, B.J.; Hilderbrand, E.R.; Satta, R.; Tai, R.; He, D.; You, C.; Chen, H.; Xu, P.; Coles, C.; Brodie, M.S.; et al. Estrogen Receptor a Regulates Ethanol Excitation of Ventral Tegmental Area Neurons and Binge Drinking in Female Mice. *J. Neurosci.* **2020**, *40*, 5196–5207. [CrossRef] [PubMed]
148. Roberts, D.C.; Bennett, S.A.; Vickers, G.J. The estrous cycle affects cocaine self-administration on a progressive ratio schedule in rats. *Psychopharmacology* **1989**, *98*, 408–411. [CrossRef] [PubMed]
149. Carroll, M.E.; Morgan, A.D.; Lynch, W.J.; Campbell, U.C.; Dess, N.K. Intravenous cocaine and heroin self-administration in rats selectively bred for differential saccharin intake: Phenotype and sex differences. *Psychopharmacology* **2002**, *161*, 304–313. [CrossRef]
150. Becker, J.B.; Hu, M. Sex differences in drug abuse. *Front. Neuroendocrinol.* **2008**, *29*, 36–47. [CrossRef]
151. Richard, J.E.; Lopez-Ferreras, L.; Anderberg, R.H.; Olandersson, K.; Skibicka, K.P. Estradiol is a critical regulator of food-reward behavior. *Psychoneuroendocrinology* **2017**, *78*, 193–202. [CrossRef]
152. Contini, A.; Sanna, F.; Maccioni, P.; Colombo, G.; Argiolas, A. Comparison between male and female rats in a model of self-administration of a chocolate-flavored beverage: Behavioral and neurochemical studies. *Behav. Brain Res.* **2018**, *344*, 28–41. [CrossRef] [PubMed]
153. Mollereau, C.; Parmentier, M.; Mailleux, P.; Butour, J.-L.; Moisand, C.; Chalon, P.; Caput, D.; Vassart, G.; Meunier, J.-C. ORL1, a novel member of the opioid receptor family. *FEBS Lett.* **1994**, *341*, 33–38. [CrossRef]
154. Meunier, J.-C.; Mollereau, C.; Toll, L.; Suaudeau, C.; Moisand, C.; Alvinerie, P.; Butour, J.-L.; Guillemot, J.-C.; Ferrara, P.; Monsarrat, B.; et al. Isolation and structure of the endogenous agonist of opioid receptor-like ORL_1 receptor. *Nature* **1995**, *377*, 532–535. [CrossRef] [PubMed]
155. Reinscheid, R.K.; Nothacker, H.-P.; Bourson, A.; Ardati, A.; Henningsen, R.A.; Bunzow, J.R.; Grandy, D.K.; Langen, H.; Monsma, F.J.; Civelli, O. Orphanin FQ: A neuropeptide that activates an opioidlike G protein-coupled receptor. *Science* **1995**, *270*, 792–794. [CrossRef]
156. Bunzow, J.R.; Saez, C.; Mortrud, M.; Bouvier, C.; Williams, J.T.; Low, M.; Grandy, D.K. Molecular cloning and tissue distribution of a putative member of the rat opioid receptor gene family that is not a μ, δ or kappa opioid receptor type. *FEBS Lett.* **1994**, *347*, 284–288. [CrossRef]
157. Mogil, J.S.; Grisel, J.E.; Reinscheid, R.K.; Civelli, O.; Belknap, J.K.; Grandy, D.K. Orphanin FQ is a functional anti-opioid peptide. *Neuroscience* **1996**, *75*, 333–337. [CrossRef]
158. Sim, L.J.; Xiao, R.Y.; Childers, S.R. Identification of opioid receptor-like (ORL1) peptide-stimulated [^{35}S]GTPgammaS binding in rat brain. *Neuroreport* **1996**, *7*, 729–733. [CrossRef]
159. Ikeda, K.; Watanabe, M.; Ichikawa, T.; Kobayashi, T.; Yano, R.; Kumanishi, T. Distribution of prepro-nociceptin/orphanin FQ mRNA and its receptor mRNA in developing and adult mouse central nervous systems. *J. Comp. Neurol.* **1998**, *399*, 139–151. [CrossRef]
160. Toll, L.; Bruchas, M.R.; Calo, G.; Cox, B.M.; Zaveri, N.T. Nociceptin/orphanin FQ receptor structure, signaling, ligands, functions, and interactions with opioid systems. *Pharmacol. Rev.* **2016**, *68*, 419–457. [CrossRef] [PubMed]
161. Zhang, N.R.; Planer, W.; Siuda, E.R.; Zhao, H.-C.; Stickler, L.; Chang, S.D.; Baird, M.A.; Cao, Y.-Q.; Bruchas, M.R. Serine 363 is required for nociceptin/orphanin FQ opioid receptor (NOPR) desensitization, internalization and arrestin signaling. *J. Biol. Chem.* **2012**, *287*, 42019–42030. [CrossRef]
162. Connor, M.; Vaughan, C.W.; Chieng, B.; Christie, M.J. Nociceptin receptor coupling to a potassium conductance in rat locus coeruleus neurones in vitro. *Br. J. Pharmacol.* **1996**, *119*, 1614–1618. [CrossRef]

163. Connor, M.; Yeo, A.; Henderson, G. The effect of nociceptin on Ca^{2+} channel current and intracellular Ca^{2+} in the SH-SY5Y human neuroblastoma cell line. *Br. J. Pharmacol.* **1996**, *118*, 205–207. [CrossRef] [PubMed]
164. Connor, M.; Christie, M.J. Modulation of Ca^{2+} channel currents of acutely dissociated rat periaqueductal grey neurons. *J. Physiol.* **1998**, *509*, 47–58. [CrossRef]
165. Fukuda, K.; Shoda, T.; Morikawa, H.; Kato, S.; Mima, H.; Mori, K. Activation of phospholipase A2 by the nociceptin receptor expressed in Chinese hamster ovary cells. *J. Neurochem.* **1998**, *71*, 2186–2192. [CrossRef]
166. Yung, L.Y.; Joshi, S.A.; Chan, R.Y.; Chan, J.S.; Pei, G.; Wong, Y.H. GalphaL1 (Galpha14) couples the opioid receptor-like1 receptor to stimulation of phospholipase C. *J. Pharmacol. Exp. Ther.* **1999**, *288*, 232–238. [PubMed]
167. Armstead, W. NOC/oFQ activates PKC and generates superoxide to impair hypotensive cerebrovasodilation after hypoxia/ischemia. *Med. Sci. Monit.* **2002**, *8*, BR0.
168. Norton, C.S.; Neal, C.R.; Kumar, S.; Akil, H.; Watson, S.J. Nociceptin/orphanin FQ and opiod receptor-like receptor mRNA expression in dopamine systems. *J. Comp. Neurol.* **2002**, *444*, 358–368. [CrossRef]
169. Sinchak, K.; Romeo, H.E.; Micevych, P.E. Site-specific estrogen and progestin regulation of orphanin FQ/nociceptin and nociceptin opioid receptor mRNA expression in the female rat limbic hypothalamic system. *J. Comp. Neurol.* **2006**, *496*, 252–268. [CrossRef]
170. Pomonis, J.D.; Billington, C.J.; Levine, A.S. Orphanin FQ, agonist of the orphan opioid receptor, stimulates feeding in rats. *Neuroreport* **1996**, *8*, 369–371. [CrossRef]
171. Polidori, C.; Calo', G.; Ciccocioppo, R.; Guerrini, R.; Regoli, D.; Massi, M. Pharmacological characterization of the nociceptin mediating hyperphagia: Identification of a selective antagonist. *Psychopharmacology* **2000**, *148*, 430–437. [CrossRef]
172. Polidori, C.; de Caro, G.; Massi, M. The hyperphagic effect of nociceptin/orphanin FQ in rats. *Peptides* **2000**, *21*, 1051–1062. [CrossRef]
173. Olszewski, P.K.; Levine, A.S. Minireview: Characterization of influence of central nociceptin/orphanin FQ on consummatory behavior. *Endocrinology* **2004**, *145*, 2627–2632. [CrossRef] [PubMed]
174. Economidou, D.; Policani, F.; Angellotti, T.; Massi, M.; Terada, T.; Ciccocioppo, R. Effect of novel NOP receptor ligands in food intake in rats. *Peptides* **2006**, *27*, 775–783. [CrossRef]
175. Matsushita, H.; Ishihara, A.; Mashiko, S.; Tanaka, T.; Kanno, T.; Iwaasa, H.; Ohta, H.; Kanatani, A. Chronic intracerebroventricular infusion of nociceptin/orphanin FQ produces body weight gain by affecting both feeding and energy metabolism. *Endocrinology* **2009**, *150*, 2668–2673. [CrossRef] [PubMed]
176. Farhang, B.; Pietruszewski, L.; Lutfy, K.; Wagner, E.J. The role of the NOP receptor in regulating food intake, meal pattern, and the excitability of proopiomelanocortin neurons. *Neuropharmacology* **2010**, *59*, 190–200. [CrossRef] [PubMed]
177. Wagner, E.J.; Rønnekleiv, O.K.; Grandy, D.K.; Kelly, M.J. The peptide orphanin FQ inhibits β-endorphin neurons and neurosecretory cells in the hypothalamic arcuate nucleus by activating an inwardly-rectifying K^{+} conductance. *Neuroendocrinology* **1998**, *67*, 73–82. [CrossRef]
178. Borgquist, A.; Kachani, M.; Tavitian, N.; Sinchak, K.; Wagner, E.J. Estradiol negatively modulates the pleiotropic actions of orphanin FQ/nociceptin at proopiomelanocortin synapses. *Neuroendocrinology* **2013**, *98*, 60–72. [CrossRef]
179. Hernandez, J.; Perez, L.; Soto, R.; Le, N.; Gastelum, C.; Wagner, E.J. Nociceptin/orphanin FQ neurons in the Arcuate Nucleus and Ventral Tegmental Area Act via Nociceptin Opioid Peptide Receptor Signaling to Inhibit Proopiomelanocortin and A 10 Dopamine Neurons and Thereby Modulate Ingestion of Palatable Food. *Physiol. Behav.* **2020**, *228*. [CrossRef]
180. Bewick, G.A.; Dhillo, W.S.; Darch, S.J.; Murphy, K.G.; Gardiner, J.V.; Jethwa, P.H.; Kong, W.M.; Ghatei, M.A.; Bloom, S.R. Hypothalamic cocaine- and amphetamine-regulated transcript (CART) and agouti-related protein (AgRP) neurons coexpress the NOP1 receptor and nociceptin alters CART and AgRP release. *Endocrinology* **2005**, *146*, 3526–3534. [CrossRef] [PubMed]
181. Jais, A.; Paeger, L.; Sotelo-Hitschfeld, T.; Bremser, S.; Prinzensteiner, M.; Klemm, P.; Mykytiuk, V.; Widdershooven, P.J.M.; Vesting, A.J.; Grzelka, K.; et al. $PNOC^{ARC}$ neurons promote hyperphagia and obesity upon high-fat feeding. *Neuron* **2020**, *106*, 1009–1025. [CrossRef]
182. Parsons, M.P.; Burt, J.; Cranford, A.; Alberto, C.; Zipperlen, K.; Hirasawa, M. Nociceptin Induces Hypophagia in the Perifornical and Lateral Hypothalamic Area. *PLoS ONE* **2012**, *7*. [CrossRef]
183. Murphy, N.P.; Ly, N.T.; Maidment, N.T. Intracerebroventricular orphanin FQ/nociceptin suppresses dopamine release in the nucleus accumbens of anaesthetized rats. *Neuroscience* **1996**, *75*, 1–4. [CrossRef]
184. Murphy, N.P.; Maidment, N.T. Orphanin FQ/nociceptin modulation of mesolimbic dopamine transmission determined by microdialysis. *J. Neurochem.* **1999**, *73*, 179–186. [CrossRef] [PubMed]
185. Murphy, N.P.; Lee, Y.; Maidment, N.T. Orphanin FQ/nociceptin blocks acquisition of morphine place preference. *Brain Res.* **1999**, *832*, 168–170. [CrossRef]
186. Di Giannuario, A.; Pieretti, S.; Catalani, A.; Loizzo, A. Orphanin FQ reduces morphine-induced dopamine release in the nucleus accumbens: A microdialysis study in rats. *Neurosci. Lett.* **1999**, *272*, 183–186. [CrossRef]
187. Zheng, F.; Grandy, D.K.; Johnson, S.W. Actions of orphanin FQ/nociceptin on rat ventral tegmental area neurons in vitro. *Br. J. Pharmacol.* **2002**, *136*, 1065–1071. [CrossRef] [PubMed]
188. Olianas, M.C.; Dedoni, S.; Boi, M.; Onali, P. Activation of nociceptin/orphanin FQ-NOP receptor system inhibits tyrosine hydroxylase phosphorylation, dopamine synthesis, and dopamine D(1) receptor signaling in rat nucleus accumbens and dorsal striatum. *J. Neurochem.* **2008**, *107*, 544–556. [CrossRef]

189. Emmerson, P.J.; Miller, R.J. Pre- and postsynaptic actions of opioid and orphan opioid agonists in the rat arcuate nucleus and ventromedial hypothalamus in vitro. *J. Physiol.* **1999**, *517*, 431–445. [CrossRef]
190. Chee, M.J.; Price, C.J.; Statnick, M.A.; Colmers, W.F. Nociceptin/orphanin FQ suppresses the excitability of neurons in the ventromedial nucleus of the hypothalamus. *J. Physiol.* **2011**, *589*, 3103–3114. [CrossRef] [PubMed]
191. Conde, K.; Meza, C.; Kelly, M.J.; Sinchak, K.; Wagner, E.J. Estradiol rapidly attenuates ORL-1 receptor-mediated inhibition of proopiomelanocortin neurons via G_q-coupled, membrane-initiated signaling. *Neuroendocrinology* **2016**, *103*, 787–805. [CrossRef] [PubMed]
192. Borgquist, A.; Rivas, V.M.; Kachani, M.; Sinchak, K.; Wagner, E.J. Gonadal steroids differentially modulate the actions of orphanin FQ/nociceptin at a physiologically relevant circuit controlling female sexual receptivity. *J. Neuroendocrinol.* **2014**, *26*, 329–340. [CrossRef]
193. Koizumi, M.; Cagniard, B.; Murphy, N.P. Endogenous nociceptin modulates diet preference independent of motivation and reward. *Physiol. Behav.* **2009**, *97*, 1–13. [CrossRef]
194. Toledo, M.A.; Pedregal, C.; Lafuente, C.; Diaz, N.; Martinez-Grau, M.A.; Jiménez, A.; Benito, A.; Torrado, A.; Mateos, C.; Joshi, E.M.; et al. Discovery of a novel series of orally active nociceptin/orphanin FQ (NOP) receptor antagonists based on a dihydrospiro(piperidine-4,7′-thieno[2,3-c]pyran) scaffold. *J. Med. Chem.* **2014**, *57*, 3418–3429. [CrossRef]
195. Statnick, M.A.; Chen, Y.; Ansonoff, M.A.; Witkin, J.M.; Rorick-Kehn, L.; Suter, T.M.; Song, M.; Hu, C.; Lafuente, C.; Jiménez, A.; et al. A Novel Nociceptin Receptor Antagonist LY2940094 Inhibits Excessive Feeding Behavior in Rodents: A Possible Mechanism for the Treatment of Binge Eating Disorder. *J. Pharmacol. Exp. Ther.* **2016**, *356*, 493–502. [CrossRef] [PubMed]
196. Hardaway, J.A.; Jensen, J.; Kim, M.; Mazzone, C.M.; Sugam, J.A.; Diberto, J.F.; Lowery-Gionta, E.G.; Hwa, L.S.; Pleil, K.E.; Bulik, C.M.; et al. Nociceptin receptor antagonist SB 61211 decreases high fat diet binge eating. *Behav. Brain Res.* **2016**, *307*, 25–34. [CrossRef]
197. Rudecki, A.P.; Gray, S.L. PACAP in the defense of energy homeostasis. *Trends Endocrinol. Metab.* **2016**, *27*, 620–632. [CrossRef] [PubMed]
198. Arimura, A.; Somogyvari-Vigh, A.; Miyata, A.; Mizuno, K.; Coy, D.H.; Kitada, C. Tissue Distribution of PACAP as Determined by RIA: Highly Abundant in the Rat Brain and Testes. *Endocrinology* **1991**, *129*, 2787–2789. [CrossRef] [PubMed]
199. Ghatei, M.A.; Takahashi, K.; Suzuki, Y.; Gardiner, J.; Jones, P.M.; Bloom, S.R. Distribution, molecular characterization of pituitary adenylate cyclase-activating polypeptide and its precursor encoding messenger RNA in human and rat tissues. *J. Endocrinol.* **1993**, *136*, 159–166. [CrossRef]
200. Vaudry, D.; Falluel-Morel, A.; Bourgault, S.; Basille, M.; Burel, D.; Wurtz, O.; Fournier, A.; Chow, B.K.C.; Hashimoto, H.; Galas, L.; et al. Pituitary adenylate cyclase-activating polypeptide and its receptors: 20 years after the discovery. *Pharmacol. Rev.* **2009**, *61*, 283–357. [CrossRef]
201. Gottschall, P.E.; Tatsuno, I.; Miyata, A.; Arimura, A. Characterization and distribution of binding sites for the hypothalamic peptide, pituitary adenylate cyclase-activating polypeptide. *Endocrinology* **1990**, *127*, 272–277. [CrossRef]
202. Lam, H.C.; Takahashi, K.; Ghatei, M.A.; Kanse, S.M.; Polak, J.M.; Bloom, S.R. Binding sites of a novel neuropeptide pituitary-adenylate-cyclase-activating polypeptide in the rat brain and lung. *Eur. J. Biochem.* **1990**, *193*, 725–729. [CrossRef]
203. Nishimoto, M.; Furuta, A.; Aoki, S.; Kudo, Y.; Miyakawa, H.; Wada, K. PACAP/PAC1 autocrine system promotes proliferation and astrogenesis in neural progenitor cells. *Glia* **2007**, *55*, 317–327. [CrossRef]
204. Lutz-Bucher, B.; Monnier, D.; Koch, B. Evidence for the presence of receptors for pituitary adenylate cyclase-activating polypeptide in the neurohypophysis that are positively coupled to cyclic AMP formation and neurohypophyseal hormone secretion. *Neuroendocrinology* **1996**, *64*, 153–161. [CrossRef]
205. Uchimura, D.; Katafuchi, T.; Hori, T.; Yanaihara, N. Facilitatory effects of pituitary adenylate cyclase activating polypeptide (PACAP) on neurons in the magnocellular portion of the rat hypothalamic paraventricular nucleus (PVN) in vitro. *J. Neuroendocrinol.* **1996**, *8*, 137–143. [CrossRef]
206. Shibuya, I.; Kabashima, N.; Tanaka, K.; Setiadji, V.S.; Noguchi, J.; Harayama, N.; Ueta, Y.; Yamashita, H. Patch-clamp analysis of the mechanism of PACAP-induced excitation in rat supraoptic neurones. *J. Neuroendocrinol.* **1998**, *10*, 759–768. [CrossRef] [PubMed]
207. Shibuya, I.; Noguchi, J.; Tanaka, K.; Harayama, N.; Inoue, U.; Kabashima, N.; Ueta, Y.; Hattori, Y.; Yamashita, H. PACAP increases the cytosolic Ca^{2+} concentration and stimulates somatodendritic vasopressin release in rat supraoptic neurons. *J. Neuroendocrinol.* **1998**, *10*, 31–42. [CrossRef]
208. Ichinose, M.; Asai, M.; Sawada, M. Activation of outward current by pituitary adenylate cyclase activating polypeptide in mouse microglial cells. *J. Neurosci. Res.* **1998**, *51*, 382–390. [CrossRef]
209. Delgado, M.; De la Fuente, M.; Martinez, C.; Gomariz, R.P. Pituitary adenylate cyclase-activating polypeptides (PACAP27 and PACAP38) inhibit the mobility of murine thymocytes and splenic lymphocytes: Comparison with VIP and implication of cAMP. *J. Neuroimmunol.* **1995**, *62*, 137–146. [CrossRef]
210. Garrido, E.; Delgado, M.; Martinez, C.; Gomariz, R.P.; De la Fuente, M. Pituitary adenylate cyclase-activating polypeptide (PACAP38) modulates lymphocyte and macrophage functions: Stimulation of adherence and opposite effect on mobility. *Neuropeptides* **1996**, *30*, 583–595. [CrossRef]

211. Martinez, C.; Delgado, M.; Pozo, D.; Leceta, J.; Calvo, J.R.; Ganea, D.; Gomariz, R.P. Vasoactive intestinal peptide and pituitary adenylate cyclase-activating polypeptide modulate endotoxin-induced IL-6 production by murine peritoneal macrophages. *J. Leukoc. Biol.* **1998**, *63*, 591–601. [CrossRef] [PubMed]
212. Delgado, M.; Martinez, C.; Pozo, D.; Calvo, J.R.; Leceta, J.; Ganea, D.; Gomariz, R.P. Vasoactive intestinal peptide (VIP) and pituitary adenylate cyclase-activation polypeptide (PACAP) protect mice from lethal endotoxemia through the inhibition of TNF-alpha and IL-6. *J. Immunol.* **1999**, *162*, 1200–1205.
213. Delgado, M.; Munoz-Elias, E.J.; Gomariz, R.P.; Ganea, D. VIP and PACAP inhibit IL-12 production in LPS-stimulated macrophages. Subsequent effect on IFNgamma synthesis by T cells. *J. Neuroimmunol.* **1999**, *96*, 167–181. [CrossRef]
214. Delgado, M.; Munoz-Elias, E.J.; Gomariz, R.P.; Ganea, D. Vasoactive intestinal peptide and pituitary adenylate cyclase-activating polypeptide enhance IL-10 production by murine macrophages: In vitro and in vivo studies. *J. Immunol.* **1999**, *162*, 1707–1716. [PubMed]
215. Villalba, M.; Bockaert, J.; Journot, L. Pituitary adenylate cyclase-activating polypeptide (PACAP-38) protects cerebellar granule neurons from apoptosis by activating the mitogen-activated protein kinase (MAP kinase) pathway. *J. Neurosci.* **1997**, *17*, 83–90. [CrossRef]
216. Vaudry, D.; Basille, M.; Anouar, Y.; Fournier, A.; Vaudry, H.; Gonzalez, B.J. The neurotrophic activity of PACAP on rat cerebellar granule cells is associated with activation of the protein kinase A pathway and c-fos gene expression. *Ann. N. Y. Acad. Sci.* **1998**, *865*, 92–99. [CrossRef]
217. Vaudry, D.; Gonzalez, B.J.; Basille, M.; Anouar, Y.; Fournier, A.; Vaudry, H. Pituitary adenylate cyclase-activating polypeptide stimulates both c-fos gene expression and cell survival in rat cerebellar granule neurons through activation of the protein kinase A pathway. *Neuroscience* **1998**, *84*, 801–812. [CrossRef]
218. Tatsuno, I.; Gottschall, P.E.; Arimura, A. Inhibition of mitogen-stimulated proliferation of murine splenocytes by a novel neuropeptide, pituitary adenylate cyclase activating polypeptide: A comparative study with vasoactive intestinal peptide. *Endocrinology* **1991**, *128*, 728–734. [CrossRef]
219. Mei, Y.A. High-voltage-activated calcium current and its modulation by dopamine D4 and pituitary adenylate cyclase activating polypeptide receptors in cerebellar granule cells. *Zhongguo Yao Li Xue Bao* **1999**, *20*, 3–9.
220. Roy, A.; Derakhshan, F.; Wilson, R.J.A. Stress peptide PACAP engages multiple signaling pathways within the carotid body to initiate excitatory responses in respiratory and sympathetic chemosensory afferents. *Am. J. Physiol. Regul. Integr. Comp. Physiol.* **2013**, *304*, R1070–R1084. [CrossRef] [PubMed]
221. Gupte, R.P.; Kadunganattil, S.; Shepherd, A.J.; Merrill, R.; Planer, W.; Bruchas, M.R.; Strack, S.; Mohapatra, D.P. Convergent phosphomodulation of the major neuronal dendritic potassium channel Kv4.2 by pituitary adenylate cyclase-activating polypeptide. *Neuropharmacology* **2016**, *101*, 291–308. [CrossRef]
222. Komatsu, M.; Schermerhorn, T.; Straub, S.G.; Sharp, G.W. Pituitary adenylate cyclase-activating peptide, carbachol, and glucose stimulate insulin release in the absence of an increase in intracellular Ca^{2+}. *Mol. Pharmacol.* **1996**, *50*, 1047–1054. [PubMed]
223. Yada, T.; Sakurada, M.; Ishihara, H.; Nakata, M.; Shioda, S.; Yaekura, K.; Hamakawa, N.; Yanagida, K.; Kikuchi, M.; Oka, Y. Pituitary adenylate cyclase-activating polypeptide (PACAP) is an islet substance serving as an intra-islet amplifier of glucose-induced insulin secretion in rats. *J. Physiol.* **1997**, *505*, 319–328. [CrossRef]
224. Yada, T.; Sakurada, M.; Nakata, M.; Yaekura, K.; Kikuchi, M. PACAP as low as 10(-13) M raises cytosolic Ca^{2+} activity in pancreatic B-cells by augmenting Ca^{2+} influx through L-type Ca^{2+} channels to trigger insulin release. *Adv. Exp. Med. Biol.* **1997**, *426*, 165–171.
225. Filipsson, K.; Sundler, F.; Ahren, B. PACAP is an islet neuropeptide which contributes to glucose-stimulated insulin secretion. *Biochem. Biophys. Res. Commun.* **1999**, *256*, 664–667. [CrossRef] [PubMed]
226. Vu, J.P.; Goyal, D.; Luong, L.; Oh, S.; Sanghu, R.; Norris, J.; Parsons, W.; Pisegna, J.R.; Germano, P.M. PACAP intraperitoneal treatment suppresses appetite and food intake via PAC1 receptor in mice by inhibiting ghrelin and increasing GLP-1 and leptin. *Am. J. Physiol. Gastrointest. Liver Physiol.* **2015**, *309*, G816–G825. [CrossRef] [PubMed]
227. Hibino, H.; Inanobe, A.; Furutani, K.; Murakami, S.; Findlay, I.; Kurachi, Y. Inwardly rectifying potassium channels: Their structure, function, and physiological roles. *Physiol. Rev.* **2010**, *90*, 291–366. [CrossRef]
228. McTaggart, J.S.; Clark, R.H.; AShcroft, F.M. The role of the KATP channel in glucose homeostasis in health and disease: More than meets the islet. *J. Physiol.* **2010**, *588*, 3201–3209. [CrossRef] [PubMed]
229. Wang, G.; Pan, J.; Tan, Y.-Y.; Sun, X.-K.; Zhang, Y.-F.; Zhou, H.-Y.; Ren, R.-J.; Wang, X.-J.; Chen, S.-D. Neuroprotective effects of PACAP27 in mice model of Parkinson's disease involved in the modulation of K(ATP) subunits and D2 receptors in the striatum. *Neuropeptides* **2008**, *42*, 267–276. [CrossRef]
230. Koide, M.; Syed, A.U.; Braas, K.M.; May, V.; Wellman, G.C. Pituitary adenylate cyclase activating polypeptide (PACAP) dilates cerebellar arteries through activation of large-conductance Ca(2+)-activated (BK) and ATP-sensitive (K ATP) K (+) channels. *J. Mol. Neurosci.* **2014**, *54*, 443–450. [CrossRef]
231. Tanida, M.; Hayata, A.; Shintani, N.; Yamamoto, N.; Kurata, Y.; Shibamoto, T.; Morgan, D.A.; Rahmouni, K.; Hashimoto, H. Central PACAP mediates the sympathetic effects of leptin in a tissue-specific manner. *Neuroscience* **2013**, *238*, 297–304. [CrossRef]
232. Mungan, Z.; Hammer, R.A.; Akarca, U.S.; Komaki, G.; Ertan, A.; Arimura, A. Effect of PACAP on gastric acid secretion in rats. *Peptides* **1995**, *16*, 1051–1056. [CrossRef]
233. Mirfendereski, S.; Tobin, G.; Hakanson, R.; Ekstrom, J. Pituitary adenylate cyclase activating peptide (PACAP) in salivary glands of the rat: Origin, and secretory and vascular effects. *Acta Physiol. Scand.* **1997**, *160*, 15–22. [CrossRef]

234. Li, P.; Chang, T.M.; Coy, D.; Chey, W.Y. Inhibition of gastric acid secretion in rat stomach by PACAP is mediated by secretin, somatostatin, and PGE(2). *Am. J. Physiol. Gastrointest. Liver Physiol.* **2000**, *278*, G121–G127. [CrossRef]
235. Kuwahara, A.; Kuwahara, Y.; Mochizuki, T.; Yanaihara, N. Action of pituitary adenylate cyclase-activating polypeptide on ion transport in guinea pig distal colon. *Am. J. Physiol.* **1993**, *264*, G433–G441. [CrossRef]
236. Takeuchi, K.; Takehara, K.; Kato, S.; Yagi, K. PACAPs stimulate duodenal bicarbonate secretion at PACAP receptors in the rat. *Am. J. Physiol.* **1997**, *272*, G646–G653. [CrossRef]
237. Katsoulis, S.; Schmidt, W.E.; Schwarzhoff, R.; Folsch, U.R.; Jin, J.G.; Grider, J.R.; Makhlouf, G.M. Inhibitory transmission in guinea pig stomach mediated by distinct receptors for pituitary adenylate cyclase-activating peptide. *J. Pharmacol. Exp. Ther.* **1996**, *278*, 199–204. [PubMed]
238. Robberecht, P.; De Neef, P.; Lefebvre, R.A. Influence of selective VIP receptor agonists in the rat gastric fundus. *Eur. J. Pharmacol.* **1998**, *359*, 77–80. [CrossRef]
239. Ozawa, M.; Aono, M.; Moriga, M. Central effects of pituitary adenylate cyclase activating polypeptide (PACAP) on gastric motility and emptying in rats. *Dig. Dis. Sci.* **1999**, *44*, 735–743. [CrossRef]
240. Schworer, H.; Katsoulis, S.; Creutzfeldt, W.; Schmidt, W.E. Pituitary adenylate cyclase activating peptide, a novel VIP-like gut-brain peptide, relaxes the guinea-pig taenia caeci via apamin-sensitive potassium channels. *Naunyn-Schmiedebergs Arch. Pharmacol.* **1992**, *346*, 511–514. [CrossRef]
241. Katsoulis, S.; Clemens, A.; Schworer, H.; Creutzfeldt, W.; Schmidt, W.E. Pituitary adenylate cyclase activating polypeptide (PACAP) is a potent relaxant of the rat ileum. *Peptides* **1993**, *14*, 587–592. [CrossRef]
242. Grider, J.R.; Katsoulis, S.; Schmidt, W.E.; Jin, J.G. Regulation of the descending relaxation phase of intestinal peristalsis by PACAP. *J. Auton. Nerv. Syst.* **1994**, *50*, 151–159. [CrossRef]
243. Ekblad, E.; Sundler, F. Distinct receptors mediate pituitary adenylate cyclase-activating peptide- and vasoactive intestinal peptide-induced relaxation of rat ileal longitudinal muscle. *Eur. J. Pharmacol.* **1997**, *334*, 61–66. [CrossRef]
244. Yokota, C.; Kawai, K.; Ohashi, S.; Watanabe, Y.; Yamashita, K. PACAP stimulates glucose output from the perfused rat liver. *Peptides* **1995**, *16*, 55–60. [CrossRef]
245. Sekiguchi, Y.; Kasai, K.; Hasegawa, K.; Suzuki, Y.; Shimoda, S. Glycogenolytic activity of pituitary adenylate cyclase activating polypeptide (PACAP) in vivo and in vitro. *Life Sci.* **1994**, *55*, 1219–1228. [CrossRef]
246. Masuo, Y.; Ohtaki, T.; Masuda, Y.; Nagai, Y.; Suno, M.; Tsuda, M.; Fujino, M. Autoradiographic distribution of pituitary adenylate cyclase activating polypeptide (PACAP) binding sites in the rat brain. *Neurosci. Lett.* **1991**, *126*, 103–106. [CrossRef]
247. Hawke, Z.; Ivanov, T.R.; Bechtold, D.A.; Dhillon, H.; Lowell, B.B.; Luckman, S.M. PACAP neurons in the hypothalamic ventromedial nucleus are targets of central leptin signaling. *J. Neurosci.* **2009**, *29*, 14828–14835. [CrossRef] [PubMed]
248. Tomimoto, S.; Ojika, T.; Shintani, N.; Hashimoto, H.; Hamagami, K.-I.; Ikeda, K.; Nakata, M.; Yada, T.; Sakurai, Y.; Shimada, T.; et al. Markedly reduced white adipose tissue and increased insulin sensitivity in adcyap1-deficient mice. *J. Pharmacol. Sci.* **2008**, *107*, 41–48. [CrossRef]
249. Segal, J.P.; Stallings, N.R.; Lee, C.E.; Zhao, L.; Socci, N.; Viale, A.; Harris, T.M.; Soares, M.B.; Childs, G.; Elmquist, J.K.; et al. Use of laser-capture microdissection for the identification of marker genes for the ventromedial hypothalamic nucleus. *J. Neurosci.* **2005**, *25*, 4181–4188. [CrossRef]
250. Mounien, L.; Do Rego, J.-C.; Bizet, P.; Boutelet, I.; Gourcerol, G.; Fournier, A.; Brabet, P.; Costentin, J.; Vaudry, H.; Jegou, S. Pituitary adenylate cyclase-activating polypeptide inhibits food intake in mice through activation of the hypothalamic melanocortin system. *Neuropsychopharmacol* **2009**, *34*, 424–435. [CrossRef] [PubMed]
251. Khodai, T.; Nunn, N.; Worth, A.A.; Feetham, C.H.; Belle, M.D.C.; Piggins, H.D.; Luckman, S.M. PACAP neurons in the ventromedial hypothalamic nucleus are glucose inhibited and their selective activation induces hyperglycemia. *Front. Endocrinol.* **2018**, *9*, 632. [CrossRef]
252. Lindberg, D.; Chen, P.; Li, C. Conditional viral tracing reveals that steroidogenic factor 1-positive neurons of the dorsomedial subdivision of the ventromedial hypothalamus project to the autonomic centers of the hypothalamus and hindbrain. *J. Comp. Neurol.* **2013**, *521*, 3167–3190. [CrossRef]
253. Resch, J.M.; Boisvert, J.P.; Hourigan, A.E.; Muller, C.R.; Yi, S.S.; Choi, S. Stimulation of the hypothalamic ventromedial nuclei by pituitary adenylate cyclase-activating polypeptide induces hypophagia and thermogenesis. *Am. J. Physiol. Regul. Integr. Comp. Physiol.* **2011**, *301*, R1625–R1634. [CrossRef]
254. Resch, J.M.; Maunze, B.; Gerhardt, A.K.; Magnuson, S.K.; Phillips, K.A.; Choi, S. Intrahypothalamic pituitary adenylate cyclase-activating polypeptide regulates energy balance via site-specific actions on feeding and metabolism. *Am. J. Physiol. Endocrinol. Metab.* **2013**, *305*, E1452–E1463. [CrossRef]
255. Mounien, L.; Bizet, P.; Boutelet, I.; Gourcerol, G.; Fournier, A.; Vaudry, H.; Jégou, S. Pituitary adenylate cyclase-activating polypeptide directly modulates the activity of proopiomelanocortin neurons in the rat arcuate nucleus. *Neuroscience* **2006**, *143*, 155–163. [CrossRef]
256. Ramikie, T.S.; Ressler, K.J. Stress-related disorders, pituitary adenylate cyclase-activating peptide (PACAP)ergic system, and sex differences. *Dialogues Clin. Neurosci.* **2016**, *18*, 403–413. [PubMed]

257. Hurley, M.M.; Maunze, B.; Block, M.E.; Frenkel, M.M.; Reily, M.J.; Kim, E.; Chen, Y.; Li, Y.; Baker, D.A.; Liu, Q.-S.; et al. Pituitary adenylate-cyclase activating polypeptide regulates hunger- and palatability-induced binge eating. *Front. Neurosci.* **2016**, *10*, 383. [CrossRef] [PubMed]
258. Hurley, M.M.; Robble, M.R.; Callan, G.; Choi, S.; Wheeler, R.A. Pituitary adenylate cyclase-activiating polypeptide (PACAP) acts in the nucleus accumbens to reduce hedonic drive. *Int. J. Obes.* **2019**, *43*, 928–932. [CrossRef] [PubMed]

International Journal of
Molecular Sciences

MDPI

Review

The Microbiota and the Gut–Brain Axis in Controlling Food Intake and Energy Homeostasis

Marina Romaní-Pérez †, Clara Bullich-Vilarrubias †, Inmaculada López-Almela, Rebeca Liébana-García, Marta Olivares and Yolanda Sanz *

Microbial Ecology, Nutrition & Health Research Unit, Institute of Agrochemistry and Food Technology, National Research Council (IATA-CSIC), 46980 Valencia, Spain; marina.romani@iata.csic.es (M.R.-P.); clarabullich@iata.csic.es (C.B.-V.); inma.lopez@iata.csic.es (I.L.-A.); reliegar@iata.csic.es (R.L.-G.); m.olivares@iata.csic.es (M.O.)

* Correspondence: yolsanz@iata.csic.es

† These authors contribute equally to this work.

Abstract: Obesity currently represents a major societal and health challenge worldwide. Its prevalence has reached epidemic proportions and trends continue to rise, reflecting the need for more effective preventive measures. Hypothalamic circuits that control energy homeostasis in response to food intake are interesting targets for body-weight management, for example, through interventions that reinforce the gut-to-brain nutrient signalling, whose malfunction contributes to obesity. Gut microbiota–diet interactions might interfere in nutrient sensing and signalling from the gut to the brain, where the information is processed to control energy homeostasis. This gut microbiota–brain crosstalk is mediated by metabolites, mainly short chain fatty acids, secondary bile acids or amino acids-derived metabolites and subcellular bacterial components. These activate gut–endocrine and/or neural-mediated pathways or pass to systemic circulation and then reach the brain. Feeding time and dietary composition are the main drivers of the gut microbiota structure and function. Therefore, aberrant feeding patterns or unhealthy diets might alter gut microbiota–diet interactions and modify nutrient availability and/or microbial ligands transmitting information from the gut to the brain in response to food intake, thus impairing energy homeostasis. Herein, we update the scientific evidence supporting that gut microbiota is a source of novel dietary and non-dietary biological products that may beneficially regulate gut-to-brain communication and, thus, improve metabolic health. Additionally, we evaluate how the feeding time and dietary composition modulate the gut microbiota and, thereby, the intraluminal availability of these biological products with potential effects on energy homeostasis. The review also identifies knowledge gaps and the advances required to clinically apply microbiome-based strategies to improve the gut–brain axis function and, thus, combat obesity.

Keywords: microbiota; gut–brain axis; nutrient sensing; food intake and obesity

Citation: Romaní-Pérez, M.; Bullich-Vilarrubias, C.; López-Almela, I.; Liébana-García, R.; Olivares, M.; Sanz, Y. The Microbiota and the Gut–Brain Axis in Controlling Food Intake and Energy Homeostasis. *Int. J. Mol. Sci.* **2021**, *22*, 5830. https://doi.org/10.3390/ijms22115830

Academic Editors: Rosalía Rodríguez-Rodríguez and Cristina Miralpeix

Received: 6 April 2021
Accepted: 26 May 2021
Published: 29 May 2021

Publisher's Note: MDPI stays neutral with regard to jurisdictional claims in published maps and institutional affiliations.

1. Gut and Brain Control of Energy Homeostasis

The gastrointestinal tract is a huge sensory organ able to transmit nutrient-related information to the brain where diverse endocrine and neural inputs converge to ultimately control feeding behaviour and whole-body energy homeostasis through efferent outputs. Nutrient sensors are receptors that bind molecules derived from the macronutrient digestion. They are located in enteroendocrine cells (EECs), which are the primary chemosensory cells in the gut, and are in direct contact with the luminal environment [1]. The activation of nutrient sensors of EECs initiates the secretion of gut hormones that in turn trigger the downstream processes required to maintain energy homeostasis postprandially. The most extensively studied gut hormones are cholecystokinin (CCK), gastric inhibitory polypeptide (GIP), mainly secreted in the upper part of the intestine, and glucagon-like peptide-1 (GLP-1) and peptide tyrosine tyrosine (PYY), mainly secreted in the distal part [2]. EECs specifically sense carbohydrates, proteins and lipids through a diverse repertoire of nutrient sensors. Among others, sodium/glucose cotransporter 1 (SGTL1) mediates carbohydrate sensing, mainly sensed in the form of glucose, and, then, induces GIP and GLP-1 secretion; calcium-sensing receptor (CaSR) senses dietary amino acids and, then, secretes CCK and GLP-1; and G-protein coupled receptor (GPR) 120 (FFAR4), GPR40 (FFAR1) and GPR119 sense products from lipid digestion and induce the secretion of CCK, GIP, GLP-1 and PYY [1,3]. EECs are in close proximity to vagal afferent nerve terminals in the lamina propria, which express intestinal hormone receptors. Enteroendocrine hormones secreted after a meal activate nutrient-sensing signalling via endocrine routes, when gut hormones reach the brain or other organs through systemic circulation, or via paracrine routes, when hormones stimulate vagal afferents nearby in the intestinal mucosa and, then, the signal reaches the brain.

The arcuate nucleus (ARC) of the hypothalamus is accessible to humoral signals since is not fully protected by the blood–brain barrier [4]. The ARC contains two subpopulations of neurons; those expressing anorexigenic propiomelanocortin (POMC), the precursor of α-melanocyte-stimulating hormone (αMSH), and the cocaine and amphetamine regulated transcript (CART); and those neurons expressing the agouti gene-related peptide (AgRP) and neuropeptide Y (NPY) [5]. The activation of the POMC/CART neurons after feeding induces the release of α-melanocyte stimulating hormone (α-MSH) that binds to melanocortin 4 receptor (MC4R) in the paraventricular nucleus (PVN) where the information is integrated to suppress food intake and regulate body weight [6]. Thus, the increased levels of gut hormones such as CCK, GLP1 or PYY after a meal reach, via endocrine routes, the ARC to suppress food intake [7]. Via paracrine routes, gut hormones can stimulate vagal afferents [8], which are bipolar neurons whose somas converge in the nodose ganglion and their proximal extensions terminate in the nucleus of the solitary tract (NTS) in the brainstem [9]. This, together with the hypothalamus, represents an important integrative gut–brain hub. The NTS also contains POMC neurons and, through monoamine neurotransmission, transmits sensory information to upstream brain areas including the PVN and the dorsal vagal complex (DVC) that send efferent outputs involved in the vago–vagal reflex [5]. In summary, there are different paths for transmitting nutrient signals to the brain. The hypothalamus senses nutrients through the action of enteroendocrine hormones released in the intestine, which reach the brain by humoral pathways [10–12], that is through the circulatory system, or by paracrine pathways, that is by activating the nerve terminals of intestinal vagal afferents, which are the main focus of the present review.

The increased intake of energy-dense and palatable foods impairs the brain circuits controlling energy homeostasis, whose deficient response to nutrient signals alters feeding behaviour, which contributes to obesity. Accordingly, the restoration of nutrient signalling via the gut–brain axis represents a promising strategy to improve the central control of energy homeostasis in response to meals and, thus, help combat obesity [13–15]. The gut microbiota is a biological factor that might directly or indirectly influence nutrient-sensing and, theoretically, its modulation could aid in the restoration of gut-to-brain communication and maintaining energy homeostasis. The role of gut microbiota in obesity has been proven through faecal transplantation experiments, which produce the metabolic phenotype of the donor in the recipient organism [16]; however, the mechanisms by which microbiota influence energy homeostasis and body-weight regulation are not yet fully understood. Western diets, rich in saturated fat and simple sugars, alter the gut ecosystem reducing bacterial diversity and increasing the abundance of potential pathogens [17]. This, in turn, could alter the metabolism of macronutrients and, thereby, the ligands available for nutrient sensors in the intestinal lumen as well as the presence of structural microbial components that might also act as ligands of sensors that mediate the gut-to-brain communication. Sensors of microbially produced metabolites and bacterial components are located in EECs, vagal afferents and, occasionally, in the hypothalamus and can be activated by ligands reaching the brain through the systemic circulation. These receptors sense microbial-derived metabolites such as short chain fatty acids (SCFAs), secondary bile acids (BAs) and amino acid-derived metabolites and subcellular bacterial components such as caseinolytic peptidase B (ClpB), lipopolysaccharide (LPS) or muramyldipeptide (MDP).

Here, we review the role of dietary factors, including feeding patterns and diet composition, in modulating gut microbiota structure and function and potentially affecting gut-to-brain nutrient sensing, mainly via endocrine and neural routes. We also compile evidence of microbial molecules able to modulate specifically hypothalamic-mediated control of the energy homeostasis, placing special emphasis on their role in regulating food intake and, thereby, obesity.

2. Circadian Rhythms, Eating Patterns and Gut Microbiota in Energy Homeostasis Control

Most of the physiological functions display circadian rhythms. At cellular level, these are governed by clock proteins that generate self-sustained daily oscillations of the biological processes that allow cells to anticipate and thus to optimally respond and adapt to environmental variations [18,19]. In mammals, most cells throughout the body express clock genes that elicit autonomous circadian oscillations [20]. The hypothalamic clock located in the suprachiasmatic nucleus acts as a master pacemaker that synchronizes the secondary clocks [21,22]. Eating behaviour and energy metabolism show a well-defined circadian pattern through the day. In humans and rodents, the timed feeding patterns are governed by metabolic hormones and nutrients [23] and shift certain secondary clocks without affecting the suprachiasmatic nucleus, whose circadian oscillations are mainly governed by light–dark cycles [24,25]. Circadian rhythmicity, specially that affecting eating behaviour, is required to maintain energy homeostasis. Indeed, humans with disrupted circadian rhythms have higher risk of developing obesity and type 2 diabetes (T2D) [26,27] and mistimed eating increases adiposity in rodents [28] and also favours obesity and increases postprandial glycaemia in humans [29,30].

Circadian rhythmicity is involved in gut microbiota–host interactions. Gut microbiota play a role in maintaining the circadian rhythms of the host, including those related to eating patterns and energy homeostasis, and vice versa. The abundance of bacterial species, as well as of their derived metabolites, vary throughout the day, suggesting associations between gut microbes and eating behaviours in humans [31]. A recent study by Reitmeier et al. [32] supports the link between diurnal fluctuations of the gut microbiota and metabolic health in humans. The authors identified these diurnal oscillations in faecal samples of three independent large-scale population studies and detected aberrant oscillations in subjects with metabolic disorders such as obesity or T2D [32].

2.1. Gut Microbiota Influences Circadian Rhythms Affecting Metabolism

The influence of gut microbiota on the host circadian rhythms and on energy metabolism has been demonstrated in peripheral tissues such as the gut, liver or white adipose tissue. In mice, gut microbiota is required to maintain the circadian rhythmicity of the expression of the Toll-like receptors (TLRs), through which the microbiota-associated molecular patterns (MAMPs) communicate with the intestinal epithelial cells (IECs) of the host [33]. The absence of gut microbiota impairs the circadian clock of the IEC leading to an increased synthesis of ileal corticosterone, which impairs systemic metabolic homeostasis. This is evidence for the role of the microbiota–IEC dialogue oscillation during the day in metabolic health maintenance [33]. Similarly, germ free mice show altered expression of circadian-related genes in liver and white adipose tissue [34]. Microbiota-derived molecules, especially those that activate aryl hydrocarbon receptor (AHR) and pregnane X receptor (PXR), are crucial to maintain the rhythmicity and sexual dimorphism of the growth hormone secretion, which maintains rhythmic gene expression and metabolome in liver in a sex-dependent manner [34].

2.2. Eating Patterns, Gut Microbial Diurnal Oscillations and Energy Homeostasis

The eating rhythms of the host cause microbiota oscillations, which seem flexible and able to adapt to the nutritional environment. Mice deficient in the clock gene *Per1/2* or mice submitted to an experimental "jet-lag" show altered eating rhythms throughout the day linked to aberrant microbiota fluctuations [35]. Time-restricted feeding of *Per1/2* knock-out mice partially restores the microbiota oscillations, proving that eating behaviour is the main driver of microbiota diurnal oscillations [35]. A fat-rich diet also blunts the cyclical changes of the gut microbiota [36,37] and, consequently, their derived-metabolites, thus affecting host metabolism. For instance, the fluctuations of species belonging to the family Lachnospiraceae and their derived metabolite butyrate disappear under a high fat diet (HFD), which affects hepatic clock genes and metabolism [36]. Additionally, the spatial distribution of microbes in the gut seems to follow daily fluctuations. In rodents, bacterial adherence to the intestinal epithelium and their proximity to the mucosal surface are higher in the dark than in the light phase and, over the course of the day, some bacterial species show oscillations of epithelial adherence, findings that are dependent on feeding time [38]. Importantly, the nutrient-induced microbiota oscillations throughout the day lead to fluctuations of luminal and serum metabolites, such as amino acids and polyamines, responsible for the circadian hepatic transcriptome required for physiological processes such as detoxification functions [38]. Eating rhythmicity persists in antibiotic-induced microbiota depletion [38] but obese-associated hyperphagia might be transmissible through gut microbiota transplants to germ-free mice; this occurs in hyperphagic TLR5 knock-out mice [39] although not in *ob/ob* mice. This suggests that eating rhythmicity is driven mainly by the host circadian clock machinery driving endocrine hormone secretion, but nutrient signalling could also be modulated by gut microbiota activated pathways.

Interestingly, the diurnal bacterial growth dynamics, which are controlled by bacterial quorum sensing, availability of nutrients, gut motility and immunity, apparently overlap with host-feeding cycles [40]. In this regard, a bacterial growth-based model of appetite control has been proposed, in which the exponential and stationary growth phases of the bacteria are associated with satiety-induced signals of the prandial and postprandial phase, respectively, while the decline growth phase is coupled with hunger-related signals, defining intermeal intervals [40]. Therefore, to understand how the gut microbiota shapes eating rhythms it is necessary to investigate the bacterial growth fluctuations parallel to the generation of dietary and bacterially derived ligands of nutrient sensors able to centrally control energy homeostasis (such as SCFAs, secondary BAs or amino acid-derived metabolites—see Section 4 for details). For instance, ClpB, an antigen mimetic of αMSH produced by *Escherichia coli* [41], activates host satiety pathways following nutrient-induced bacterial growth [42]. Similarly, nutrient-induced bacterial growth might enhance the generation and release of bacterial ligands, such as cell wall and membrane components (e.g., lipopolysaccharide, LPS, or muramyldipeptide, MDP) or quorum sensing molecules, which in turn activate nutrient sensing pathways in the gut. Although the causal link between the diurnal oscillations of these molecules and feeding rhythms has yet to be proven, LPS and MDP are known to suppress food intake through immune sensing pathways linked to satiety and sickness behaviour [43–45]. Through an immune-related cascade in which LPS and MDP bind to TLR4 and NOD2, respectively, both molecules seem to modulate GLP-1-mediated signalling [46–48]. Nevertheless, the extent to which nutrient-induced bacterial growth and the consequent release of LPS or MDP synchronize meal-related host rhythms under physiological conditions is still unknown. Indeed, the effects of LPS and MDP on feeding behaviour have mainly been described in the context of bacterial infection, where these bacterial molecules reach the systemic circulation. However, their role as modulators of nutrient signalling in the gut, for instance as GLP-1 secretagogues, remains mostly unexplored. In this regard, LPS could act as an inducer of GLP-1 secretion when administered orally only if the gut barrier integrity is impaired [46]

3. Diet Composition Influences Gut Microbiota and Gut-to-Brain Nutrient-Sensing

Adherence to a particular dietary pattern, such as a Western, vegetarian or Mediterranean diet, or dietary interventions characterized by large variations in the macronutrients proportions, such as high protein or high fat (ketogenic) diets, impacts gut microbiota composition and function [49–51]. The composition in macronutrients of these diets exerts an important effect on the availability of microbially derived ligands of dietary and non-dietary nature in the luminal content able trigger gut-to-brain sensing routes and controlling food intake and energy metabolism (summarized in Table 1). The availability of these ligands depends on multiple biological processes, including the microbiota-mediated catabolism of ingested nutrients and their absorption by enterocytes.

Most of the simple carbohydrates, proteins and fats are absorbed in the proximal regions of the small intestine while indigestible complex carbohydrates, the preferred carbon source of gut microbiota, reach the colon, favouring the growth of anaerobic bacteria and species diversity. The gut microbiota structure and composition are flexible, showing rapid adaptations to macronutrient shifts (24–48 h) that remain for short periods while more permanent changes might require longer adherence to a dietary pattern [52,53]. When the dietary macronutrient composition is high in non- or low-fermentable nutrients (e.g., lipids or proteins) these macronutrients overwhelm their intestinal absorption and reach the colon, where they are used by the best adapted microbes with the subsequent shift in their ability to activate gut nutrient sensing routes.

Although the diet per se modulates the energy metabolism, here we focus on effects possibly mediated by the interactions between the gut microbiota and the main macronutrients of different diets for which there is substantial scientific evidence. A special emphasis is placed on those interactions that potentially affect gut-to-brain nutrient sensing and, thereby, energy homeostasis.

3.1. Western Diets

Over the last two decades, the dietary patterns of modern societies have exhibited a shift towards a diet low in fibre and high in saturated fats, simple sugars and refined foods, termed the "Western diet" [54]. The Western diet is frequently associated with altered eating patterns, leading to hyperphagia and obesity onset. Specifically, the overconsumption of saturated fats and simple sugars in Western diets contributes to impair eating behaviour as a consequence of an overstimulation of the nutrient sensing routes that disturbs how brain senses these nutrient-related signals to control food intake patterns. Particularly, the intake of dietary fats amplifies meal sizes in humans; this phenomenon is exacerbated in obese subjects [55] that also showed reduced PYY secretion following a lipid load [56,57]. Studies in obese individuals reported increased postprandial levels of GLP-1 and CCK and occasional reductions of GLP-1 secretion [58,59] while investigations in rodents demonstrated that HFD diminished the sensitivity of GLP-1 and CCK contributing to alter eating behaviour and energy homeostasis [60,61]. In addition to the effects on eating behaviour, mouse studies also demonstrate that diets rich in saturated fats impair the enteric detection of glucose required to induce glycogen depots in muscle through a GLP-1 receptor mechanism in the arcuate NPY-expressing neurons [62].

Sugar-enriched foods also contribute to increasing weight gain as demonstrated by the meta-analyses of randomised controlled intervention trials and observational studies [63]. Excessive intake of sugar is associated with overeating due to impaired hedonic and homeostatic brain circuits [64–66] related to defective gut-to brain glucose sensing. In lean humans, calories from dietary sugars negatively correlate with glucose-induced GLP-1 secretion and positively with dorsal striatal food cue reactivity to palatable food [67]. Obesity in combination with increased dietary sugars has an additive effect reducing circulating levels of GLP-1 after a glucose load [68]. Functional magnetic resonance imaging also reveals that, compared with lean individuals, obese subjects show altered sucrose-related functional connectivity of lateral hypothalamus and NTS with reward-related brain areas resulting in reduced sucrose-associated hedonic responses [69] while studies in mice indicate that the Western diet impairs glucose sensing in POMC-expression neurons [70].

Together with the altered feeding behaviour and energy homeostasis, the Western diet also decreases the bacterial diversity in the human gut, partly due to its reduced content of complex carbohydrates [71] and especially if diets are rich in high saturated fatty acids (SFAs) [72].

Compared with children from rural areas of Africa, European children eating a typical Western diet showed a higher abundance of Firmicutes with an overrepresentation of *Faecalibacterium*, while genera from Bacteroidetes, such as *Alistipes* and *Prevotella*, are increased and decreased, respectively [50]. These associations can reflect a metabolic adaptation of the intestinal bacteria to a new nutritional environment. How gut microbiota, through their interactions with Western diet-derived nutrients (i.e., saturated fats and simple sugars), influences the pool of bacterial-derived metabolites potentially activating gut nutrient sensing routes to control eating patterns and energy homeostasis is in an early stage of investigation. Herein, we compile associative studies in humans and mechanistic studies in rodents which provide evidence of these interactions affecting the central control of energy homeostasis.

In humans, short-term adaptations of the gut microbiota to an animal-based diet consist of a predominance of bile-tolerant microorganisms such as *Alistipes* or *Bilophila* and a decrease in bacteria adapted to metabolize dietary plant polysaccharide such *as Roseburia, Eubacterium rectale* and *Ruminococcus bromii* [53]. In vitro human gut simulator assays have demonstrated the predominance of lipid degradation-related genes and the downregulation of carbohydrate degradation genes, along with an increase in Gammaproteobacteria and the genera *Alistipes* and *Bilophila* in bacterial communities exposed to a fat-based medium [73]. In mice, a SFAs-enriched diet, especially high in palmitic acid, boosts overgrowth of *Bilophila wadsworthia*, which aggravates the HFD-induced metabolic disturbances [74]. The increased abundance of *B. wadsworthia* seems to be a secondary consequence of the SFAs on BAs metabolism [75]. Specifically, the SFAs-enriched diet increases the host production of BAs conjugated with taurine, which increase the fitness and growth of *B. wadsworthia*, while inhibiting other bacteria. *B. wadsworthia* also exacerbates the HFD-induced decrease in deoxycholic acid (DCA) and hyodeoxycholic acid (HDCA), secondary BAs that mediate gut nutrient sensing routes [75] (see Section 4 for details).

Studies in mice fed saturated fat-enriched diets combined with microbiota transplants or antibiotic treatments have provided insights into the causal relationship between specific bacterial genera or species and nutrient sensing routes involved in the central control of glucose homeostasis. For instance, the abundance of *Lactobacillus gasseri* (which is reduced by Western diets) helps to sense oleic and linoleic acid in the small intestine to ultimately reduce hepatic glucose production through downregulation of the bile acids receptor farnesoid acid receptor (FXR) and upregulation of the long-chain acyl-CoA synthetase-3 (ACSL3) [76]. In addition, the Western diet-induced lactobacilli decrease seems to be involved in the defective GLP-1 receptor/nitric oxide synthase (nNOS)-mediated signalling in response to oral glucose load to centrally control insulin secretion and gastric emptying, a process initiated by LPS and MDP [47].

Western diet also provides an excess of non-absorbable glucose and fructose to the small intestine and subsequently reach the colon. Gavage of [13C]fructose in mice reveals that high doses of fructose saturate the fructose-to-glucose conversion in the small intestine, enhancing fructose utilization by the microbiota in the colon via the hexokinase pathway to further generate tricarboxylic acid cycle (TCA) intermediates, such as essential amino acids (valine and leucine) and SCFAs (succinate, butyrate and acetate) [77,78]. In addition, sucrose-derived glucose and fructose in the colon downregulate a gene involved in gut bacteria colonization, reducing the fitness of bacterial species such as *Bacteroides thetaiotaomicron* associated with a lean phenotype by some studies [79]. The Western diet-induced decrease in the abundance of *Lactobacillus* spp. in the small intestine also impairs the glucose transporter SGLT1 and the GLP-1 receptor-mediated glucose sensing in the gut required to centrally control energy homeostasis in rats [80].

3.2. Vegetarian Diets

Vegetarian diets largely vary in composition according to the interindividual choice of foods. Commonly, they are devoid of meat and can include eggs and dairy products (lacto-ovo-vegetarian, lacto-vegetarian or ovo-vegetarian); exclude eggs and dairy products (vegan diets) or only include vegetables, fruit, nuts, seeds, legumes and sprouted grains (raw vegan diet) [81].

Compared with omnivorous diets, individuals adhering to vegan or vegetarian diets show less uncontrolled eating and emotional eating [82]. In the short-term, compared with a processed meat meal, a plant-based tofu meal enhances the secretion of GLP-1 in T2D individuals with a concomitant increase in satiety and a reduction of triglycerides in plasma [83,84].

Deciphering which are the vegetarian/vegan diet-associated nutrients and molecules that facilitate the gut-to brain nutrient sensing would provide valuable knowledge for designing effective anti-obesity dietary interventions for subjects who chose this dietary pattern. Overall, well-implemented vegetarian diets meet or exceed the recommended fibre intake [81,85]. A comparative study demonstrated that while the daily intake of sugars does not differ between vegan, vegetarian and omnivorous diets and the intake of proteins shows slight differences, subjects who adhered to a vegetarian/vegan diet consumed fewer calories and saturated and monounsaturated fats and higher amounts of dietary fibre [86], pointing out a main role of fibres in the ability of plant-based diets to modulate the gut–brain axis.

Dietary fibres, complex carbohydrates unable to be digested and absorbed in the upper part of the human intestine and which pass to the distal part, serve as substrate for intestinal microbes [87]. These complex polysaccharides, highly abundant in plant-based diets, per se display protective effects against the progression of obesity; moreover, new investigations suggest that the gut microbiota might be an intermediate player of the fibre benefits. Fibre fermentation generates diverse molecules including lactate, pyruvate and succinate as well as the SCFAs including acetate, propionate and butyrate in a molar ratio of 60:20:20, approximately [88]. Specifically, acetate is produced by phosphate acetyltransferase and acetate kinase; propionate is catalysed via the succinate, acrylate and propanediol pathways; and butyrate is produced by phosphate butyryltransferase and butyrate kinase [89].

Overall, an overrepresentation of the genera *Prevotella* and *Ruminococcus* is present in faeces of humans who adhere to plant-based diets [90]. Interventional studies in humans also show that different types of fibres favour the growth of bacteria with the enzymatic machinery directly involved in their fermentation. For instance, the abundance of *Bifidobacterium* spp., whose genome encodes transporters and enzymes involved in complex carbohydrate catabolism [91], are increased in the human colon as a result of dietary supplementation with inulin-type fructans [92], resistant starch [93], galactoologosacharides [94] or arabyno-oligosaccharides, which also increase *Prevotella* [95,96]. Some studies also associate the bifidogenic effect of the inulin-type fructans supplementation with improvements in oral glucose tolerance in obese women and reductions in body weight z-score in obese children [92]. Nonetheless, most of the human studies only show associations between specific bacterial taxa and metabolic benefits, and further studies are needed to demonstrate causality. Thus, it is also possible that the microbiota-mediated metabolic benefits of prebiotic fibres could depend on the initial microbiota configuration of the human subject and its capability to enhance the production of enough fermented products (e.g., SCFAs or others) and that this variation explains the inconsistency of the results across intervention trials [96,97].

Nevertheless, some studies in germ-free mice intentionally colonized with specific microbiotas have already demonstrated that, for example, plant-based diets induce microbiota-dependent benefits. In particular, *Prevotella copri* seems to mediate the fibre-induced glucose tolerance improvement as revealed by a study conducted in germ-free mice colonized with the gut microbiota from individuals that favourably respond to barley kernel-based bread consumption [98]. An important route of communication between the gut and the brain is the intestinal gluconeogenesis that, via portal glucose sensing, is essential for maintaining the postprandial regulation of the hypothalamic energy homeostasis control, including food intake and endogenous production of glucose [99]. It seems that *P. copri* could modulate the gut–brain axis to centrally control glucose homeostasis antagonizing the effects of other intestinal bacteria. Particularly, *P. copri* seems to limit the effects of *Bacteroides thetaiotaomicron* on energy metabolism impairment, possibly due to the ability of the latter bacterium to reduce colonic gluconeogenesis [98]. This idea was later demonstrated by de Vadder et al. 2016 [100] who evidenced the capacity of the *P. copri* colonization in mice to increase the intraluminal content of succinate, which could be used as substrate for the intestinal gluconeogenesis to then reduce hepatic glucose production via portal glucose sensing.

3.3. Mediterranean Diet

The Mediterranean diet is characterized by a high-level consumption of fruits, vegetables, grains, and fish and seafood as the main animal protein [101,102]. Scientific evidence supports that most of the benefits, mainly cardiovascular, of the Mediterranean diet are attributed to the high content of monounsaturated fatty acids (MUFAs), mainly oleic acid; ω-3 polyunsaturated fatty acids (PUFAs), mainly α-linolenic acid, eicosapentaenoic acid and docosahexaenoic acid; polyphenols, flavonoids and non-flavonoids, and fibre [103,104].

Contrary to SFAs, unsaturated fats are associated with metabolic benefits [105,106] and, accordingly, the Mediterranean diet is considered adequate to promote metabolic health without restricting total fat intake. Oleic and α-linolenic acid are GLP-1 secretagogues [107–109] and initiate lipid sensing routes to centrally control the endogenous production of glucose [110].

Dietary polyphenols present in vegetables and fruits might also display protective effects against obesity by the modulation of the hypothalamic function. Polyphenols can reach the brain and also initiate gut nutrient sensing routes. In the gut, polyphenols are able to induce the secretion of GLP-1, and PYY [111] in postprandial periods favouring the insulin-mediated glucose-lowering effects [112,113]. Whether or not these effects are mediated by the central nervous system should be further ascertain.

The Mediterranean diet also shapes the gut microbiota and the associated metabolome. Obese individuals adhere to a Mediterranean diet for 2 years show increases in the genera *Bacteroides*, *Prevotella*, and *Faecalibacterium*, and most importantly of *Roseburia* and *Ruminococcus* and the species *Parabacteroides distasonis* and *Faecalibacterium prausnitzii* [114]. Compared with Western diet non-human primate's consumers, Mediterranean diet consumers also show higher levels of genera *Lactobacillus*, *Clostridium* and *Oscillospira* [115]. The overrepresentation of these gut bacteria species could represent an adaptation of the microbial ecosystem to the higher preabsorptive abundance of PUFAs, polyphenols and complex carbohydrates, not absorbed in the upper gut, and thus able to interact with intestinal bacteria to modulate the pool of nutrient and microbe-derived metabolites activating the gut–brain axis.

Studies in mice reveal that, compared with saturated fats, intake of polyunsaturated fats limit the progression of obesity and induce different changes to the gut microbiota composition [116,117]. Overall lipids are not primarily digested by intestinal microbes but lipid-induced gut microbiota changes can equally influence how the gut senses nutrients. *Lactobacillus* spp. might be overrepresented under a diet rich in linoleic acid since they specifically develop resistance to its toxicity [118] and also use this ω6 fatty acid to produce PUFAs-derived metabolites such as 10-hydroxy-cis-12-octadecenoic acid (HYA) [119]. HYA improves metabolic health, suppressing food intake through lipid sensing-mediated signalling in L cells, in which GPR40 and/or GPR120 activation induces GLP-1 secretion in mice [119]. In addition, gut bacteria might influence lipid sensing routes and the exportation of fatty acids to extra-intestinal tissues by influencing the fatty acid storage in the enterocyte. For instance, *Lactobacillus paracasei* and *Escherichia coli* decrease the intestinal secretion of the absorbed oleic acid but through different metabolites derived from the fermentation of complex carbohydrates [120]. Specifically, intestinal storage of oleic acid is promoted by *L. paracasei*-produced L-lactate by inhibiting malonyl-CoA-induced fatty-acid beta oxidation, a route activated by *E.coli*-produced acetate, which increases oleic acid degradation [120].

Additionally, gut microbiota favours the bioavailability of polyphenols from food in the intestinal lumen through complex multienzymatic reactions [121] modulating the polyphenol-associated gut nutrient sensing. To date, the identification of bacteria species specifically modulating polyphenol metabolic routes is under an early stage of investigation. Nevertheless, some evidence is available indicating, for instance, that *Bacillus subtilis*, which protects against obesity in mice [122], may produce protocatechuic acid from dietary quercetin in the human gut, which can virtually bind to the GLP-1 receptor as predicted by molecular docking simulations [123].

3.4. Diets Based on Macronutrients Ratio Variations

Herein we review the effects of diets used for weight loss, which are based on variations of the macronutrients proportions, such as high protein diets and low carbohydrate-high fat diets. We also focus on the potential role of these diets in the postprandial modulation of the gut–brain axis by changes in the pool of microbial and nutritional ligands in the intestinal lumen derived from the dietary macronutrients.

High protein diets are characterized by an increased intake of food rich in proteins (25–35% of energy compared with 12–18% of the standard protein diets) [124] and frequently associated with reduced carbohydrate consumption. These type of diets seem to be appropriate to rapidly induce weight loss [125]. Indeed, high protein diets positively regulate energy metabolism since, compared with other macronutrients, proteins strongly induce satiety and stimulate intestinal gluconeogenesis and thermogenesis [99,126]. Nevertheless, in the long-term, diets with different macronutrients ratios but the same energy-restricted content have similar effects on body weight maintenance, which could also be due to a poor long-term adherence to all diets [127–129].

In normal weight or obese human subjects, high protein meals induce the greatest satiety compared with isocaloric diets with high content of carbohydrates or fats, an effect that is dependent on PYY, the secretion of which is preferentially enhanced by proteins [130]. In addition to gut hormones, high protein diets modulate the gut–brain axis to control food intake and energy metabolism by stimulating the intestinal gluconeogenesis from postprandial to postabsorptive periods [99]. Digested peptides in the upper gut antagonize the μ-opioid receptors in the spinal and vagal afferents of the portal vein, a signalling that centrally activates the intestinal gluconeogenesis involved in the glucose sensing-induced food intake suppression [131,132]. Additionally, enriched-protein meals are a source of gluconeogenesis substrates for the gut such as glutamine and glutamate [99]. Compared with carbohydrates and fats, proteins also have the highest effect inducing thermogenesis [126]. Although the underlying mechanisms needs to be clarified specifically for proteins, some studies have identified thermogenesis-dependent gut–brain axis mechanisms mediated by gut hormones. For instance, GLP-1 centrally enhances thermogenesis through sympathetic efferents [133] and the duodenal hormone secretin postprandially activates the thermogenesis to induce satiety [134].

High protein diets also increase the amount of amino acids that can be fermented by gut microbiota in the colon to obtain energy and to produce nutrient sensing ligands. These include amino acid-derived SCFAs, branched chain fatty acids (BCFAs: isobutyrate, 2-methylbutyrate and isovalerate) and other molecules derived from tryptophan or glutamate, among others.

Compared with carbohydrates, the fermentation of proteins produces fewer SCFAs, although it still contributes substantially to microbial organic acid production. Amino acid-derived acetate is produced from glycine, alanine, threonine, glutamate, lysine and aspartate; butyrate is produced from glutamate and lysine and propionate from threonine [135].

Tryptophan is also biochemically transformed by gut microbiota, leading to the production of either serotonin (5-HT), indole, kynurenine or other derivative compounds [136]. Intestinal bacteria modulate the production of 5-HT in the gut directly or indirectly through microbe–host interactions. Members of the gut microbiota also possess tryptophanase activity, mediating the conversion of tryptophan into indole, which serve as interspecies signalling molecule that control bacterial physiology [137]. Indole-producing bacteria include, among others, species belonging to *Bacillus, Clostridium, Enterococcus, Bacteroides, Enterobacter, E. coli, Prevotella, Shigella* and *Vibrio* [138]. In addition, the gut microbiota modulates the expression of the rate-limiting host enzyme involved in the conversion of tryptophan to kynurenine, indoleamine 2,3-dioxygenase (IDO1), thus influencing the levels of kynurenine and its derivatives such as kynurenic acid [139]. Although some of these compounds might be toxic, such as indoxyl sulphate and quinolinic acid [140,141], others can potentially activate 5-HT receptors, aryl hydrocarbon receptor (AhR), and GPR35, which affect energy homeostasis (see Section 4 for details).

Intestinal bacteria can also decarboxylase glutamate, producing γ-aminobutyric acid (GABA) via the enzymatic activity of the glutamate decarboxylase, which helps to maintain the intracellular pH of the bacteria [142]. Strains belonging to the genera *Lactobacillus, Bifidobacterium, Lactococcus, Streptococcus, Escherichia, Listeria,* and *Aspergillus* have been reported to produce GABA [143–145], which also modulate nutrient sensing in the gut (see Section 4 for details).

Nevertheless, few studies address associative or causative links between the gut microbiota and high-protein diets on the control of food intake and energy homeostasis via the gut–brain axis.

Overall, diets high in proteins and low in carbohydrates reduce faecal abundance of SCFAs; i.e. butyrate [146], acetate and propionate [147,148], while increasing branch chain fatty acids; i.e., 2-methylbutyrate [146], isobutyrate and isovalerate [53,148,149]. Additionally, high intake of dietary proteins often increases the levels of kynurenic acid and indoxyl sulphate in plasma [150].

Concerning the effects on gut microbiota composition, interventional studies in humans with high protein diets initially suggest that variations (such as reductions of *Bifidobacterium, Roseburia* and *Eubacterium rectale*), were a consequence of the reduced intake of dietary fibres or the caloric restriction associated with these interventions [151]. More recent studies comparing the effects of different protein sources suggest that proteins also play a direct role in driving at least microbial-mediated metabolite changes with potential health impacts [146]. This study shows associations between bacterial abundance and faecal metabolites in overweight subjects conducting a 3-week high-protein diet intervention (either soy- or casein-based diet). For example, the study identified a positive correlation between *Oscillospira* and *Odoribacter* and amino acid-derived bacterial metabolites measured by targeted metabolomics and 1H nuclear magnetic resonance [146].

In humans, protein intake is positively associated with ClpB-like gene function [152] and in vitro studies demonstrate that, compared with other macronutrients (D-fructose and oleic acid), *E.coli* requires protein supplementation (bovine serum albumin) to increase the mRNA and protein levels of ClpB, which induces a dose-dependent stimulating effect on PYY secretion [153]. However, future investigations are needed to assess whether postprandial increases in ClpB coupled with bacterial growth account for the PYY-mediated food intake suppression in high protein diets in humans [130].

Larger-scale human interventions are needed to elucidate to what extent changes in microbiota-derived metabolites from dietary proteins affect nutrient sensing and, ultimately, control energy homeostasis, with an especial emphasis on intestinal gluconeogenesis and gut-mediated enhancement of thermogenesis. This will shed light on the link between protein-derived microbial metabolites and high protein-associated weight loss. Furthermore, additional studies are needed to assess the risk and benefits of high-protein dietary interventions in improving metabolic health, since these reduce butyrate production (the main energy source for colonic enterocytes) and increase levels of mucosal and renal toxic compounds (e.g., tyrosine-derivate p-cresol and indoxyl sulphate) [151,154].

Low carbohydrate diets are frequently defined as diets with less than 20% of calories from carbohydrates with high content of fats (55–65%) and occasionally high in proteins (25–30%) [155]. These types of diets include ketogenic diets, frequently characterized by high content of fats, which were originally used to treat epilepsy due to their associated anticonvulsant effect driven by the increased production of ketone bodies and the modulation of GABA neurotransmission and mitochondrial metabolism [156]. Currently, these diets are also conceived for weight loss purposes. The metabolic benefits of ketogenic diets are based on their reduced capacity to postprandially increase glycemia and insulinemia and the concomitant enhancement of cellular catabolic routes using fats as the main source of energy. Accordingly, fat depots are reduced and gluconeogenesis, followed by ketogenesis, are also enhanced to supply glucose to cells [157]. These diets are effective to rapidly lose weight although long-term adherence poses challenges due to their associated contraindications [158]. Some studies reveal that ketogenic diets also induce less hunger and reduce the desire to eat in humans when comparing the appetite assessments before and during adherence to the diet [159]. Low carbohydrate diets also induce satiety in T2D patients [160] and, compared with diets with high content in carbohydrates, have lower capacity to stimulate food intake-related brain areas [161]. Nevertheless, the exact underlying mechanisms of the satiety effects caused by these diets remained elusive, especially those possibly dependent on gut microbiota.

Some studies indicate that diet-induced ketogenesis seems to mediate the reduced circulating levels of ghrelin associated with lower appetite in overweigh/obese individuals following a low energy diet [162] although ketone bodies inhibit the GLP-1 release by the EECs [163] and directly activate orexigenic hypothalamic routes in the brain [164]. Similar to SCFAs, ketone bodies also initiate GPR41 and GPR43 signalling to control energy metabolism. In particular, the acetate binds to GPR43 to induce lipid utilization in plasma [119] and β-hydroxybutyrate antagonizes GPR41 in sympathetic neurons to attenuate the sympathetic mediated metabolism [165]. To date, how these ketone bodies impact on the hypothalamus via the GPR41/43 has not been explored.

Ketogenic diets also change the gut microbiota structure and function in obese individuals with an overall decrease in butyrate and butyrate-producing bacteria, such as *Roseburia* spp. and *Eubacterium rectale* and *Bifidobacterium* spp. These changes are mainly attributed to the reduced content of the diet in complex carbohydrates and the concomitant reduction of SCFAs production [148].

Since ketogenic diets are also diets rich in fats, the identification of the main mechanisms through which these diets impact on the gut–brain axis through gut microbiota-dependent mechanisms is complex. These mechanisms could be related to the reduction of SCFAs generated in the gut by the intestinal microbiota. Nevertheless, the role of other specific gut microbiota adaptations to these diets in shaping the specific pool of bioactive ligands and the existence of potential interactions with the levels of ketone bodies and their effects remain to be investigated.

In this regard, it is unknown whether the ketogenic diet-associated gut microbiota play a causative role in the postprandial control of the gut–brain axis. A recent study conducted in humans and mice identified that, compared with a conventional HFD, a ketogenic diet, high in fats and low in proteins, specifically reduces species from the genus *Bifidobacterium* by a direct effect of intestinal ketone bodies inhibiting their growth [166]. Additionally, the ketogenic diet displays metabolic improvements in mice, although these were attributed to a protective effect on the intestinal immunity rather than to the modulation of the gut–brain axis [166]. A direct role of ketogenic diets modulating the gut–brain axis was evidenced in another study in mice demonstrating that a carbohydrate-restricted diet with high content of fats favours cross-feeding between *Akkermansia muciniphila* and *Parabacteroides*, resulting in a reduced gamma-glutamyltranspeptidase activity in faeces driving an increased GABA/glutamate ratio in the hippocampus that in turn confers protection against refractory epilepsy [167]. Although how ketogenic diets centrally control energy homeostasis through gut microbiota-mediated mechanisms remains uncertain, given the role of GABAergic neurotransmission on the central control of energy homeostasis and food intake (see Section 4 for details) it seems plausible that ketogenic diet modulates the gut–brain axis in postprandial periods by a similar gut microbiota-dependent mechanism operating in the hypothalamus. On the other hand, the gut microbiota, by modulating the lipid metabolism, seems to influence the circulating levels of β-hydroxybutyrate, which exerts neuroprotective effects on the brain, although this effect has not been explored for its relationship with metabolic health [168].

4. Microbial Ligands Mediating Gut–Brain Communication and Energy Homeostasis

Here, we review the microbial products, including bacterial metabolites and bacterial cell components, that might impact on brain functions by modulating nutrient sensing signalling through enteroendocrine humoral and neural pathways, and could contribute to controlling energy homeostasis (Figure 1, Table 1). Beyond the control of food intake, gut microbiota might influence the whole-body energy metabolism by modulating the parasympathetic and sympathetic efferent tone [47,169], although this mechanism has been the subject of far fewer studies. Thus, this section focuses on the role of gut microbiota in controlling food intake and energy homeostasis, mainly through effects on the hypothalamus-mediated food-intake suppression and, specially, in the postprandial periods.

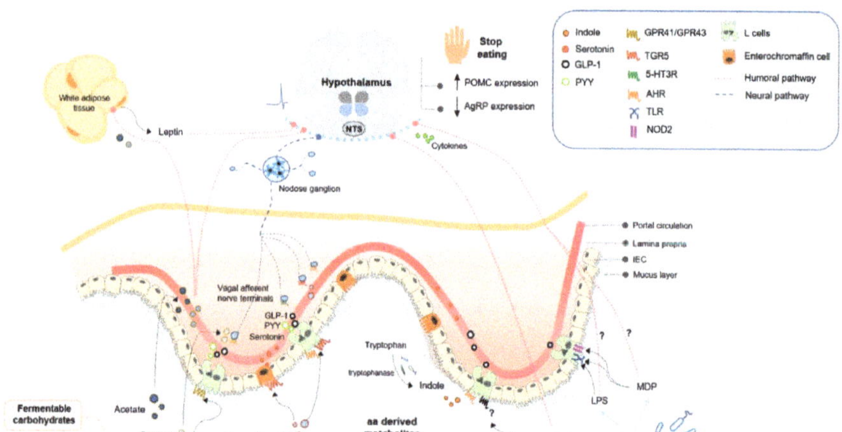

Figure 1. Bacterially produced metabolites from dietary nutrients and structural components of non-dietary origin modulate food intake in the brain (hypothalamus) through humoral and/or enterodocrine and neural signalling pathways. Here, we represent the pathways by which bacterial metabolites and non-dietary bacterial components (LPS, MDP and ClpB) induce an anorexigenic response in postprandial periods and a long-term food intake control. 5-HT3R, 5-hydroxytryptamine type 3 receptor; aa, amino acid; AgRP, agouti gene-related peptide; AHR, aryl hydrocarbon receptor; BA, bile acids; ClpB, caseinolytic peptidase B; GLP-1, glucagonlike peptide-1; GPR41/FFAR3, free fatty acid receptor 3; GPR43/FFAR2, free fatty acid receptor 2; IEC, intestinal epithelial cells; LPS, lipopolysaccharide; MDP, muramyldipeptide; NOD2, Nucleotide-binding oligomerization domain 2; NTS, nucleus tractus solitarius; PYY, peptide YY; POMC, proopiomelanocortin; SCFA, short-chain fatty acids; TGR5, takeda G protein-coupled receptor 5; TLR, Toll-like receptor.

Table 1. Main microbially derived ligands of dietary and non-dietary nature involved in gut-to brain nutrient sensing and control energy homeostasis.

Dietary Nutrients	Gut Bacterial-Derived Ligand	Bacterial Producers	Bacterial-Producing Enzyme	Receptor	Pathway	Function	References
Fermentable carbohydrates	SCFAs (acetate, propionate, butyrate)	*Prevotella* [90], *Ruminococcus* [90], *Bifidobacterium* sp. [91], *Prevotella* [95,96]	Phosphate acetyltransferase and acetate kinase for acetate	FFAR2/GPR43 (L cells)	Humoral pathway	Food intake suppression, ARC neuronal activation, increase in acetyl-CoA carboxylase activity and AMPK inducing an increase in POMC and reduction in AgRP expression, leptin release from adipocytes	[89,170–173]
			Enzymes involved in succinate, acrylate and propanediol pathways for propionate	FFAR3/GPR41 (L cells, enteric neurons, nodose ganglion neurons)	Humoral pathway, gut nutrient sensing pathways (GLP-1, PYY)	Food intake suppression, leptin release from adipocytes, control of postprandial glucose, control of intestinal gluconeogenesis	[89,132,173–177]
			Phosphate butyryltransferase and butyrate kinase for butyrate	FFAR3/GPR41 (L cells, enteric neurons, nodose ganglion neurons)	Gut nutrient sensing pathways (GLP-1, GIP, vagal afferents)	Food intake suppression, stimulation of POMC expression, suppression of AgRP expression, suppression of orexigenic neurons activity	[89,178–181]
Bile acids (BAs) (involved in lipid digestion)	Secondary BAs	Members of the genera: *Lactobacillus* [182–184], *Bifidobacterium* [182,185], *Enterococcus* [186,187], *Clostridium* [182,188], *Listeria* [182,189], *Bacteroides* [182]	Bacterial bile salt hydrolases (BSH) (deconjugation of primary BA to secondary BA)	TGR5 (L cells, vagal afferents, nodose ganglion neurons, hypothalamic neurons)	Humoral pathway, gut nutrient sensing pathways (GLP-1, PYY, 5-HT, vagal afferents)	Food intake suppression in synergy with CCK1R activation, activation of POMC/CART-expressing hypothalamic neurons, glucose homeostasis, 5-HT3R activation in intestinal vagal afferent terminals (probably modulating food intake)	[190–200]
Proteins	Indole	Members of the genera: *Bacillus*, *Clostridium*, *Enterococcus*, *Bacteroides*, *Enterobacter*, *Escherichia*, *Prevotella*, *Shigella* and *Vibrio* [138]	Tryptophanase (tryptophan to indole)	AHR (L cells)	Gut nutrient sensing pathways (GLP-1)	Contribution to eating patterns unknown	[136,201–203]
	GABA	Members of the genera: *Lactobacillus*, *Bifidobacterium*, *Lactococcus*, *Streptococcus*, *Escherichia*, *Listeria*, and *Aspergillus* [143–145]	Glutamate decarboxylase (glutamate to GABA)	$GABA_A$, $GABA_B$ (L cells, vagal afferents)	Gut nutrient sensing pathways (potentially through vagal afferents)	Contribution to nutrient sensing in the brain unknown	[142,204–207]

Table 1. *Cont.*

Dietary Nutrients	Gut Bacterial-Derived Ligand	Bacterial Producers	Bacterial-Producing Enzyme	Receptor	Pathway	Function	References
Bacterial cellular components	ClpB (mimetic of α-MSH)	Order *Enterobacteriales*, including *E. coli* strains and *Hafnia* genus [208]	-	Unidentified	Humoral pathway, gut nutrient sensing pathways (PYY)	Food intake suppression by increasing POMC and decreasing AgRP expression, enhancement of POMC neuronal activity	[41,42,154,209–211]
	LPS	Gram-negative bacteria [212]	-	CD14/TLR4 (enteric neurons, nodose ganglion neurons)	Humoral pathway, gut nutrient sensing pathways (GLP-1 and potentially through vagal afferents)	Reduction of food intake, enhancement of GLP-1-induced NO production in enteric neurons (possibly contributing to an anorexigenic shift in neuropeptides expression), satiogenic effect probably by changes in hypothalamic cytokine expression, increase nodose ganglion neurons excitability	[43,45,47,213–218]
	MDP	Gram-positive bacteria (minor component in Gram-negative bacteria) [219]	-	CD14/NOD2/TLR2 (L-cells, enteric neurons)	Humoral pathway, gut nutrient sensing pathways (GLP-1)	Reduction in food intake, enhancement of GLP-1-induced NO production in enteric neurons, glucose tolerance	[43,45,47,213–217,220]

4.1. Short Chain Fatty Acids

The function of SCFAs in protecting against metabolic alterations of diet-induced obesity is relatively well-established. Supplementing diet with fermentable carbohydrates, such as inulin, oligofructose or pectin, reduces food intake and improves HFD-induced glucose intolerance and weight gain in rodents and humans [170,221–225]. These effects are associated with SCFA production.

Although the three main SCFAs activate both GPR43/FFAR2 and GPR41/FFAR3, acetate is a potent agonist of FFAR2 [226], whereas propionate and butyrate have higher affinity for FFAR3 [226]. Both receptors are expressed in L-cells [227,228], and FFAR3 has been detected in enteric neurons [229] and in nodose ganglia cells [230]. Besides being agonists of FFAR2 and FFAR3, all SCFAs and lactate can potentially affect food intake by inhibiting the signalling of the orexigenic hormone ghrelin. In Hek293a cells stably expressing human growth hormone secretagogue receptor type 1a (GHSR1a), the three SCFAS (acetate, propionate, butyrate) and lactate reduced the Ca^{2+} influx in presence of ghrelin and showed antagonistic GHSR-1a properties as they attenuated the ghrelin-mediated receptor internalization [231]. As this SCFAs-mediated signalling is still under an early stage of investigation (revised by [232]), here we focus on the role of SCFAs in food intake suppression, acting as FFAR2/FFAR3 agonists.

SCFAs are known to regulate food intake by modulating hypothalamic function, either reaching systemic circulation to the brain or via nutrient signalling mediated directly by GLP-1 and PYY generated in EECs or via vagal afferents. Among SCFAs, acetate seems to reach the brain through systemic routes, while propionate and butyrate mainly activate gut nutrient sensing pathways.

The analysis of biodistribution of 11C-acetate infused in the mouse colon indicates that acetate crosses the blood–brain barrier and reaches the hypothalamus, suppressing food intake short-term [171]. Moreover, acetate from fermentable fibre increases the hypothalamic neuronal activation in the ARC, but not in the ventral medial hypothalamus (VMH) or the PVN [170,171]. These changes are accompanied by an increase in hypothalamic activity of acetyl-CoA carboxylase and AMP-activated protein kinase (AMPK), subsequently inducing a downstream acute rise in POMC expression and reduction in AgRP expression [171,172]. In addition, systemic acetate enhances leptin release from adipocytes, possibly via an as yet undetermined FFAR2-dependent mechanism [173], thus contributing to a shift towards an anorexigenic neuronal activation pattern in the ARC [233]. Contrary to the abovementioned control diet-based research, a study in rats indicates that, under an energy-dense diet, gut microbiota enhance the production of acetate from glucose and fatty acids, contributing to diet-induced obesity [234]. Indeed, HFD increases acetate levels in the intestinal lumen and also systemically results in hyperphagia and an enhanced parasympathetic tone in β-cells, which impairs glucose stimulating insulin secretion [234]. The role of acetate in signalling through gut nutrient sensing routes is less evident. Some studies reveal that acetate does not stimulate GLP-1 or PYY secretion [171,235] and others report acetate-induced PYY secretion in the distal, but not proximal colon in a small cohort of obese humans without exploring its effects on food intake [236,237].

In overweight humans, the acute administration of inulin-propionate ester, which enables propionate delivery in the colon specifically, increases the postprandial secretion of GLP-1 and PYY along with reduced food intake, together with a non-significant trend to decrease long-term energy intake [225]. In mice, propionate also robustly stimulates both GLP-1 and PYY secretion either in vivo or in primary murine colonic cultures, probably via a FFAR2-dependent mechanism [174]. Additionally, propionate acts as a food intake suppressor through independent mechanisms of the endocrine actions of gut hormones. Indeed, independently of changes of the circulating levels of GLP-1 and PYY, non-obese healthy men receiving inulin-propionate showed reduced *ad libitum* food intake and lower oxygen level-dependent signal in reward-related brain structures [238]. Thus propionate can impact on brain function through vagal afferents innervating the gut and the portal vein. In this regard De Vadder and colleagues demonstrated that luminal propionate controls the postprandial levels of glucose through a gut–brain loop [175]. In particular, this process is initiated by the propionate-induced activation of FFAR3 expressed in the afferents of the portal vein, which leads to the activation of neurons in brain regions receiving vagal and spinal inputs, i.e., DVC, spinal C1 segment and the parabrachial nucleus (PBN), respectively, as well as in the hypothalamic areas receiving inputs from DVC and PBN, i.e., PVN, the lateral, hypothalamus (LH) and the ARC. Apparently, propionate centrally controls the intestinal gluconeogenesis, a process that induces metabolic benefits [132,176], including the reduction of the endogenous production of glucose independently of insulin [175].

Furthermore, like acetate, elevated plasma propionate levels induce leptin release from adipocytes via the somewhat controversial FFAR2- or FFAR3-dependent mechanism [173,177], promoting hypothalamic anorexigenic neuronal activation [233].

Among the three main SCFAs, butyrate may be the strongest stimulator of anorexigenic peptides [178] and the most potent suppressor of food intake [179]. In obese and healthy humans, all three main SCFAs separately administered increase plasma levels of PYY in both fasting and postprandial conditions [236]. In mice, oral, but not intravenous, administration of butyrate reduces food intake, according to decreased neuronal activity in NTS and DVC [180]. Additionally, butyrate influences the hypothalamic circuitry, suppressing the activity of orexigenic neurons [180], decreasing AgRP expression [181] and enhancing POMC expression [181]. In addition, subdiaphragmatic vagotomy eradicates the anorectic effects of butyrate, suggesting that the vagal nerve is necessary to convey the short- and long-term satiety signalling of acute or chronic administration of butyrate in the context of obesity [180]. Moreover, intraperitoneal butyrate induces activation of nodose ganglia neurons [179]. Intestinal butyrate may transmit satietogenic signals by stimulating GLP-1 and GIP secretion in L-cells and K-cells, respectively [178], and these effects are probably reinforced by ghrelin inhibition [181]. It is not clear whether butyrate induces PYY secretion, since studies report no changes [181], decrease [235] or slight increase [178] in PYY release as a result of butyrate increase in the caecal content of animals orally receiving a probiotic, a pea protein supplementation or oral butyrate administration. Furthermore, it is not yet clear whether butyrate-induced gut hormone secretion is mediated by a FFAR3-dependent mechanism [178,181]. Therefore, the evidence about the mechanism through which SCFA are involved in satiety signalling is not fully consistent, probably due to experimental differences, for instance, in hormone measurement timings, types of dietary fibre supplementation, administration site and the use of different animal models.

4.2. Microbial Metabolites of Bile Acids

Bile acids (BA) are steroid acids synthesized in the liver from cholesterol, conjugated to either taurine or glycine and postprandially released in the duodenum to facilitate the absorption of dietary lipids and fat-soluble vitamins [239–241]. The majority of primary BA secreted in the intestine are actively reabsorbed in the ileum and transported back through the portal circulation to the liver (enterohepatic circulation) [242]. The remaining small portion of primary BA are deconjugated and dehydroxylated in the ileum and colon by intestinal bacteria into secondary BAs [243,244], mainly DCA and lithocholic acid (LCA) in humans and rodents, and murideoxycholic acid in rodents [239,245]. Bacterial bile salt hydrolases (BSH) are essential enzymes for deconjugation of primary BA to secondary BAs [190]. In humans, BSH genes and enzymes have been characterized in Gram-positive bacterial species of the genera *Lactobacillus* [182,183,190], *Bifidobacterium* [184,190], *Enterococcus* [185,186], *Clostridium* [187,190] and *Listeria* [188,190] and in Gram-negative bacteria of the genus *Bacteroides* [189,190].

Besides their role in lipid digestion, BAs have recently been characterized as ligands of takeda G protein-coupled receptor 5 (TGR5), regulating lipid and glucose metabolism once activated [246–248]. Dietary macronutrients and feeding patterns influence the composition and secretion of bile acids, respectively, and thus the microbiota-mediated bile acids' effects on the host. Interventional studies in humans reveal that a fibre-rich low-fat diet is associated with low levels of all secondary BAs, and both glycine- and taurine-conjugated bile acids in faeces [249]. By contrast, a high fat diet increases the abundance of unconjugated and secondary BAs (DCA, TDCA, 12keto-LCA, 3b-DCA and TLCA) coupled with changes in species belonging to the genera *Bacteroides*, *Clostridium*, *Bifidobacterium* and *Lactobacillus* and a positive correlation between *Bacteroides* and DCA [250].

Studies in humans and rodents report that microbial BAs modulate energy homeostasis directly by reaching the brain through systemic circulation and/or activating gut nutrient sensing routes. The main functional effects of BAs reported are on feeding behaviour or glucose homeostasis. Plasma BAs levels notably rise postprandially in humans [251,252] and rodents [253], thus fluctuating in the systemic circulation along with the circadian rhythm upon food intake, suggesting a role of BA as lipid-sensing molecules and short-term satiety signals that reach the brain through a humoral route. Diet fat content is positively associated with plasma BAs concentrations in humans [252] and mice [191], and with hypothalamic BAs concentrations in mice [191]. Interestingly, DCA and other BAs have been detected in the rat brain [254,255], and their levels are positively correlated with plasmatic levels [254], indicating that secondary BAs may reach the brain from the intestine by diffusion across the blood–brain barrier [254]. Accordingly, TGR5 expression has been detected in neurons [192] and more recently in the hypothalamus related to glucose metabolism [191]. Altogether, this evidence suggests secondary BAs generated by gut microbiota activity potentially play a role in the direct control of hypothalamic function.

In addition, secondary BAs activate enteroendocrine and/or neural gut nutrient sensing routes. Studies in mice demonstrate that downregulation of TGR5 in the context of diet-induced obesity impairs glucose tolerance [193] and increases food intake coupled with a substantial decrease in the plasma levels of GLP-1 and PYY [194,195]. By contrast, TGR5 overexpression in mice improves oral glucose tolerance by inducing a marked enhancement of the postprandial secretion of GLP-1, probably due to increased BAs flow after a highly lipidic meal, [193]. Gut bile acid sensing is also supported by other in vitro [196] and in vivo [195,197] studies in mice showing that secondary BAs trigger GLP-1 and PYY secretion from EECs via TGR5 activation. Human colonic GLP-1-producing EECs also express TGR5, the activation of which by TDCA increases the secretion of GLP-1 [256]. Importantly, obese subjects with or without diabetes show an impaired BA metabolic pathway [256] which might contribute to aberrant GLP-1 secretion [257]. Similarly, experiments in mice demonstrate that secondary BAs induce the secretion of 5-HT [198] subsequently activating 5-HT3 receptors on vagal afferent terminals [199]. The impact of secondary BAs resulting from gut microbiota activity on the brain is still understudied [247] but secondary BAs, through TGR5-GLP-1 and/or 5-HT gut sensing pathways, can potentially modulate food intake and energy homeostasis via efferent routes. Specifically, the role of secondary BAs in inducing satiety through TGR5-activation in vagal afferent neurons has been demonstrated. Indeed, TGR5 expression has been detected in the nodose ganglia cells colocalizing with CCK-1R in rats [200]. Moreover, vagal TGR5 mediates DCA-specific activation of POMC and CART-expressing neurons, but not orexigenic neurons, in the hypothalamus resulting in reduced spontaneous postprandial food intake [200]. Additionally, silencing TGR5 and CCK-1R in the nodose ganglia has an additive effect increasing spontaneous food intake and suggesting a synergistic effect of BA and CCK in hypothalamic short-term satiety signalling via the vagal pathway [200].

Therefore, the bacterial BAs bioconversion capacity may have a strong impact on the intestinal availability of secondary BAs [258,259]. Subsequently, the postprandial enhancement of the TGR5-mediated signalling may be affected by these secondary BAs generated in the intestine as well as by the BAs pool that reaches the brain [247], contributing to the modulation of hypothalamic function. However, to our knowledge, there are currently no studies reporting the effects of specific bacterial species or strains on BA sensing to centrally control energy homeostasis.

4.3. Amino Acid Derived Metabolites

In the gut, as well as in the brain, 5-HT is synthesized from tryptophan [260], an essential amino acid obtained from dietary proteins [261]. Studies in rats demonstrate that 5-HT is released by enterochromaffin cells (ECs) in response to nutrient stimuli such as carbohydrates, glucose or lipids [262,263]. 5-HT binds to 5-HT3 receptors in the vagal afferents, acting thus as sensory transducers [262]. Functionality, the activation of 5-HT signalling in vagal afferents mediates glucose induced inhibition of gastric emptying and lipid-induced food intake suppression, which occurs simultaneously with CCK1 signalling [263,264]. Similarly, in humans, inhibition of 5-HT3 receptor increases the liquid meal ingested [265].

Hafnia alvei, *E. coli* K-12 and species and strains of the genera *Lactococcus*, *Lactobacillus*, *Streptococcus* and *Klebsiella* have been shown in vitro to synthesize 5-HT from tryptophan [266]. The mechanisms by which gut bacteria directly synthesize 5-HT are not clearly established as bacteria lack tryptophan hydroxylase 1 (TPH1), an essential enzyme for 5-HT biosynthesis. Therefore, an alternative mechanism has been proposed, in which tryptophan is decarboxylated to tryptamine [267]. The role of gut microbiota as an enhancer of 5-HT biosynthesis and release from ECs may partly occur through host–microbe interactions mediated by bacterial structural components or metabolites according to certain reports. For example, *Escherichia coli* modulates the host tryptophan hydroxylase-1 activity, enhancing the synthesis of the 5-HT precursor 5-hydroxytryptophan (5-HTP) and the extra-cellular levels of 5-HT through interaction with secreted host derived factors [268]. This strain also contributes to 5-HT clearance by its uptake into enterocytes through the serotonin transporter SERT, which is required for completion of the serotonin circuit synthesis. Other studies show that spore-forming microbes from a healthy human increase 5-HT in colon and blood, through a process that seems to involve host interactions with bacterially produced metabolites such as deoxycholate and SCFAs [269,270]. Probably acting through SCFAs, gut microbiota upregulates the expression of THP1 in colonocytes as demonstrated by the colonization of germ-free mice with mouse or human gut microbiota [270].

A recent study reveals that ECs differentially express chemical sensors according to their spatial location in the gut. Surprisingly, in the small intestine, ECs do not express nutrient sensors but might indirectly release 5-HT in response to nutrient stimuli through the paracrine action of GLP-1, from the neighbour EECs [271]. By contrast, compared with small intestine ECs, colonic ECs present higher expression of microbial sensors such as TGR5, which binds to secondary BAs; GPR132, to acyl amides and lactate; GPR35, to aromatic acids, OLFR558, to isovalerate and FFAR2, to SCFAs, all able to trigger 5-HT release [271]. Accordingly, gut microbiota, through their metabolites, would induce the release of 5-HT in colon and indirectly in the small intestine by stimulating GLP-1 secretion. In fact, although 5-HT cannot cross the blood–brain barrier [139], peripheral 5-HT has been demonstrated to reduce food intake [272–274], suggesting an alternative signalling pathway for the effects of gut-derived 5-HT on the regulation of appetite [275,276].

Studies indicate that, in contrast to 5-HT, its precursor 5-HTP can cross the blood–brain barrier, probably leading to 5-HT production in the brain [139,277,278] and accounting for the role of gut-derived 5-HTP in eliciting satiety. In fact, 5-HTP biosynthesis is enhanced in ECs by gut microbiota-derived SCFAs by upregulating expression of TPH1 as indicated above [278]. Thus, these findings support a role of the gut microbiota-induced 5-HT and 5-HTP in food intake, although further investigations are needed to specifically identify the postprandial microbially produced metabolites and structural products potentially influencing feeding behaviour via this route.

Tryptophan can be fermented into indole [201] which is a ligand of aryl hydrocarbon receptor (AHR), a transcription factor that regulates gene expression. In vitro experiments with GLUTag cell lines show that activation of AHR with an agonist increases the expression of proglucagon as well as GLP-1 secretion [202]. In line with this, experiments with primary mouse colonic L cells demonstrate that indole elicits rapid GLP-1 release during short exposure but that release is inhibited over longer periods and at lower doses [203]. How these actions are coordinated with eating patterns remains unknown; however, the authors proposed that indole at low concentration suppress GLP-1 from L colonic cells but induces an opposite effect in response to high-protein diet. Although indole-mediated nutrient sensing signalling is in an early research stage, the evidence above suggests that AHR might act as a sensor of microbially produced metabolites capable of triggering GLP-1-mediated nutrient signalling from the L cells of the gut to the brain.

In mice, obesity has been associated with increased intestinal IDO1, an enzyme catalysing tryptophan degradation via the kynurenine pathway, thus limiting the bacterial production of indole from tryptophan and increasing kynurenine and its derivatives such as kynurenic acid [279]. In mice, intraperitoneal administration of kynurenic acid induces energy expenditure without affecting locomotion or food intake [280]. This effect seems to be dependent on the activation of GPR35 in the adipose tissue [280], a receptor that has also been detected in sensory neurons of the extrinsic intestinal innervations [281], suggesting that kynurenic acid could play a role in gut-to-brain sensory transmission, which is as yet unexplored.

To our knowledge, so far, studies are lacking on the impact of GABA produced by gut microbes from the dietary amino acid glutamate on the gut–brain communication and, thus, on the control of energy homeostasis. Therefore, the pathways by which microbially produced GABA may influence the hypothalamus function can only be hypothesized from unconnected observations. GABA is the main inhibitory neurotransmitter in the central nervous system. Peripheral GABA cannot cross the blood–brain barrier [282], but activates gut nutrient sensing signalling pathways. First, GABA has been demonstrated to stimulate GLP-1 release from the EECs line GLUTag [204]. Moreover, GABAB receptors are expressed along the gastrointestinal tract [283,284], and co-localized with 5-HT-producing cells identified as morphologically similar to EECs [284], probably inhibiting 5-HT-release by ECs [285]. Second, GABA transmits sensory information by activating GABAA [205] and GABAB receptors [206] expressed in vagal afferents or by secreting exosomes from GABA-stimulated intestinal cells that, in turn, activate neurons [207]. Nonetheless, specific studies are needed to understand the contribution of gut microbiota-derived GABA to nutrient sensing from the gut to the brain and, thus, to the control energy homeostasis in both lean and obese subjects.

4.4. Cellular Components of Gut Bacteria

Certain cellular components of intestinal bacteria stimulate gut-to-brain routes of communication, which may be involved in the central control of energy homeostasis, especially through food intake modulation. This is the case for protein fragments of gut bacteria displaying molecular mimicry with human appetite-regulatory peptides and neuropeptides; i.e., protein fragments of the human α-MSH sequence being identical to gut bacterial-derived proteins from *E. coli*, *Bifidobacterium longum*, *Bacillus cereus*, and certain enteropathogenic bacterial strains [286]. In particular, ClpB produced by *E. coli* K12 is a mimetic of the anorexigenic peptide α-MSH [41], which has a discontinuous five-amino acid overlap containing part of the α-MSH sequence [287]. ClpB is a heat-shock protein with ATPase and chaperon activity mediating the resolubilization of heat-denaturated protein aggregates, thus having a protective function for bacterial cell-induced damage. More recently, the α-MSH-like motif identified within ClpB protein has been confirmed to be specific to the order Enterobacteriales and conserved among its taxa, including *E. coli* strains and *Hafnia* genus [209].

Recent preclinical and clinical studies have demonstrated the effectiveness of ClpB protein on modulating energy homeostasis. In humans, shotgun metagenomic analysis of faeces revealed that gut microbiota ClpB gene, which is associated with decreased weight gain [152], was lower in obese compared to lean individuals [152]. In obese mice, oral administration of the native *E.coli* K12 strain, unlike the ClpB-deficient strain, decreased body weight gain [209], while its food-intake suppression activity was lost when the ClpB-deficient *E. coli* was administrated [41]. In addition, intragastric administration of *Hafnia alvei* HA4597, a ClpB-producer, reduces caloric intake and increases lipolytic effects in *ob/ob* mice coupled with reduced fat mass and body weight gain in both genetically and diet-induced obese mice [209]. Along with the aforementioned effects, Lucas et al. found a decrease in glycaemia, plasma cholesterol and alanine aminotransferase, a marker of obesity-induced steatohepatitis [208]. Mechanistically, ClpB modulates the hypothalamic circuit that controls food intake. In particular, AgRP expression is decreased in obese mice administered the ClpB-producer *H. alvei* [209], while in an animal model of anorexia increased ClpB plasma levels were associated with increased POMC expression [210]. Moreover, *bdnf* mRNA levels increased in mice receiving *E. coli* proteins collected at stationary phase, which include ClpB [42]. These findings prove the contribution of ClpB in upregulating the expression of hypothalamic anorexigenic neuropeptides.

In addition, it is known that ClpB transmits nutrient information from the gut to the brain, directly reaching the hypothalamus when traveling in the systemic circulation, or indirectly by the activation of gut nutrient sensing signalling. The presence of ClpB has been reported in the plasma and in the hypothalamus of rodents and healthy humans [211,288]. Moreover, ClpB protein plasma levels in obese mice are increased by treatment with the ClpB-producer *H. alvei* HA4597, in accordance with food intake reduction [209] and patients with eating disorders show elevated ClpB plasma concentrations compared to healthy individuals [288]. Ex vivo electrophysiological experiments also reveal that ClpB from the stationary stage of *E.coli* directly stimulated the firing rate of POMC neurons, while systemic administration of *E.coli* in the stationary stage increased cFos immunolabelled POMC neurons in the ARC and VMH [42].

ClpB has also been found in the colon of mice, rats, and healthy humans [211]. ClpB stimulates PYY secretion in cultured rat intestinal mucosa in a dose-dependent manner [211] and the ClpB-induced PYY secretion may be enhanced by nutrient-induced bacterial growth [42]. Accordingly, in response to food intake, ClpB initiates a PYY-mediated endocrine or neural nutrient sensing signalling that ultimately regulate the hypothalamic function in suppressing food intake. However, the receptor linking the ClpB agonist with PYY secretion has yet to be identified [153].

A synthetic fragment of ClpB has partial agonist activity on MC3R and MC1R, but not on MC4R [287], whose activation by α-MSH induces PYY secretion [289]. However, the ability of the ClpB-α-MSH-like motif to activate MCRs [287] supports a spatial complementarity of bacterial-derived ClpB and MCRs and, therefore, further investigations regarding the receptors by which ClpB drives its satiating effects are required.

Overall, ClpB is suggested to have a satietogenic effect by systemic and/or neural/endocrine routes. By a systemic route, plasmatic ClpB depends on the quantity of ClpB-producing bacteria in the intestine and, therefore, plasmatic ClpB might be a long-term satiety signal by modulating hypothalamic neuropeptide expression, ultimately influencing meal patterns. By a neural/endocrine route, ClpB might also be a short-term meal termination signal by stimulating release of satiety hormones in the intestine.

Some studies also indicate that LPS and MDP may play a role as food intake modulators. LPS is the outer membrane's major component of Gram-negative bacteria [290]. In obesity, LPS plasma levels are increased due to elevated gut permeability [212], leading to the low-grade inflammation characterizing this metabolic disorder [291]. MDP is a minor component of the peptidoglycan of the cell wall of Gram-negative and more abundant in Gram-positive bacteria [219]. Under normal conditions, MDP is released constantly from degraded gut bacteria [45]. LPS and MDP are CD14/TLR4 and NOD2 agonists, respectively, and both enhance GLP-1-induced nitric oxide production in enteric neurons [47], which may contribute to satiety signalling to the hypothalamus by promoting a shift towards anorexigenic neuropeptides expression [213]. Moreover, MDP activates NOD2 expressed in L-cells to induce GLP-1 release in healthy, but not diet-induced obese mice [220]. Although these effects have been related to glucose tolerance [220] MDP-induced GLP-1 release should also be studied for its effects on food intake.

Interestingly, several studies with rodents report clear effects of LPS and MDP on satiety. Acute or chronic intraperitoneal injections of LPS or MDP reduce food intake [43,45,214,215]. Some studies have found LPS to induce stronger anorexigenic effects in the brain compared to MDP [44,216,292]. The satiety-producing potential between both bacterial components is probably explained by their different abilities to enhance cytokine expression in the hypothalamus [216], since no changes have been detected in the expression of the neuropeptides POMC, NPY or leptin after LPS or MDP administration to the brain [216].

The anorexigenic effects of LPS and MDP seem to be mediated by CD14 for both and by TLR4 for LPS and TLR2 for MDP [217]. The LPS from *E. coli* isolated from healthy humans has been demonstrated to increase murine nodose ganglia neurons excitability in vitro via a TLR4-dependent mechanism [218], suggesting that the intestinal-derived LPS effects on hypothalamic regulation of food intake is mediated by vagal afferent neurons. In contrast to these findings, subdiaphragmatic vagal deafferentation in rats did not suppress food intake reduction after intraperitoneal injection of LPS or MDP, suggesting that peripheral administration of both bacterial components enhances appetite reduction via a humoral but not vagal pathway [214]. Moreover, MDP gavage in rats has no effects in MDP circulating levels nor in feeding patterns, while intraperitoneal MDP reduces food intake, indicating the importance of the humoral pathway for MDP satiety signalling [45]. However, intestinal MDP translocation should be specifically investigated to better define the underlying mechanisms of intestinal-derived MDP on appetite regulation.

Lugarini et al. found that LPS and MDP suppress food intake in obese rats to a similar extent as in non-obese rats [215], suggesting that both bacterial components exert anorexigenic effects regardless obesity. However, the doses of injected LPS and MDP in this study may not reflect the amount of LPS and MDP crossing the intestinal epithelial barrier, whose integrity depends on the metabolic state [212]. Further investigation is required to fully unravel the mechanisms and pathways through which intestinal structural bacterial components reach the hypothalamic centre of appetite regulation under healthy and pathological conditions in vivo.

5. Tackling Obesity with Gut Microbes Mediating in Gut–Brain Communication

Hyperphagia and obesity are caused by defective sensitivity in the hypothalamus or in food reward brain areas to peripheral signals reflecting the nutritional status of the body. Current antiobesity drugs mainly target the central nervous system to suppress appetite. Monotherapies including the agonist of 5-HT2C receptor, lorcaserin, or the protease-resistant long-acting GLP-1, liraglutide, as well as combinational therapies of stimulators of noradrenaline or serotonine-release combined with sympathomimetic anticonvulsant or opiod receptor antagonist are currently approved for treating obesity and T2D, combined with exercise and a balanced diet [14,15,293]. As emerging pharmacotherapies, gut peptides acting as potential agonists of lipid sensing and bile acids receptors and, thereby, as secretagogues are being preclinically and clinically tested as potential drugs to improve metabolic disturbance, especially T2D [294]. Notably, bariatric surgery is the most effective therapy for treating obesity, producing important and sustained weight-loss. This is conducted only in a limited number of subjects with a body mass index (BMI) > 40 and also implies surgical risks [295,296]. It is hypothesized that, rather than mechanical alterations, bariatric surgery alters gut signals sensed by the brain, including amplification of postprandial secretion of gut hormones, resulting in a beneficial impact on energy homeostasis [13]. Research is underway to advance the clinical management of obesity with less invasive pharmacotherapies, which mimic the molecular adaptations of bariatric surgery, including those based on gut–peptide combinatorial strategies [13]. However, the applicability of new drugs is frequently limited by unwanted side effects and safer alternatives are being considered, such as devices to modulate the vagal nerve [14].

In the light of this review, the gut microbiota represents a potential target for favourably regulating gut–brain communication and thereby controlling food intake, energy homeostasis and obesity. Indeed, the supplementation of key intestinal bacteria could increase the microbiota-derived bioactive molecules acting as enhancers of gut hormone secretion and as vagal afferent stimulators, optimizing the response to nutritional clues via gut–brain signalling in obesity. Notably, promising studies have revealed that the gut microbiota might act in harmony with the postprandial processes in the gut that control food intake, as occurs for bacterial-produced ClpB.

Currently, microbiome-based approaches, including faecal microbiota transplantation (FMT) and the administration of intestinal bacteria (probiotics or bio-therapeutic products), are being explored for their impact on gut microbiota structure and function for therapeutic or preventive purposes.

The clinical applicability of FMT to treat obesity is being explored in experimental clinical trials. To date, a couple of studies have been completed and indicate that the microbial shift towards a lean-like profile induced by the oral administration of capsules containing the faecal microbiota from lean donors was not associated with metabolic benefits in the recipients [297,298]. By contrast, an improvement in insulin resistance was observed in male volunteers with metabolic syndrome after a duodenal infusion of gut microbiota from lean donors obtained from small intestine biopsy species [299]. Nevertheless, investigations in this field provide limited data regarding how the microbiota-based approaches specifically control the gut–brain axis in obesity [300].

New indigenous intestinal bacterial species are also being identified and tested for their potential use as next generation probiotics or live biotherapeutic products [301] to maintain metabolic health and combat obesity. This strategy may also contribute to progressing towards the development of safer and more effective microbiome-based strategies for the clinical management of obesity as compared to FMT [302]. Some examples of these new bacterial species and strains are *Eubacterium hallii* L2–7, *Akkermansia muciniphila* ATCC® BAA-835™ or *Bacteroides uniformis* CECT 7771. These were initially selected based on associations between an increased abundance of the bacterial species and a healthy metabolic phenotype in a substantial number of human studies. To date, the clinical efficacy in overweight volunteers has already been demonstrated for *A. muciniphila* (clinical trial no. NCT02637115). The pasteurized *A. muciniphila* significantly improved metabolic parameters (insulin sensitivity, insulinemia and cholesterol) and tended to reduce body weight and fat mass [303]. Nonetheless, the investigations on the mechanisms by which these bacteria induce their benefits have mainly focused on the regulation of immune pathways involved in obesity-associated inflammation, generally resulting in metabolic improvements in preclinical trials [304]. Although evidence of the possible action of these bacteria through the modulation of the gut–brain axis is scarce [305–307], preclinical studies demonstrated that *A. muciniphila* administration to obese mice increases the intestinal production of the endocannabinoid, 2-arachidonoylglycerol (2-AG); and the endocannabinoid analogue, oleoylglycerol (2-OG) [305]. These ligands can modulate the secretory function of L cells and, theoretically, the activity of vagal afferents [308,309].

Other emerging strategies to identify new probiotics or live bio-therapeutic products tend to mimic drug discovery approaches. In this case, the selection of effector intestinal bacteria is based on in silico predictions (computational molecular modelling) of their capacity to produce bioactive molecules [310]. Based on this strategy, the strain *Hafnia alvei* 4597 was selected to provide the protein ClpB, a bacterial mimetic of α-MSH, which induced satiety. This discovery was based on the initial detection of α-MSH reactive autoantibodies in plasma of humans and rodents [311] and in the in silico screening of the presence of ClpB in members of the family Enterobacteriaceae, including the genus *Hafnia* [312]. Studies in obese mice demonstrated that the administration of *Hafnia alvei* 4597 suppressed food intake and improved the metabolic disturbances associated to obesity [209]. Clinical studies in humans to show the efficacy are underway (clinical trial no. NCT03657186).

In spite of this progress, further advances are still needed to increase the effectiveness and clinical applicability of microbiome-based strategies. In fact, interindividual variability of the human host and its gut microbiota could change the response to these therapies [313]. Therefore, efforts are also underway to develop precision probiotics through a deep understanding of their mode of action and the factors influencing the individual host response [310]. This requires the integration of the person-specific genotypic and phenotypic variables, including microbiome data (strain-level composition, transcriptomics, metabolomics, etc.), that may help to predict the health outcomes of a specific intervention. This strategy may also be applied to the identification of personalized microbiome-based strategies that improve nutrient sensing routes and, thus, control energy homeostasis. To this end, further efforts are needed for the identification of cross-feeding pathways between different intestinal bacteria of a specific host and the resulting dietary and bacterial-derived effector molecules, considering the person's meal timings and dietary composition.

Author Contributions: Writing—Original Draft: M.R.-P., C.B.-V., I.L.-A., Y.S.; Writing—Review and Editing: Y.S., M.O., R.L.-G.; Figures: M.O., R.L.-G.; Funding Acquisition: Y.S. All authors have read and agreed to the published version of the manuscript.

Funding: This work is supported by grant AGL2017-88801-P from the Spanish Ministry of Science and Innovation (MICINN, Spain). The Marie Sklodowska-Curie grant agreement No. 797297 (M.R-P), the FPI grant of I.L.-A., the FPU grant of R.L. and the Juan de la Cierva contract of M.O. and the contract of C. B-V for promotion of youth employment in R+D+I from MICINN (Spain) are fully acknowledged.

Institutional Review Board Statement: Not applicable.

Informed Consent Statement: Not applicable.

Data Availability Statement: Not applicable.

Conflicts of Interest: The authors declare no conflict of interest.

References

1. Reimann, F.; Tolhurst, G.; Gribble, F.M. G-Protein-Coupled Receptors in Intestinal Chemosensation. *Cell Metab.* **2012**, *15*, 421–431. [CrossRef] [PubMed]
2. Gribble, F.M.; Reimann, F. Function and mechanisms of enteroendocrine cells and gut hormones in metabolism. *Nat. Rev. Endocrinol.* **2019**, *15*, 226–237. [CrossRef]
3. Psichas, A.; Reimann, F.; Gribble, F.M. Gut chemosensing mechanisms. *J. Clin. Investig.* **2015**, *125*, 908–917. [CrossRef] [PubMed]
4. Bennett, L.; Yang, M.; Enikolopov, G.; Iacovitti, L. Circumventricular organs: A novel site of neural stem cells in the adult brain. *Mol. Cell. Neurosci.* **2009**, *41*, 337–347. [CrossRef]
5. Cone, R.D. Anatomy and regulation of the central melanocortin system. *Nat. Neurosci.* **2005**, *8*, 571–578. [CrossRef]
6. Dietrich, M.; Horvath, T.L. Hypothalamic control of energy balance: Insights into the role of synaptic plasticity. *Trends Neurosci.* **2013**, *36*, 65–73. [CrossRef] [PubMed]
7. Strader, A.D.; Woods, S.C. Gastrointestinal hormones and food intake. *Gastroenterology* **2005**, *128*, 175–191. [CrossRef]
8. Wang, Y.B.; De Lartigue, G.; Page, A.J. Dissecting the Role of Subtypes of Gastrointestinal Vagal Afferents. *Front. Physiol.* **2020**, *11*, 11. [CrossRef]
9. Berthoud, H.-R.; Neuhuber, W.L. Functional and chemical anatomy of the afferent vagal system. *Auton. Neurosci.* **2000**, *85*, 1–17. [CrossRef]
10. Cruciani-Guglielmacci, C.; Fioramonti, X. Editorial: Brain Nutrient Sensing in the Control of Energy Balance: New Insights and Perspectives. *Front. Physiol.* **2019**, *10*. [CrossRef]
11. Duca, F.A.; Yue, J.T. Fatty acid sensing in the gut and the hypothalamus: In Vivo and in vitro perspectives. *Mol. Cell. Endocrinol.* **2014**, *397*, 23–33. [CrossRef]
12. Blouet, C.; Schwartz, G.J. Hypothalamic nutrient sensing in the control of energy homeostasis. *Behav. Brain Res.* **2010**, *209*, 1–12. [CrossRef] [PubMed]
13. Clemmensen, C.; Müller, T.D.; Woods, S.C.; Berthoud, H.-R.; Seeley, R.J.; Tschöp, M.H. Gut-Brain Cross-Talk in Metabolic Control. *Cell* **2017**, *168*, 758–774. [CrossRef] [PubMed]
14. Gautron, L.; Elmquist, J.K.; Williams, K.W. Neural Control of Energy Balance: Translating Circuits to Therapies. *Cell* **2015**, *161*, 133–145. [CrossRef]
15. Drucker, D.J. Mechanisms of Action and Therapeutic Application of Glucagon-like Peptide-1. *Cell Metab.* **2018**, *27*, 740–756. [CrossRef] [PubMed]
16. Ridaura, V.K.; Faith, J.J.; Rey, F.E.; Cheng, J.; Duncan, A.E.; Kau, A.L.; Griffin, N.W.; Lombard, V.; Henrissat, B.; Bain, J.R.; et al. Gut Microbiota from Twins Discordant for Obesity Modulate Metabolism in Mice. *Science* **2013**, *341*, 1241214. [CrossRef] [PubMed]
17. Sonnenburg, J.L.; Bäckhed, F. Diet–microbiota interactions as moderators of human metabolism. *Nat. Cell Biol.* **2016**, *535*, 56–64. [CrossRef] [PubMed]

18. Takahashi, J.S. Transcriptional architecture of the mammalian circadian clock. *Nat. Rev. Genet.* **2017**, *18*, 164–179. [CrossRef] [PubMed]
19. Pittendrigh, C.S. Temporal Organization: Reflections of a Darwinian Clock-Watcher. *Annu. Rev. Physiol.* **1993**, *55*, 17–54. [CrossRef]
20. Yoo, S.-H.; Yamazaki, S.; Lowrey, P.L.; Shimomura, K.; Ko, C.H.; Buhr, E.D.; Siepka, S.M.; Hong, H.-K.; Oh, W.J.; Yoo, O.J.; et al. PERIOD2:LUCIFERASE real-time reporting of circadian dynamics reveals persistent circadian oscillations in mouse peripheral tissues. *Proc. Natl. Acad. Sci. USA* **2004**, *101*, 5339–5346. [CrossRef]
21. Sakamoto, K.; Nagase, T.; Fukui, H.; Horikawa, K.; Okada, T.; Tanaka, H.; Sato, K.; Miyake, Y.; Ohara, O.; Kako, K.; et al. Multitissue Circadian Expression of Rat periodHomolog (rPer2) mRNA Is Governed by the Mammalian Circadian Clock, the Suprachiasmatic Nucleus in the Brain. *J. Biol. Chem.* **1998**, *273*, 27039–27042. [CrossRef]
22. Akhtar, R.; Reddy, A.B.; Maywood, E.S.; Clayton, J.D.; King, V.M.; Smith, A.G.; Gant, T.W.; Hastings, M.H.; Kyriacou, C.P. Circadian Cycling of the Mouse Liver Transcriptome, as Revealed by cDNA Microarray, Is Driven by the Suprachiasmatic Nucleus. *Curr. Biol.* **2002**, *12*, 540–550. [CrossRef]
23. Challet, E. The circadian regulation of food intake. *Nat. Rev. Endocrinol.* **2019**, *15*, 393–405. [CrossRef]
24. Wehrens, S.M.; Christou, S.; Isherwood, C.; Middleton, B.; Gibbs, M.A.; Archer, S.N.; Skene, D.; Johnston, J.D. Meal Timing Regulates the Human Circadian System. *Curr. Biol.* **2017**, *27*, 1768–1775.e3. [CrossRef]
25. Stokkan, K.-A.; Yamazaki, S.; Tei, H.; Sakaki, Y.; Menaker, M. Entrainment of the Circadian Clock in the Liver by Feeding. *Science* **2001**, *291*, 490–493. [CrossRef]
26. Knutsson, A.; Kempe, A. Shift work and diabetes—A systematic review. *Chronobiol. Int.* **2014**, *31*, 1146–1151. [CrossRef]
27. McHill, A.W.; Wright, K.P. Role of sleep and circadian disruption on energy expenditure and in metabolic predisposition to human obesity and metabolic disease. *Obes. Rev.* **2017**, *18*, 15–24. [CrossRef] [PubMed]
28. Ramirez-Plascencia, O.D.; Saderi, N.; Escobar, C.; Salgado-Delgado, R.C. Feeding during the rest phase promotes circadian conflict in nuclei that control energy homeostasis and sleep-wake cycle in rats. *Eur. J. Neurosci.* **2017**, *45*, 1325–1332. [CrossRef]
29. Reutrakul, S.; Hood, M.; Crowley, S.J.; Morgan, M.K.; Teodori, M.; Knutson, K. The Relationship Between Breakfast Skipping, Chronotype, and Glycemic Control in Type 2 Diabetes. *Chronobiol. Int.* **2013**, *31*, 64–71. [CrossRef] [PubMed]
30. Jakubowicz, D.; Wainstein, J.; Landau, Z.; Raz, I.; Ahren, B.; Chapnik, N.; Ganz, T.; Menaged, M.; Barnea, M.; Bar-Dayan, Y.; et al. Influences of Breakfast on Clock Gene Expression and Postprandial Glycemia in Healthy Individuals and Individuals With Diabetes: A Randomized Clinical Trial. *Diabetes Care* **2017**, *40*, 1573–1579. [CrossRef] [PubMed]
31. Kaczmarek, J.L.; Musaad, S.M.; Holscher, H.D. Time of day and eating behaviors are associated with the composition and function of the human gastrointestinal microbiota. *Am. J. Clin. Nutr.* **2017**, *106*, ajcn156380. [CrossRef]
32. Reitmeier, S.; Kiessling, S.; Clavel, T.; List, M.; Almeida, E.L.; Ghosh, T.S.; Neuhaus, K.; Grallert, H.; Linseisen, J.; Skurk, T.; et al. Arrhythmic Gut Microbiome Signatures Predict Risk of Type 2 Diabetes. *Cell Host Microbe* **2020**, *28*, 258–272.e6. [CrossRef] [PubMed]
33. Mukherji, A.; Kobiita, A.; Ye, T.; Chambon, P. Homeostasis in Intestinal Epithelium Is Orchestrated by the Circadian Clock and Microbiota Cues Transduced by TLRs. *Cell* **2013**, *153*, 812–827. [CrossRef]
34. Weger, B.D.; Gobet, C.; Yeung, J.; Martin, E.; Jimenez, S.; Betrisey, B.; Foata, F.; Berger, B.; Balvay, A.; Foussier, A.; et al. The Mouse Microbiome Is Required for Sex-Specific Diurnal Rhythms of Gene Expression and Metabolism. *Cell Metab.* **2019**, *29*, 362–382.e8. [CrossRef]
35. Thaiss, C.A.; Zeevi, D.; Levy, M.; Zilberman-Schapira, G.; Suez, J.; Tengeler, A.C.; Abramson, L.; Katz, M.N.; Korem, T.; Zmora, N.; et al. Transkingdom Control of Microbiota Diurnal Oscillations Promotes Metabolic Homeostasis. *Cell* **2014**, *159*, 514–529. [CrossRef] [PubMed]
36. Leone, V.; Gibbons, S.M.; Martinez, K.; Hutchison, A.L.; Huang, E.Y.; Cham, C.M.; Pierre, J.F.; Heneghan, A.F.; Nadimpalli, A.; Hubert, N.; et al. Effects of Diurnal Variation of Gut Microbes and High-Fat Feeding on Host Circadian Clock Function and Metabolism. *Cell Host Microbe* **2015**, *17*, 681–689. [CrossRef] [PubMed]
37. Zarrinpar, A.; Chaix, A.; Yooseph, S.; Panda, S. Diet and Feeding Pattern Affect the Diurnal Dynamics of the Gut Microbiome. *Cell Metab.* **2014**, *20*, 1006–1017. [CrossRef] [PubMed]
38. Thaiss, C.A.; Levy, M.; Korem, T.; Dohnalová, L.; Shapiro, H.; Jaitin, D.A.; David, E.; Winter, D.R.; Gury-BenAri, M.; Tatirovsky, E.; et al. Microbiota Diurnal Rhythmicity Programs Host Transcriptome Oscillations. *Cell* **2016**, *167*, 1495–1510.e12. [CrossRef]
39. Vijay-Kumar, M.; Aitken, J.D.; Carvalho, F.A.; Cullender, T.C.; Mwangi, S.; Srinivasan, S.; Sitaraman, S.V.; Knight, R.; Ley, R.E.; Gewirtz, A.T. Metabolic Syndrome and Altered Gut Microbiota in Mice Lacking Toll-Like Receptor 5. *Science* **2010**, *328*, 228–231. [CrossRef]
40. Fetissov, S.O. Role of the gut microbiota in host appetite control: Bacterial growth to animal feeding behaviour. *Nat. Rev. Endocrinol.* **2017**, *13*, 11–25. [CrossRef]
41. Tennoune, N.; Chan, P.; Breton, J.; Legrand, R.; Chabane, Y.N.; Akkermann, K.; Järv, A.; Ouelaa, W.; Takagi, K.; Ghouzali, I.; et al. Bacterial ClpB heat-shock protein, an antigen-mimetic of the anorexigenic peptide α-MSH, at the origin of eating disorders. *Transl. Psychiatry* **2014**, *4*, e458. [CrossRef]
42. Breton, J.; Tennoune, N.; Lucas, N.; Francois, M.; Legrand, R.; Jacquemot, J.; Goichon, A.; Guérin, C.; Peltier, J.; Pestel-Caron, M.; et al. Gut Commensal E. coli Proteins Activate Host Satiety Pathways following Nutrient-Induced Bacterial Growth. *Cell Metab.* **2016**, *23*, 324–334. [CrossRef] [PubMed]

43. Langhans, W.; Harlacher, R.; Balkowski, G.; Scharrer, E. Comparison of the effects of bacterial lipopolysaccharide and muramyl dipeptide on food intake. *Physiol. Behav.* **1990**, *47*, 805–813. [CrossRef]
44. Langhans, W.; Balkowski, G.; Savoldelli, D. Differential feeding responses to bacterial lipopolysaccharide and muramyl dipeptide. *Am. J. Physiol. Integr. Comp. Physiol.* **1991**, *261*, R659–R664. [CrossRef]
45. Fosset, S.; Fromentin, G.; Rampin, O.; Lang, V.; Mathieu, F.; Tomé, D. Pharmacokinetics and feeding responses to muramyl dipeptide in rats. *Physiol. Behav.* **2003**, *79*, 173–182. [CrossRef]
46. Lebrun, L.J.; Lenaerts, K.; Kiers, D.; de Barros, J.-P.P.; Le Guern, N.; Plesnik, J.; Thomas, C.; Bourgeois, T.; Dejong, C.H.; Kox, M.; et al. Enteroendocrine L Cells Sense LPS after Gut Barrier Injury to Enhance GLP-1 Secretion. *Cell Rep.* **2017**, *21*, 1160–1168. [CrossRef] [PubMed]
47. Grasset, E.; Puel, A.; Charpentier, J.; Collet, X.; Christensen, J.E.; Tercé, F.; Burcelin, R. A Specific Gut Microbiota Dysbiosis of Type 2 Diabetic Mice Induces GLP-1 Resistance through an Enteric NO-Dependent and Gut-Brain Axis Mechanism. *Cell Metab.* **2017**, *25*, 1075–1090.e5. [CrossRef]
48. Nguyen, A.T.; Mandard, S.; Dray, C.; Deckert, V.; Valet, P.; Besnard, P.; Drucker, D.J.; Lagrost, L.; Grober, J. Lipopolysaccharides-Mediated Increase in Glucose-Stimulated Insulin Secretion: Involvement of the GLP-1 Pathway. *Diabetes* **2013**, *63*, 471–482. [CrossRef] [PubMed]
49. Cabrera-Mulero, A.; Tinahones, A.; Bandera, B.; Moreno-Indias, I.; Macías-González, M.; Tinahones, F.J. Keto microbiota: A powerful contributor to host disease recovery. *Rev. Endocr. Metab. Disord.* **2019**, *20*, 415–425. [CrossRef]
50. De Filippo, C.; Cavalieri, D.; Di Paola, M.; Ramazzotti, M.; Poullet, J.B.; Massart, S.; Collini, S.; Pieraccini, G.; Lionetti, P. Impact of diet in shaping gut microbiota revealed by a comparative study in children from Europe and rural Africa. *Proc. Natl. Acad. Sci. USA* **2010**, *107*, 14691–14696. [CrossRef]
51. Martín-Peláez, S.; Fito, M.; Castaner, O. Mediterranean Diet Effects on Type 2 Diabetes Prevention, Disease Progression, and Related Mechanisms. A Review. *Nutrients* **2020**, *12*, 2236. [CrossRef] [PubMed]
52. Wu, G.D.; Chen, J.; Hoffmann, C.; Bittinger, K.; Chen, Y.Y.; Keilbaugh, S.A.; Bewtra, M.; Knights, D.; Walters, W.A.; Knight, R.; et al. Linking Long-Term Dietary Patterns with Gut Microbial Enterotypes. *Science* **2011**, *334*, 105–108. [CrossRef]
53. David, L.A.; Maurice, C.F.; Carmody, R.N.; Gootenberg, D.; Button, J.E.; Wolfe, B.E.; Ling, A.V.; Devlin, A.S.; Varma, Y.; Fischbach, M.; et al. Diet rapidly and reproducibly alters the human gut microbiome. *Nat. Cell Biol.* **2014**, *505*, 559–563. [CrossRef] [PubMed]
54. Popkin, B.M.; Gordon-Larsen, P. The nutrition transition: Worldwide obesity dynamics and their determinants. *Int. J. Obes.* **2004**, *28 Suppl 3*, S2–S9. [CrossRef]
55. Speechly, D.; Buffenstein, R. Appetite dysfunction in obese males: Evidence for role of hyperinsulinaemia in passive overconsumption with a high fat diet. *Eur. J. Clin. Nutr.* **2000**, *54*, 225–233. [CrossRef] [PubMed]
56. Le Roux, C.W.; Batterham, R.L.; Aylwin, S.J.B.; Patterson, M.; Borg, C.-M.; Wynne, K.; Kent, A.; Vincent, R.; Gardiner, J.; Ghatei, M.A.; et al. Attenuated Peptide YY Release in Obese Subjects Is Associated with Reduced Satiety. *Endocrinology* **2006**, *147*, 3–8. [CrossRef]
57. Mittelman, S.D.; Klier, K.; Braun, S.; Azen, C.; Geffner, M.E.; Buchanan, T.A. Obese Adolescents Show Impaired Meal Responses of the Appetite-Regulating Hormones Ghrelin and PYY. *Obesity* **2010**, *18*, 918–925. [CrossRef]
58. Hira, T.; Pinyo, J.; Hara, H. What Is GLP-1 Really Doing in Obesity? *Trends Endocrinol. Metab.* **2020**, *31*, 71–80. [CrossRef]
59. French, S.J.; Murray, B.; Rumsey, R.D.E.; Fadzlin, R.; Read, N.W. Adaptation to high-fat diets: Effects on eating behaviour and plasma cholecystokinin. *Br. J. Nutr.* **1995**, *73*, 179–189. [CrossRef] [PubMed]
60. A Duca, F.; Swartz, T.D.; Sakar, Y.; Covasa, M. Decreased intestinal nutrient response in diet-induced obese rats: Role of gut peptides and nutrient receptors. *Int. J. Obes.* **2013**, *37*, 375–381. [CrossRef]
61. Duca, F.A.; Zhong, L.; Covasa, M. Reduced CCK signaling in obese-prone rats fed a high fat diet. *Horm. Behav.* **2013**, *64*, 812–817. [CrossRef]
62. Knauf, C.; Cani, P.D.; Kim, N.-H.; Iglesias, M.A.; Chabo, C.; Waget, A.; Colom, A.; Rastrelli, S.; Delzenne, N.M.; Drucker, D.J.; et al. Role of Central Nervous System Glucagon-Like Peptide-1 Receptors in Enteric Glucose Sensing. *Diabetes* **2008**, *57*, 2603–2612. [CrossRef] [PubMed]
63. Morenga, L.T.; Mallard, S.; Mann, J. Dietary sugars and body weight: Systematic review and meta-analyses of randomised controlled trials and cohort studies. *BMJ* **2012**, *346*, e7492. [CrossRef] [PubMed]
64. Ochoa, M.; Lallès, J.-P.; Malbert, C.-H.; Val-Laillet, D. Dietary sugars: Their detection by the gut-brain axis and their peripheral and central effects in health and diseases. *Eur. J. Nutr.* **2014**, *54*, 1–24. [CrossRef] [PubMed]
65. I Domingos, A.; Sordillo, A.; Dietrich, M.; Liu, Z.-W.; A Tellez, L.; Vaynshteyn, J.; Ferreira, J.; I Ekstrand, M.; Horvath, T.L.; E De Araujo, I.; et al. Hypothalamic melanin concentrating hormone neurons communicate the nutrient value of sugar. *eLife* **2013**, *2*, e01462. [CrossRef]
66. Steinbusch, L.; Labouèbe, G.; Thorens, B. Brain glucose sensing in homeostatic and hedonic regulation. *Trends Endocrinol. Metab.* **2015**, *26*, 455–466. [CrossRef]
67. Dorton, H.M.; Luo, S.; Monterosso, J.R.; Page, K.A. Influences of Dietary Added Sugar Consumption on Striatal Food-Cue Reactivity and Postprandial GLP-1 Response. *Front. Psychiatry* **2018**, *8*, 297. [CrossRef]
68. Jones, S.; Luo, S.; Dorton, H.M.; Yunker, A.G.; Angelo, B.; Defendis, A.; Monterosso, J.R.; Page, K.A. Obesity and Dietary Added Sugar Interact to Affect Postprandial GLP-1 and Its Relationship to Striatal Responses to Food Cues and Feeding Behavior. *Front. Endocrinol.* **2021**, *12*. [CrossRef]

69. Kilpatrick, L.A.; Coveleskie, K.; Connolly, L.; Labus, J.S.; Ebrat, B.; Stains, J.; Jiang, Z.; Suyenobu, B.Y.; Raybould, H.E.; Tillisch, K.; et al. Influence of Sucrose Ingestion on Brainstem and Hypothalamic Intrinsic Oscillations in Lean and Obese Women. *Gastroenterology* **2014**, *146*, 1212–1221. [CrossRef]
70. Parton, L.E.; Ye, C.P.; Coppari, R.; Enriori, P.J.; Choi, B.; Zhang, C.-Y.; Xu, C.; Vianna, C.R.; Balthasar, N.; Lee, C.E.; et al. Glucose sensing by POMC neurons regulates glucose homeostasis and is impaired in obesity. *Nature* **2007**, *449*, 228–232. [CrossRef]
71. Sonnenburg, E.D.; Smits, S.A.; Tikhonov, M.; Higginbottom, S.K.; Wingreen, N.S.; Sonnenburg, J.L. Diet-induced extinctions in the gut microbiota compound over generations. *Nat. Cell Biol.* **2016**, *529*, 212–215. [CrossRef]
72. Wolters, M.; Ahrens, J.; Pérez, M.R.; Watkins, C.; Sanz, Y.; Benítez-Páez, A.; Stanton, C.; Günther, K. Dietary fat, the gut microbiota, and metabolic health—A systematic review conducted within the MyNewGut project. *Clin. Nutr.* **2019**, *38*, 2504–2520. [CrossRef]
73. Agans, R.; Gordon, A.; Kramer, D.L.; Perez-Burillo, S.; Rufián-Henares, J.A.; Paliy, O. Dietary Fatty Acids Sustain the Growth of the Human Gut Microbiota. *Appl. Environ. Microbiol.* **2018**, *84*, 84. [CrossRef] [PubMed]
74. Devkota, S.; Wang, Y.; Musch, M.W.; Leone, V.; Fehlner-Peach, H.; Nadimpalli, A.; Antonopoulos, D.A.; Jabri, B.; Chang, E.B. Dietary-fat-induced taurocholic acid promotes pathobiont expansion and colitis in Il10−/− mice. *Nat. Cell Biol.* **2012**, *487*, 104–108. [CrossRef] [PubMed]
75. Natividad, J.M.; Lamas, B.; Pham, H.P.; Michel, M.-L.; Rainteau, D.; Bridonneau, C.; DA Costa, G.; Vlieg, J.V.H.; Sovran, B.; Chamignon, C.; et al. Bilophila wadsworthia aggravates high fat diet induced metabolic dysfunctions in mice. *Nat. Commun.* **2018**, *9*, 1–15. [CrossRef]
76. Bauer, P.V.; Duca, F.A.; Waise, T.Z.; Dranse, H.J.; Rasmussen, B.A.; Puri, A.; Rasti, M.; O'Brien, C.A.; Lam, T.K. Lactobacillus gasseri in the Upper Small Intestine Impacts an ACSL3-Dependent Fatty Acid-Sensing Pathway Regulating Whole-Body Glucose Homeostasis. *Cell Metab.* **2018**, *27*, 572–587.e6. [CrossRef] [PubMed]
77. Jang, C.; Hui, S.; Lu, W.; Cowan, A.J.; Morscher, R.J.; Lee, G.; Liu, W.; Tesz, G.J.; Birnbaum, M.J.; Rabinowitz, J.D. The Small Intestine Converts Dietary Fructose into Glucose and Organic Acids. *Cell Metab.* **2018**, *27*, 351–361.e3. [CrossRef]
78. Zhao, S.; Jang, C.; Liu, J.; Uehara, K.; Gilbert, M.; Izzo, L.; Zeng, X.; Trefely, S.; Fernandez, S.; Carrer, A.; et al. Dietary fructose feeds hepatic lipogenesis via microbiota-derived acetate. *Nat. Cell Biol.* **2020**, *579*, 586–591. [CrossRef]
79. Townsend, G.E.; Han, W.; Schwalm, N.; Raghavan, V.; Barry, N.A.; Goodman, A.L.; Groisman, E.A. Dietary sugar silences a colonization factor in a mammalian gut symbiont. *Proc. Natl. Acad. Sci. USA* **2019**, *116*, 233–238. [CrossRef]
80. Bauer, P.V.; Duca, F.A.; Waise, T.Z.; Rasmussen, B.A.; Abraham, M.A.; Dranse, H.J.; Puri, A.; O'Brien, C.A.; Lam, T.K. Metformin Alters Upper Small Intestinal Microbiota that Impact a Glucose-SGLT1-Sensing Glucoregulatory Pathway. *Cell Metab.* **2018**, *27*, 101–117.e5. [CrossRef]
81. Melina, V.; Craig, W.; Levin, S. Position of the Academy of Nutrition and Dietetics: Vegetarian Diets. *J. Acad. Nutr. Diet.* **2016**, *116*, 1970–1980. [CrossRef] [PubMed]
82. Brytek-Matera, A. Restrained Eating and Vegan, Vegetarian and Omnivore Dietary Intakes. *Nutrients* **2020**, *12*, 2133. [CrossRef] [PubMed]
83. Belinova, L.; Kahleova, H.; Malinska, H.; Topolcan, O.; Vrzalova, J.; Oliyarnyk, O.; Kazdova, L.; Hill, M.; Pelikanova, T. Differential Acute Postprandial Effects of Processed Meat and Isocaloric Vegan Meals on the Gastrointestinal Hormone Response in Subjects Suffering from Type 2 Diabetes and Healthy Controls: A Randomized Crossover Study. *PLoS ONE* **2014**, *9*, e107561. [CrossRef]
84. Klementova, M.; Thieme, L.; Haluzik, M.; Pavlovicova, R.; Hill, M.; Pelikanova, T.; Kahleova, H. A Plant-Based Meal Increases Gastrointestinal Hormones and Satiety More Than an Energy- and Macronutrient-Matched Processed-Meat Meal in T2D, Obese, and Healthy Men: A Three-Group Randomized Crossover Study. *Nutrients* **2019**, *11*, 157. [CrossRef] [PubMed]
85. Karlsen, M.C.; Rogers, G.; Miki, A.; Lichtenstein, A.H.; Folta, S.C.; Economos, C.D.; Jacques, P.F.; Livingston, K.A.; McKeown, N.M. Theoretical Food and Nutrient Composition of Whole-Food Plant-Based and Vegan Diets Compared to Current Dietary Recommendations. *Nutrients* **2019**, *11*, 625. [CrossRef] [PubMed]
86. Clarys, P.; Deliens, T.; Huybrechts, I.; Deriemaeker, P.; Vanaelst, B.; De Keyzer, W.; Hebbelinck, M.; Mullie, P. Comparison of Nutritional Quality of the Vegan, Vegetarian, Semi-Vegetarian, Pesco-Vegetarian and Omnivorous Diet. *Nutrients* **2014**, *6*, 1318–1332. [CrossRef]
87. Stephen, A.M.; Champ, M.M.-J.; Cloran, S.J.; Fleith, M.; Van Lieshout, L.; Mejborn, H.; Burley, V.J. Dietary fibre in Europe: Current state of knowledge on definitions, sources, recommendations, intakes and relationships to health. *Nutr. Res. Rev.* **2017**, *30*, 149–190. [CrossRef]
88. Berggren, A.M.; Björck, I.M.E.; Nyman, E.M.G.L.; Eggum, B.O. Short-chain fatty acid content and pH in caecum of rats given various sources of carbohydrates. *J. Sci. Food Agric.* **1993**, *63*, 397–406. [CrossRef]
89. Zhao, C.; Dong, H.; Zhang, Y.; Li, Y. Discovery of potential genes contributing to the biosynthesis of short-chain fatty acids and lactate in gut microbiota from systematic investigation in *E. coli*. *NPJ Biofilms Microbiomes* **2019**, *5*, 1–8. [CrossRef]
90. Tomova, A.; Bukovsky, I.; Rembert, E.; Yonas, W.; Alwarith, J.; Barnard, N.D.; Kahleova, H. The Effects of Vegetarian and Vegan Diets on Gut Microbiota. *Front. Nutr.* **2019**, *6*, 47. [CrossRef]
91. Ventura, M.; Canchaya, C.; Tauch, A.; Chandra, G.; Fitzgerald, G.F.; Chater, K.F.; Van Sinderen, D. Genomics of Actinobacteria: Tracing the Evolutionary History of an Ancient Phylum. *Microbiol. Mol. Biol. Rev.* **2007**, *71*, 495–548. [CrossRef]
92. Delzenne, N.M.; Olivares, M.; Neyrinck, A.M.; Beaumont, M.; Kjølbæk, L.; Larsen, T.M.; Benítez-Páez, A.; Romaní-Pérez, M.; Garcia-Campayo, V.; Bosscher, D.; et al. Nutritional interest of dietary fiber and prebiotics in obesity: Lessons from the MyNewGut consortium. *Clin. Nutr.* **2020**, *39*, 414–424. [CrossRef]

93. Venkataraman, A.; Sieber, J.R.; Schmidt, A.W.; Waldron, C.; Theis, K.R.; Schmidt, T.M. Variable responses of human microbiomes to dietary supplementation with resistant starch. *Microbiome* **2016**, *4*, 1–9. [CrossRef] [PubMed]
94. Davis, L.M.G.; Martínez, I.; Walter, J.; Goin, C.; Hutkins, R.W. Barcoded Pyrosequencing Reveals That Consumption of Galactooligosaccharides Results in a Highly Specific Bifidogenic Response in Humans. *PLoS ONE* **2011**, *6*, e25200. [CrossRef] [PubMed]
95. Benítez-Páez, A.; Kjølbæk, L.; Del Pulgar, E.M.G.; Brahe, L.K.; Astrup, A.; Matysik, S.; Schött, H.-F.; Krautbauer, S.; Liebisch, G.; Boberska, J.; et al. A Multi-omics Approach to Unraveling the Microbiome-Mediated Effects of Arabinoxylan Oligosaccharides in Overweight Humans. *mSystems* **2019**, *4*, e00209–e00219. [CrossRef]
96. Kjølbæk, L.; Benítez-Páez, A.; del Pulgar, E.M.G.; Brahe, L.K.; Liebisch, G.; Matysik, S.; Rampelli, S.; Vermeiren, J.; Brigidi, P.; Larsen, L.H.; et al. Arabinoxylan oligosaccharides and polyunsaturated fatty acid effects on gut microbiota and metabolic markers in overweight individuals with signs of metabolic syndrome: A randomized cross-over trial. *Clin. Nutr.* **2020**, *39*, 67–79. [CrossRef]
97. Canfora, E.E.; van der Beek, C.M.; Hermes, G.D.; Goossens, G.H.; Jocken, J.W.; Holst, J.J.; van Eijk, H.M.; Venema, K.; Smidt, H.; Zoetendal, E.G.; et al. Supplementation of Diet With Galacto-oligosaccharides Increases Bifidobacteria, but Not Insulin Sensitivity, in Obese Prediabetic Individuals. *Gastroenterology* **2017**, *153*, 87–97.e3. [CrossRef] [PubMed]
98. Kovatcheva-Datchary, P.; Nilsson, A.; Akrami, R.; Lee, Y.S.; De Vadder, F.; Arora, T.; Hallen, A.; Martens, E.; Björck, I.; Bäckhed, F. Dietary Fiber-Induced Improvement in Glucose Metabolism Is Associated with Increased Abundance of Prevotella. *Cell Metab.* **2015**, *22*, 971–982. [CrossRef] [PubMed]
99. Soty, M.; Gautier-Stein, A.; Rajas, F.; Mithieux, G. Gut-Brain Glucose Signaling in Energy Homeostasis. *Cell Metab.* **2017**, *25*, 1231–1242. [CrossRef] [PubMed]
100. De Vadder, F.; Kovatcheva-Datchary, P.; Zitoun, C.; Duchampt, A.; Bäckhed, F.; Mithieux, G. Microbiota-Produced Succinate Improves Glucose Homeostasis via Intestinal Gluconeogenesis. *Cell Metab.* **2016**, *24*, 151–157. [CrossRef]
101. Products, N.A.A. (Nda) E.P.O.D. Scientific Opinion on the substantiation of health claims related to fruits and/or vegetables (ID 1212, 1213, 1214, 1217, 1218, 1219, 1301, 1425, 1426, 1427, 1428, 1429, 1430) and to the "Mediterranean diet" (ID 1423) pursuant to Article 13(1) of Regulati. *EFSA J.* **2011**, *9*, 2245. [CrossRef]
102. Willett, W.C.; Sacks, F.; Trichopoulou, A.; Drescher, G.; Ferro-Luzzi, A.; Helsing, E.; Trichopoulos, D. Mediterranean diet pyramid: A cultural model for healthy eating. *Am. J. Clin. Nutr.* **1995**, *61*, 1402S–1406S. [CrossRef] [PubMed]
103. Estruch, R.; Ros, E.; Salas-Salvadó, J.; Covas, M.-I.; Corella, D.; Arós, F.; Gómez-Gracia, E.; Ruiz-Gutiérrez, V.; Fiol, M.; Lapetra, J.; et al. Primary Prevention of Cardiovascular Disease with a Mediterranean Diet Supplemented with Extra-Virgin Olive Oil or Nuts. *N. Engl. J. Med.* **2018**, *378*, e34. [CrossRef] [PubMed]
104. Mazzocchi, A.; Leone, L.; Agostoni, C.; Pali-Schöll, I. The Secrets of the Mediterranean Diet. Does [Only] Olive Oil Matter? *Nutr.* **2019**, *11*, 2941. [CrossRef] [PubMed]
105. Bray, G.A.; Lovejoy, J.C.; Smith, S.R.; Delany, J.P.; Lefevre, M.; Hwang, D.; Ryan, D.; York, D.A. The Influence of Different Fats and Fatty Acids on Obesity, Insulin Resistance and Inflammation. *J. Nutr.* **2002**, *132*, 2488–2491. [CrossRef] [PubMed]
106. Beulen, Y.; Martínez-González, M.A.; Van De Rest, O.; Salas-Salvadó, J.; Sorlí, J.V.; Gómez-Gracia, E.; Fiol, M.; Estruch, R.; Santos-Lozano, J.M.; Schröder, H.; et al. Quality of Dietary Fat Intake and Body Weight and Obesity in a Mediterranean Population: Secondary Analyses within the PREDIMED Trial. *Nutrients* **2018**, *10*, 2011. [CrossRef] [PubMed]
107. Iakoubov, R.; Ahmed, A.; Lauffer, L.M.; Bazinet, R.P.; Brubaker, P.L. Essential Role for Protein Kinase Cζ in Oleic Acid-Induced Glucagon-Like Peptide-1 Secretion in Vivo in the Rat. *Endocrinology* **2011**, *152*, 1244–1252. [CrossRef]
108. Poreba, M.A.; Dong, C.X.; Li, S.K.; Stahl, A.; Miner, J.H.; Brubaker, P.L. Role of fatty acid transport protein 4 in oleic acid-induced glucagon-like peptide-1 secretion from murine intestinal L cells. *Am. J. Physiol. Metab.* **2012**, *303*, E899–E907. [CrossRef]
109. Hirasawa, A.; Tsumaya, K.; Awaji, T.; Katsuma, S.; Adachi, T.; Yamada, M.; Sugimoto, Y.; Miyazaki, S.; Tsujimoto, G. Free fatty acids regulate gut incretin glucagon-like peptide-1 secretion through GPR120. *Nat. Med.* **2005**, *11*, 90–94. [CrossRef] [PubMed]
110. Duca, F.A.; Waise, T.M.Z.; Peppler, W.T.; Lam, T.K.T. The metabolic impact of small intestinal nutrient sensing. *Nat. Commun.* **2021**, *12*, 1–12. [CrossRef]
111. Panickar, K.S. Effects of dietary polyphenols on neuroregulatory factors and pathways that mediate food intake and energy regulation in obesity. *Mol. Nutr. Food Res.* **2012**, *57*, 34–47. [CrossRef] [PubMed]
112. Avila, J.A.D.; García, J.R.; Aguilar, G.A.G.; De La Rosa, L.A. The Antidiabetic Mechanisms of Polyphenols Related to Increased Glucagon-Like Peptide-1 (GLP1) and Insulin Signaling. *Molecules* **2017**, *22*, 903. [CrossRef]
113. Hlebowicz, J.; Hlebowicz, A.; Lindstedt, S.; Björgell, O.; Höglund, P.; Holst, J.J.; Darwiche, G.; Almér, L.-O. Effects of 1 and 3 g cinnamon on gastric emptying, satiety, and postprandial blood glucose, insulin, glucose-dependent insulinotropic polypeptide, glucagon-like peptide 1, and ghrelin concentrations in healthy subjects. *Am. J. Clin. Nutr.* **2009**, *89*, 815–821. [CrossRef]
114. Haro, C.; García, A.C.; Rangel-Zúñiga, O.A.; Alcalá-Díaz, J.F.; Landa, B.B.; Clemente, J.C.; Pérez-Martínez, P.; López-Miranda, J.; Pérez-Jiménez, F.; Camargo, A. Consumption of Two Healthy Dietary Patterns Restored Microbiota Dysbiosis in Obese Patients with Metabolic Dysfunction. *Mol. Nutr. Food Res.* **2017**, *61*. [CrossRef] [PubMed]
115. Nagpal, R.; Shively, C.A.; Appt, S.A.; Register, T.C.; Michalson, K.T.; Vitolins, M.Z.; Yadav, H. Gut Microbiome Composition in Non-human Primates Consuming a Western or Mediterranean Diet. *Front. Nutr.* **2018**, *5*, 28. [CrossRef]
116. Ms, E.Y.H.; Leone, V.A.; Devkota, S.; Wang, Y.; Brady, M.J.; Chang, E.B. Composition of Dietary Fat Source Shapes Gut Microbiota Architecture and Alters Host Inflammatory Mediators in Mouse Adipose Tissue. *J. Parenter. Enter. Nutr.* **2013**, *37*, 746–754. [CrossRef]

117. Caesar, R.; Tremaroli, V.; Kovatcheva-Datchary, P.; Cani, P.D.; Bäckhed, F. Crosstalk between Gut Microbiota and Dietary Lipids Aggravates WAT Inflammation through TLR Signaling. *Cell Metab.* **2015**, *22*, 658–668. [CrossRef]
118. Di Rienzi, S.C.; Jacobson, J.; A Kennedy, E.; E Bell, M.; Shi, Q.; Waters, J.L.; Lawrence, P.; Brenna, J.T.; A Britton, R.; Walter, J.; et al. Resilience of small intestinal beneficial bacteria to the toxicity of soybean oil fatty acids. *eLife* **2018**, *7*. [CrossRef] [PubMed]
119. Miyamoto, J.; Igarashi, M.; Watanabe, K.; Karaki, S.-I.; Mukouyama, H.; Kishino, S.; Li, X.; Ichimura, A.; Irie, J.; Sugimoto, Y.; et al. Gut microbiota confers host resistance to obesity by metabolizing dietary polyunsaturated fatty acids. *Nat. Commun.* **2019**, *10*, 1–15. [CrossRef] [PubMed]
120. Araújo, J.R.; Tazi, A.; Burlen-Defranoux, O.; Vichier-Guerre, S.; Nigro, G.; Licandro, H.; Demignot, S.; Sansonetti, P.J. Fermentation Products of Commensal Bacteria Alter Enterocyte Lipid Metabolism. *Cell Host Microbe* **2020**, *27*, 358–375.e7. [CrossRef]
121. Catalkaya, G.; Venema, K.; Lucini, L.; Rocchetti, G.; Delmas, D.; Daglia, M.; De Filippis, A.; Xiao, H.; Quiles, J.L.; Xiao, J.; et al. Interaction of dietary polyphenols and gut microbiota: Microbial metabolism of polyphenols, influence on the gut microbiota, and implications on host health. *Food Front.* **2020**, *1*, 109–133. [CrossRef]
122. Zhang, Y.; Sun, Q.; Li, Z.; Wang, H.; Li, J.; Wan, X. Fermented soybean powder containing Bacillus subtilis SJLH001 protects against obesity in mice by improving transport function and inhibiting angiogenesis. *J. Funct. Foods* **2019**, *59*, 60–70. [CrossRef]
123. Kartinah, N.T.; Fadilah, F.; Ibrahim, E.I.; Suryati, Y. The Potential of Hibiscus sabdariffa Linn in Inducing Glucagon-Like Peptide-1 via SGLT-1 and GLPR in DM Rats. *BioMed Res. Int.* **2019**, *2019*, 8724824. [CrossRef] [PubMed]
124. Clifton, P.M.; Keogh, J. Metabolic effects of high-protein diets. *Curr. Atheroscler. Rep.* **2007**, *9*, 472–478. [CrossRef]
125. Waliłko, E.; Napierała, M.; Bryśkiewicz, M.; Fronczyk, A.; Majkowska, L. High-Protein or Low Glycemic Index Diet—Which Energy-Restricted Diet Is Better to Start a Weight Loss Program? *Nutrients* **2021**, *13*, 1086. [CrossRef]
126. Hall, K.D.; Guo, J. Obesity Energetics: Body Weight Regulation and the Effects of Diet Composition. *Gastroenterology* **2017**, *152*, 1718–1727.e3. [CrossRef] [PubMed]
127. Franz, M.J.; VanWormer, J.J.; Crain, A.L.; Boucher, J.L.; Histon, T.; Caplan, W.; Bowman, J.D.; Pronk, N.P. Weight-Loss Outcomes: A Systematic Review and Meta-Analysis of Weight-Loss Clinical Trials with a Minimum 1-Year Follow-Up. *J. Am. Diet. Assoc.* **2007**, *107*, 1755–1767. [CrossRef] [PubMed]
128. Sacks, F.M.; Bray, G.A.; Carey, V.J.; Smith, S.R.; Ryan, D.; Anton, S.D.; McManus, K.; Champagne, C.M.; Bishop, L.M.; Laranjo, N.; et al. Comparison of Weight-Loss Diets with Different Compositions of Fat, Protein, and Carbohydrates. *N. Engl. J. Med.* **2009**, *360*, 859–873. [CrossRef]
129. Freedhoff, Y.; Hall, K.D. Weight loss diet studies: We need help not hype. *Lancet* **2016**, *388*, 849–851. [CrossRef]
130. Batterham, R.L.; Heffron, H.; Kapoor, S.; Chivers, J.E.; Chandarana, K.; Herzog, H.; Le Roux, C.W.; Thomas, E.L.; Bell, J.D.; Withers, D.J. Critical role for peptide YY in protein-mediated satiation and body-weight regulation. *Cell Metab.* **2006**, *4*, 223–233. [CrossRef]
131. Duraffourd, C.; De Vadder, F.; Goncalves, D.; Delaere, F.; Penhoat, A.; Brusset, B.; Rajas, F.; Chassard, D.; Duchampt, A.; Stefanutti, A.; et al. Mu-Opioid Receptors and Dietary Protein Stimulate a Gut-Brain Neural Circuitry Limiting Food Intake. *Cell* **2012**, *150*, 377–388. [CrossRef]
132. Mithieux, G.; Misery, P.; Magnan, C.; Pillot, B.; Gautier-Stein, A.; Bernard, C.; Rajas, F.; Zitoun, C. Portal sensing of intestinal gluconeogenesis is a mechanistic link in the diminution of food intake induced by diet protein. *Cell Metab.* **2005**, *2*, 321–329. [CrossRef]
133. Beiroa, D.; Imbernon, M.; Gallego, R.; Senra, A.; Herranz, D.; Villarroya, F.; Serrano, M.; Fernø, J.; Salvador, J.; Escalada, J.; et al. GLP-1 Agonism Stimulates Brown Adipose Tissue Thermogenesis and Browning Through Hypothalamic AMPK. *Diabetes* **2014**, *63*, 3346–3358. [CrossRef]
134. Li, Y.; Schnabl, K.; Gabler, S.-M.; Willershäuser, M.; Reber, J.; Karlas, A.; Laurila, S.; Lahesmaa, M.; U-Din, M.; Bast-Habersbrunner, A.; et al. Secretin-Activated Brown Fat Mediates Prandial Thermogenesis to Induce Satiation. *Cell* **2018**, *175*, 1561–1574.e12. [CrossRef]
135. Davila, A.-M.; Blachier, F.; Gotteland, M.; Andriamihaja, M.; Benetti, P.-H.; Sanz, Y.; Tomé, D. Intestinal luminal nitrogen metabolism: Role of the gut microbiota and consequences for the host. *Pharmacol. Res.* **2013**, *68*, 95–107. [CrossRef]
136. Krautkramer, K.A.; Fan, J.; Bäckhed, F. Gut microbial metabolites as multi-kingdom intermediates. *Nat. Rev. Microbiol.* **2021**, *19*, 77–94. [CrossRef] [PubMed]
137. Kim, J.; Park, W. Indole: A signaling molecule or a mere metabolic byproduct that alters bacterial physiology at a high concentration? *J. Microbiol.* **2015**, *53*, 421–428. [CrossRef]
138. Lee, J.-H.; Lee, J. Indole as an intercellular signal in microbial communities. *FEMS Microbiol. Rev.* **2010**, *34*, 426–444. [CrossRef]
139. Agus, A.; Planchais, J.; Sokol, H. Gut Microbiota Regulation of Tryptophan Metabolism in Health and Disease. *Cell Host Microbe* **2018**, *23*, 716–724. [CrossRef] [PubMed]
140. Leong, S.C.; Sirich, T.L. Indoxyl Sulfate—Review of Toxicity and Therapeutic Strategies. *Toxins* **2016**, *8*, 358. [CrossRef] [PubMed]
141. Sundaram, G.; Brew, B.J.; Jones, S.P.; Adams, S.; Lim, C.K.; Guillemin, G.J. Quinolinic acid toxicity on oligodendroglial cells: Relevance for multiple sclerosis and therapeutic strategies. *J. Neuroinflammation* **2014**, *11*, 1–11. [CrossRef] [PubMed]
142. Feehily, C.; Karatzas, K. Role of glutamate metabolism in bacterial responses towards acid and other stresses. *J. Appl. Microbiol.* **2012**, *114*, 11–24. [CrossRef] [PubMed]
143. Mazzoli, R.; Pessione, E. The Neuro-endocrinological Role of Microbial Glutamate and GABA Signaling. *Front. Microbiol.* **2016**, *7*, 1934. [CrossRef]

144. Diez-Gutiérrez, L.; Vicente, L.S.; Barrón, L.J.R.; Villarán, M.D.C.; Chavarri, M. Gamma-aminobutyric acid and probiotics: Multiple health benefits and their future in the global functional food and nutraceuticals market. *J. Funct. Foods* **2020**, *64*, 103669. [CrossRef]
145. Barrett, E.; Ross, R.P.; O'Toole, P.W.; Fitzgerald, G.F.; Stanton, C. γ-Aminobutyric acid production by culturable bacteria from the human intestine. *J. Appl. Microbiol.* **2012**, *113*, 411–417. [CrossRef] [PubMed]
146. Beaumont, M.; Portune, K.J.; Steuer, N.; Lan, A.; Cerrudo, V.; Audebert, M.; Dumont, F.; Mancano, G.; Khodorova, N.; Andriamihaja, M.; et al. Quantity and source of dietary protein influence metabolite production by gut microbiota and rectal mucosa gene expression: A randomized, parallel, double-blind trial in overweight humans. *Am. J. Clin. Nutr.* **2017**, *106*, 1005–1019. [CrossRef] [PubMed]
147. Windey, K.; De Preter, V.; Louat, T.; Schuit, F.; Herman, J.; VanSant, G.; Verbeke, K. Modulation of Protein Fermentation Does Not Affect Fecal Water Toxicity: A Randomized Cross-Over Study in Healthy Subjects. *PLoS ONE* **2012**, *7*, e52387. [CrossRef]
148. Duncan, S.H.; Belenguer, A.; Holtrop, G.; Johnstone, A.M.; Flint, H.J.; Lobley, G.E. Reduced Dietary Intake of Carbohydrates by Obese Subjects Results in Decreased Concentrations of Butyrate and Butyrate-Producing Bacteria in Feces. *Appl. Environ. Microbiol.* **2006**, *73*, 1073–1078. [CrossRef] [PubMed]
149. Russell, W.R.; Gratz, S.W.; Duncan, S.H.; Holtrop, G.; Ince, J.; Scobbie, L.; Duncan, G.; Johnstone, A.M.; E Lobley, G.; Wallace, R.J.; et al. High-protein, reduced-carbohydrate weight-loss diets promote metabolite profiles likely to be detrimental to colonic health. *Am. J. Clin. Nutr.* **2011**, *93*, 1062–1072. [CrossRef]
150. Poesen, R.; Mutsaers, H.A.M.; Windey, K.; Broek, P.H.V.D.; Verweij, V.; Augustijns, P.; Kuypers, D.; Jansen, J.; Evenepoel, P.; Verbeke, K.; et al. The Influence of Dietary Protein Intake on Mammalian Tryptophan and Phenolic Metabolites. *PLoS ONE* **2015**, *10*, e0140820. [CrossRef]
151. Blachier, F.; Beaumont, M.; Portune, K.J.; Steuer, N.; Lan, A.; Audebert, M.; Khodorova, N.; Andriamihaja, M.; Airinei, G.; Benamouzig, R.; et al. High-protein diets for weight management: Interactions with the intestinal microbiota and consequences for gut health. A position paper by the my new gut study group. *Clin. Nutr.* **2019**, *38*, 1012–1022. [CrossRef]
152. Rodríguez, M.A.; Mayneris-Perxachs, J.; Burokas, A.; Pérez-Brocal, V.; Moya, A.; Portero-Otin, M.; Ricart, W.; Maldonado, R.; Fernández-Real, J.-M. Gut bacterial ClpB-like gene function is associated with decreased body weight and a characteristic microbiota profile. *Microbiome* **2020**, *8*, 1–10. [CrossRef]
153. Dominique, M.; Breton, J.; Guérin, C.; Bole-Feysot, C.; Lambert, G.; Déchelotte, P.; Fetissov, S. Effects of Macronutrients on the In Vitro Production of ClpB, a Bacterial Mimetic Protein of α-MSH and Its Possible Role in Satiety Signaling. *Nutrients* **2019**, *11*, 2115. [CrossRef] [PubMed]
154. Portune, K.; Beaumont, M.; Davila, A.-M.; Tomé, D.; Blachier, F.; Sanz, Y. Gut microbiota role in dietary protein metabolism and health-related outcomes: The two sides of the coin. *Trends Food Sci. Technol.* **2016**, *57*, 213–232. [CrossRef]
155. Barber, T.; Hanson, P.; Kabisch, S.; Pfeiffer, A.; Weickert, M. The Low-Carbohydrate Diet: Short-Term Metabolic Efficacy Versus Longer-Term Limitations. *Nutrients* **2021**, *13*, 1187. [CrossRef]
156. Rogawski, M.A.; Löscher, W.; Rho, J.M. Mechanisms of Action of Antiseizure Drugs and the Ketogenic Diet. *Cold Spring Harb. Perspect. Med.* **2016**, *6*, a022780. [CrossRef] [PubMed]
157. Brouns, F. Overweight and diabetes prevention: Is a low-carbohydrate–high-fat diet recommendable? *Eur. J. Nutr.* **2018**, *57*, 1301–1312. [CrossRef] [PubMed]
158. Puchalska, P.; Crawford, P.A. Multi-dimensional Roles of Ketone Bodies in Fuel Metabolism, Signaling, and Therapeutics. *Cell Metab.* **2017**, *25*, 262–284. [CrossRef] [PubMed]
159. A Gibson, A.; Seimon, R.V.; Lee, C.M.Y.; Ayre, J.; Franklin, J.; Markovic, T.P.; Caterson, I.D.; Sainsbury, A. Do ketogenic diets really suppress appetite? A systematic review and meta-analysis. *Obes. Rev.* **2014**, *16*, 64–76. [CrossRef] [PubMed]
160. Skytte, M.J.; Samkani, A.; Astrup, A.; Frystyk, J.; Rehfeld, J.F.; Holst, J.J.; Madsbad, S.; Burling, K.; Fenger, M.; Thomsen, M.N.; et al. Effects of carbohydrate restriction on postprandial glucose metabolism, β-cell function, gut hormone secretion, and satiety in patients with Type 2 diabetes. *Am. J. Physiol. Metab.* **2021**, *320*, E7–E18. [CrossRef]
161. Holsen, L.M.; Hoge, W.S.; Lennerz, B.S.; Cerit, H.; Hye, T.; Moondra, P.; Goldstein, J.M.; Ebbeling, C.B.; Ludwig, D.S. Diets Varying in Carbohydrate Content Differentially Alter Brain Activity in Homeostatic and Reward Regions in Adults. *J. Nutr.* **2021**. [CrossRef]
162. Sumithran, P.; Prendergast, L.A.; Delbridge, E.; Purcell, K.; Shulkes, A.; Kriketos, A.D.; Proietto, J. Ketosis and appetite-mediating nutrients and hormones after weight loss. *Eur. J. Clin. Nutr.* **2013**, *67*, 759–764. [CrossRef]
163. Wallenius, V.; Elias, E.; Elebring, E.; Haisma, B.; Casselbrant, A.; Larraufie, P.; Spak, E.; Reimann, F.; Le Roux, C.W.; Docherty, N.G.; et al. Suppression of enteroendocrine cell glucagon-like peptide (GLP)-1 release by fat-induced small intestinal ketogenesis: A mechanism targeted by Roux-en-Y gastric bypass surgery but not by preoperative very-low-calorie diet. *Gut* **2020**, *69*, 1423–1431. [CrossRef]
164. Carneiro, L.; Geller, S.; Fioramonti, X.; Hébert, A.; Repond, C.; Leloup, C.; Pellerin, L. Evidence for hypothalamic ketone body sensing: Impact on food intake and peripheral metabolic responses in mice. *Am. J. Physiol. Metab.* **2016**, *310*, E103–E115. [CrossRef]
165. Kimura, I.; Inoue, D.; Maeda, T.; Hara, T.; Ichimura, A.; Miyauchi, S.; Kobayashi, M.; Hirasawa, A.; Tsujimoto, G. Short-chain fatty acids and ketones directly regulate sympathetic nervous system via G protein-coupled receptor 41 (GPR41). *Proc. Natl. Acad. Sci. USA* **2011**, *108*, 8030–8035. [CrossRef]

166. Ang, Q.Y.; Alexander, M.; Newman, J.C.; Tian, Y.; Cai, J.; Upadhyay, V.; Turnbaugh, J.A.; Verdin, E.; Hall, K.D.; Leibel, R.L.; et al. Ketogenic Diets Alter the Gut Microbiome Resulting in Decreased Intestinal Th17 Cells. *Cell* **2020**, *181*, 1263–1275.e16. [CrossRef]
167. Olson, C.A.; Vuong, H.E.; Yano, J.M.; Liang, Q.Y.; Nusbaum, D.; Hsiao, E.Y. The Gut Microbiota Mediates the Anti-Seizure Effects of the Ketogenic Diet. *Cell* **2018**, *173*, 1728–1741.e13. [CrossRef] [PubMed]
168. Leclercq, S.; Le Roy, T.; Furgiuele, S.; Coste, V.; Bindels, L.B.; Leyrolle, Q.; Neyrinck, A.M.; Quoilin, C.; Amadieu, C.; Petit, G.; et al. Gut Microbiota-Induced Changes in β-Hydroxybutyrate Metabolism Are Linked to Altered Sociability and Depression in Alcohol Use Disorder. *Cell Rep.* **2020**, *33*, 108238. [CrossRef] [PubMed]
169. Muller, P.A.; Schneeberger, M.; Matheis, F.; Wang, P.; Kerner, Z.; Ilanges, A.; Pellegrino, K.; Del Mármol, J.; Castro, T.B.R.; Furuichi, M.; et al. Microbiota modulate sympathetic neurons via a gut–brain circuit. *Nat. Cell Biol.* **2020**, *583*, 441–446. [CrossRef]
170. Anastasovska, J.; Arora, T.; Canon, G.J.S.; Parkinson, J.R.; Tuohy, K.; Gibson, G.R.; Nadkarni, N.A.; So, P.-W.; Goldstone, A.; Thomas, E.L.; et al. Fermentable Carbohydrate Alters Hypothalamic Neuronal Activity and Protects Against the Obesogenic Environment. *Obesity* **2012**, *20*, 1016–1023. [CrossRef] [PubMed]
171. Frost, G.; Sleeth, M.L.; Sahuri-Arisoylu, M.; Lizarbe, B.; Cerdan, S.; Brody, L.; Anastasovska, J.; Ghourab, S.; Hankir, M.; Zhang, S.; et al. The short-chain fatty acid acetate reduces appetite via a central homeostatic mechanism. *Nat. Commun.* **2014**, *5*, 3611. [CrossRef]
172. Shen, L.; Keenan, M.J.; Martin, R.J.; Tulley, R.T.; Raggio, A.M.; McCutcheon, K.L.; Zhou, J. Dietary Resistant Starch Increases Hypothalamic POMC Expression in Rats. *Obesity* **2009**, *17*, 40–45. [CrossRef] [PubMed]
173. Zaibi, M.S.; Stocker, C.J.; O'Dowd, J.; Davies, A.; Bellahcene, M.; Cawthorne, M.A.; Brown, A.J.; Smith, D.M.; Arch, J.R. Roles of GPR41 and GPR43 in leptin secretory responses of murine adipocytes to short chain fatty acids. *FEBS Lett.* **2010**, *584*, 2381–2386. [CrossRef] [PubMed]
174. Psichas, A.; Sleeth, M.L.; Murphy, K.G.; Brooks, L.; A Bewick, G.; Hanyaloglu, A.C.; A Ghatei, M.; Bloom, S.R.; Frost, G. The short chain fatty acid propionate stimulates GLP-1 and PYY secretion via free fatty acid receptor 2 in rodents. *Int. J. Obes.* **2015**, *39*, 424–429. [CrossRef] [PubMed]
175. De Vadder, F.; Kovatcheva-Datchary, P.; Goncalves, D.; Vinera, J.; Zitoun, C.; Duchampt, A.; Bäckhed, F.; Mithieux, G. Microbiota-Generated Metabolites Promote Metabolic Benefits via Gut-Brain Neural Circuits. *Cell* **2014**, *156*, 84–96. [CrossRef] [PubMed]
176. Troy, S.; Soty, M.; Ribeiro, L.; Laval, L.; Migrenne, S.; Fioramonti, X.; Pillot, B.; Fauveau, V.; Aubert, R.; Viollet, B.; et al. Intestinal Gluconeogenesis Is a Key Factor for Early Metabolic Changes after Gastric Bypass but Not after Gastric Lap-Band in Mice. *Cell Metab.* **2008**, *8*, 201–211. [CrossRef]
177. Xiong, Y.; Miyamoto, N.; Shibata, K.; Valasek, M.A.; Motoike, T.; Kedzierski, R.M.; Yanagisawa, M. Short-chain fatty acids stimulate leptin production in adipocytes through the G protein-coupled receptor GPR41. *Proc. Natl. Acad. Sci. USA* **2004**, *101*, 1045–1050. [CrossRef] [PubMed]
178. Lin, H.V.; Frassetto, A.; Jr, E.J.K.; Nawrocki, A.R.; Lu, M.M.; Kosinski, J.R.; Hubert, J.A.; Szeto, D.; Yao, X.; Forrest, G.; et al. Butyrate and Propionate Protect against Diet-Induced Obesity and Regulate Gut Hormones via Free Fatty Acid Receptor 3-Independent Mechanisms. *PLoS ONE* **2012**, *7*, e35240. [CrossRef]
179. Goswami, C.; Iwasaki, Y.; Yada, T. Short-chain fatty acids suppress food intake by activating vagal afferent neurons. *J. Nutr. Biochem.* **2018**, *57*, 130–135. [CrossRef]
180. Li, Z.; Yi, C.-X.; Katiraei, S.; Kooijman, S.; Zhou, E.; Chung, C.K.; Gao, Y.; van den Heuvel, J.K.; Meijer, O.C.; Berbée, J.F.P.; et al. Butyrate reduces appetite and activates brown adipose tissue via the gut-brain neural circuit. *Gut* **2018**, *67*, 1269–1279. [CrossRef] [PubMed]
181. Yadav, H.; Lee, J.-H.; Lloyd, J.; Walter, P.; Rane, S.G. Beneficial Metabolic Effects of a Probiotic via Butyrate-induced GLP-1 Hormone Secretion. *J. Biol. Chem.* **2013**, *288*, 25088–25097. [CrossRef]
182. Wang, Z.; Zeng, X.; Mo, Y.; Smith, K.; Guo, Y.; Lin, J. Identification and Characterization of a Bile Salt Hydrolase from Lactobacillus salivarius for Development of Novel Alternatives to Antibiotic Growth Promoters. *Appl. Environ. Microbiol.* **2012**, *78*, 8795–8802. [CrossRef] [PubMed]
183. Elkins, C.A.; Moser, S.A.; Savage, D.C. Genes encoding bile salt hydrolases and conjugated bile salt transporters in Lactobacillus johnsonii 100-100 and other Lactobacillus species. *Microbiology* **2001**, *147*, 3403–3412. [CrossRef]
184. Kim, G.-B.; Miyamoto, C.M.; Meighen, E.A.; Lee, B.H. Cloning and Characterization of the Bile Salt Hydrolase Genes (bsh) from Bifidobacterium bifidum Strains. *Appl. Environ. Microbiol.* **2004**, *70*, 5603–5612. [CrossRef]
185. Wijaya, A.; Hermann, A.; Abriouel, H.; Specht, I.; Yousif, N.M.K.; Holzapfel, W.H.; Franz, C.M.A.P. Cloning of the Bile Salt Hydrolase (bsh) Gene from Enterococcus faecium FAIR-E 345 and Chromosomal Location of bsh Genes in Food Enterococci. *J. Food Prot.* **2004**, *67*, 2772–2778. [CrossRef]
186. Chand, D.; Ramasamy, S.; Suresh, C. A highly active bile salt hydrolase from Enterococcus faecalis shows positive cooperative kinetics. *Process. Biochem.* **2016**, *51*, 263–269. [CrossRef]
187. Kang, D.-J.; Ridlon, J.M.; Moore, D.R.; Barnes, S.; Hylemon, P.B. Clostridium scindens baiCD and baiH genes encode stereo-specific 7α/7β-hydroxy-3-oxo-Δ4-cholenoic acid oxidoreductases. *Biochim. Biophys. Acta (BBA) Mol. Cell Biol. Lipids* **2008**, *1781*, 16–25. [CrossRef] [PubMed]
188. Begley, M.; Hill, C.; Gahan, C. Bile Salt Hydrolase Activity in Probiotics. *Appl. Environ. Microbiol.* **2006**, *72*, 1729–1738. [CrossRef]
189. Yao, L.; Seaton, S.C.; Ndousse-Fetter, S.; A Adhikari, A.; DiBenedetto, N.; I Mina, A.; Banks, A.S.; Bry, L.; Devlin, A.S. A selective gut bacterial bile salt hydrolase alters host metabolism. *eLife* **2018**, *7*. [CrossRef]

190. Ridlon, J.M.; Kang, D.-J.; Hylemon, P.B. Bile salt biotransformations by human intestinal bacteria. *J. Lipid Res.* **2006**, *47*, 241–259. [CrossRef]
191. La Frano, M.R.; Hernandez-Carretero, A.; Weber, N.; Borkowski, K.; Pedersen, T.L.; Osborn, O.; Newman, J.W. Diet-induced obesity and weight loss alter bile acid concentrations and bile acid–sensitive gene expression in insulin target tissues of C57BL/6J mice. *Nutr. Res.* **2017**, *46*, 11–21. [CrossRef]
192. Keitel, V.; Görg, B.; Bidmon, H.J.; Zemtsova, I.; Spomer, L.; Zilles, K.; Häussinger, D. The bile acid receptor TGR5 (Gpbar-1) acts as a neurosteroid receptor in brain. *Glia* **2010**, *58*, 1794–1805. [CrossRef]
193. Thomas, C.; Gioiello, A.; Noriega, L.; Strehle, A.; Oury, J.; Rizzo, G.; Macchiarulo, A.; Yamamoto, H.; Mataki, C.; Pruzanski, M.; et al. TGR5-Mediated Bile Acid Sensing Controls Glucose Homeostasis. *Cell Metab.* **2009**, *10*, 167–177. [CrossRef]
194. Bensalem, A.; Murtaza, B.; Hichami, A.; Khan, A.S.; Oulamara, H.; Merlen, G.; Berrichi, M.; Agli, A.-N.; Tordjmann, T.; Khan, N.A. Bile acid receptor TGR5 is critically involved in preference for dietary lipids and obesity. *J. Nutr. Biochem.* **2020**, *76*, 108298. [CrossRef] [PubMed]
195. Kuhre, R.E.; Albrechtsen, N.J.W.; Larsen, O.; Jepsen, S.L.; Balk-Møller, E.; Andersen, D.B.; Deacon, C.F.; Schoonjans, K.; Reimann, F.; Gribble, F.M.; et al. Bile acids are important direct and indirect regulators of the secretion of appetite- and metabolism-regulating hormones from the gut and pancreas. *Mol. Metab.* **2018**, *11*, 84–95. [CrossRef]
196. Katsuma, S.; Hirasawa, A.; Tsujimoto, G. Bile acids promote glucagon-like peptide-1 secretion through TGR5 in a murine enteroendocrine cell line STC-1. *Biochem. Biophys. Res. Commun.* **2005**, *329*, 386–390. [CrossRef]
197. Ullmer, C.; Sanchez, R.A.; Sprecher, U.; Raab, S.; Mattei, P.; Dehmlow, H.; Sewing, S.; Iglesias, A.; Beauchamp, J.; Conde-Knape, K. Systemic bile acid sensing by G protein-coupled bile acid receptor 1 (GPBAR1) promotes PYY and GLP-1 release. *Br. J. Pharmacol.* **2013**, *169*, 671–684. [CrossRef] [PubMed]
198. Alemi, F.; Poole, D.P.; Chiu, J.; Schoonjans, K.; Cattaruzza, F.; Grider, J.R.; Bunnett, N.W.; Corvera, C.U. The Receptor TGR5 Mediates the Prokinetic Actions of Intestinal Bile Acids and Is Required for Normal Defecation in Mice. *Gastroenterology* **2013**, *144*, 145–154. [CrossRef] [PubMed]
199. Yu, Y.; Villalobos-Hernandez, E.C.; Pradhananga, S.; Baker, C.C.; Keating, C.; Grundy, D.; Lomax, A.E.; Reed, D.E. Deoxycholic acid activates colonic afferent nerves via 5-HT3 receptor-dependent and -independent mechanisms. *Am. J. Physiol. Liver Physiol.* **2019**, *317*, G275–G284. [CrossRef] [PubMed]
200. Wu, X.; Li, J.-Y.; Lee, A.; Lu, Y.-X.; Zhou, S.-Y.; Owyang, C. Satiety induced by bile acids is mediated via vagal afferent pathways. *JCI Insight* **2020**, *5*, 5. [CrossRef]
201. Smith, E.; Macfarlane, G. Dissimilatory Amino Acid Metabolism in Human Colonic Bacteria. *Anaerobe* **1997**, *3*, 327–337. [CrossRef]
202. Natividad, J.M.; Agus, A.; Planchais, J.; Lamas, B.; Jarry, A.C.; Martin, R.; Michel, M.-L.; Chong-Nguyen, C.; Roussel, R.; Straube, M.; et al. Impaired Aryl Hydrocarbon Receptor Ligand Production by the Gut Microbiota Is a Key Factor in Metabolic Syndrome. *Cell Metab.* **2018**, *28*, 737–749.e4. [CrossRef]
203. Chimerel, C.; Emery, E.; Summers, D.K.; Keyser, U.; Gribble, F.M.; Reimann, F. Bacterial Metabolite Indole Modulates Incretin Secretion from Intestinal Enteroendocrine L Cells. *Cell Rep.* **2014**, *9*, 1202–1208. [CrossRef]
204. Gameiro, A.; Reimann, F.; Habib, A.M.; O'Malley, D.; Williams, L.; Simpson, A.K.; Gribble, F.M. The neurotransmitters glycine and GABA stimulate glucagon-like peptide-1 release from the GLUTag cell line. *J. Physiol.* **2005**, *569*, 761–772. [CrossRef]
205. Ashworth-Preece, M.; Krstew, E.; Jarrott, B.; Lawrence, A.J. Functional GABAA receptors on rat vagal afferent neurones. *Br. J. Pharmacol.* **1997**, *120*, 469–475. [CrossRef]
206. Andrews, P.L.R. Abdominal vagal afferent neurones: Aan important target for the treatment of gastrointestinal dysfunction. *Curr. Opin. Pharmacol.* **2002**, *2*, 650–656. [CrossRef]
207. Inotsuka, R.; Uchimura, K.; Yamatsu, A.; Kim, M.; Katakura, Y. γ-Aminobutyric acid (GABA) activates neuronal cells by inducing the secretion of exosomes from intestinal cells. *Food Funct.* **2020**, *11*, 9285–9290. [CrossRef] [PubMed]
208. Lucas, N.; Legrand, R.; Deroissart, C.; Dominique, M.; Azhar, S.; Le Solliec, M.-A.; Léon, F.; Rego, J.-C.D.; Déchelotte, P.; Fetissov, S.O.; et al. *Hafnia alvei* HA4597 Strain Reduces Food Intake and Body Weight Gain and Improves Body Composition, Glucose, and Lipid Metabolism in a Mouse Model of Hyperphagic Obesity. *Microorganism* **2019**, *8*, 35. [CrossRef] [PubMed]
209. Legrand, R.; Lucas, N.; Dominique, M.; Azhar, S.; Deroissart, C.; Le Solliec, M.-A.; Rondeaux, J.; Nobis, S.; Guérin, C.; Léon, F.; et al. Commensal *Hafnia alvei* strain reduces food intake and fat mass in obese mice—a new potential probiotic for appetite and body weight management. *Int. J. Obes.* **2020**, *44*, 1041–1051. [CrossRef] [PubMed]
210. Dominique, M.; Legrand, R.; Galmiche, M.; Azhar, S.; Deroissart, C.; Guérin, C.; Rego, J.-L.D.; Leon, F.; Nobis, S.; Lambert, G.; et al. Changes in Microbiota and Bacterial Protein Caseinolytic Peptidase B During Food Restriction in Mice: Relevance for the Onset and Perpetuation of Anorexia Nervosa. *Nutrients* **2019**, *11*, 2514. [CrossRef] [PubMed]
211. Dominique, M.; Boulete, I.; Bole-Feysot, C.; Leon, F.; Rego, J.-C.D.; Fetissov, S.; Déchelotte, P.; Lambert, G.; Legrand, R.; Lucas, N. Rôle de la protéine bactérienne ClpB et d'un de ses fragments peptidiques dans la régulation de la prise alimentaire. *Nutr. Clin. Métabolisme* **2019**, *33*, 23–24. [CrossRef]
212. Cani, P.; Bibiloni, R.; Knauf, C.; Waget, A.; Neyrinck, A.M.; Delzenne, N.; Burcelin, R. Changes in Gut Microbiota Control Metabolic Endotoxemia-Induced Inflammation in High-Fat Diet-Induced Obesity and Diabetes in Mice. *Diabetes* **2008**, *57*, 1470–1481. [CrossRef]
213. Wellhauser, L.; Chalmers, J.A.; Belsham, D.D. Nitric Oxide Exerts Basal and Insulin-Dependent Anorexigenic Actions in POMC Hypothalamic Neurons. *Mol. Endocrinol.* **2016**, *30*, 402–416. [CrossRef] [PubMed]

214. Porter, M.H.; Hrupka, B.J.; Langhans, W.; Schwartz, G.J. Vagal and splanchnic afferents are not necessary for the anorexia produced by peripheral IL-1β, LPS, and MDP. *Am. J. Physiol. Integr. Comp. Physiol.* **1998**, *275*, R384–R389. [CrossRef]
215. Lugarini, F.; Hrupka, B.; Schwartz, G.; Plata-Salaman, C.; Langhans, W. Acute and chronic administration of immunomodulators induces anorexia in Zucker rats. *Physiol. Behav.* **2005**, *84*, 165–173. [CrossRef] [PubMed]
216. Gayle, D.; E Ilyin, S.; Flynn, M.C.; Plata-Salamán, C.R. Lipopolysaccharide (LPS)- and muramyl dipeptide (MDP)-induced anorexia during refeeding following acute fasting: Characterization of brain cytokine and neuropeptide systems mRNAs. *Brain Res.* **1998**, *795*, 77–86. [CrossRef]
217. Von Meyenburg, C.; Hrupka, B.H.; Arsenijevic, D.; Schwartz, G.J.; Landmann, R.; Langhans, W. Role for CD14, TLR2, and TLR4 in bacterial product-induced anorexia. *Am. J. Physiol. Integr. Comp. Physiol.* **2004**, *287*, R298–R305. [CrossRef] [PubMed]
218. Pradhananga, S.; Tashtush, A.A.; Allen-Vercoe, E.; Petrof, E.O.; Lomax, A.E. Protease-dependent excitation of nodose ganglion neurons by commensal gut bacteria. *J. Physiol.* **2020**, *598*, 2137–2151. [CrossRef] [PubMed]
219. Liu, Y.-J.; Kobayashi, K.S. Muramyl Dipeptide and its Derivatives: Peptide Adjuvant in Immunological Disorders and Cancer Therapy. *Curr. Bioact. Compd.* **2011**, *7*, 180–197. [CrossRef]
220. Williams, L.; AlShehri, A.; Robichaud, B.; Cudmore, A.; Gagnon, J. The Role of the Bacterial Muramyl Dipeptide in the Regulation of GLP-1 and Glycemia. *Int. J. Mol. Sci.* **2020**, *21*, 5252. [CrossRef]
221. Reimer, R.A.; Maurer, A.D.; Eller, L.K.; Hallam, M.C.; Shaykhutdinov, R.; Vogel, H.J.; Weljie, A.M. Satiety Hormone and Metabolomic Response to an Intermittent High Energy Diet Differs in Rats Consuming Long-Term Diets High in Protein or Prebiotic Fiber. *J. Proteome Res.* **2012**, *11*, 4065–4074. [CrossRef] [PubMed]
222. Arora, T.; Loo, R.L.; Anastasovska, J.; Gibson, G.R.; Tuohy, K.; Sharma, R.K.; Swann, J.R.; Deaville, E.R.; Sleeth, M.L.; Thomas, E.L.; et al. Differential Effects of Two Fermentable Carbohydrates on Central Appetite Regulation and Body Composition. *PLoS ONE* **2012**, *7*, e43263. [CrossRef]
223. Delmée, E.; Cani, P.D.; Gual, G.; Knauf, C.; Burcelin, R.; Maton, N.; Delzenne, N.M. Relation between colonic proglucagon expression and metabolic response to oligofructose in high fat diet-fed mice. *Life Sci.* **2006**, *79*, 1007–1013. [CrossRef]
224. Cani, P.; Neyrinck, A.M.; Maton, N.; Delzenne, N.M. Oligofructose Promotes Satiety in Rats Fed a High-Fat Diet: Involvement of Glucagon-Like Peptide-1. *Obes. Res.* **2005**, *13*, 1000–1007. [CrossRef] [PubMed]
225. Chambers, E.S.; Viardot, A.; Psichas, A.; Morrison, D.J.; Murphy, K.G.; Zac-Varghese, S.E.K.; MacDougall, K.; Preston, T.; Tedford, C.; Finlayson, G.S.; et al. Effects of targeted delivery of propionate to the human colon on appetite regulation, body weight maintenance and adiposity in overweight adults. *Gut* **2014**, *64*, 1744–1754. [CrossRef]
226. Brown, A.J.; Goldsworthy, S.M.; Barnes, A.A.; Eilert, M.M.; Tcheang, L.; Daniels, D.; Muir, A.I.; Wigglesworth, M.J.; Kinghorn, I.; Fraser, N.J.; et al. The Orphan G Protein-coupled Receptors GPR41 and GPR43 Are Activated by Propionate and Other Short Chain Carboxylic Acids. *J. Biol. Chem.* **2003**, *278*, 11312–11319. [CrossRef]
227. Tolhurst, G.; Heffron, H.; Lam, Y.S.; Parker, H.E.; Habib, A.M.; Diakogiannaki, E.; Cameron, J.; Grosse, J.; Reimann, F.; Gribble, F.M. Short-Chain Fatty Acids Stimulate Glucagon-Like Peptide-1 Secretion via the G-Protein-Coupled Receptor FFAR2. *Diabetes* **2011**, *61*, 364–371. [CrossRef] [PubMed]
228. Karaki, S.-I.; Mitsui, R.; Hayashi, H.; Kato, I.; Sugiya, H.; Iwanaga, T.; Furness, J.B.; Kuwahara, A. Short-chain fatty acid receptor, GPR43, is expressed by enteroendocrine cells and mucosal mast cells in rat intestine. *Cell Tissue Res.* **2006**, *324*, 353–360. [CrossRef]
229. Nøhr, M.K.; Pedersen, M.H.; Gille, A.; Egerod, K.L.; Engelstoft, M.S.; Husted, A.S.; Sichlau, R.M.; Grunddal, K.V.; Poulsen, S.S.; Han, S.; et al. GPR41/FFAR3 and GPR43/FFAR2 as Cosensors for Short-Chain Fatty Acids in Enteroendocrine Cells vs FFAR3 in Enteric Neurons and FFAR2 in Enteric Leukocytes. *Endocrinology* **2013**, *154*, 3552–3564. [CrossRef]
230. Nøhr, M.; Egerod, K.; Christiansen, S.; Gille, A.; Offermanns, S.; Schwartz, T.; Møller, M. Expression of the short chain fatty acid receptor GPR41/FFAR3 in autonomic and somatic sensory ganglia. *Neuroscience* **2015**, *290*, 126–137. [CrossRef]
231. Torres-Fuentes, C.; Golubeva, A.V.; Zhdanov, A.; Wallace, S.; Arboleya, S.; Papkovsky, D.B.; El Aidy, S.; Ross, P.; Roy, B.L.; Stanton, C.; et al. Short-chain fatty acids and microbiota metabolites attenuate ghrelin receptor signaling. *FASEB J.* **2019**, *33*, 13546–13559. [CrossRef]
232. Schalla, M.A.; Stengel, A. Effects of microbiome changes on endocrine ghrelin signaling—A systematic review. *Peptides* **2020**, *133*, 170388. [CrossRef] [PubMed]
233. Sahu, A. Leptin signaling in the hypothalamus: Emphasis on energy homeostasis and leptin resistance. *Front. Neuroendocr.* **2003**, *24*, 225–253. [CrossRef] [PubMed]
234. Perry, R.J.; Peng, L.; Barry, N.A.; Cline, G.W.; Zhang, D.; Cardone, R.L.; Petersen, K.F.; Kibbey, R.G.; Goodman, N.A.B.A.L.; Shulman, G.I. Acetate mediates a microbiome–brain–β-cell axis to promote metabolic syndrome. *Nat. Cell Biol.* **2016**, *534*, 213–217. [CrossRef]
235. Adam, C.L.; Gratz, S.W.; Peinado, D.I.; Thomson, L.M.; Garden, K.E.; Williams, P.A.; Richardson, A.J.; Ross, A.W. Effects of Dietary Fibre (Pectin) and/or Increased Protein (Casein or Pea) on Satiety, Body Weight, Adiposity and Caecal Fermentation in High Fat Diet-Induced Obese Rats. *PLoS ONE* **2016**, *11*, e0155871. [CrossRef] [PubMed]
236. Canfora, E.E.; Van Der Beek, C.M.; Jocken, J.W.E.; Goossens, G.; Holst, J.J.; Damink, S.W.M.O.; Lenaerts, K.; DeJong, C.H.C.; Blaak, E.E. Colonic infusions of short-chain fatty acid mixtures promote energy metabolism in overweight/obese men: A randomized crossover trial. *Sci. Rep.* **2017**, *7*, 1–12. [CrossRef] [PubMed]

237. Van Der Beek, C.M.; Canfora, E.E.; Lenaerts, K.; Troost, F.J.; Damink, S.W.O.; Holst, J.J.; Masclee, A.A.; Dejong, C.H.; Blaak, E.E. Distal, not proximal, colonic acetate infusions promote fat oxidation and improve metabolic markers in overweight/obese men. *Clin. Sci.* **2016**, *130*, 2073–2082. [CrossRef]
238. Byrne, C.; Chambers, E.S.; Alhabeeb, H.; Chhina, N.; Morrison, D.J.; Preston, T.; Tedford, C.; Fitzpatrick, J.; Irani, C.; Busza, A.; et al. Increased colonic propionate reduces anticipatory reward responses in the human striatum to high-energy foods. *Am. J. Clin. Nutr.* **2016**, *104*, 5–14. [CrossRef]
239. Ringseis, R.; Gessner, D.K.; Eder, K. The Gut–Liver Axis in the Control of Energy Metabolism and Food Intake in Animals. *Annu. Rev. Anim. Biosci.* **2020**, *8*, 295–319. [CrossRef]
240. Houten, S.; Watanabe, M.; Auwerx, J. Endocrine functions of bile acids. *EMBO J.* **2006**, *25*, 1419–1425. [CrossRef]
241. Russell, D.W. Fifty years of advances in bile acid synthesis and metabolism. *J. Lipid Res.* **2009**, *50*, S120–S125. [CrossRef] [PubMed]
242. Dawson, P.A.; Karpen, S.J. Intestinal transport and metabolism of bile acids. *J. Lipid Res.* **2015**, *56*, 1085–1099. [CrossRef] [PubMed]
243. Narushima, S.; Itoh, K.; Miyamoto, Y.; Park, S.-H.; Nagata, K.; Kuruma, K.; Uchida, K. Deoxycholic acid formation in gnotobiotic mice associated with human intestinal bacteria. *Lipids* **2006**, *41*, 835–843. [CrossRef] [PubMed]
244. Gustafsson, B.E.; Midtvedt, T.; Norman, A. Metabolism of Cholic Acid in Germfree Animals after the Establishment in the Intes-tinal Tract of Deconjugating and 7 Alpha-Dehydroxylating Bacteria. *Acta Pathol. Microbiol. Scand.* **2009**, *72*, 433–443. [CrossRef]
245. Mertens, K.L.; Kalsbeek, A.; Soeters, M.R.; Eggink, H.M. Bile Acid Signaling Pathways from the Enterohepatic Circulation to the Central Nervous System. *Front. Neurosci.* **2017**, *11*, 617. [CrossRef]
246. Chiang, J.Y.L.; Ferrell, J.M. Bile Acids as Metabolic Regulators and Nutrient Sensors. *Annu. Rev. Nutr.* **2019**, *39*, 175–200. [CrossRef]
247. Monteiro-Cardoso, V.F.; Corlianò, M.; Singaraja, R.R. Bile Acids: A Communication Channel in the Gut-Brain Axis. *NeuroMolecular Med.* **2021**, *23*, 99–117. [CrossRef] [PubMed]
248. Chen, X.; Lou, G.; Meng, Z.; Huang, W. TGR5: A Novel Target for Weight Maintenance and Glucose Metabolism. *Exp. Diabetes Res.* **2011**, *2011*, 853501. [CrossRef]
249. Trefflich, I.; Marschall, H.-U.; di Giuseppe, R.; Ståhlman, M.; Michalsen, A.; Lampen, A.; Abraham, K.; Weikert, C. Associations between Dietary Patterns and Bile Acids—Results from a Cross-Sectional Study in Vegans and Omnivores. *Nutrients* **2019**, *12*, 47. [CrossRef]
250. Wan, Y.; Yuan, J.; Li, J.; Li, H.; Zhang, J.; Tang, J.; Ni, Y.; Huang, T.; Wang, F.; Zhao, F.; et al. Unconjugated and secondary bile acid profiles in response to higher-fat, lower-carbohydrate diet and associated with related gut microbiota: A 6-month randomized controlled-feeding trial. *Clin. Nutr.* **2020**, *39*, 395–404. [CrossRef]
251. Glicksman, C.; Pournaras, D.J.; Wright, M.; Roberts, R.; Mahon, D.; Welbourn, R.; Sherwood, R.; Alaghband-Zadeh, J.; Le Roux, C. Postprandial plasma bile acid responses in normal weight and obese subjects. *Ann. Clin. Biochem. Int. J. Lab. Med.* **2010**, *47*, 482–484. [CrossRef]
252. Sonne, D.P.; Van Nierop, F.S.; Kulik, W.; Soeters, M.R.; Vilsbøll, T.; Knop, F.K. Postprandial Plasma Concentrations of Individual Bile Acids and FGF-19 in Patients with Type 2 Diabetes. *J. Clin. Endocrinol. Metab.* **2016**, *101*, 3002–3009. [CrossRef]
253. Eggink, H.M.; Oosterman, J.E.; De Goede, P.; De Vries, E.M.; Foppen, E.; Koehorst, M.; Groen, A.K.; Boelen, A.; Romijn, J.A.; La Fleur, S.E.; et al. Complex interaction between circadian rhythm and diet on bile acid homeostasis in male rats. *Chronobiol. Int.* **2017**, *34*, 1339–1353. [CrossRef] [PubMed]
254. Higashi, T.; Watanabe, S.; Tomaru, K.; Yamazaki, W.; Yoshizawa, K.; Ogawa, S.; Nagao, H.; Minato, K.; Maekawa, M.; Mano, N. Unconjugated bile acids in rat brain: Analytical method based on LC/ESI-MS/MS with chemical derivatization and estimation of their origin by comparison to serum levels. *Steroids* **2017**, *125*, 107–113. [CrossRef] [PubMed]
255. Zheng, X.; Chen, T.; Zhao, A.; Wang, X.; Xie, G.; Huang, F.; Liu, J.; Zhao, Q.; Wang, S.; Wang, C.; et al. The Brain Metabolome of Male Rats across the Lifespan. *Sci. Rep.* **2016**, *6*, 24125. [CrossRef]
256. Calderon, G.; McRae, A.; Rievaj, J.; Davis, J.; Zandvakili, I.; Linker-Nord, S.; Burton, D.; Roberts, G.; Reimann, F.; Gedulin, B.; et al. Ileo-colonic delivery of conjugated bile acids improves glucose homeostasis via colonic GLP-1-producing enteroendocrine cells in human obesity and diabetes. *EBioMedicine* **2020**, *55*, 102759. [CrossRef]
257. Vilsbøll, T.; Krarup, T.; Deacon, C.F.; Madsbad, S.; Holst, J.J. Reduced Postprandial Concentrations of Intact Biologically Active Glucagon-Like Peptide 1 in Type 2 Diabetic Patients. *Diabetes* **2001**, *50*, 609–613. [CrossRef]
258. Li, F.; Jiang, C.; Krausz, K.W.; Li, Y.; Albert, I.; Hao, H.; Fabre, K.M.; Mitchell, J.B.; Patterson, A.; Gonzalez, F.J. Microbiome remodelling leads to inhibition of intestinal farnesoid X receptor signalling and decreased obesity. *Nat. Commun.* **2013**, *4*, 1–10. [CrossRef]
259. Wahlström, A.; Sayin, S.I.; Marschall, H.-U.; Bäckhed, F. Intestinal Crosstalk between Bile Acids and Microbiota and Its Impact on Host Metabolism. *Cell Metab.* **2016**, *24*, 41–50. [CrossRef] [PubMed]
260. Le Floc'H, N.; Otten, W.; Merlot, E. Tryptophan metabolism, from nutrition to potential therapeutic applications. *Amino Acids* **2011**, *41*, 1195–1205. [CrossRef]
261. Friedman, M.; Levin, C. Nutritional and medicinal aspects of d-amino acids. *Amino Acids* **2011**, *42*, 1553–1582. [CrossRef]
262. Zhu, J.X.; Wu, X.Y.; Owyang, C.; Li, Y. Intestinal serotonin acts as a paracrine substance to mediate vagal signal transmission evoked by luminal factors in the rat. *J. Physiol.* **2001**, *530*, 431–442. [CrossRef]

263. Raybould, H.E.; Glatzle, J.; Robin, C.; Meyer, J.H.; Phan, T.; Wong, H.; Sternini, C. Expression of 5-HT3 receptors by extrinsic duodenal afferents contribute to intestinal inhibition of gastric emptying. *Am. J. Physiol. Liver Physiol.* **2003**, *284*, G367–G372. [CrossRef] [PubMed]
264. Savastano, D.M.; Covasa, M. Intestinal nutrients elicit satiation through concomitant activation of CCK1 and 5-HT3 receptors. *Physiol. Behav.* **2007**, *92*, 434–442. [CrossRef]
265. Janssen, P.; Vos, R.; Van Oudenhove, L.; Tack, J. Influence of the 5-HT3 receptor antagonist ondansetron on gastric sensorimotor function and nutrient tolerance in healthy volunteers. *Neurogastroenterol. Motil.* **2010**, *23*, 444-e175. [CrossRef] [PubMed]
266. O'Mahony, S.; Clarke, G.; Borre, Y.; Dinan, T.; Cryan, J. Serotonin, tryptophan metabolism and the brain-gut-microbiome axis. *Behav. Brain Res.* **2015**, *277*, 32–48. [CrossRef]
267. Williams, B.B.; Van Benschoten, A.H.; Cimermancic, P.; Donia, M.S.; Zimmermann, M.; Taketani, M.; Ishihara, A.; Kashyap, P.C.; Fraser, J.S.; Fischbach, M.A. Discovery and Characterization of Gut Microbiota Decarboxylases that Can Produce the Neurotransmitter Tryptamine. *Cell Host Microbe* **2014**, *16*, 495–503. [CrossRef]
268. Nzakizwanayo, J.; Dedi, C.; Standen, G.; Macfarlane, W.M.; Patel, B.A.; Jones, B.V. Escherichia coli Nissle 1917 enhances bioavailability of serotonin in gut tissues through modulation of synthesis and clearance. *Sci. Rep.* **2015**, *5*, 17324. [CrossRef] [PubMed]
269. Yano, J.M.; Yu, K.; Donaldson, G.P.; Shastri, G.G.; Ann, P.; Ma, L.; Nagler, C.R.; Ismagilov, R.F.; Mazmanian, S.K.; Hsiao, E.Y. Indigenous Bacteria from the Gut Microbiota Regulate Host Serotonin Biosynthesis. *Cell* **2015**, *161*, 264–276. [CrossRef]
270. Reigstad, C.S.; Salmonson, C.E.; Iii, J.F.R.; Szurszewski, J.H.; Linden, D.R.; Sonnenburg, J.L.; Farrugia, G.; Kashyap, P.C. Gut microbes promote colonic serotonin production through an effect of short-chain fatty acids on enterochromaffin cells. *FASEB J.* **2015**, *29*, 1395–1403. [CrossRef]
271. Lund, M.L.; Egerod, K.L.; Engelstoft, M.S.; Dmytriyeva, O.; Theodorsson, E.; Patel, B.A.; Schwartz, T.W. Enterochromaffin 5-HT cells—A major target for GLP-1 and gut microbial metabolites. *Mol. Metab.* **2018**, *11*, 70–83. [CrossRef]
272. Fletcher, P.; Burton, M. Effects of manipulations of peripheral serotonin on feeding and drinking in the rat. *Pharmacol. Biochem. Behav.* **1984**, *20*, 835–840. [CrossRef]
273. Pollock, J.D.; Rowland, N. Peripherally administered serotonin decreases food intake in rats. *Pharmacol. Biochem. Behav.* **1981**, *15*, 179–183. [CrossRef]
274. Voigt, J.-P.; Fink, H. Serotonin controlling feeding and satiety. *Behav. Brain Res.* **2015**, *277*, 14–31. [CrossRef] [PubMed]
275. Meguid, M.M.; O Fetissov, S.; Varma, M.; Sato, T.; Zhang, L.; Laviano, A.; Rossi-Fanelli, F. Hypothalamic dopamine and serotonin in the regulation of food intake. *Nutrients* **2000**, *16*, 843–857. [CrossRef]
276. Fetissov, S.O.; Meguid, M.M. Serotonin delivery into the ventromedial nucleus of the hypothalamus affects differently feeding pattern and body weight in obese and lean Zucker rats. *Appetite* **2010**, *54*, 346–353. [CrossRef]
277. Agren, H.; Reibring, L.; Hartvig, P.; Tedroff, J.; Bjurling, P.; Hörnfeldt, K.; Andersson, Y.; Lundqvist, H.; Långström, B. Low brain uptake of L-[11C]5-hydroxytryptophan in major depression: A positron emission tomography study on patients and healthy volunteers. *Acta Psychiatr. Scand.* **1991**, *83*, 449–455. [CrossRef]
278. Tian, P.; O'Riordan, K.J.; Lee, Y.-K.; Wang, G.; Zhao, J.; Zhang, H.; Cryan, J.F.; Chen, W. Towards a psychobiotic therapy for depression: Bifidobacterium breve CCFM1025 reverses chronic stress-induced depressive symptoms and gut microbial abnormalities in mice. *Neurobiol. Stress* **2020**, *12*, 100216. [CrossRef]
279. Laurans, L.; Venteclef, N.; Haddad, Y.; Chajadine, M.; Alzaid, F.; Metghalchi, S.; Sovran, B.; Denis, R.G.P.; Dairou, J.; Cardellini, M.; et al. Genetic deficiency of indoleamine 2,3-dioxygenase promotes gut microbiota-mediated metabolic health. *Nat. Med.* **2018**, *24*, 1113–1120. [CrossRef] [PubMed]
280. Agudelo, L.Z.; Ferreira, D.M.S.; Cervenka, I.; Bryzgalova, G.; Dadvar, S.; Jannig, P.R.; Klein, A.; Lakshmikanth, T.; Sustarsic, E.G.; Porsmyr-Palmertz, M.; et al. Kynurenic Acid and Gpr35 Regulate Adipose Tissue Energy Homeostasis and Inflammation. *Cell Metab.* **2018**, *27*, 378–392.e5. [CrossRef] [PubMed]
281. Egerod, K.L.; Petersen, N.; Timshel, P.N.; Rekling, J.C.; Wang, Y.; Liu, Q.; Schwartz, T.W.; Gautron, L. Profiling of G protein-coupled receptors in vagal afferents reveals novel gut-to-brain sensing mechanisms. *Mol. Metab.* **2018**, *12*, 62–75. [CrossRef]
282. Kuriyama, K.; Sze, P. Blood-brain barrier to H3-γ-aminobutyric acid in normal and amino oxyacetic acid-treated animals. *Neuropharmacology* **1971**, *10*, 103–108. [CrossRef]
283. Hyland, N.P.; Cryan, J.F. A Gut Feeling about GABA: Focus on GABAB Receptors. *Front. Pharmacol.* **2010**, *1*, 124. [CrossRef]
284. Nakajima, K.; Tooyama, I.; Kuriyama, K.; Kimura, H. Immunohistochemical demonstration of GABAB receptors in the rat gastrointestinal tract. *Neurochem. Res.* **1996**, *21*, 211–215. [CrossRef] [PubMed]
285. Schwörer, H.; Racké, K.; Kilbinger, H. GABA receptors are involved in the modulation of the release of 5-hydroxytryptamine from the vascularly perfused small intestine of the guinea-pig. *Eur. J. Pharmacol.* **1989**, *165*, 29–37. [CrossRef]
286. Fetissov, S.O.; Sinno, M.H.; Coëffier, M.; Bole-Feysot, C.; Ducrotté, P.; Hökfelt, T.; Déchelotte, P. Autoantibodies against appetite-regulating peptide hormones and neuropeptides: Putative modulation by gut microflora. *Nutrients* **2008**, *24*, 348–359. [CrossRef] [PubMed]
287. Ericson, M.D.; Schnell, S.M.; Freeman, K.T.; Haskell-Luevano, C. A fragment of the Escherichia coli ClpB heat-shock protein is a micromolar melanocortin 1 receptor agonist. *Bioorganic Med. Chem. Lett.* **2015**, *25*, 5306–5308. [CrossRef]
288. Breton, J.; Legrand, R.; Akkermann, K.; Järv, A.; Harro, J.; Déchelotte, P.; Fetissov, S.O. Elevated plasma concentrations of bacterial ClpB protein in patients with eating disorders. *Int. J. Eat. Disord.* **2016**, *49*, 805–808. [CrossRef]

289. Panaro, B.L.; Tough, I.R.; Engelstoft, M.S.; Matthews, R.T.; Digby, G.J.; Møller, C.L.; Svendsen, B.; Gribble, F.; Reimann, F.; Holst, J.J.; et al. The Melanocortin-4 Receptor Is Expressed in Enteroendocrine L Cells and Regulates the Release of Peptide YY and Glucagon-like Peptide 1 In Vivo. *Cell Metab.* **2014**, *20*, 1018–1029. [CrossRef]
290. Schletter, J.; Heine, H.; Ulmer, A.J.; Rietschel, E.T. Molecular mechanisms of endotoxin activity. *Arch. Microbiol.* **1995**, *164*, 383–389. [CrossRef] [PubMed]
291. Cani, P.D.; Amar, J.; Iglesias, M.A.; Poggi, M.; Knauf, C.; Bastelica, D.; Neyrinck, A.M.; Fava, F.; Tuohy, K.; Chabo, C.; et al. Metabolic Endotoxemia Initiates Obesity and Insulin Resistance. *Diabetes* **2007**, *56*, 1761–1772. [CrossRef]
292. Plata-Salamán, C. Cytokines and feeding. *Int. J. Obes.* **2001**, *25*, S48–S52. [CrossRef] [PubMed]
293. A Bray, G.; Frühbeck, G.; Ryan, D.; Wilding, J.P.H. Management of obesity. *Lancet* **2016**, *387*, 1947–1956. [CrossRef]
294. Gimeno, R.E.; Briere, D.A.; Seeley, R.J. Leveraging the Gut to Treat Metabolic Disease. *Cell Metab.* **2020**, *31*, 679–698. [CrossRef] [PubMed]
295. Toplak, H.; Woodward, E.; Yumuk, V.; Oppert, J.-M.; Halford, J.C.; Frühbeck, G. 2014 EASO Position Statement on the Use of Anti-Obesity Drugs. *Obes. Facts* **2015**, *8*, 166–174. [CrossRef]
296. Schutz, D.D.; Busetto, L.; Dicker, D.; Farpour-Lambert, N.; Pryke, R.; Toplak, H.; Widmer, D.; Yumuk, V.; Schutz, Y. European Practical and Patient-Centred Guidelines for Adult Obesity Management in Primary Care. *Obes. Facts* **2019**, *12*, 40–66. [CrossRef]
297. Yu, E.W.; Gao, L.; Stastka, P.; Cheney, M.C.; Mahabamunuge, J.; Soto, M.T.; Ford, C.B.; Bryant, J.A.; Henn, M.R.; Hohmann, E.L. Fecal microbiota transplantation for the improvement of metabolism in obesity: The FMT-TRIM double-blind placebo-controlled pilot trial. *PLoS Med.* **2020**, *17*, e1003051. [CrossRef] [PubMed]
298. Allegretti, J.R.; Kassam, Z.; Mullish, B.; Chiang, A.; Carrellas, M.; Hurtado, J.; Marchesi, J.R.; McDonald, J.A.; Pechlivanis, A.; Barker, G.F.; et al. Effects of Fecal Microbiota Transplantation With Oral Capsules in Obese Patients. *Clin. Gastroenterol. Hepatol.* **2020**, *18*, 855–863.e2. [CrossRef]
299. Vrieze, A.; Van Nood, E.; Holleman, F.; Salojärvi, J.; Kootte, R.S.; Bartelsman, J.F.; Dallinga–Thie, G.M.; Ackermans, M.T.; Serlie, M.J.; Oozeer, R.; et al. Transfer of Intestinal Microbiota From Lean Donors Increases Insulin Sensitivity in Individuals With Metabolic Syndrome. *Gastroenterology* **2012**, *143*, 913–916.e7. [CrossRef]
300. Hartstra, A.V.; Schüppel, V.; Imangaliyev, S.; Schrantee, A.; Prodan, A.; Collard, D.; Levin, E.; Dallinga-Thie, G.; Ackermans, M.T.; Winkelmeijer, M.; et al. Infusion of donor feces affects the gut–brain axis in humans with metabolic syndrome. *Mol. Metab.* **2020**, *42*, 101076. [CrossRef] [PubMed]
301. FDA Early Clinical Trials with Live Biotherapeutic Products: Chemistry, Manu-facturing, and Control Information: Guidance for Industry. 2016. Available online: www.fda.gov (accessed on 26 May 2021).
302. O'Toole, P.W.; Marchesi, J.; Hill, C. Next-generation probiotics: The spectrum from probiotics to live biotherapeutics. *Nat. Microbiol.* **2017**, *2*, 17057. [CrossRef] [PubMed]
303. Depommier, C.; Everard, A.; Druart, C.; Plovier, H.; Van Hul, M.; Vieira-Silva, S.; Falony, G.; Raes, J.; Maiter, D.; Delzenne, N.; et al. Supplementation with Akkermansia muciniphila in overweight and obese human volunteers: A proof-of-concept exploratory study. *Nat. Med.* **2019**, *25*, 1096–1103. [CrossRef] [PubMed]
304. Pérez, M.R.; Agusti, A.; Sanz, Y. Innovation in microbiome-based strategies for promoting metabolic health. *Curr. Opin. Clin. Nutr. Metab. Care* **2017**, *20*, 484–491. [CrossRef] [PubMed]
305. Everard, A.; Belzer, C.; Geurts, L.; Ouwerkerk, J.P.; Druart, C.; Bindels, L.B.; Guiot, Y.; Derrien, M.; Muccioli, G.G.; Delzenne, N.M.; et al. Cross-talk between Akkermansia muciniphila and intestinal epithelium controls diet-induced obesity. *Proc. Natl. Acad. Sci. USA* **2013**, *110*, 9066–9071. [CrossRef]
306. Gauffin-Cano, P.; Santacruz, A.; Moya, Á.; Sanz, Y. Bacteroides uniformis CECT 7771 Ameliorates Metabolic and Immunological Dysfunction in Mice with High-Fat-Diet Induced Obesity. *PLoS ONE* **2012**, *7*, e41079. [CrossRef]
307. Udayappan, S.; Manneras-Holm, L.; Chaplin, A.; Belzer, C.; Herrema, H.; Dallinga-Thie, G.M.; Duncan, S.H.; Stroes, E.S.G.; Groen, A.K.; Flint, H.J.; et al. Oral treatment with Eubacterium hallii improves insulin sensitivity in db/db mice. *NPJ Biofilms Microbiomes* **2016**, *2*, 16009. [CrossRef]
308. Hansen, K.B.; Rosenkilde, M.M.; Knop, F.K.; Wellner, N.; Diep, T.A.; Rehfeld, J.F.; Andersen, U.B.; Holst, J.J.; Hansen, H.S. 2-Oleoyl Glycerol Is a GPR119 Agonist and Signals GLP-1 Release in Humans. *J. Clin. Endocrinol. Metab.* **2011**, *96*, E1409–E1417. [CrossRef]
309. DiPatrizio, N. Endocannabinoids and the Gut-Brain Control of Food Intake and Obesity. *Nutrients* **2021**, *13*, 1214. [CrossRef]
310. Veiga, P.; Suez, J.; Derrien, M.; Elinav, E. Moving from probiotics to precision probiotics. *Nat. Microbiol.* **2020**, *5*, 878–880. [CrossRef]
311. Fetissov, S.O.; Hallman, J.; Oreland, L.; Klinteberg, B.A.; Grenbäck, E.; Hulting, A.-L.; Hökfelt, T. Autoantibodies against -MSH, ACTH, and LHRH in anorexia and bulimia nervosa patients. *Proc. Natl. Acad. Sci. USA* **2002**, *99*, 17155–17160. [CrossRef]
312. Fetissov, S.O.; Legrand, R.; Lucas, N. Bacterial Protein Mimetic of Peptide Hormone as a New Class of Protein- based Drugs. *Curr. Med. Chem.* **2019**, *26*, 546–553. [CrossRef] [PubMed]
313. Maruvada, P.; Leone, V.; Kaplan, L.M.; Chang, E.B. The Human Microbiome and Obesity: Moving beyond Associations. *Cell Host Microbe* **2017**, *22*, 589–599. [CrossRef] [PubMed]

International Journal of
Molecular Sciences

MDPI

Review

Hypothalamic Actions of SIRT1 and SIRT6 on Energy Balance

Mar Quiñones [1,2,*], Eva Martínez-Grobas [1], Johan Fernø [3], Raquel Pérez-Lois [4], Luisa María Seoane [2,4] and Omar Al Massadi [2,4,*]

1 Department of Physiology, CIMUS, University of Santiago de Compostela-Instituto de Investigación Sanitaria de Santiago de Compostela, 15782 Santiago de Compostela, Spain; eva.martinez.grobas@rai.usc.es
2 CIBER Fisiopatología de la Obesidad y Nutrición (CIBERobn), 15706 Santiago de Compostela, Spain; luisamaria.seoane@usc.es
3 Hormone Laboratory, Department of Medical Biochemistry and Pharmacology, Haukeland University Hospital, N-5021 Bergen, Norway; johan.ferno@uib.no
4 Instituto de Investigación Sanitaria de Santiago de Compostela, Complexo Hospitalario Universitario de Santiago (CHUS/SERGAS), Travesía da Choupana s/n, 15706 Santiago de Compostela, Spain; raquelpl.93@gmail.com
* Correspondence: mar.quinones@usc.es (M.Q.); omar.al-massadi.iglesias@sergas.es (O.A.M.)

Abstract: Sirtuins are NAD+ dependent deacetylases that regulate a large number of physiological processes. These enzymes are highly conserved and act as energy sensors to coordinate different metabolic responses in a controlled manner. At present, seven mammalian sirtuins (SIRT 1-7) have been identified, with SIRT1 and SIRT6 shown to exert their metabolic actions in the hypothalamus, both with crucial roles in eliciting responses to dampen metabolic complications associated with obesity. Therefore, our aim is to compile the current understanding on the role of SIRT1 and SIRT6 in the hypothalamus, especially highlighting their actions on the control of energy balance.

Keywords: SIRT1; SIRT6; energy balance; food intake; body weight; adiposity; type 2 diabetes; obesity

Citation: Quiñones, M.; Martínez-Grobas, E.; Ferno, J.; Pérez-Lois, R.; Seoane, L.M.; Al Massadi, O. Hypothalamic Actions of SIRT1 and SIRT6 on Energy Balance. *Int. J. Mol. Sci.* **2021**, *22*, 1430. https://doi.org/10.3390/ijms22031430

Academic Editor: Cristina Miralpeix
Received: 23 December 2020
Accepted: 28 January 2021
Published: 31 January 2021

1. Introduction

1.1. Obesity

The incidence of obesity has reached pandemic proportions and become a major cause of various social, economic and health problems in our society [1]. In fact, worldwide obesity has nearly tripled since 1975 and the World Health Organization has declared that overweight is among the top ten conditions of major health risk in the world and among the top five in developed countries [2]. Obesity is defined as a state on excess of fat produced as a consequence of a high energy intake and a low energy expenditure that leads to serious health problems [3,4]. While food intake is tightly regulated by homeostatic and hedonic drives [5], changes in energy expenditure can be explained by differences in basal metabolism (energy expended to keep the body functioning at rest), physical activity and adaptive thermogenesis [6]. Therefore, obesity is a complex and multifactorial disease involving genetic, biological but also behavioral factors.

The first line of treatment of obese people is to follow a healthy diet and do regular exercise. However, these guidelines are difficult to adhere to for the obese population and most people with obesity that lose weight will regain it over time, ultimately with a higher body weight than before starting the weight loss [7,8]. Bariatric surgery is currently the most efficient method for long-term weight loss. However, this is a costly and invasive intervention not without risk and side-effects, and is only recommended to a small proportion of morbidly obese people [9]. The currently available pharmacological treatments to combat obesity have limited effectiveness and more efficient drugs are urgently needed. Therefore, it is necessary to ascertain in depth the molecular mechanisms underlying this disease and to identify novel therapeutic targets for treatment.

1.2. The Hypothalamus as the Master Regulator of Energy Balance

The central nervous system (CNS) is a key organ in energy balance and body weight regulation. The CNS elicits these functions through three mechanism: (a) controlling food-related behaviors, including food intake (b) controlling the autonomic nervous system, which regulate the energy expenditure and other metabolic processes; and (c) controlling the neuroendocrine system by modulating hormone secretion. The modulation and coordination of this complex systems occurs in different brain areas [10,11]. However, it is well known that the hypothalamic neuronal networks integrate peripheral information (i.e., nutrients and hormones) to modulate energy balance [12,13]. The hypothalamus, a central structure composed of anatomically distinct nuclei interconnected via axonal projections, is at large the most studied area in term of its regulation of food intake and body weight. In this neuronal network, we can distinguish the hypothalamic arcuate nucleus (ARC) as the best positioned nucleus to receive signals from the periphery and to mediate a subsequent homeostatic response to peripheral tissues. This nucleus is considered the "master hypothalamic center" for feeding control [13,14].

The ARC is composed by two antagonist neuronal populations, neuropeptide Y/Agouti related peptide (NPY/AgRP) neurons that induces a positive energy balance and proopiomelanocortin (POMC) neurons and cocaine and amphetamine related transcript (CART) that when activated induces negative energy balance. Both neuron types regulate food intake, energy expenditure and nutrient partitioning. When energy intake exceeds expenditure, the expression of orexigenic neuropeptides, such as AgRP and NPY decreases, whereas the expression of anorexigenic neuropeptides, such as CART and POMC increases [3,10]. Opposite changes occur when energy expenditure exceeds intake.

Hypothalamic neurons in the ARC respond to peripheral nutrients, such as glucose and fatty acids as well as hormones, such as leptin and ghrelin, the two most important hormones in the regulation of food intake and body weight [11,12]. Leptin, secreted by, and in proportion to white adipose tissue (WAT), informs the hypothalamus of energy buildup and activates POMC/CART neurons and inhibits NPY/AgRP neurons, resulting in an inhibition of feeding and an increase in energy expenditure. In contrast, ghrelin is a stomach-derived peptide that increases food intake through increased NPY and AgRP expression. Therefore, both leptin and ghrelin use these pathways in opposite ways to exert their roles on energy balance.

Adjacent to the ARC we find the ventromedial nucleus (VMH), a hypothalamic area that receives projections mainly from NPY/AgRP and POMC neurons. These VMH neurons express steroidogenic factor 1 (SF1) and project their axons at the same time to the ARC and to secondary hypothalamic nuclei, as well as regions of the brain stem [15]. These neurons are activated in respond to changes in circulating leptin and other metabolic signals, and due to their projections to the ARC, an elaborate neuronal circuitry underlying metabolic homeostasis is formed [15,16].

Another relevant metabolic nucleus is the lateral hypothalamic area (LHA), a region of the brain that has for long been known for its role in the regulation of eating and considered the most interconnected area of the hypothalamus. This hypothalamic area is mainly composed of two types orexigenic neuronal populations, namely orexin-producing neurons and melanin-concentrating hormone (MCH)-producing neurons [17–19].

Taking into account that several mechanisms are involved in the development and maintenance of obesity within the hypothalamus, in the next section we introduce the role of a highly conserved family of class III deacetylases denominated Sirtuins, (SIRTs) as one of these relevant metabolic regulators.

2. Sirtuins: Metabolic Sensors

Mammalian SIRTs are homologous to the *Sir2* gene of *Saccharomyces cerevisiae*. Seven *Sir2* orthologues, members of the SIRT family (SIRT1-7), have been described in mammals. All members share the conserved SIRT domain, but they differ in subcellular location and function [20–22]. SIRTs actively participate in the regulation of energy metabolism related to diet/caloric restriction, but they also have an important role in other metabolic processes, such as aging, cell survival, and DNA repair [23,24].

The aim of this review is to describe the hypothalamic actions of SIRTs in a metabolic context. Until now, only SIRT1 and SIRT6 have been shown to act in the hypothalamus to control energy balance. Therefore, in this review, we will describe in detail some of the most important hypothalamic mechanisms by which these two SIRTs regulate energy metabolism.

2.1. Hypothalamic SIRT1 in the Control of Energy Balance

SIRT1 is expressed in a wide range of tissues and organs, such as the liver, pancreas, heart, muscle, and adipose tissue (Reviewed in [20–22]). SIRT1 is also expressed in important metabolic centers of the brain, including ARC, VMH, dorsomedial nucleus and paraventricular nucleus of the hypothalamus (PVN), the area postrema and the nucleus of the solitary tract in the hindbrain (For review see [20–22]). SIRT1 is involved in the control of metabolic processes that regulate body weight, such as control of food intake [25–28], adiposity [26,29], energy expenditure [29], thermogenesis of brown adipose tissue (BAT) and "browning" of WAT [29] (Figure 1).

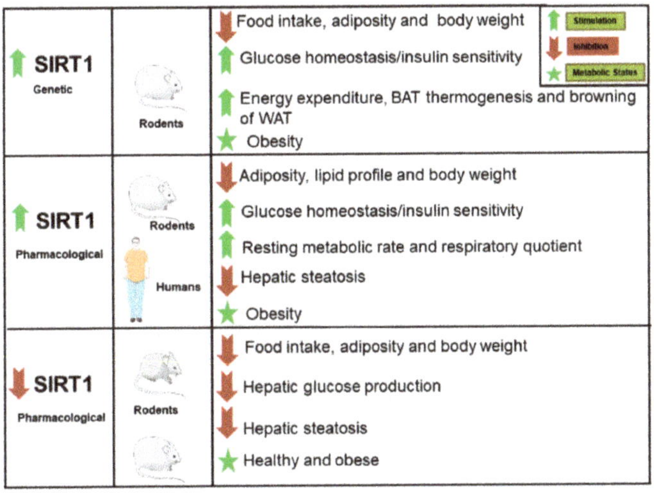

Figure 1. Description of the metabolic actions elicited by SIRT1 in mice and humans. Pharmacological or genetic modulation of SIRT1 regulates body weight, adiposity in WAT and liver, food intake, energy expenditure, resting metabolic rate, respiratory quotient, BAT thermogenesis, browning of WAT, glucose homeostasis and insulin sensitivity. green arrow: stimulation, red arrow: inhibition, green star:metabolic state. Abbreviations used-BAT; brown adipose tissue; SIRT1: sirtuin 1; WAT; white adipose tissue. The figures were generated by using materials from Servier Medical Art (Servier) under consideration of a Creative Commons Attribution 3.0 Unported License.

With regard to genetic animal models for SIRT1, the global knock out (KO) mice for SIRT1 are not viable or have several metabolic complications [30], whereas moderate overexpression of SIRT1 in mice improves several metabolic parameters associated with obesity [31] (Figure 1). Consistently, the pharmacological activation of SIRT1 produces a variety of beneficial metabolic actions in vivo (Figure 1). In this sense, natural activators of SIRT1, such as the polyphenol resveratrol (RSV) [32,33] or synthetic molecules [34–36], have been used for this purpose. These pharmacological activators improve insulin sensitivity, increase insulin secretion, enhance mitochondrial function, decrease adiposity and lower glucose levels [37–41]. Intriguingly, a negative energy balance was also observed in mice following pharmacological inhibition of SIRT1 [26,27]. Therefore, a controversy exists on the role of SIRT1 in regulation of energy metabolism and more studies are needed to ascertain the mechanism of action of these SIRT1 modulators.

2.2. SIRT1 and Food Intake

Several studies have shown that pharmacological inhibition of SIRT1 in the hypothalamus reduces the caloric intake and body weight in rodents [26,27], which occur through the modulation of the central melanocortin system [25,27,28]. In fact, central administration of a melanocortin receptor antagonist (SHU9119) reversed the anorectic effect of hypothalamic SIRT1 inhibition [27]. Indeed, this anorexigenic action is mediated by the repression of AgRP and stimulation of POMC levels. In addition, SIRT1 requires the Forkhead Box O1 transcription factor (FOXO1) to exert these actions, since FOXO1 inhibition blocks the actions of SIRT1 on AgRP and POMC neurons [27]. However, SIRT1 has other downstream effectors, e.g., the actions of SIRT1 on food intake are mediated by the mitochondrial redox machinery, because the decrease in feeding induced by the action of EX-527, a pharmacological SIRT1 inhibitor, is impaired in uncoupling protein-2 KO animals [28].

In agreement with these data, hypothalamic SIRT1 can act as a mediator of several nutritional hormones (reviewed in [42]). In fact, the pharmacological blockade of SIRT1 in the CNS or the genetic inhibition of p53, a downstream effector of SIRT1, prevents the orexigenic action of ghrelin, by impairing the regulation of this hormone on its target neuropeptides NPY/AgRP and its target transcription factors namely FOXO1, cAMP response element-binding protein and brain-specific homeobox transcription factor [25]. Subsequent studies with conditional KO mice models for SIRT1 and p53 in specific neuronal populations show that the food intake induced by ghrelin rely on the action of these energy sensors specifically in AgRP neurons [28,43]. These data highlight the relevance of SIRT1/p53 pathway in the orexigenic action of ghrelin.

Consistently with these findings, it has been shown that SIRT1 mediates the metabolic actions regulated by other orexigenic hormone such as MCH. In this study, it was shown that the action of MCH on food intake, glucose homeostasis and adiposity rely on SIRT1 in POMC neurons, due to the specific deletion of SIRT1 on a subset of these neurons comprise the metabolic actions of MCH [26]. These data reveal for the first time the neuronal basis of the metabolic effects of MCH on body weight and food intake.

Another study demonstrated the importance of SIRT1 in food selection, and that SIRT1 mediates the regulation of macronutrient choice [44]. In this study, targeting central SIRT1 by gain-of function and loss-of-function mice models, it was shown that the preference for high sucrose diet was reduced when SIRT1 was stimulated at neuronal level, while it was increased when SIRT1 was inhibited at this level. These effects of diet preference over sucrose were shown to be mediated by SIRT1 specifically in oxytocin neurons, since overexpression or deletion of SIRT1 in oxytocin PVN neurons displayed the same results on sucrose preference as the neuronal manipulation of SIRT1. However, when standard diet (SD) vs. high fat diet (HFD) was offered to the animals in the latter mentioned mice model, no differences were found, which show that SIRT1 in oxytocin neurons regulates the preference for sucrose, but not for fat. The fibroblast growth hormone-21 (FGF-21) is a peripheral signal targeting oxytocin neurons to regulate macronutrient preference. In the same study SIRT1 was identified as a regulator of sucrose preference by promoting FGF21 sensitivity in oxytocin neurons.

2.3. SIRT1 in Hypothalamic Neurons

Studies with conditional mice models have shown that the effects of SIRT1 in the hypothalamus may depend on the neuronal type in which it acts (Figure 2).

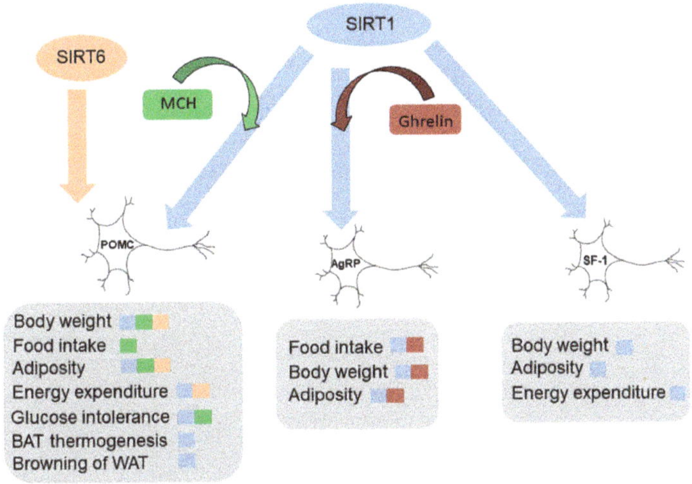

Figure 2. SIRT1 and SIRT6 in hypothalamic neurons: The effects of SIRT1 and SIRT6 in the hypothalamus may depend on the neuronal type. Abbreviations used-AgRP: Agouti-related protein neurons; BAT; brown adipose tissue; MCH: melanin concentrating hormone; POMC: Proopiomelanocortin neurons; SF-1: steroidogenic factor 1; WAT; white adipose tissue. The figures were generated by using materials from Servier Medical Art (Servier) under consideration of a Creative Commons Attribution 3.0 Unported License.

It has been reported that inhibition of SIRT1 specifically in POMC neurons does not significantly alter body weight or adiposity in SD-fed mice [29]. However, these mice are more vulnerable to diet-induced obesity (DIO) due to a reduction in energy expenditure and a reorganization of WAT through the sympathetic nervous system. Consistent with these data, overexpression of SIRT1 in POMC neurons has the opposite effect, and these mice showed a lean phenotype due to an increase in energy expenditure and activation of sympathetic nervous system in adipose tissue [45].

Additionally, SIRT1 is essential for the metabolic functions of AgRP neurons. The inhibition of SIRT1 specifically in AgRP neurons decreases the consumption of food, which leads to a decrease in adiposity and consequently a reduced body weight [28]. There are, however, also opposing experimental data, demonstrating that overexpression of SIRT1 in AgRP neurons improves weight gain and reduces the food intake associated with age in mice [45].

Overexpression of SIRT1 in SF1 neurons shows protective effects against DIO, with reduction in body weight, fat mass, and leptin levels, and increased energy expenditure [46]. Accordingly, the specific inhibition of SIRT1 in SF1 neurons makes HFD fed mice more susceptible to obesity and diabetes mellitus [46].

These data indicate that SIRT1 plays an essential role in hypothalamic neurons that are associated with the metabolic adaptations that determine body weight gain in obesity.

2.4. SIRT1 a Link between Metabolism and Reproduction

SIRT1 has been shown to be the link between metabolism and reproduction through hypothalamic neurons [47]. First, it has been observed how female animals that overexpress SIRT1 have a normal body weight and show a delay in the development of puberty. Using virogenetic techniques, it was demonstrated that the overexpression of SIRT1 specifically in the ARC promoted this delay. Moreover, in this work it was also shown that SIRT1 is expressed in kisspeptin, neurokinin B and dynorphin (KNDy) neurons, which express the *Kiss1* pubertal activator gene. It is important to highlight that the content of SIRT1 in the KNDy neurons of the ARC and the recruitment of SIRT1 towards the *Kiss1* promoter in vivo change depending on the nutritional status of the animal. In states of overnutrition, the neuronal content of SIRT1 and its association with *Kiss1* decrease, which accelerates and promotes puberty [47]. In contrast, malnutrition raises SIRT1 levels, prolongs the repression of *Kiss1* by SIRT1, and delays puberty. These results identify SIRT1-mediated *Kiss1* inhibition as a key epigenetic mechanism by which nutritional cues and obesity influence puberty in mammals.

In summary, hypothalamic SIRT1 signaling appears to be a key mediator of energy metabolism and the physiological response to obesity.

3. Hypothalamic SIRT6 in the Control of Energy Balance

SIRT6 is a nuclear protein associated with physiological and pathological processes, regulating obesity, insulin resistance, inflammation and energy metabolism [48]. It is expressed in various tissues, with the highest level of expression in adipose tissue, skeletal muscle and heart [49–51]. SIRT6 is also highly expressed in the CNS, and its expression is regulated by the availability of nutrients, showing low levels in the hypothalamus and specifically in POMC neurons in obesity [52]. Therefore, SIRT6 is postulated as a relevant energy sensor at the central level and a promising pharmacological target in the regulation of energy metabolism.

A seminal work in the study of SIRT6 described that global SIRT6 KO mice suffer a severe multisystemic phenotype, with severe hypoglycemia and a short life expectancy [49]. A remarkable characteristic of these mice is that they show a delay in postnatal growth associated with low levels of insulin-like growth factor 1. In agreement with these results, mice in which SIRT6 was inactivated at the neural level (NSIRT6-KO) display a significantly reduced growth during development, smaller adult phenotype and lower body weight compared to controls [50]. These results demonstrate that neural inactivation of SIRT6 is sufficient to cause growth retardation. However, elderly male NSIRT6 KO mice, but not female mice, were significantly heavier than controls [50].

Mice deficient for SIRT6, specifically in POMC neurons (SIRT6 POMC KO), showed no changes in body size or differences in glucose levels [52]. Therefore, hypoglycemia in SIRT6 KO mice and growth retardation in NSIRT6 KO mice were not due to SIRT6 deficiency in POMC neurons. On the contrary, SIRT6 POMC KO mice showed a modest, but significant, increase in body weight under SD that was greater when the mice were fed HFD [52]. This increase in body weight is associated with a higher adiposity in WAT and liver, caloric intake, glucose intolerance and less energy expenditure. Consistently, it was also found that the activity of the leptin signaling pathway was significantly reduced in SIRT6 POMC KO mice, showing high levels of leptin in the serum of these mice both under SD and HFD [52].

Interestingly, and in line with these previous results the global overexpression of SIRT6 decrease body weight, adiposity and glucose tolerance [51]. Moreover, and consistent with these data, the study of Tang et al. has demonstrated that overexpression of SIRT6 in the ARC of obese mice decreases body weight, adiposity and food intake [52]. These data suggest that the activation of SIRT6 in POMC neurons could be the responsible of this effect, however this possibility was not tested yet.

Altogether, these studies postulate to SIRT6 as an essential player in the regulation of body energy homeostasis.

4. Therapeutic Options

The positive effects of SIRTs in animal models on prevention and treatment of obesity and metabolic syndrome suggested their use as therapeutic agents. However, the knowledge about the involvement of SIRTs on clinical trials for the treatment of metabolic disease is limited.

In this sense, the administration of RSV, the most studied natural SIRT1 activator, reduces adipocyte size, blood glucose, preserves pancreatic β-cells and improves insulin action in rhesus monkeys fed a high-fat, high sugar diet for 2 years [53,54]. Importantly, treatment with RSV during 30 days in obese men confirmed these initial reports in non-human primates [55]. Other epidemiological studies in humans showed positive effects of RSV in patients with diabetes or obesity-mediated insulin resistance [40,56,57]. Accordingly, RSV improves plasma triglyceride concentration [37,41], lowers circulating cytokine levels [37–39], elicits better metabolic flexibility with lower homeostatic model assessment for insulin resistance index [37], suppresses postprandial glucagon responses, decreases resting metabolic rate and improves respiratory quotient in obese subjects [37,58]. Interestingly, all of these effects elicited by RSV on energy balance, glucose homeostasis and insulin sensitivity were not observed in non-obese healthy humans [59].

It is important to highlight that in spite of these results, the use of the natural activator RVS as a therapeutic target bear some controversies. For example, the use of RVS shows low tolerability and specificity, inconsistent dosages activity and interactions with other compounds (for a discussion of this issue see [60]). These shortcomings have stimulated to the development of a number of new small synthetic molecules activators. However, although these molecules are 1000 times more potent and their bioavailability are improved compared to the natural compounds, they do not show a superior efficacy in preclinical or clinical studies [34–36]. For example, some epidemiological observations with the synthetic SIRT1 activator, SRT2104, showed that the administration of this compound during 28 days produces a mild reduction in the glucose levels and slightly improvement in the body weight [61]. Other studies in elderly or type 2 diabetic patients shows that the administration of SRT2104 reduces the lipid profile and the body weight [62] without changes in glucose levels [36,61].

Despite these controversies, SIRT1 activators often show an acceptable efficacy in treating metabolic disturbances in clinical trials, where improvements are restricted to subjects with a pathological condition. Thus, more studies with larger cohorts are necessary to obtain a consensus for the actions, specificity, dosage, and effectiveness of these SIRT1 activators.

5. Conclusions and Remarks

In this review we cover the most relevant studies focusing on the metabolic actions of SIRT1 and SIRT6 at the hypothalamic level. In this sense, it is important to highlight that the effect of hypothalamic SIRT1 on energy balance depends on the neuronal type where it acts, which has been demonstrated for a variety of metabolic effects. E.g., the specific deletion of SIRT1 in POMC and SF1 neurons from HFD-fed mice shows that SIRT1 regulates metabolism through energy expenditure but does not affect food intake, whereas specific modulation of SIRT1 in AgRP neurons specifically affects eating behavior. Other examples of the relevance of this deacetylase in the regulation of metabolic processes came from studies that describe its involvement in the signaling pathways of some relevant metabolic hormones. More specifically, it has recently been shown that SIRT1, in addition to mediate the orexigenic action of ghrelin, is essential for the metabolic actions of MCH. Importantly in both cases, the ARC melanocortin system plays a fundamental role in the action of SIRT1 in these processes.

SIRT1 may also regulate diet preference via actions in the PVN Oxitocin neurons. This is relevant since the effectiveness of the diet and its impact on body weight depends both on the amount of caloric intake that is consumed and the balance of macronutrients of the ingested meal. It is also important to highlight the effects of SIRT1 on conditions related to obesity, such as puberty. SIRT1 has been shown to be a key factor in the development of puberty. Specifically, it has been identified that the ARC is the main integrator area of these effects, with epigenetic inhibition of *Kiss1* by SIRT1 as a key mechanism by which nutritional signals and obesity influence puberty of children.

In addition to the effects of SIRT1, this review also provides a comprehensive overview on the involvement of hypothalamic SIRT6 in the control of energy metabolism. It has been shown recently that the specific deletion of SIRT6 in POMC neurons increases food intake and decreases energy expenditure, promoting an obesogenic phenotype in mice. These results suggest that SIRT6 has a protective effect against DIO by its action in POMC neurons. In this sense, future studies should focus on the possible effects of SIRT6 modulation in other hypothalamic neurons such as AgRP or SF1 neurons. Therefore, all these results strongly support SIRT6 as a new hypothalamic molecular mediator in the regulation of energy homeostasis.

The data described in this review reveal SIRT1 and SIRT6 as multifaceted mediators of energy metabolism, affecting processes such as food intake, food preference, puberty, body weight, adiposity, glucose homeostasis and insulin resistance.

Apart from these considerations there are controversial results regarding the central actions of these SIRTs that we have discussed. While it seems clear that SIRT1 activators improve several metabolic parameters, in some circumstances this is also true for SIRT1 inhibitors. For example, neural deletion of SIRT1 improves insulin sensitivity and glucose intolerance in mice, while the administration of the pharmacological inhibitor EX-527 improves body weight, food intake, hepatic steatosis and fibrosis in diabetic rats [25–27,63]. In this sense, the healthy phenotype observed in these rodent models following SIRT1 inhibition may at first seem contradictory, but there are other previous examples in the literature involving other mammalian SIRTs that show similar results. Indeed, SIRT6 overexpression protected against DIO [51] in the same way that SIRT6 deletion did [49]. There is currently no clear explanation for the controversial results regarding the central actions of these SIRTs, but it is believed that the processes that involves metabolic regulation are highly complex and that this may sometimes lead to counterintuitive findings. Therefore, these data imply that the implications for these variety of effects are that their levels must be under a fine-tuning regulation and that their alterations may lead to unpredictable side effects without an evident dose-response relationship.

In spite of these somewhat contradictory results, we believe that SIRT1 and SIRT6 emerge as important targets for the development of novel therapies in the control of obesity and its comorbidities. Importantly, it should be noted that the fact that these proteins regulate energy balance in specific high-calorie regimens is of paramount importance for the development of new strategies to combat metabolic diseases. Finally, we should not forget the potential involvement of the other SIRTs different from SIRT1 and SIRT6 in the control of energy balance at the hypothalamic level, which should be the focus of future studies.

Author Contributions: M.Q. and O.A.M. wrote the manuscript. All authors made substantial contributions to discussion of the content and reviewed/edited the manuscript before submission. All authors have read and agreed to the published version of the manuscript.

Funding: This work is supported by grants of Xunta de Galicia (MQ:2018-PG013). Centro de Investigación Biomédica en Red (CIBER) de Fisiopatología de la Obesidad y Nutrición (CIBERobn). CIBERobn is an initiative of the Instituto de Salud Carlos III (ISCIII) of Spain which is supported by FEDER funds. Western Norway Regional Health Authority (Helse Vest RHF).M.Q. is a recipient of a Postdoctoral contract from Galician Government (Xunta de Galicia ED481B2018/004). O.A.M. was funded by a research contract Miguel Servet (CP20/00146) from the ISCIII. The figures were generated by using materials from Servier Medical Art (Servier) under consideration of a Creative Commons Attribution 3.0 Unported License.

Institutional Review Board Statement: Not applicable.

Informed Consent Statement: Not applicable.

Data Availability Statement: Not applicable.

Conflicts of Interest: The authors declare no conflict of interest.

References

1. Williams, E.P.; Mesidor, M.; Winters, K.; Dubbert, P.M.; Wyatt, S.B. Overweight and obesity: Prevalence, consequences, and causes of a growing public health problem. *Curr. Obes. Rep.* **2015**, *4*, 363–370. [CrossRef] [PubMed]
2. NCD Risk Factor Collaboration. Worldwide trends in body-mass index, underweight, overweight, and obesity from 1975 to 2016: A pooled analysis of 2416 population-based measurement studies in 128.9 million children, adolescents, and adults. *Lancet* **2017**, *390*, 2627–2642. [CrossRef]
3. Morton, G.J.; Cummings, D.E.; Baskin, D.G.; Barsh, G.S.; Schwartz, M.W. Central nervous system control of food intake and body weight. *Nature* **2006**, *443*, 289–295. [CrossRef]
4. Schneeberger, M.; Gomis, R.; Claret, M. Hypothalamic and brainstem neuronal circuits controlling homeostatic energy balance. *J. Endocrinol.* **2014**, *220*, T25–T46. [CrossRef] [PubMed]
5. Al Massadi, O.; Nogueiras, R.; Dieguez, C.; Girault, J.A. Ghrelin and food reward. *Neuropharmacology* **2019**, *148*, 131–138. [CrossRef]
6. Gonzalez-Garcia, I.; Milbank, E.; Dieguez, C.; Lopez, M.; Contreras, C. Glucagon, GLP-1 and Thermogenesis. *Int. J. Mol. Sci.* **2019**, *20*, 3445. [CrossRef]
7. MacLean, P.S.; Higgins, J.A.; Jackman, M.R.; Johnson, G.C.; Fleming-Elder, B.K.; Wyatt, H.R.; Melanson, E.L.; Hill, J.O. Peripheral metabolic responses to prolonged weight reduction that promote rapid, efficient regain in obesity-prone rats. *Am. J. Physiol. Regul. Integr. Comp. Physiol.* **2006**, *290*, R1577–R1588. [CrossRef]
8. Crujeiras, A.B.; Goyenechea, E.; Abete, I.; Lage, M.; Carreira, M.C.; Martinez, J.A.; Casanueva, F.F. Weight regain after a diet-induced loss is predicted by higher baseline leptin and lower ghrelin plasma levels. *J. Clin. Endocrinol. Metab.* **2010**, *95*, 5037–5044. [CrossRef]
9. Evers, S.S.; Sandoval, D.A.; Seeley, R.J. The physiology and molecular underpinnings of the effects of bariatric surgery on obesity and diabetes. *Annu. Rev. Physiol.* **2017**, *79*, 313–334. [CrossRef]
10. Friedman, J.M. Modern science versus the stigma of obesity. *Nat. Med.* **2004**, *10*, 563–569. [CrossRef]
11. Farooqi, S.; O'Rahilly, S. Genetics of obesity in humans. *Endocr. Rev.* **2006**, *27*, 710–718. [CrossRef] [PubMed]
12. Al Massadi, O.; Lopez, M.; Tschop, M.; Dieguez, C.; Nogueiras, R. Current understanding of the hypothalamic ghrelin pathways inducing appetite and adiposity. *Trends Neurosci.* **2017**, *40*, 167–180. [CrossRef] [PubMed]
13. Timper, K.; Bruning, J.C. Hypothalamic circuits regulating appetite and energy homeostasis: Pathways to obesity. *Dis. Models Mech.* **2017**, *10*, 679–689. [CrossRef] [PubMed]
14. Waterson, M.J.; Horvath, T.L. Neuronal regulation of energy homeostasis: Beyond the hypothalamus and feeding. *Cell Metab.* **2015**, *22*, 962–970. [CrossRef]
15. Sternson, S.M.; Shepherd, G.M.; Friedman, J.M. Topographic mapping of VMH→Arcuate nucleus microcircuits and their reorganization by fasting. *Nat. Neurosci.* **2005**, *8*, 1356–1363. [CrossRef]
16. Dhillon, H.; Zigman, J.M.; Ye, C.; Lee, C.E.; McGovern, R.A.; Tang, V.; Kenny, C.D.; Christiansen, L.M.; White, R.D.; Edelstein, E.A.; et al. Leptin directly activates SF1 neurons in the VMH, and this action by leptin is required for normal body-weight homeostasis. *Neuron* **2006**, *49*, 191–203. [CrossRef]
17. Kawauchi, H.; Kawazoe, I.; Tsubokawa, M.; Kishida, M.; Baker, B.I. Characterization of melanin-concentrating hormone in chum salmon pituitaries. *Nature* **1983**, *305*, 321–323. [CrossRef]
18. De Lecea, L.; Kilduff, T.S.; Peyron, C.; Gao, X.B.; Foye, P.E.; Danielson, P.E.; Fukuhara, C.; Battenberg, E.L.F.; Gautvik, V.T.; Barlett, F.S., III; et al. The hypocretins: Hypothalamus-specific peptides with neuroexcitatory activity. *Proc. Natl. Acad. Sci. USA* **1998**, *95*, 322–327. [CrossRef]
19. Sakurai, T.; Amemiya, A.; Ishii, M.; Matsuzaki, I.; Chemelli, R.M.; Tanaka, H.; Williams, S.C.; Richardson, J.A.; Kozlowski, G.P.; Wilson, S.; et al. Orexins and orexin receptors: A family of hypothalamic neuropeptides and G protein-coupled receptors that regulate feeding behavior. *Cell* **1998**, *92*, 573–585. [CrossRef]
20. Coppari, R. Metabolic actions of hypothalamic SIRT1. *Trends Endocrinol. Metab.* **2012**, *23*, 179–185. [CrossRef]
21. Nogueiras, R.; Habegger, K.M.; Chaudhary, N.; Finan, B.; Banks, A.S.; Dietrich, M.O.; Horvath, T.L.; Sinclair, D.A.; Pfluger, P.T.; Tschöp, M.H. Sirtuin 1 and sirtuin 3: Physiological modulators of metabolism. *Physiol. Rev.* **2012**, *92*, 1479–1514. [CrossRef] [PubMed]
22. Al Massadi, O.; Quinones, M.; Lear, P.; Dieguez, C.; Nogueiras, R. The brain: A new organ for the metabolic actions of SIRT1. *Horm. Metab. Res.* **2013**, *45*, 960–966. [CrossRef] [PubMed]
23. Chang, H.C.; Guarente, L. SIRT1 and other sirtuins in metabolism. *Trends Endocrinol. Metab.* **2014**, *25*, 138–145. [CrossRef] [PubMed]
24. Carafa, V.; Rotili, D.; Forgione, M.; Cuomo, F.; Serretiello, E.; Hailu, G.S.; Jahro, E.; Lahtela-Kakkonen, M.; Mai, A.; Altucci, L. Sirtuin functions and modulation: From chemistry to the clinic. *Clin. Epigenet.* **2016**, *8*, 61. [CrossRef] [PubMed]
25. Velasquez, D.A.; Martinez, G.; Romero, A.; Vázquez, M.J.; Boit, K.D.; Dopeso-Reyes, I.G.; López, M.; Vidal, A.; Nogueiras, R.; Diéguez, C. The central Sirtuin 1/p53 pathway is essential for the orexigenic action of ghrelin. *Diabetes* **2011**, *60*, 1177–1185. [CrossRef]
26. Al-Massadi, O.; Quinones, M.; Clasadonte, J.; Hernandez-Bautista, R.; Romero-Picó, A.; Folgueira, C.; Morgan, D.A.; Kalló, I.; Heras, V.; Senra, A.; et al. MCH regulates SIRT1/FoxO1 and reduces POMC neuronal activity to induce hyperphagia, adiposity, and glucose intolerance. *Diabetes* **2019**, *68*, 2210–2222. [CrossRef]

27. Cakir, I.; Perello, M.; Lansari, O.; Messier, N.J.; Vaslet, C.A.; Nillni, E.A. Hypothalamic Sirt1 regulates food intake in a rodent model system. *PLoS ONE* **2009**, *4*, e8322. [CrossRef]
28. Dietrich, M.O.; Antunes, C.; Geliang, G.; Liu, Z.W.; Borok, E.; Nie, Y.; Xu, A.W.; Souza, D.O.; Gao, Q.; Diano, S.; et al. Agrp neurons mediate Sirt1's action on the melanocortin system and energy balance: Roles for Sirt1 in neuronal firing and synaptic plasticity. *J. Neurosci.* **2010**, *30*, 11815–11825. [CrossRef]
29. Ramadori, G.; Fujikawa, T.; Fukuda, M.; Anderson, J.; Morgan, D.A.; Mostoslavsky, R.; Stuart, R.C.; Vianna, C.R.; Nillni, E.A.; Rahmouni, K.; et al. SIRT1 deacetylase in POMC neurons is required for homeostatic defenses against diet-induced obesity. *Cell Metab.* **2010**, *12*, 78–87. [CrossRef]
30. Cheng, H.L.; Mostoslavsky, R.; Saito, S.; Manis, J.P.; Gu, Y.; Patel, P.; Bronson, R.; Apella, E.; Alt, F.W.; Chua, K.F. Developmental defects and p53 hyperacetylation in Sir2 homolog (SIRT1)-deficient mice. *Proc. Natl. Acad. Sci. USA* **2003**, *100*, 10794–10799. [CrossRef]
31. Pfluger, P.T.; Herranz, D.; Velasco-Miguel, S.; Serrano, M.; Tschop, M.H. Sirt1 protects against high-fat diet-induced metabolic damage. *Proc. Natl. Acad. Sci. USA* **2008**, *105*, 9793–9798. [CrossRef] [PubMed]
32. Lagouge, M.; Argmann, C.; Gerhart-Hines, Z.; Meziane, H.; Lerin, C.; Daussin, F.; Messadeq, N.; Milne, J.; Lambert, P.; Elliott, P.; et al. Resveratrol improves mitochondrial function and protects against metabolic disease by activating SIRT1 and PGC-1alpha. *Cell* **2006**, *127*, 1109–1122. [CrossRef] [PubMed]
33. Baur, J.A.; Pearson, K.J.; Price, N.L.; Jamieson, H.A.; Lerin, C.; Kalra, A.; Prabhu, V.V.; Allard, J.S.; Lopez-Lluch, G.; Lewis, K.; et al. Resveratrol improves health and survival of mice on a high-calorie diet. *Nature* **2006**, *444*, 337–342. [CrossRef] [PubMed]
34. Milne, J.C.; Lambert, P.D.; Schenk, S.; Carney, D.P.; Smith, J.J.; Gagne, D.J.; Jin, L.; Boss, O.; Perni, R.B.; Vu, C.B.; et al. Small molecule activators of SIRT1 as therapeutics for the treatment of type 2 diabetes. *Nature* **2007**, *450*, 712–716. [CrossRef] [PubMed]
35. Venkatasubramanian, S.; Noh, R.M.; Daga, S.; Langrish, J.P.; Joshi, N.V.; Mills, N.L.; Hoffmann, E.; Jacobson, E.W.; Vlasuk, G.P.; Waterhouse, B.R.; et al. Cardiovascular effects of a novel SIRT1 activator, SRT2104, in otherwise healthy cigarette smokers. *J. Am. Heart Assoc.* **2013**, *2*, e000042. [CrossRef]
36. Libri, V.; Brown, A.P.; Gambarota, G.; Haddad, J.; Shields, G.S.; Dawes, H.; Pinato, D.J.; Hoffmann, E.; Elliot, P.J.; Vlasuk, G.P.; et al. A pilot randomized, placebo controlled, double blind phase I trial of the novel SIRT1 activator SRT2104 in elderly volunteers. *PLoS ONE* **2012**, *7*, e51395. [CrossRef]
37. Timmers, S.; Konings, E.; Bilet, L.; Houtkooper, R.H.; van de Weijer, T.; Goossens, G.H.; Hoeks, J.; van der Krieken, S.; Ryu, D.; Kersten, S.; et al. Calorie restriction-like effects of 30 days of resveratrol supplementation on energy metabolism and metabolic profile in obese humans. *Cell Metab.* **2011**, *14*, 612–622. [CrossRef]
38. Ghanim, H.; Sia, C.L.; Abuaysheh, S.; Korzeniewski, K.; Patnaik, P.; Marumganti, A.; Chaudhuri, A.; Dandona, P. An antiinflammatory and reactive oxygen species suppressive effects of an extract of Polygonum cuspidatum containing resveratrol. *J. Clin. Endocrinol. Metab.* **2010**, *95*, E1–E8. [CrossRef]
39. Ghanim, H.; Sia, C.L.; Korzeniewski, K.; Lohano, T.; Abuaysheh, S.; Marumganti, A.; Chaudhuri, A.; Dandona, P. A resveratrol and polyphenol preparation suppresses oxidative and inflammatory stress response to a high-fat, high-carbohydrate meal. *J. Clin. Endocrinol. Metab.* **2011**, *96*, 1409–1414. [CrossRef]
40. Liu, K.; Zhou, R.; Wang, B.; Mi, M.T. Effect of resveratrol on glucose control and insulin sensitivity: A meta-analysis of 11 randomized controlled trials. *Am. J. Clin. Nutr.* **2014**, *99*, 1510–1519. [CrossRef]
41. Bo, S.; Ciccone, G.; Castiglione, A.; Gambino, R.; De Michieli, F.; Villois, P.; Durazzo, M.; Cavallo-Perin, P.; Cassader, M. Anti-inflammatory and antioxidant effects of resveratrol in healthy smokers a randomized, double-blind, placebo-controlled, cross-over trial. *Curr. Med. Chem.* **2013**, *20*, 1323–1331. [CrossRef] [PubMed]
42. Quinones, M.; Al-Massadi, O.; Ferno, J.; Nogueiras, R. Cross-talk between SIRT1 and endocrine factors: Effects on energy homeostasis. *Mol. Cell. Endocrinol.* **2014**, *397*, 42–50. [CrossRef] [PubMed]
43. Quinones, M.; Al-Massadi, O.; Folgueira, C.; Bremser, S.; Gallego, R.; Torres-Leal, L.; Haddad-Tóvolli, R.; García-Caceres, C.; Hernandez-Bautista, R.; Lam, B.Y.H.; et al. p53 in AgRP neurons is required for protection against diet-induced obesity via JNK1. *Nat. Commun.* **2018**, *9*, 3432. [CrossRef] [PubMed]
44. Matsui, S.; Sasaki, T.; Kohno, D.; Yaku, K.; Inutsuka, A.; Yokota-Hashimoto, H.; Kikuchi, O.; Suga, T.; Kobayashi, M.; Yamanaka, A.; et al. Neuronal SIRT1 regulates macronutrient-based diet selection through FGF21 and oxytocin signalling in mice. *Nat. Commun.* **2018**, *9*, 4604. [CrossRef] [PubMed]
45. Sasaki, T.; Kikuchi, O.; Shimpuku, M.; Susanti, V.Y.; Yokota-Hashimoto, H.; Taguchi, R.; Shibusawa, N.; Sato, T.; Tang, L.; Amano, K.; et al. Hypothalamic SIRT1 prevents age-associated weight gain by improving leptin sensitivity in mice. *Diabetologia* **2014**, *57*, 819–831. [CrossRef]
46. Ramadori, G.; Fujikawa, T.; Anderson, J.; Berglund, E.D.; Frazao, R.; Michán, S.; Vianna, C.R.; Sinclair, D.A.; Elias, C.F.; Coppari, R. SIRT1 deacetylase in SF1 neurons protects against metabolic imbalance. *Cell Metab.* **2011**, *14*, 301–312. [CrossRef]
47. Vazquez, M.J.; Toro, C.A.; Castellano, J.M.; Ruiz-Pino, F.; Roa, J.; Beiroa, D.; Heras, V.; Velasco, I.; Dieguez, C.; Pinilla, L.; et al. SIRT1 mediates obesity- and nutrient-dependent perturbation of pubertal timing by epigenetically controlling Kiss1 expression. *Nat. Commun.* **2018**, *9*, 4194. [CrossRef]
48. Kuang, J.; Chen, L.; Tang, Q.; Zhang, J.; Li, Y.; He, J. The role of Sirt6 in obesity and diabetes. *Front. Physiol.* **2018**, *9*, 135. [CrossRef]

49. Mostoslavsky, R.; Chua, K.F.; Lombard, D.B.; Pang, W.W.; Fischer, M.R.; Gellon, L.; Liu, P.; Mostoslavsky, G.; Franco, S.; Murphy, M.M.; et al. Genomic instability and aging-like phenotype in the absence of mammalian SIRT6. *Cell* **2006**, *124*, 315–329. [CrossRef]
50. Schwer, B.; Schumacher, B.; Lombard, D.B.; Xiao, C.; Kurtev, M.V.; Gao, J.; Schneider, J.I.; Chai, H.; Bronson, R.T.; Tsai, L.-H.; et al. Neural sirtuin 6 (Sirt6) ablation attenuates somatic growth and causes obesity. *Proc. Natl. Acad. Sci. USA* **2010**, *107*, 21790–21794. [CrossRef]
51. Kanfi, Y.; Peshti, V.; Gil, R.; Naiman, S.; Nahum, L.; Levin, E.; Kronfeld-Schor, N.; Cohen, H.Y. SIRT6 protects against pathological damage caused by diet-induced obesity. *Aging Cell* **2010**, *9*, 162–173. [CrossRef] [PubMed]
52. Tang, Q.; Gao, Y.; Liu, Q.; Yang, X.; Wu, T.; Huang, C.; Huang, Y.; Zhang, J.; Zhang, Z.; Li, R.; et al. Sirt6 in pro-opiomelanocortin neurons controls energy metabolism by modulating leptin signaling. *Mol. Metab.* **2020**, *37*, 100994. [CrossRef] [PubMed]
53. Jimenez-Gomez, Y.; Mattison, J.A.; Pearson, K.J.; Martin-Montalvo, A.; Palacios, H.H.; Sossong, A.M.; Ward, T.M.; Younts, C.M.; Lewis, K.; Allard, J.S.; et al. Resveratrol improves adipose insulin signaling and reduces the inflammatory response in adipose tissue of rhesus monkeys on high-fat, high-sugar diet. *Cell Metab.* **2013**, *18*, 533–545. [CrossRef] [PubMed]
54. Fiori, J.L.; Shin, Y.K.; Kim, W.; Krzysik-Walker, S.M.; González-Mariscal, I.; Carlson, O.D.; Sanghvi, M.; Moaddel, R.; Farhang, K.; Gadkaree, S.K.; et al. Resveratrol prevents beta-cell dedifferentiation in nonhuman primates given a high-fat/high-sugar diet. *Diabetes* **2013**, *62*, 3500–3513. [CrossRef] [PubMed]
55. Konings, E.; Timmers, S.; Boekschoten, M.V.; Goossens, G.H.; Jocken, J.W.; Afman, L.A.; Müller, M.; Schrauwen, P.; Mariman, E.C.; Blaak, E.E. The effects of 30 days resveratrol supplementation on adipose tissue morphology and gene expression patterns in obese men. *Int. J. Obes.* **2014**, *38*, 470–473. [CrossRef]
56. Poulsen, M.M.; Vestergaard, P.F.; Clasen, B.F.; Radko, Y.; Christensen, L.P.; Stødkilde-Jørgensen, H.; Møller, N.; Jessen, N.; Pedersen, S.B.; Jørgensen, O.L. High-dose resveratrol supplementation in obese men: An investigator-initiated, randomized, placebo-controlled clinical trial of substrate metabolism, insulin sensitivity, and body composition. *Diabetes* **2013**, *62*, 1186–1195. [CrossRef] [PubMed]
57. Hausenblas, H.A.; Schoulda, J.A.; Smoliga, J.M. Resveratrol treatment as an adjunct to pharmacological management in type 2 diabetes mellitus—Systematic review and meta-analysis. *Mol. Nutr. Food Res.* **2015**, *59*, 147–159. [CrossRef]
58. Knop, F.K.; Konings, E.; Timmers, S.; Schrauwen, P.; Holst, J.J.; Blaak, E.E. Thirty days of resveratrol supplementation does not affect postprandial incretin hormone responses, but suppresses postprandial glucagon in obese subjects. *Diabet. Med.* **2013**, *30*, 1214–1218. [CrossRef]
59. Yoshino, J.; Conte, C.; Fontana, L.; Mittendorfer, B.; Imai, S.I.; Schechtman, K.B.; Gu, C.; Kunz, I.; Fanelli, F.R.; Patterson, B.W.; et al. Resveratrol supplementation does not improve metabolic function in nonobese women with normal glucose tolerance. *Cell Metab.* **2012**, *16*, 658–664. [CrossRef]
60. Novelle, M.G.; Wahl, D.; Dieguez, C.; Bernier, M.; de Cabo, R. Resveratrol supplementation: Where are we now and where should we go? *Ageing Res. Rev.* **2015**, *21*, 1–15. [CrossRef]
61. Baksi, A.; Kraydashenko, O.; Zalevkaya, A.; Stets, R.; Elliott, P.; Haddad, J.; Hoffmann, E.; Vlasuk, G.P.; Jacobson, E.W. A phase II, randomized, placebo-controlled, double-blind, multi-dose study of SRT2104, a SIRT1 activator, in subjects with type 2 diabetes. *Br. J. Clin. Pharm.* **2014**, *78*, 69–77. [CrossRef] [PubMed]
62. Noh, R.M.; Venkatasubramanian, S.; Daga, S.; Langrish, J.; Mills, N.L.; Lang, N.N.; Hoffmann, E.; Waterhouse, B.; Newby, D.E.; Frier, B.M. Cardiometabolic effects of a novel SIRT1 activator, SRT2104, in people with type 2 diabetes mellitus. *Open Heart* **2017**, *4*, e000647. [CrossRef] [PubMed]
63. Kundu, A.; Dey, P.; Park, J.H.; Kim, I.S.; Kwack, S.J.; Kim, H.S. EX-527 Prevents the progression of high-fat diet-induced hepatic steatosis and fibrosis by upregulating SIRT4 in Zucker rats. *Cells* **2020**, *9*, 1101. [CrossRef] [PubMed]

International Journal of
Molecular Sciences

MDPI

Review

New Insights of SF1 Neurons in Hypothalamic Regulation of Obesity and Diabetes

Anna Fosch [1], Sebastián Zagmutt [1], Núria Casals [1,2] and Rosalía Rodríguez-Rodríguez [1,*]

[1] Basic Sciences Department, Faculty of Medicine and Health Sciences, Universitat Internacional de Catalunya, 08195 Sant Cugat del Vallès, Spain; afosch@uic.es (A.F.); szagmutt@uic.es (S.Z.); ncasals@uic.es (N.C.)
[2] Centro de Investigación Biomédica en Red de Fisiopatología de la Obesidad y la Nutrición (CIBEROBN), Instituto de Salud Carlos III, 28029 Madrid, Spain
* Correspondence: rrodriguez@uic.es; Tel.: +34-935-042-002

Abstract: Despite the substantial role played by the hypothalamus in the regulation of energy balance and glucose homeostasis, the exact mechanisms and neuronal circuits underlying this regulation remain poorly understood. In the last 15 years, investigations using transgenic models, optogenetic, and chemogenetic approaches have revealed that SF1 neurons in the ventromedial hypothalamus are a specific lead in the brain's ability to sense glucose levels and conduct insulin and leptin signaling in energy expenditure and glucose homeostasis, with minor feeding control. Deletion of hormonal receptors, nutritional sensors, or synaptic receptors in SF1 neurons triggers metabolic alterations mostly appreciated under high-fat feeding, indicating that SF1 neurons are particularly important for metabolic adaptation in the early stages of obesity. Although these studies have provided exciting insight into the implications of hypothalamic SF1 neurons on whole-body energy homeostasis, new questions have arisen from these results. Particularly, the existence of neuronal sub-populations of SF1 neurons and the intricate neurocircuitry linking these neurons with other nuclei and with the periphery. In this review, we address the most relevant studies carried out in SF1 neurons to date, to provide a global view of the central role played by these neurons in the pathogenesis of obesity and diabetes.

Keywords: SF1 neurons; ventromedial hypothalamus nucleus; obesity; diabetes; energy homeostasis; glucose homeostasis

Citation: Fosch, A.; Zagmutt, S.; Casals, N.; Rodríguez-Rodríguez, R. New Insights of SF1 Neurons in Hypothalamic Regulation of Obesity and Diabetes. *Int. J. Mol. Sci.* **2021**, *22*, 6186. https://doi.org/10.3390/ijms22126186

Academic Editor: Clara Balsano

Received: 29 April 2021
Accepted: 5 June 2021
Published: 8 June 2021

1. Introduction

Obesity is a multifactorial chronic disease associated with a higher risk of developing cardiovascular diseases, diabetes, cancer, and, more recently, COVID-19 infection. According to the World Health Organization (WHO), in 2016, worldwide obesity had nearly tripled since 1975, with 39% of adults and 18% of children and adolescents overweight or obese [1]. The main metabolic comorbidity of obesity is type-2 diabetes that occurs when body tissues become resistant to insulin and is estimated to be the seventh leading cause of death [2]. Therefore, understanding the molecular and physiological mechanisms underlying the control of feeding behavior, energy balance and glucose homeostasis is crucial for the prevention and treatment of obesity and diabetes.

The regulation of peripheral metabolism and glucose homeostasis not only resides in the tissue's autonomous responses to nutrients and hormonal signals but also in some brain regions, particularly the hypothalamus. The hypothalamus integrates multiple metabolic inputs from peripheral organs with afferent stimuli coming from other brain regions and coordinates a diversity of efferent responses to control food intake, fat metabolism, hormone secretion, body temperature, locomotion, and behavior in order to maintain energy balance and blood glucose levels. Within the hypothalamus, the ventromedial nucleus (VMH) located above the arcuate nucleus (ARC) and the median eminence, was identified in the mid-1900s as the satiety center because its injury produced hyperphagia, insulin resistance, and body weight gain [3,4]. At that point, VMH was demonstrated to play a key role in the control of energy expenditure and glucose homeostasis [3,5]. Since then, intensive research has been done on VMH and it is currently known that this hypothalamic area encompasses a heterogeneous set of neurons, which are differentiated by the genes they are expressing (Figure 1). Many of the genes highly expressed in the VMH have been identified and their functions have been explored (Figure 1) [6–8].

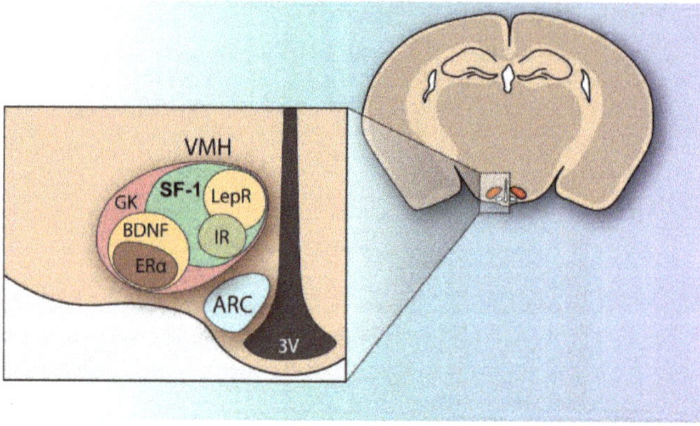

Figure 1. Schematic illustration of the pattern of genes highly expressed in the VMH. The majority of VMH cells, especially in the dorsomedial and central regions of VMH, express the nuclear receptor steroidogenic factor 1 (SF1). Leptin receptor (LEPR) positive cells mainly converge in the dorsomedial part, whereas insulin receptor (IR) maps the central region. Cells expressing brain-derived neurotrophic factor (BDNF) are mainly distributed in central and lateral areas of the VMH, estrogen receptor (ERα)-expressing cells are limited to the lateral region, and glucokinase (GK)-positive cells are present throughout the VMH. This schematic diagram is based on previous articles from Choi et al. [8] and Yi et al. [6].

The majority of VMH cells, especially in the dorsomedial and central regions of VMH (VMHdm and VMHc, respectively) [9], express the nuclear receptor steroidogenic factor 1 (SF1) (Figure 1), also known as NR5A1, which is essential for VMH development and function [10–12] and is considered a key marker of that nucleus since SF1 expression is specific and selective to VMH in the brain. Outside the brain, SF1 can be found in the adrenal gland and gonads with differential roles during development [13]. For many years, SF1 was classified as an orphan nuclear receptor, but between 2005 and 2007 several pieces of evidence demonstrated that phospholipids can be ligands of SF1, such as phosphatidic acid being an activator and sphingosine an inhibitor of SF1 transcriptional activity [13]. Interestingly, SF1 can suffer different post-translational modifications, which regulate its stability and transcriptional activity [14], but also control the expression of numerous downstream target genes, including CB1, BDNF, and Crhr2 [13]. Considering this, in order to explore the importance of SF1 neurons, transgenic mice lacking this nuclear receptor were studied by different researchers. Mice lacking SF1 were not viable due to a failure in the proper development of adrenal glands and gonads [11,13,15]. However, when rescued from lethality by adrenal transplantation from WT littermates and corticosteroid injections, mutant mice displayed robust weight gain resulting from both hyperphagia and reduced energy expenditure [16]. Interestingly, a similar phenotype was observed in humans with mutations of the SF1 gene, who often show mild to severe obesity [17,18]. To avoid the confounding metabolic effects of glucocorticoid injections and adrenal transplantations to global SF1 KO mice, Elmquist's group generated an alternative mouse model in which SF1 was specifically deleted in the VMH after completion of the nucleus development using CamKII-Cre [19,20]. These postnatal VMH-specific SF1 KO mice showed increased weight gain and impaired thermogenesis in response to a high-fat diet (HFD), being the first demonstration that the transcription factor SF1 is postnatally required in the VMH for normal energy homeostasis, especially under the HFD condition.

In an attempt to clarify the contribution of this specific population of neurons to hypothalamic regulation of obesity and diabetes, in the last 15 years, several transgenic models have been developed by deleting specific targets in SF1 neurons related to energy balance and glucose homeostasis. In 2013, a profound review article was published by Choi and colleagues [8] summarizing the last updates of SF1 neurons in energy homeostasis. Since then, new neuronal-based approaches (i.e., optogenetic and chemogenetic technology) and the generation of new transgenic mice in key target proteins have provided exciting insight into the implication of SF1 neurons on whole-body energy balance, particularly thermogenesis and glucose homeostasis, that are compiled in the present review.

Despite these studies contributing to understanding the mechanisms by which SF1 neurons regulate energy homeostasis, new questions arose from these results. An important issue to clarify is the heterogeneity of SF1 neurons in terms of glucose, leptin, and insulin sensing or non-sensing neurons, indicating the operation complexity of SF1 neurons and the intricate mechanisms of afferent and efferent neurocircuitry involved in the control of energy balance. In this review, we address the most relevant studies carried out on SF1 neurons to date, from transgenic mice to neurocircuitry studies, in order to discuss the most relevant findings of these investigations and provide an exhaustive overview of the role played by SF1 neurons in the hypothalamic regulation of energy expenditure and glucose homeostasis, and the potential mechanisms involved.

2. Unraveling the Functions of SF1 Neurons in Energy Balance by Optogenetic and Chemogenetic Approaches

In order to selectively manipulate the SF1 neuronal activity in a physiological context, optogenetic and chemogenetic approaches have emerged [21,22]. These tools use channels that are activated by light and engineered G-protein coupled receptors controlled by exogenous molecules, respectively [21,22]. The incorporation of these approaches into animal models has greatly advanced our understanding of the SF1 role and neuronal circuits.

Mice engineered to activate SF1 neurons by optogenetics were designed through the injection of adeno-associated virus (AAV) particles expressing a Cre-dependent channelrhodopsin (ChRs) into the VMH of SF1-Cre mice to produce SF1-ChRs animals [23,24]. The advantage of this technology is the light-controllable activation of SF1 neurons in a spatiotemporal manner.

One of the first results obtained from this model revealed that SF1 neurons are closely associated with defensive reactions. David J. Anderson and colleagues in 2015, demonstrated that the optogenetic stimulation of SF1 neurons applying a frequency of 20 Hz induced freezing or activity burst, while no response on feeding behavior or energy balance was reported [25]. However, the electrophysiological record of the firing pattern observed in these studies was generated by applying a high-frequency burst of spiking, while the physiological spiking of the neurons in steady-state ranged between 3.5–6.3 Hz [26]. Considering this information, two years later, other researchers showed that SF1 neurons exert a differential effect depending on the frequency of activation. They confirmed that high-frequency activation (>20 Hz) evokes a profound defensive response which includes freezing and escape attempts, but low-frequency activation (2 Hz) suppresses feeding after fasting and reduces the time that mice spend near to the food [24]. These novel results suggest that SF1 neurons dynamically modulate feeding and anxiety-related behaviors by changing the firing pattern and also indicate that this subset of hypothalamic neurons is involved in the fight or flight response.

The chemogenetic modulation of SF1 neurons under potentially more physiological firing patterns has been performed using hM3Dq and hM4Di designer receptors exclusively activated by designer drugs (DREADDs). The expression of DREADDs specifically in SF1 neurons allows to excite or inhibit them in response to clozapine (CNO) or JHU37160 dihydrochloride (J60) which is an in vivo DREADD agonist with high affinity and potency for hM3Dq and hM4Di receptors [27,28]. In agreement with the feeding behavior-related results obtained from optogenetic approaches, the administration of CNO in fasted SF1-hM3Dq mice revealed a reduction in food consumption. On the contrary, inhibition of SF1 neurons in SF1-hM4Di mice increased cumulative food intake in ad libitum-fed mice. Authors manifested that none of the dramatic defensive behaviors seen with high-frequency optogenetic activation of SF1 neurons was observed [24]. At peripheral levels, they also demonstrated that the chronic chemogenetic modulation of SF1 neurons modified the body fat mass since the continuous administration of CNO (3 weeks) in SF1-hM3Dq and SF1-hM4Di animals reduced and increased the fat mass content, respectively [24]. These changes in fat mass could be explained by the SF1 modulation of fat oxidation. Very recently, it was established that SF1-hM3Dq mice increased energy expenditure and fat oxidation independent of the locomotion activity within 2-h post-activation [23]. Although this study did not evaluate the energy expenditure profile in SF1-hM4Di mice, it was expected that the inactivation of SF1 neurons reduced energy expenditure. This hypothesis is supported by mice expressing tetanus toxin (TT) in SF1 neurons. Since TT prevents neurotransmitter release, SF^{TT} mice displayed reduced energy expenditure and increased body weight [29].

Besides their implication in energy balance, the use of optogenetics has highlighted the role of SF1 neurons in the hypothalamic control of systemic glucose levels. For a long time, it was known that VMH triggered the counterregulatory response (CRR) induced by hypoglycemia [30,31] but it was not clear the contribution of SF1 neurons to this feedback response. In order to elucidate whether SF1 neurons are linked to this effect, an elegant experiment using optogenetic and chemogenetic tools was done by Gregory J Morton and colleagues, showing that selective inhibition of SF1 neurons blocked recovery from insulin-induced hypoglycemia. Conversely, activation of SF1 neurons caused diabetes-range hyperglycemia [32]. This evidence is concordant with those obtained from transgenic models such as mice lacking vesicular glutamate transporter 2 (VGLUT2) specifically in SF1 neurons since this genetic disruption attenuated recovery from insulin-induced hypoglycemia [33].

Taken all together, these genetic approaches have revealed the specific involvement of SF1 neurons in many aspects of metabolic regulation due to their direct or indirect role in the maintenance of the energy balance and glucose levels, confirming the classification of VMH as a primary satiety center [34,35].

3. Manipulation of Key Targets in SF1 Neurons: Lessons from Transgenic Mice

A particularly powerful strategy developed for the exploration of SF1 neurons in obesity and diabetes has been the design of the SF1 Cre mice. Several groups have generated different SF1 Cre transgenic lines in which the expression of Cre recombinase is derived by *Sf1* regulatory elements [20,36]. These lines allow for ablating general factors or targets known to be associated with energy homeostasis by crossing them with *floxed* strains. In the following sub-sections, we discuss different studies of SF1-CRE transgenic mice organized by the type of molecular target under investigation (Table 1) as well as the sex-specific effect of SF1 neurons in energy balance.

3.1. Hormone Receptors and Related Signaling Pathways

Due to the importance of the anorectic hormone leptin in the central control of energy homeostasis, the physiological effects after deletion of leptin receptor (LEPR) in SF1 neurons have been thoroughly investigated. The first and most representative study was performed by Bradford Lowell and colleagues, in which genetic deletion of LEPR selectively from hypothalamic SF1 neurons triggered an increase in body weight gain without changes in food intake, leaving these mice unable to adapt to HFD or to activate energy expenditure [20]. To reinforce the key role of the leptin pathway in SF1 neurons, selective inactivation of *Socs3*, a negative mediator of central leptin—pSTAT3 signaling, in SF1 neurons (Figure 2) was developed. Conversely to the deletion of LEPR, mice lacking *Socs3* showed improved weight-reducing effects of leptin, with a decrease in food intake and an enhanced energy expenditure under chow diet or HFD condition [37]. The importance of leptin signaling in energy balance through SF1 neurons was also reinforced by the specific deletion of the G protein α-subunit Gsα [38]. Gsα couple receptors for hormones, neurotransmitters, and other factors to activate adenyl-cyclase leading to cAMP generation, which is a negative regulator of leptin action (Figure 2). Then, the lack of Gsα in SF1 neurons increases leptin sensitivity [38].

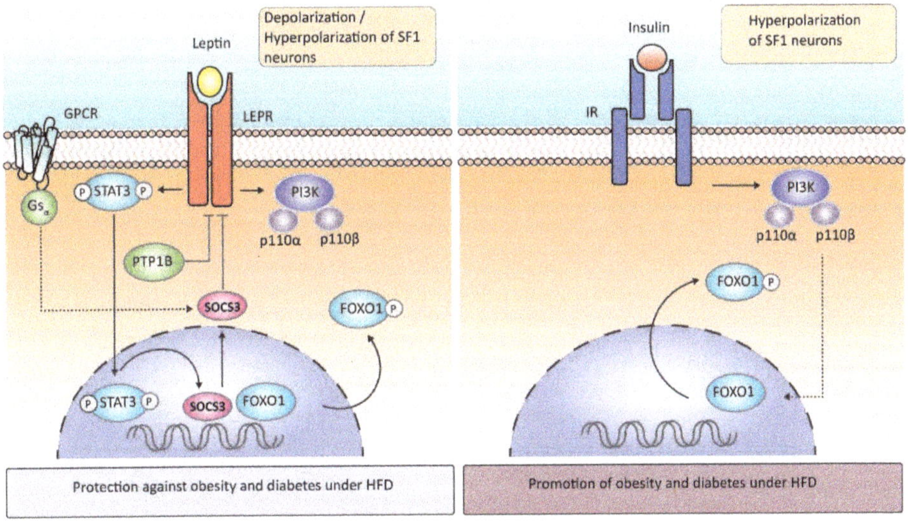

Figure 2. Leptin and insulin signaling in SF1 neurons. Leptin binds LEPR to regulate body weight gain, energy expenditure, and insulin sensitivity in the VMH via STAT3 phosphorylation which is translocated into the nucleus and induces the expression of genes such as SOCS3, a negative regulator of LEPR. Leptin also stimulates a PI3K-dependent pathway, particularly involving subunits p110α, which contributes to energy expenditure regulation, and p110β, which contributes to both brown fat thermogenesis and glucose metabolism regulation in SF1 neurons. The Gsα subunit of G protein-coupled receptors (GPCR) of hormones and the protein-tyrosine phosphatase 1B (PTP1B) negatively regulates leptin action. The activation of this leptin signaling pathway triggers protection against obesity and diabetes under a high-fat diet (HFD) feeding. According to insulin signaling in SF1 neurons, exposure to HFD drives hyperinsulinemia, insulin binds insulin receptor (IR), which in turn activates PI3K that indirectly phosphorylates FOXO1 to induce transcription factors and promote obesity and glucose metabolism disruption.

The protein-tyrosine phosphatase 1B (PTP1B) is another negative regulator of leptin signaling in SF1 neurons (Figure 2). Its action is mediated through selective dephosphorylation of the two signaling molecules JAK2 and STAT3. In vivo studies have demonstrated that whole-brain deletion of PTP1B resulted in leanness, hypersensitivity to leptin, and resistance to HFD-induced obesity, a phenotype partly associated with increased hypothalamic activation of STAT3 [39]. Surprisingly, its specific deletion in SF1 neurons resulted in increased adiposity in female mice exposed to HFD due to low energy expenditure, whereas leptin sensitivity was enhanced, and food intake was attenuated, findings that were likely explained by increased STAT3 activation [40]. Mice lacking PTP1B in SF1 neurons also had improved leptin and insulin signaling in VMH, suggesting that increased insulin responsiveness in SF1 neurons could overcome leptin hypersensitivity and promote adiposity [39,40].

A more recent study tried to rescue native LEPR in SF1 neurons in *LepR*-deficient mice. They concluded that LEPR signaling in the VMH is not sufficient to protect against obesity in this null mouse [41]. This finding could explain that this neuronal population expressing LEPR works in conjunction with other types of neurons expressing the same receptor, and SF1 neurons by themselves cannot compensate for all receptor deficiency. Summing up, leptin signaling in SF1 neurons plays a key role in energy homeostasis regulation and mediates the proper physiological adaptation to HFD to avoid or delay the onset of obesity.

According to glucose metabolism, leptin has been long related to glucose homeostasis improving insulin sensitivity, since intra-VMH injection of leptin increases glucose uptake in peripheral tissues [42] and normalizes hyperglycemia [43]. Very recently it has been demonstrated that central leptin infusion in mice with SF1 neuron-specific LEPR deficiency corrected diabetic hyperglycemia [43]. Despite this result, re-expressing the receptor in SF1 neurons of null mice showed that SF1 neurons were not sufficient to mediate the antidiabetic action of leptin [44]. The essential action of leptin in SF1 to correct diabetic hyperglycemia was clarified by further investigations. Particularly, in the specific knock—out of *Socs3* in SF1 neurons, where leptin signaling is over-activated, Ren Zhang and colleagues observed improved glucose homeostasis, showing protection against hyperglycemia and hyperinsulinemia caused by HFD feeding [37]. These studies demonstrate that leptin in VMH neurons improves glucose and insulin metabolism, although this area is not essential as there are redundant neuronal circuits regulating it. Optogenetic activation of SF1 neurons has the same output as leptin increasing glucose uptake but it does not normalize blood glucose levels, which leads to an understanding of two different subsets of SF1 neurons, one subset increasing insulin sensitivity and the second one increasing blood glucose levels.

Another key hormone implicated in the balancing of energy metabolism is insulin. It is known that insulin acutely suppresses food intake and decreases fat mass in both rodents and humans [45,46] (Figure 2). Mice lacking insulin receptors (IR) in SF1 neurons did not show any differences in body weight when fed a chow diet but under HFD conditions, mutant mice were protected against obesity and showed an enhanced leptin sensitivity and glucose homeostasis [26]. Interestingly, exposure to HFD led to the overactivation of insulin in the VMH, leading to a reduction in SF1 neurons firing frequency, in comparison to the insulin resistance induced in ARC neurons. The differential dysregulation of insulin action in these hypothalamic nuclei under HFD could indicate cooperation between these responses to drive obesity [26]. These findings also suggest that hyperinsulinemia evoked by HFD feeding controls SF1 neuronal activity, which leads to changes in the synaptic inputs to other neuronal populations such as POMC, resulting in obesity and diabetes. However, the specific contribution of insulin signaling in SF1 neurons and its relationship to peripheral insulin resistance and glucose levels needs further investigation.

Leptin depolarizes or hyperpolarizes SF1 neurons, depending on the subpopulation, while insulin only hyperpolarizes them, as both actions are in a downstream PI3K-dependent manner [20,47–49] (Figure 2). PI3K is formed by two different subunits, p85 and p110, and specifically, the subunit p110β is necessary for depolarizing and hyperpolarizing SF1 neurons, while p110α is only needed in the hyperpolarization process [47]. Mice lacking p110α in SF1 neurons had reduced energy expenditure in response to hypercaloric feeding and, therefore, displayed an obesogenic phenotype. Moreover, this p110α subunit was not required to regulate glucose metabolism in SF1 neurons [49]. Mice lacking p110β in the same neuronal population had also decreased energy expenditure (reduced thermogenesis) leading to increased susceptibility to obesity, whereas, in contrast to the p110α subunit, p110β involved changes in peripheral insulin sensitivity [50]. In line with this evidence, deletion of FOXO1, a downstream transcription factor of insulin-PI3K (Figure 2), in SF1 neurons resulted in a lean phenotype with high energy expenditure, even in fasting, and these null mice presented an enhanced insulin sensitivity and glucose tolerance, in concordance with genetic deletion of IR in SF1neurons [51]. Although leptin and insulin can inhibit SF1 neurons using the same molecular cascade, they are anatomically segregated within the VMH (neurons expressing LEPR receptor are located in the VMHdm when depolarizing and scattered throughout the nucleus when hyperpolarizing, whereas those expressing IR are in the VMHc close to the ventricle) [47], which would explain the different effects observed when deleting their receptors.

Other hormones studied in SF1 neurons are estrogens. Female mice lacking the estrogenic receptor α (ERα) in SF1 neurons were obese due to a reduced energy expenditure [52]. Ablation of the ERα led to abdominal obesity with adipocyte hypertrophy in females, but not in male mice [53]. Despite the fact that most of the studies on SF1 neurons until now were performed only in male mice, these last results described, and others discussed later [40,54], reinforce the notion that SF1 neurons may have a sex-specific effect on energy balance and glucose metabolism.

Growth hormone (GH) also plays a role in glucose metabolism via SF1 neurons. GH is secreted in a metabolic stress situation such as hypoglycemia. Deletion of its receptor in SF1 neurons resulted in an impaired capacity for recovery from hypoglycemia [55]. These mice showed an altered CRR due to changes in the neurocircuit that regulates the parasympathetic nervous system. This result supports the importance of SF1 neurons in the proper functionality of glucose homeostasis.

Altogether, these findings identify SF1 neurons (the predominant VMH population) as a key player in the regulation of energy expenditure and glucose homeostasis, being particularly important in the adaptive response to HFD feeding. Most of the mutant mice with deletion of several hormone receptors and associated proteins in SF1 neurons have no changes or mild metabolic alterations under chow diet, but they show substantial metabolic variations under HFD exposure. The action of hormones and related proteins in SF1 neurons is also involved in the CRR to hypoglycemia to maintain glucose balance between the brain and the periphery. Future studies are needed to describe the specific molecular mechanisms and subsets of SF1 neurons underlying the effects of hormones in glucose homeostasis and energy expenditure.

Table 1. Genetic models developed to study SF1 neurons in energy balance.

Type of Target	Target	Mice Model Name	Sex	Challenge	BW	FI	EE	Adiposity	Glycemia	Glucose Tolerance	Insulin Sensitivity	Leptin Sensitivity	SNS Activity	Ref.
Hormone receptors and related signaling pathways	LEPR	Sf1-Cre, *Lepr*$^{\text{flox/flox}}$	M	SD	↑	n.s.	n.s.	↑	n.s.	-	-	-	-	[20]
				HFD	↑	↑	↓	↑	n.s.	-	-	-	-	
	SOCS3	Sf1-Cre, *Socs3*$^{\text{flox/flox}}$	M	SD	n.s.	↓	↓	-	↓	↑	↑	↑	-	[37]
				HF-HS	n.s.	↓	↓	-	↓	↑	↑	↑	-	
				Leptin (a)	↓	↓	-	-	-	-	-	↑	-	
	$G_s\alpha$	VMHGsKO	M (b)	SD	n.s.	n.s.	n.s.	-	n.s.	n.s.	n.s.	n.s.	-	[38]
				HFD	n.s.	n.s.	n.s.	-	↓	↑	↑	↑	-	
	PTP1B	Sf1-*Ptpn1*$^{-/-}$	F	HFD	↑	↓	↓	↑	-	-	↑	↑	↓	[39]
			M	HFD	n.s.	-	-	n.s.	-	-	-	-	-	
	IR	SF-1$^{\Delta IR}$		SD	n.s.	n.s.	n.s.	n.s.	-	-	-	-	-	[26]
				HFD	↓	↓	n.s.	↓	n.s.	↑	↑	↑	-	
	p110α	p110α$^{\text{lox/lox}}$/SF1-Cre	M	SD	n.s.	n.s.	n.s.	n.s.	n.s.	-	n.s.	↓	-	[49]
				HFD	↑	n.s.	↓	↑	-	-	-	-	-	
	p110β	p110β KO$^{\text{sf1}}$	M	SD	n.s.	n.s.	↓ BAT th. (c)	n.s.	n.s.	↓	↓	-	-	[50]
				HFD	↑	n.s.	↓	↑	↑	-	-	-	-	
	FOXO1	*Foxo1* KO$^{\text{Sf1}}$	M	SD	↓	n.s.	↑	↓	-	-	-	-	-	[51]
			F	SD	↓	n.s.	↑	↓	-	-	-	-	-	
			M	HFD	↓	n.s.	↑	↓	↓	↑	↑	↑	-	
	ERα	ERα$^{\text{lox/lox}}$/SF1-Cre	F	SD	↑	n.s.	↓	↑	n.s.	↓	-	-	-	[52]
			F	HFD	↑	n.s.	↓	↑	-	-	-	-	↓	
Nutrient sensors	AMPK	SF1-Cre AMPKα1$^{\text{flox/flox}}$	M	SD	↓	n.s.	↑	↓	-	-	-	-	↑ (d)	[56]
			M	HFD	↓	n.s.	↑	↓	↓	↑	n.s.	-	-	
	SIRT1	Sf1-Cre; *Sirt1*$^{\text{loxP/loxP}}$	M/F	SD	n.s.	n.s.	n.s.	n.s.	n.s.	-	-	-	-	[57]
				HFD	↑	n.s.	↓	↑	↑	↓	↓	↓	-	

Table 1. *Cont.*

Type of Target	Target	Mice Model Name	Sex	Challenge	BW	FI	EE	Adiposity	Glycemia	Glucose Tolerance	Insulin Sensitivity	Leptin Sensitivity	SNS Activity	Ref.
Glutamatergic neurotransmission and synaptic receptors	VGLUT2	Sf1-Cre; *Vglut2*$^{flox/flox}$	M/F	SD	n.s.	-	-	-	↓	-	-	-	-	[33]
			M/F	HFD	↑	↑	n.s.	↑	-	-	-	-	-	
	mGluR5	mGluR5$^{2L/2L:SF1\text{-}Cre}$	F	SD	n.s.	n.s.	n.s.	-	n.s.	↓	↓	-	↓	[54]
			M	SD	n.s.	n.s.	n.s.	-	n.s.	n.s.	n.s.	n.s.	n.s.	
			M/F	HFD	n.s.	n.s.	n.s.	-	-	-	-	-	-	
	α2δ-1	α2δ-1$^{2L/2L:SF1\text{-}Cre}$	M	SD	n.s.	n.s.	n.s.	n.s.	n.s.	↓	↓	-	↓	[58,59]
			M	HFD	n.s.	n.s.	n.s.	n.s.	n.s.	n.s.	n.s.	-	-	
			F	SD	n.s.	n.s.	n.s.	n.s.		↓	↓	-	-	
			F	HFD	↑	n.s.	n.s.	n.s.	n.s.	↓	↓	-	↑	
	CB1	SF1-CB$_1$-KO	M	SD	n.s.	n.s.	↑ BAT th.	↓	n.s.	↑	↑	↑	↑	[60]
			M	HFD	↑	↑	n.s.	↑	-	↓	n.s.	↓	↓	
Modulators of autophagy, mitochondrial and primary cilia function	*Atfg7*	Sf1-Cre; *Atg7*$^{loxP/loxP}$	M	Fasting	n.s.	↓	↓	n.s.	n.s.	n.s.	n.s.	↓	-	[61]
	UCP2	*Ucp2*KOKI^{Sf1}	M	Chow diet	n.s.	n.s.	n.s.	n.s.	n.s.	↑	↑	-	-	[62]
	IFT88	IFT88-KO^{SF-1}	M/F	Chow diet	↑	n.s.	↓	↑	↑	↓	↓	↓	↓	[63]
			M/F	HFD	↑	↑	↓	↑	↑	-	-	-	-	

n.s.: No significant changes appreciated; -: not studied/unknown; M: male; F: female. SD: standard diet; HFD: high fat diet; HF-HS: high fat-high sucrose diet; BAT th.: brown fat thermogenesis. [(a)] Subcutaneously implanted osmotic minipumps, infusing for 14 days at 0,5 μg/h. [(b)] The study was performed in both male and female, but the metabolic alterations were only appreciated in male mice. [(c)] No changes in EE but significant decrease in BAT thermogenesis. [(d)] Increased sympathetic activity in brown fat.

3.2. Nutrient Sensors: AMPK and SIRT1

The regulation of energy homeostasis is importantly regulated by proteins acting as nutrient and energy sensors, such as AMPK and SIRT1, which restore energy balance during metabolic challenges both at the cellular and physiological levels (Figure 3). A link between both AMPK and SIRT1 in hypothalamic SF1 neurons with central control of obesity and diabetes has been established.

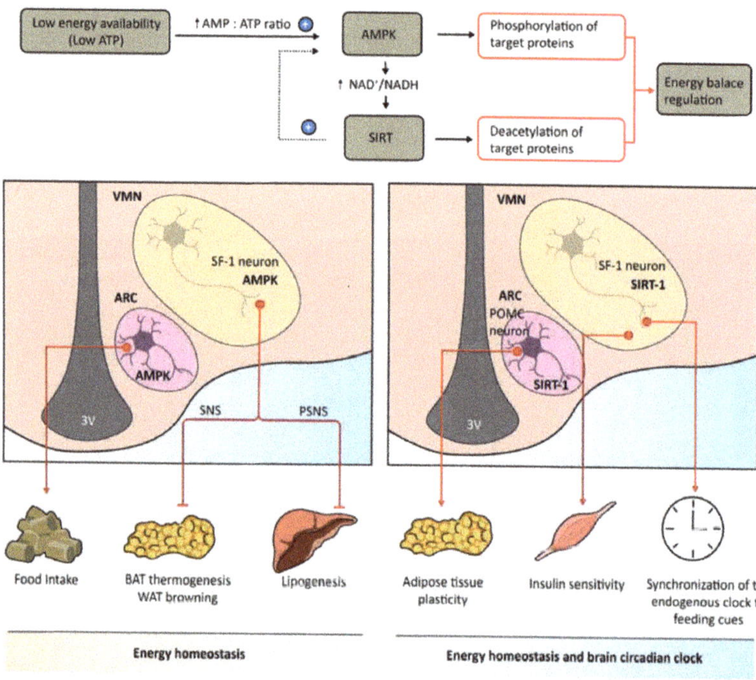

Figure 3. Nutrient sensors in SF1 neurons regulate energy homeostasis. AMPK and SIRT1 are key nutrient sensors that restore energy balance under low energy availability. The role of both AMPK and SIRT1 in the central regulation of energy balance depends on their specific location in hypothalamic nuclei. AMPK in SF1 neurons of the VMH particularly regulates energy expenditure by acting in BAT thermogenesis and subcutaneous WAT browning (via sympathetic output) and liver lipogenesis (via parasympathetic output), whereas AMPK effects in feeding control and glucose balance mostly rely on ARC neurons. SIRT1 in SF1 VMH neurons controls insulin sensitivity in skeletal muscle protecting against diet-induced obesity, but in POMC neurons of the ARC, SIRT1 regulates normal BAT-like remodeling in specific fat depots. SIRT1 in SF1 neurons also contributes to the synchronization of the circadian clock in the brain by nutritional inputs. The arrow means increase and + means activation.

AMPK is a highly conserved master regulator of metabolism, which is activated under low energy conditions, increasing energy production and reducing energy waste. Current evidence demonstrates the critical role of hypothalamic AMPK in the regulation of food intake and energy expenditure, as well as glucose and lipid homeostasis at the whole-body level [56,64–67]. Interestingly, the regulation of these effects depends on the anatomical location of AMPK in the hypothalamus: AMPK effects in feeding control and glucose homeostasis mostly rely on ARC neurons [64,66,67], whereas its effects in energy expenditure arise from the VMH [56,65,68], particularly in SF1 neurons (Figure 3).

The identification of SF1 neurons as the main neuronal AMPK-expressing population in VMH to be targeted in obesity was demonstrated by several studies of Miguel Lopez lab.

They generated an SF1 specific AMPKα_1 null mouse by Cre-Lox recombination. Ablation of AMPKα_1 in these neurons led to feeding-independent weight loss associated with an increase in energy expenditure [56]. In accordance with this, SF1-Cre AMPK$\alpha_1^{flox/flox}$ mice displayed BAT activation, which was confirmed by increased BAT temperature and UCP1 protein expression, elevated sympathetic activity, and higher ^{18}F-FDG uptake in brown fat [56]. SF1-Cre AMPK$\alpha_1^{flox/flox}$ mice fed with a long-term HFD had a feeding-independent decrease in body weight and adiposity, associated with increased energy expenditure and VO_2, higher BAT thermogenesis activation, and browning of subcutaneous WAT. These findings point to AMPKα_1, but not AMPKα_2, in SF1 neurons as the main catalytic AMPK subunit in the VMH that regulate thermogenic control, particularly under HFD exposure.

Other studies demonstrating that the stimulatory effect of thyroid hormones in BAT thermogenesis [65], WAT browning, and lipid metabolism are mediated by AMPKα_1 attenuation in SF1 neurons [68] strongly support the notion that specific targeting of the discrete neuronal population in the VMH impacts obesity in a feeding-independent, but the thermogenic-dependent manner [56]. These studies also suggest that simultaneous targeting of AMPK in SF1 neurons of the VMH and neurons in the ARC would allow for controlling both food intake and energy expenditure by inhibiting a common protein.

Although the protection of AMPKα_1 deletion in SF1 neurons against HFD-induced obesity was mainly associated with increased energy expenditure, slight alterations in glucose balance (decreased glycemia and improved glucose tolerance without changes in insulin sensitivity and insulin levels) found in mutant mice fed with an HFD, could be involved [56]. The moderate role of AMPK in glucose sensing by SF1 neurons in the VMH was also reported in a recent study in which AMPK activity suppression led to selective depletion of SF1 glucose inhibitory (GI) neurons, and activated CRR without affecting the presence of glucose excited (GE) neurons [69]. The authors suggested that the primary role of AMPK in SF1 GI neurons is to control the expression of *Txn2* (encoding a mitochondrial redox enzyme), providing protection against ROS produced during hypoglycemia. Conversely, under hyperglycemia, a reduction in ROS production is mediated by UCP2 in GE SF1 neurons [62]. These findings support the idea of two different subsets of SF1 neurons: the one regulating hyperglycemia enhancing insulin sensitivity and the second one regulating hypoglycemia by activating CRR.

In addition to AMPK, the metabolic sensor of nicotinamide adenine dinucleotide-dependent deacetylase SIRT1, which is highly expressed in the VMH, has shown to contribute to both control of energy balance and clock function from SF1 neurons [70–72]. SIRT1 deacetylates several proteins in central and peripheral tissues to provide adaptation against redox/nutrient challenges. SIRT1 drives lipid mobilization from adipose tissue, lipid oxidation in skeletal muscle and liver, and increased glucose production, all of these responses protect against diminished energy availability [70]. In the hypothalamus, SIRT1 has been revealed as an attractive target against obesity and type 2 diabetes in both POMC [73] and SF1 [57] neurons. SIRT1 activation in these hypothalamic neurons triggers a reduction in body adiposity and insulin resistance in obesogenic environments, while the deleterious effects, such as increased glucose and lipid output from the liver, would be avoided.

As observed with AMPK, the anatomical location of SIRT1 in the hypothalamus is relevant for the protection against diet-induced obesity: although SIRT in POMC neurons is required for normal BAT-like remodeling in the perigonadal fat depot [73], SIRT1 in SF1 neurons controls insulin sensitivity in the skeletal muscle [57] (Figure 3). Mice lacking SIRT1 in SF1 neurons showed hypersensitivity to hypercaloric diets, gaining more bodyweight and lower energy expenditure compared to control mice after long-term HFD, whereas food intake remained unchanged [57].

In addition, to protect against dietary obesity, SIRT1 in SF1 neurons is relevant for homeostatic responses that can protect against dietary diabetes, since mice overexpressing SIRT1 exclusively in SF1 neurons were protected from developing diet-induced insulin resistance in skeletal muscle and hyperglycemia, whereas mice lacking SIRT1 in the same neurons were more prone to develop insulin resistance in skeletal muscle after HFD [57]. It is important to note that these studies indicate that SIRT1 in SF1 neurons does not regulate food intake in either chow or HF feeding and that the regulatory effects on energy expenditure and glucose homeostasis are predominantly observed in a hypercaloric diet but not in a normal diet context. The latter evidence is intriguing since SIRT1 is known to be activated in the low-energy state. According to Coppari [70], the possible explanations are (i) HFD increases SIRT1 activity in SF1 neurons, and (ii) in chow-fed mice, compensatory mechanisms (i.e., enhancement of insulin sensitivity in the liver to provide normal glycemia) counterbalance the defects related to SIRT1 deficiency in these hypothalamic neurons. These hypotheses could explain that the metabolic imbalance is mainly observed in transgenic mice exposed to an HFD when the compensatory mechanisms are unable to prevent the whole-body effects produced to the lack of SIRT1 in hypothalamic neurons (i.e., insulin sensitivity is reduced in the skeletal muscle but is not increased in the liver).

In addition to these investigations, SIRT1 also contributes to the synchronization of the circadian clock in the brain, by nutritional inputs, which is crucial to maintaining cellular and body homeostasis [71]. Mice with targeted ablation of SIRT1 in SF1 neurons of the VMH showed a reduction in the period length in the light-entrainable activity under constant darkness and scheduled feeding, but not in ad libitum feeding [71]. Furthermore, these effects were more pronounced in KO mice under scheduled high-caloric diet feeding. This investigation suggests that SIRT1 in SF1 neurons contributes to the synchronization and/or adaptation of the endogenous clock to feeding cues (Figure 3), a function that would partially explain the metabolic alterations displayed by mice deficient in SIRT1 in SF1 neurons previously reported [57]. Accordingly, deletion of BMAL1 in SF1 neurons of the VMH in mice also drove a substantial alteration of the circadian clock in the brain, with an impact in BAT activation, supporting the importance of SF1 neurons in the regulation of clock function in the brain and the whole-body energy homeostasis [72]. The implication of SIRT1 in SF1 neurons in healthy aging or lifespan via circadian clock modulation would be also interesting to explore and would provide insight into the hypothalamic mechanisms mediated by SF1 neurons regulating energy balance.

Overall, specific targeting of metabolic sensors such as AMPK and SIRT1 in SF1 neurons of the VMH would be enough to modulate body weight gain and energy expenditure, in a feeding-independent manner, but associated with synchronization of the central clock with the periphery. This is a challenging approach but with significant translational relevance in obesity management.

3.3. Glutamatergic Neurotransmission and Synaptic Receptors

The expression and activity of synaptic receptors and glutamatergic neurotransmission in SF1 neurons of the VMH have been identified to contribute to energy balance dysregulation in obesity and diabetes.

Since neurons in the VMH are largely glutamatergic, the importance of glutamate neurotransmission versus other neurotransmitters and neuropeptides in mediating the functions of these nuclei seem to be higher, not only in energy homeostasis but also in sexual and defensive behaviors [8,33]. The first study reporting this evidence in SF1 neurons was provided by Bradford Lowell and colleagues [33]. They generated a mouse lacking the VGLUT2—a transporter mediating uptake of L-glutamate into synaptic vesicles—specifically in SF1 neurons, driving to a disruption in transsynaptic communication by glutamate in this specific population of neurons. The phenotype study was particularly focused on glucose and insulin homeostasis: these mice displayed hypoglycemia in the fasted state secondary to impaired fasting-induced glucagon increased and impaired gluconeogenesis induction in the liver [33] (Figure 4). VGLUT2 deficient mice in SF1 neurons also had defective CRR to insulin-induced hypoglycemia and central 2-deoxyglucose, as this greater degree of hypoglycemia again linked to an impaired glucagon response. It is important to mention that SF1-Cre;Vglut2$^{flox/flox}$ mice did not become obese when exposed to a chow diet and had mild obesity when fed with an HFD [33]. A partial contribution of GABA release, from a small number of SF1 neurons, in body weight regulation in these mice could explain this mild obesogenic phenotype [33]. Importantly, this study was the first demonstration of the importance of glutamate release in VMH neurocircuitry to prevent hypoglycemia.

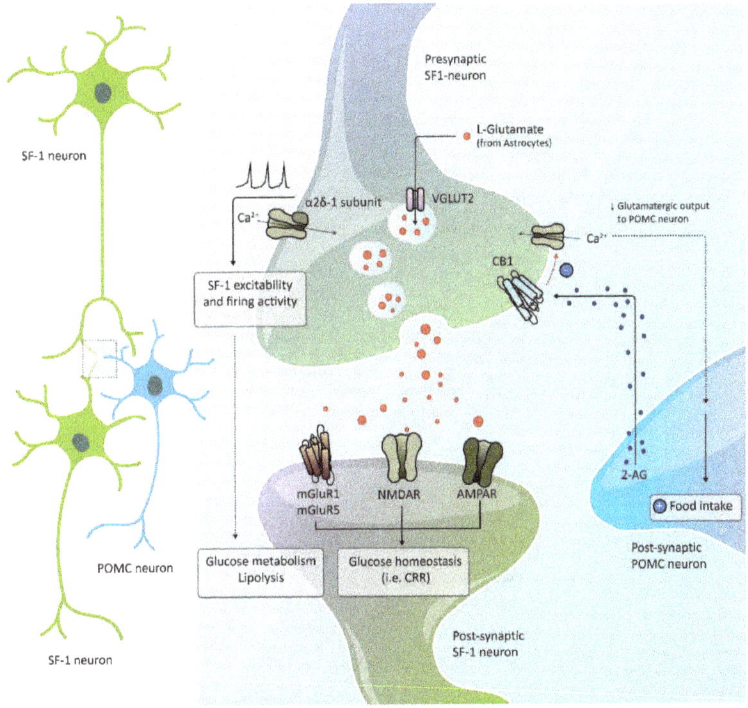

Figure 4. Glutamatergic neurotransmission and synaptic receptors in SF1 neurons and energy balance. Lack of the glutamate vesicular transporter VGLUT2 in SF1 neurons disrupts the glutamate effect in the postsynaptic receptors, leading to glucose metabolism dysfunction and a defective counterregulatory response (CRR). In line with this, the glutamate receptor mGluR5 in SF1 neurons also regulates glucose balance. The calcium channel subunit α2δ-1 is essential for excitability and firing activity of SF1 neurons which in turn regulate glucose metabolism in adipose tissue and skeletal muscle and lipolysis in white adipose tissue with no alterations in food intake. Finally, the endocannabinoid 2-arachidonoylglycerol (2-AG) released in the post-synaptic neuron, leads to retrograde activation of the CB1 receptor located in the presynaptic glutamatergic SF1 neurons, blocking calcium entry with subsequent inhibition of glutamatergic output and attenuation of POMC neurons activation, promoting food intake in diet-induced obesity.

The relevant role of glutamatergic neurotransmission in SF1 neurons to regulate energy homeostasis was also recently established by the development of mice lacking the metabotropic glutamate receptor mGluR5 in these neurons [54]. The most remarkable finding of this study was that mGluR5 depletion in SF1 neurons did not affect energy balance, but it significantly impaired glucose balance control (Figure 4), as this dysfunction was only observed in female but not in male mice. The sex-specific impairment of insulin sensitivity and glucose tolerance was associated with a reduction in intrinsic excitability and the firing rate of SF1 neurons. This suggests a functional indirect interaction of mGluR5 with estrogen receptors to control the effects of estradiol on SF1 neurons activity and glycemic control, which switch from facilitatory to detrimental in the absence of the glutamate receptor in these neurons [54]. These results are in line with the previously described study reporting that ERα depletion in SF1 neurons drives metabolic disturbances only in females [52]. However, the lack of impairment in insulin-induced hypoglycemia in mice deficient in mGluR5 is in contrast to the previous investigation in which inhibition of glutamate release by SF1 neurons impaired the CRR to hypoglycemia [33].

Other receptors involved in neuronal activity have shown their contribution to the regulation of obesity and diabetes by SF1 neurons in the VMH. Among them, the calcium channel subunit α2δ-1 has a non-canonical role in SF1 neurons to control glucose and lipid homeostasis (Figure 4). This neuronal receptor facilitates cell surface trafficking of calcium channels mediating calcium current and neurotransmitters release [74] and also acts as a receptor for thrombospondins, promoting excitatory synapse assembly [75]. The authors show that α2δ-1 is essential for the excitability and firing activity of SF1 neurons. Deletion of this receptor in SF1 neurons in mice led to neuronal hypoactivity, sympathetic tone reduction to WAT and skeletal muscle, glucose intolerance, and insulin resistance, and lipolysis alteration in WAT [58], as these metabolic alterations mainly observed in female mice [59]. Despite these findings, mice did not display alterations in food intake and body weight in response to HFD [58]. The reduction in SF1 neuronal activity observed in this transgenic mouse did not compromise CRR to hypoglycemia, in contrast to previous studies in which chemogenic or optogenetic silencing of these neurons impaired recovery from hypoglycemia [32,76]. These conflicting results could be related to the heterogeneity of SF1 neurons, with functional diversity in the VMH (e.g., leptin-activated and leptin-inhibited SF1 neurons) [47].

The endocannabinoid receptor CB1 located in SF1 neurons of the VMH has been also involved in glutamatergic output regulating feeding and in the control of metabolic flexibility in response to dietary changes. Hypothalamic slice preparations exposed to a CB1 receptor-specific agonist, displayed inhibition of the electrical activity of SF1 neurons [77]. In line with this, endocannabinoids release in postsynaptic neurons, led to retrograde activation of CB1 receptors in pre-synaptic glutamatergic SF1 neurons of the VMH, blocking calcium entry with subsequent inhibition of glutamatergic output and attenuation of POMC neurons activation, promoting feeding in diet-induced obesity [78] (Figure 4). The physiological relevance of CB1 receptors in SF1 neurons was also reported by Daniela Cota and colleagues, showing that mice lacking CB1 receptor specifically in these neurons showed a diet-dependent bidirectional metabolic phenotype: under a chow diet, deletion of CB1 decreased adiposity by increasing sympathetic activity and WAT lipolysis, whether, conversely, under HFD, lack of this receptor in SF1 neurons blunted peripheral use of lipids and produced leptin resistance, hyperphagia, body weight gain and glucose intolerance [60]. Therefore, CB1 synaptic receptor in SF1 neurons seems to act as a molecular switch for correct metabolic flexibility in the hypothalamus, protecting from the development of HFD-induced leptin resistance.

Altogether, although these studies provide substantial evidence of the role of glutamatergic neurotransmission and neuronal receptors in SF1 neurons to control glucose homeostasis and metabolic adaptation to dietary challenges, they raise important questions that still need to be clarified such as (i) the identification of the exact afferent and efferent components of the neurocircuitry linking SF1 neurons with other brain areas and the periphery (e.g., for the regulation of glucagon release in the pancreas), (ii) the evaluation of sex-dependent function, (iii) the exploration of SF1 neuronal subpopulations in response to different stimuli (e.g., important for the compensatory effects on feeding or the CRR to hypoglycemia).

3.4. Modulators of Autophagy, Mitochondrial and Primary Cilia Function

Autophagy plays a critical role in several physiological processes such as metabolic regulations. Deletion of essential autophagy genes, such as the autophagy-related gene (Atg) 7 in peripheral tissues drives significant changes in body weight and glucose balance [79,80]. More recent investigations have demonstrated that autophagy is important in the hypothalamic regulation of energy homeostasis and feeding behavior, with these effects affected differently depending on the neuronal types engaged. In the ARC, *Atg7* deletion in POMC neurons implied an obesogenic phenotype, especially when feeding an HFD [81,82], whereas loss of this gene in AgRP neurons promoted leanness [83]. A more recent study reported the metabolic importance of autophagy in the VMH, by deleting *Atg7* in SF1 VMH neurons of mice [61]. The authors found that in fed conditions, loss of *Atg7* in this neuronal population has no effect on body weight, food intake, or energy expenditure. However, when fasted overnight, mice lacking *Atg7* displayed reduced O_2 consumption, CO_2 and heat production, and reduced food intake after refeeding. *Atg7* deletion in SF1 neurons only implied moderate changes in fed plasma insulin levels and insulin resistance, without changes in glycemia and glucose tolerance.

This study demonstrates autophagy activation in the VMH, in addition to other regions such as the ARC, in response to fasting. This was supported not only by increased expression of autophagy-related proteins in the area but also by a 2- to 3-fold increase in the density of cFos-immunoreactive cells in the VMH, ARC, and DMH of control mice in response to fasting, an induction that was abrogated in these three regions in *Atg7* deficient mice in SF1 VMH neurons [61]. Then, the lack of autophagy in SF1 VMH neurons significantly impacts neurons outside of the VMH. This is in line with previous investigations showing that POMC neurons in the ARC receive excitatory inputs from the VMH, and the strength of this VMH-ARC input is reduced by fasting [84].

Loss of autophagy in SF1 VMH neurons also caused alterations in mitochondrial function and morphology, which was associated with inadequate metabolic response to fasting [61]. During fasting, mitochondrial efficiency must increase, and autophagy is important to ensure the degradation of mitochondria that are damaged beyond repair, that is, the process called mitophagy [85]. Mice lacking *Atg7* in SF1 neurons showed a reduced number of mitochondria-autophagosome contacts with alterations in mitochondria morphology and respiration, indicating abnormal mitophagy [61]. Since UCP2 expression in SF1 VMH neurons has been linked to mitophagy [62,86], the authors explored the expression of this protein, and they found it was dysregulated in mutant mice [61]. In agreement with these findings, UCP2 in SF1 VMH regulates mitochondria fission and the excitability of GE neurons in the VMH, which in turn allows the appropriate response to increased glycemia, enhancing insulin sensitivity in peripheral tissues [62]. However, the moderate effects of autophagy loss in SF1 neurons observed in glucose homeostasis do not correspond to the abnormal expression of UCP2 detected. The fact that UCP2-induced mitochondrial fission in the VMH only affects GE neurons, but not GI neurons [62], could explain the moderate phenotype of the mice deficient in *Atg7* in glucose metabolism.

Very recently, the SF1 neurons' primary cilia, a solitary antenna-like extension of the plasma membrane, was demonstrated as an important organelle for the regulation of energy homeostasis [63]. In this study, the authors deleted the intraflagellar transport 88 (IFT88), which is a critical protein in primary cilium biogenesis, specifically in SF1 neurons. Dysfunction of VMH primary cilia resulted in impaired activation of sympathetic tone, central leptin resistance, and higher body weight gain under both chow and HFD feeding. Obesity in the transgenic mice IFT88-KO^{SF-1} was caused by a marked decrease in energy expenditure, hyperphagia (only under HFD feeding), blunted BAT function, insulin resistance, and glucose intolerance. In addition, deletion of primary cilia in SF1 neurons led to an impairment in bone homeostasis. The downregulation of the sympathetic activity in IFT88-KO^{SF-1} mice was linked to the reduction of the metabolic rate and increase in bone density, as this effect considered was independent of the obese phenotype [63]. Considering the fact that primary cilia are required for proper neuronal circuit formation, deletion of this organelle in SF1 neurons may therefore influence local neuronal circuits [63].

4. The Sex-Specific Effect of SF1 Neurons on Energy Balance

It is known that the VMH is sexually dimorphic, showing females higher ERα concentration than males [87]. Additionally, as described in Section 3, selective deletion of this receptor in SF1 neurons resulted in increased abdominal obese phenotype with adipocyte hypertrophy in females, but not in males [52]. Obesity in females was caused by decreased energy expenditure as they had reduced basal metabolic rate and impaired BAT thermogenesis [52]. It has been described also that estrogens regulate the activity of GI neurons of the ventrolateral portion of VMH, since females showed an attenuated response to hypoglycemia compared to male mice, although in this study it is not specified if these neurons were SF1 positive cells [88]. As has been already specified, there are other genetic deletions that presented different outputs in a sex-specific manner. Expression of mGluR5 in SF1 neurons was necessary for estradiol protective effects in glucose balance in female but not in male mice, and mGluR5 deletion resulted in reduced electrical activity only in female mice [54]. Deletion of $\alpha 2\delta$-1 expression in SF1 neurons also presented sexually dimorphic effects depending on diet conditions. Particularly, female mice lacking $\alpha 2\delta$-1 displayed glucose intolerance and insulin resistance under chow or HFD, as this phenotype much moderate in male mice [58,59]. Cheung et al. also observed sex-dependent changes when deleting VGLUT2 in SF1 neurons, since female but not male mice presented attenuation of DIO, and transgenic male mice showed behavioral changes not observed in female mice [89].

Estrogens could affect and change some intracellular signaling cascades leading to these differences observed between males and females, as female mice have greater expression of ERα. This would explain why the presence or absence of different patterns of receptors impacts the estrogenic effect on SF1 neurons, as some of the sex-specific signaling cascades are being altered. Other sex-specific effects related to SF1 neurons can be found in their synapses with POMC neurons, as estradiol attenuated the retrograde endocannabinoid signaling from POMC to SF1 neurons, increasing the glutamatergic inputs to POMC [78]. This study opens another possibility based on the fact hypothesis that estrogens do not interact directly with the specific SF1 neurons targeted in each study, but their effects fall on other cells acting on SF1 neurons.

Then, estrogen signaling in SF1 neurons is a must for metabolic health in female mice. Despite the fact that many studies presented in this review were performed only in male mice, these last results obtained in both sexes reinforce the idea of SF1 neurons having a sex-specific effect on energy balance and glucose metabolism.

5. Exploring the Neurocircuitry That Links SF1 Neurons to Other Brain Areas in Energy Balance

Remarkable progress in the neuroanatomy of VMH projections has been obtained from novel biological tools. Some notions on how the VMH connects with other brain centers come from stereotaxic injection with anterograde axonal tracers in adult rats [90,91]. While the results from these studies established that VMH is organized in subregions depending on their projection, the inherent limitations of this method make it difficult to assess the specific neuronal network involved in SF1 neurons. To overcome these limitations, *Sf-1*^{TauGFP} and *Z/EG*$^{Sf1:Cre}$ mice models were originally designed to trace the major VMH axonal projections during embryonic and postnatal stages [9].

The first of these models is a knock-in line that contains the wheat germ agglutinin (WGA) and Tau-green fluorescent protein (TauGFP) under the control of *Sf-1* regulatory elements. In the second model, the SF1 Cre mouse was crossed with a *Z/EG* reporter mouse, resulting in constitutive expression of eGFP (enhanced GFP) after Cre-mediated recombination [9]. The analysis of the results obtained from both models indicated the efferent SF1 projection in ascending and descending tracts that innervate different structures such as the hypothalamus, thalamus, the basal forebrain, and the brainstem. Interestingly, SF1 projections targeting the general vicinity of gonadotropin-releasing hormone (GnRH) neurons suggest the potential role in fertility physiology [9].

As expected, experiments using these models showed that SF1 neurons projected to areas implicated in body weight regulation, including the paraventricular nucleus of the hypothalamus (PVN) [9,92]. However, it was not possible to detect any GFP$^+$ fibers within the ARC. Thus, additional studies using synaptophysin or rabies virus would be helpful to explore the specific network between SF1 and hypothalamic nuclei involved in energy balance. Recently, Yunglei Yang and colleagues have optogenetically identified the downstream target underlying the SF1 suppression of food intake (Figure 5). The authors used the SF1 ChR2 mice, but the fiber optic cannula was implanted above the PVN, therefore the photostimulation of this area would only activate the SF1-PVN projections. They showed that high-frequency stimulation of this circuit potently reduced food intake even in 24 h food-deprived mice [23].

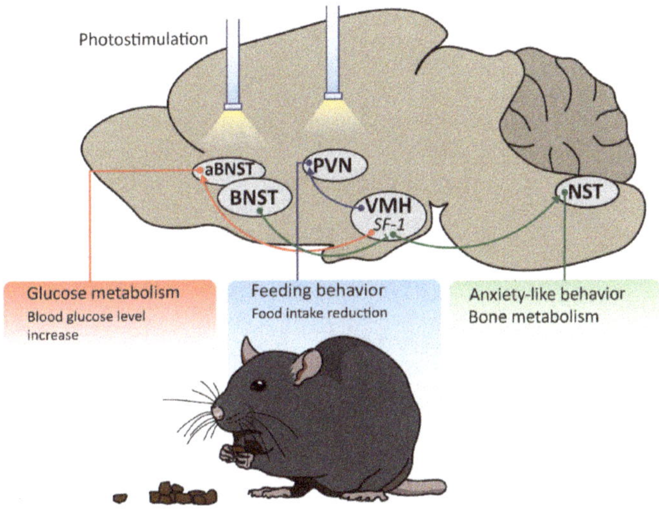

Figure 5. Neurocircuitry linking SF1 VMH neurons to other brain areas in energy balance. Using optogenetic tools, the paraventricular nucleus of the hypothalamus (PVN) has been identified as a downstream target nucleus underlying the SF1 suppression of food intake. Downstream projections of SF1 in the VMH involved in the control of glycemia are the bed nucleus of the stria terminalis (aBNST) since photostimulation of the VMH^{Sf1}-aBNST circuit leads to increases in blood glucose levels. Finally, SF1 neurons are innervated by upstream BNST neurons and send projection to the nucleus tractus solitarius (NST) to regulate anxiety-like behavior and bone metabolism.

Downstream projections of SF1 neurons regulating glycemia were identified through the same novel strategy derived from optogenetics. The researchers administrated fluorescently tagged ChRs into the VMH of SF1-Cre mice to visualize the projection fields using histological imaging to detect the EYFP reporter in axonal projections [32]. Once identified this target site, a fiber optic was implanted to depolarize the final projection by laser stimulation. With this strategy, it was possible to demonstrate that $VMH^{SF1} \rightarrow$aBNST is the most relevant circuit involved in controlling glycemia since the photostimulation of this circuit triggered an increase of the blood glucose levels [32] (Figure 5). Interestingly, other functions have been attributed to this specific circuit: Fan Yang and colleagues recently demonstrated that SF1 neurons are innervated by upstream BNST neurons and send projections to the NST nucleus to regulate anxiety-like behavior and bone metabolism [93] (Figure 5).

The use of current and novel biological tools provides several clues about the neurocircuits regulating blood glucose and energy balance, and the afferent and efferent connections linking SF1 neurons with other neuronal populations. An improving understanding of the functional organization of SF1 neurons may help to identify future strategies for metabolic diseases such as obesity and diabetes.

6. Concluding Remarks and Future Perspectives

In the last years, the critical role played by the hypothalamus in the regulation of energy balance and glucose homeostasis has gained substantial importance. However, the exact mechanisms and neuronal circuits underlying this regulation and the pathogenesis of obesity and insulin resistance remain poorly understood. The growing literature is demonstrating that the origin of these diseases is beyond feeding and pancreatic insulin secretion, and needs further definition. Here we extensively review that SF1 neurons in the hypothalamus provide a "central role" in the control of blood glucose levels, insulin sensitivity in peripheral tissues, adipose tissue plasticity, and thermogenesis activation. SF1 neurons are a specific lead in the brain's ability to sense glucose levels and conduct insulin and leptin signaling in energy expenditure and glucose homeostasis, with minor feeding control. Interestingly, transgenic mice lacking different targets in SF1 neurons show an altered metabolic phenotype mostly under HFD feeding, but not under chow diet, indicating that SF1 neurons are particularly important for metabolic adaptation in the early stages of obesity. These investigations also demonstrate the sex-specific effects of SF1 neurons in the VMH, and the importance of the analysis of the phenotype in both male and female mice when exploring energy balance and metabolism in transgenic models.

Although optogenetic and chemogenetic tools have clearly demonstrated SF1 function in glucose homeostasis, adiposity, and energy expenditure, there are still contradictory results in these functions after deletion of specific receptors in SF1 neurons (e.g., mGluR5). These controversies could be associated with the existence of several sub-populations of SF1 neurons that respond differently to insulin, leptin, and glucose. For instance, whereas some SF1 neurons seem to be specialized in the regulation of blood glucose levels, some others are responsible for insulin sensitivity in the periphery.

When investigating the hypothalamic regulation of obesity and diabetes, in addition to the identification of SF1 sub-populations, it is important to explore the coordination of SF1 neurons with other neurons outside the VMH to trigger metabolic functions. For instance, HFD-induced hyperinsulinemia drives SF1 neurons hyperpolarization, leading in turn to functional changes in the synaptic output onto POMC neurons, causing obesity and glucose intolerance. This evidence suggests that the nucleus-specific responses upon HFD feeding can cooperate to cause obesity and diabetes, and disruption of this cooperation could act as a link between both diseases.

Despite the advances in the neurocircuitry connecting SF1 VMH neurons with other brain areas, the exact efferent circuits leading to changes in peripheral tissues need further investigation. How mice lacking a protein only in SF1 neurons present a strong phenotype in the periphery in terms of glucose homeostasis and energy expenditure under nutritional challenges? Are these circuits restricted to SNS and PSNS? A new point of view also emerges from these studies: a disruption in a specific type of neurons in the hypothalamus could trigger the pathogenesis of type 2 diabetes and obesity. How this neuronal dysfunction drives peripheral insulin resistance, adiposity, and thermogenesis alteration remains unclear. Reconsidering the pathogenesis of obesity and type 2 diabetes and now including the essential role of SF1 neurons, new strategies to treat these diseases could emerge. Targeting this specific population of neurons in the VMH, and even another cooperating population outside the VMH at the same time could modulate specific metabolic functions, a challenging approach but with significant translational impacts in the management of obesity and diabetes.

Author Contributions: All authors (A.F., S.Z., N.C. and R.R.-R.) have participated in writing and editing. All authors (A.F., S.Z., N.C. and R.R.-R.) have read and agreed to the published version of the manuscript.

Funding: This work was supported by the Spanish Ministerio de Economía, Industria y Competitividad (MINECO), Agencia Estatal de Investigación (AEI), and Fondo Europeo de Desarrollo Regional (FEDER) (Grants SAF2017-82813-C3-3-R to N.C. and R.R.-R. and PCI2018-092997/AEI to R.R.-R.).

A.F. is granted with a fellowship from the Agència de Gestió d'Ajuts Universitaris i de la Recerca (AGAUR) in Catalonia.

Institutional Review Board Statement: Not applicable.

Informed Consent Statement: Not applicable.

Data Availability Statement: Not applicable.

Conflicts of Interest: The authors declare no conflict of interest.

References

1. World Health Organization. Obesity and Overweight. Available online: https://www.who.int/news-room/fact-sheets/detail/obesity-and-overweight (accessed on 19 April 2021).
2. World Health Organization. Diabetes. Available online: https://www.who.int/news-room/fact-sheets/detail/diabetes (accessed on 19 April 2021).
3. Hetherington, A.W.; Ranson, S.W. The relation of various hypothalamic lesions to adiposity in the rat. *J. Comp. Neurol.* **1942**, *76*, 475–499. [CrossRef]
4. Hetherington, A.W.; Ranson, S.W. Hypothalamic lesions and adiposity in the rat. *Anat. Rec.* **1940**, *78*, 149–172. [CrossRef]
5. Borg, W.P.; During, M.J.; Sherwin, R.S.; Borg, M.A.; Brines, M.L.; Shulman, G.I. Ventromedial hypothalamic lesions in rats suppress counterregulatory responses to hypoglycemia. *J. Clin. Investig.* **1994**, *93*, 1677–1682. [CrossRef]
6. Yi, C.X.; Scherer, T.; Tschöp, M.H. Cajal revisited: Does the VMH make us fat? *Nat. Neurosci.* **2011**, *14*, 806–808. [CrossRef]
7. McClellan, K.M.; Parker, K.L.; Tobet, S. Development of the ventromedial nucleus of the hypothalamus. *Front. Neuroendocrinol.* **2006**, *27*, 193–209. [CrossRef]
8. Choi, Y.-H.; Fujikawa, T.; Lee, J.; Reuter, A.; Kim, K.W. Revisiting the Ventral Medial Nucleus of the Hypothalamus: The Roles of SF-1 Neurons in Energy Homeostasis. *Front. Neurosci.* **2013**, *7*, 71. [CrossRef] [PubMed]
9. Cheung, C.C.; Kurrasch, D.M.; Liang, J.K.; Ingraham, H.A. Genetic labeling of steroidogenic factor-1 (SF-1) neurons in mice reveals ventromedial nucleus of the hypothalamus (VMH) circuitry beginning at neurogenesis and development of a separate non-SF-1 neuronal cluster in the ventrolateral VMH. *J. Comp. Neurol.* **2013**, *521*, 1268–1288. [CrossRef] [PubMed]
10. Zhao, L.; Ki, W.K.; Ikeda, Y.; Anderson, K.K.; Beck, L.; Chase, S.; Tobet, S.A.; Parker, K.L. Central nervous system-specific knockout of steroidogenic factor 1 results in increased anxiety-like behavior. *Mol. Endocrinol.* **2008**, *22*, 1403–1415. [CrossRef] [PubMed]
11. Ikeda, Y.; Luo, X.; Abbud, R.; Nilson, J.H.; Parker, K.L. The nuclear receptor steroidogenic factor 1 is essential for the formation of the ventromedial hypothalamic nucleus. *Mol. Endocrinol.* **1995**, *9*, 478–486. [CrossRef] [PubMed]
12. Parker, K.L.; Rice, D.A.; Lala, D.S.; Ikeda, Y.; Luo, X.; Wong, M.; Bakke, M.; Zhao, L.; Frigeri, C.; Hanley, N.A.; et al. Steroidogenic factor 1: An essential mediator of endocrine development. *Recent Prog. Horm. Res.* **2002**, *57*, 19–36. [CrossRef]
13. Hoivik, E.A.; Lewis, A.E.; Aumo, L.; Bakke, M. Molecular aspects of steroidogenic factor 1 (SF-1). *Mol. Cell. Endocrinol.* **2010**, *315*, 27–39. [CrossRef] [PubMed]
14. Lee, J.; Yang, D.J.; Lee, S.; Hammer, G.D.; Kim, K.W.; Elmquist, J.K. Nutritional conditions regulate transcriptional activity of SF-1 by controlling sumoylation and ubiquitination. *Sci. Rep.* **2016**, *6*. [CrossRef] [PubMed]
15. Sadovsky, Y.; Crawford, P.A.; Woodson, K.G.; Polish, J.A.; Clements, M.A.; Tourtellotte, L.M.; Simburger, K.; Milbrandt, J. Mice deficient in the orphan receptor steroidogenic factor 1 lack adrenal glands and gonads but express P450 side-chain-cleavage enzyme in the placenta and have normal embryonic serum levels of corticosteroids. *Proc. Natl. Acad. Sci. USA* **1995**, *92*, 10939–10943. [CrossRef] [PubMed]
16. Majdic, G.; Young, M.; Gomez-Sanchez, E.; Anderson, P.; Szczepaniak, L.S.; Dobbins, R.L.; McGarry, J.D.; Parker, K.L. Knockout Mice Lacking Steroidogenic Factor 1 Are a Novel Genetic Model of Hypothalamic Obesity. *Endocrinology* **2002**, *143*, 607–614. [CrossRef]
17. Achermann, J.C.; Ito, M.; Ito, M.; Hindmarsh, P.C.; Jameson, J.L. A mutation in the gene encoding steroidogenic factor-1 causes XY sex reversal and adrenal failure in humans [1]. *Nat. Genet.* **1999**, *22*, 125–126. [CrossRef] [PubMed]
18. Liu, W.; Liu, M.; Fan, W.Q.; Nawata, H.; Yanase, T. The Gly146Ala variation in human SF-1 gene: Its association with insulin resistance and type 2 diabetes in Chinese. *Diabetes Res. Clin. Pract.* **2006**, *73*, 322–328. [CrossRef] [PubMed]
19. Kim, K.W.; Zhao, L.; Donato, J.; Kohno, D.; Xu, Y.; Elias, C.F.; Lee, C.; Parker, K.L.; Elmquist, J.K. Steroidogenic factor 1 directs programs regulating diet-induced thermogenesis and leptin action in the ventral medial hypothalamic nucleus. *Proc. Natl. Acad. Sci. USA* **2011**, *108*, 10673–10678. [CrossRef]
20. Dhillon, H.; Zigman, J.M.; Ye, C.; Lee, C.E.; McGovern, R.A.; Tang, V.; Kenny, C.D.; Christiansen, L.M.; White, R.D.; Edelstein, E.A.; et al. Leptin directly activates SF1 neurons in the VMH, and this action by leptin is required for normal body-weight homeostasis. *Neuron* **2006**, *49*, 191–203. [CrossRef]
21. Berglund, K.; Tung, J.K.; Higashikubo, B.; Gross, R.E.; Moore, C.I.; Hochgeschwender, U. Combined Optogenetic and Chemogenetic Control of Neurons. *Methods Mol. Biol.* **2016**, *1408*, 207–225. [CrossRef]
22. Forcelli, P.A. Applications of optogenetic and chemogenetic methods to seizure circuits: Where to go next? *J. Neurosci. Res.* **2017**, *95*, 2345–2356. [CrossRef]

23. Zhang, J.; Chen, D.; Sweeney, P.; Yang, Y. An excitatory ventromedial hypothalamus to paraventricular thalamus circuit that suppresses food intake. *Nat. Commun.* **2020**, *11*, 6326. [CrossRef] [PubMed]
24. Viskaitis, P.; Irvine, E.E.; Smith, M.A.; Choudhury, A.I.; Alvarez-Curto, E.; Glegola, J.A.; Hardy, D.G.; Pedroni, S.M.A.; Paiva Pessoa, M.R.; Fernando, A.B.P.; et al. Modulation of SF1 Neuron Activity Coordinately Regulates Both Feeding Behavior and Associated Emotional States. *Cell Rep.* **2017**, *21*, 3559–3572. [CrossRef] [PubMed]
25. Kunwar, P.S.; Zelikowsky, M.; Remedios, R.; Cai, H.; Yilmaz, M.; Meister, M.; Anderson, D.J. Ventromedial hypothalamic neurons control a defensive emotion state. *Elife* **2015**, *4*, e06633. [CrossRef]
26. Klöckener, T.; Hess, S.; Belgardt, B.F.; Paeger, L.; Verhagen, L.A.W.; Husch, A.; Sohn, J.W.; Hampel, B.; Dhillon, H.; Zigman, J.M.; et al. High-fat feeding promotes obesity via insulin receptor/PI3K-dependent inhibition of SF-1 VMH neurons. *Nat. Neurosci.* **2011**, *14*, 911–918. [CrossRef] [PubMed]
27. Coutinho, E.A.; Okamoto, S.; Ishikawa, A.W.; Yokota, S.; Wada, N.; Hirabayashi, T.; Saito, K.; Sato, T.; Takagi, K.; Wang, C.C.; et al. Activation of SF1 neurons in the ventromedial hypothalamus by DREADD technology increases insulin sensitivity in peripheral tissues. *Diabetes* **2017**, *66*, 2372–2386. [CrossRef] [PubMed]
28. Bonaventura, J.; Eldridge, M.A.G.; Hu, F.; Gomez, J.L.; Sanchez-Soto, M.; Abramyan, A.M.; Lam, S.; Boehm, M.A.; Ruiz, C.; Farrell, M.R.; et al. High-potency ligands for DREADD imaging and activation in rodents and monkeys. *Nat. Commun.* **2019**, *10*, 4627. [CrossRef] [PubMed]
29. Flak, J.N.; Goforth, P.B.; Dell'Orco, J.; Sabatini, P.V.; Li, C.; Bozadjieva, N.; Sorensen, M.; Valenta, A.; Rupp, A.; Affinati, A.H.; et al. Ventromedial hypothalamic nucleus neuronal subset regulates blood glucose independently of insulin. *J. Clin. Investig.* **2020**, *130*, 2943–2952. [CrossRef]
30. Borg, W.P.; Sherwin, R.S.; During, M.J.; Borg, M.A.; Shulman, G.I. Local ventromedial hypothalamus glucopenia triggers counterregulatory hormone release. *Diabetes* **1995**, *44*, 180–184. [CrossRef]
31. Borg, M.A.; Borg, W.P.; Tamborlane, W.V.; Brines, M.L.; Shulman, G.I.; Sherwin, R.S. Chronic hypoglycemia and diabetes impair counterregulation induced by localized 2-deoxy-glucose perfusion of the ventromedial hypothalamus in rats. *Diabetes* **1999**, *48*, 584–587. [CrossRef]
32. Meek, T.H.; Nelson, J.T.; Matsen, M.E.; Dorfman, M.D.; Guyenet, S.J.; Damian, V.; Allison, M.B.; Scarlett, J.M.; Nguyen, H.T.; Thaler, J.P.; et al. Functional identification of a neurocircuit regulating blood glucose. *Proc. Natl. Acad. Sci. USA* **2016**, *113*, E2073–E2082. [CrossRef]
33. Tong, Q.; Ye, C.P.; McCrimmon, R.J.; Dhillon, H.; Choi, B.; Kramer, M.D.; Yu, J.; Yang, Z.; Christiansen, L.M.; Lee, C.E.; et al. Synaptic Glutamate Release by Ventromedial Hypothalamic Neurons Is Part of the Neurocircuitry that Prevents Hypoglycemia. *Cell Metab.* **2007**, *5*, 383–393. [CrossRef] [PubMed]
34. Nisbett, R.E. Hunger, obesity, and the ventromedial hypothalamus. *Psychol. Rev.* **1972**, *79*, 433–453. [CrossRef] [PubMed]
35. Avrith, D.; Mogenson, G.J. Reversible hyperphagia and obesity following intracerebral microinjections of colchicine into the ventromedial hypothalamus of the rat. *Brain Res.* **1978**, *153*, 99–107. [CrossRef]
36. Bingham, N.C.; Verma-Kurvari, S.; Parada, L.F.; Parker, K.L. Development of a steroidogenic factor 1/Cre transgenic mouse line. *Genesis* **2006**, *44*, 419–424. [CrossRef]
37. Zhang, R.; Dhillon, H.; Yin, H.; Yoshimura, A.; Lowell, B.B.; Maratos-Flier, E.; Flier, J.S. Selective inactivation of Socs3 in SF1 neurons improves glucose homeostasis without affecting body weight. *Endocrinology* **2008**, *149*, 5654–5661. [CrossRef]
38. Berger, A.; Kablan, A.; Yao, C.; Ho, T.; Podyma, B.; Weinstein, L.S.; Chen, M. Gsα deficiency in the ventromedial hypothalamus enhances leptin sensitivity and improves glucose homeostasis in mice on a high-fat diet. *Endocrinology* **2016**, *157*, 600–610. [CrossRef]
39. Bence, K.K.; Delibegovic, M.; Xue, B.; Gorgun, C.Z.; Hotamisligil, G.S.; Neel, B.G.; Kahn, B.B. Neuronal PTP1B regulates body weight, adiposity and leptin action. *Nat. Med.* **2006**, *12*, 917–924. [CrossRef]
40. Chiappini, F.; Catalano, K.J.; Lee, J.; Peroni, O.D.; Lynch, J.; Dhaneshwar, A.S.; Wellenstein, K.; Sontheimer, A.; Neel, B.G.; Kahn, B.B. Ventromedial hypothalamus-specific Ptpn1 deletion exacerbates diet-induced obesity in female mice. *J. Clin. Investig.* **2014**, *124*, 3781–3792. [CrossRef]
41. Senn, S.S.; Le Foll, C.; Whiting, L.; Tarasco, E.; Duffy, S.; Lutz, T.A.; Boyle, C.N. Unsilencing of native LepRs in hypothalamic SF1 neurons does not rescue obese phenotype in LepR-deficient mice. *Am. J. Physiol. Regul. Integr. Comp. Physiol.* **2019**, *317*, R451–R460. [CrossRef]
42. Minokoshi, Y.; Haque, M.S.; Shimazu, T. Microinjection of leptin into the ventromedial hypothalamus increases glucose uptake in peripheral tissues in rats. *Diabetes* **1999**, *48*, 287–291. [CrossRef] [PubMed]
43. Meek, T.H.; Matsen, M.E.; Dorfman, M.D.; Guyenet, S.J.; Damian, V.; Nguyen, H.T.; Taborsky, G.J.; Morton, G.J. Leptin action in the ventromedial hypothalamic nucleus is sufficient, but not necessary, to normalize diabetic hyperglycemia. *Endocrinology* **2013**, *154*, 3067–3076. [CrossRef]
44. Gonçalves, G.H.M.; Li, W.; Garcia, A.V.C.G.; Figueiredo, M.S.; Bjørbæk, C. Hypothalamic agouti-related peptide neurons and the central melanocortin system are crucial mediators of leptin's antidiabetic actions. *Cell Rep.* **2014**, *7*, 1093–1103. [CrossRef] [PubMed]
45. Air, E.L.; Benoit, S.C.; Blake Smith, K.A.; Clegg, D.J.; Woods, S.C. Acute third ventricular administration of insulin decreases food intake in two paradigms. *Pharmacol. Biochem. Behav.* **2002**, *72*, 423–429. [CrossRef]

46. Jauch-Chara, K.; Friedrich, A.; Rezmer, M.; Melchert, U.H.; Scholand-Engler, H.G.; Hallschmid, M.; Oltmanns, K.M. Intranasal insulin suppresses food intake via enhancement of brain energy levels in humans. *Diabetes* **2012**, *61*, 2261–2268. [CrossRef]
47. Sohn, J.W.; Oh, Y.; Kim, K.W.; Lee, S.; Williams, K.W.; Elmquist, J.K. Leptin and insulin engage specific PI3K subunits in hypothalamic SF1 neurons. *Mol. Metab.* **2016**, *5*, 669–679. [CrossRef]
48. Xu, A.W.; Kaelin, C.B.; Takeda, K.; Akira, S.; Schwartz, M.W.; Barsh, G.S. PI3K integrates the action of insulin and leptin on hypothalamic neurons. *J. Clin. Investig.* **2005**, *115*, 951–958. [CrossRef]
49. Xu, Y.; Hill, J.W.; Fukuda, M.; Gautron, L.; Sohn, J.W.; Kim, K.W.; Lee, C.E.; Choi, M.J.; Lauzon, D.A.; Dhillon, H.; et al. PI3K signaling in the ventromedial hypothalamic nucleus is required for normal energy homeostasis. *Cell Metab.* **2010**, *12*, 88–95. [CrossRef] [PubMed]
50. Fujikawa, T.; Choi, Y.H.; Yang, D.J.; Shin, D.M.; Donato, J.; Kohno, D.; Lee, C.E.; Elias, C.F.; Lee, S.; Kim, K.W. P110β in the ventromedial hypothalamus regulates glucose and energy metabolism. *Exp. Mol. Med.* **2019**, *51*. [CrossRef] [PubMed]
51. Kim, K.W.; Donato, J.; Berglund, E.D.; Choi, Y.H.; Kohno, D.; Elias, C.F.; DePinho, R.A.; Elmquist, J.K. FOXO1 in the ventromedial hypothalamus regulates energy balance. *J. Clin. Investig.* **2012**, *122*, 2578–2589. [CrossRef]
52. Xu, Y.; Nedungadi, T.P.; Zhu, L.; Sobhani, N.; Irani, B.G.; Davis, K.E.; Zhang, X.; Zou, F.; Gent, L.M.; Hahner, L.D.; et al. Distinct hypothalamic neurons mediate estrogenic effects on energy homeostasis and reproduction. *Cell Metab.* **2011**, *14*, 453–465. [CrossRef]
53. Correa, S.M.; Newstrom, D.W.; Warne, J.P.; Flandin, P.; Cheung, C.C.; Lin-Moore, A.T.; Pierce, A.A.; Xu, A.W.; Rubenstein, J.L.; Ingraham, H.A. An estrogen-responsive module in the ventromedial hypothalamus selectively drives sex-specific activity in females. *Cell Rep.* **2015**, *10*, 62–74. [CrossRef]
54. Fagan, M.P.; Ameroso, D.; Meng, A.; Rock, A.; Maguire, J.; Rios, M. Essential and sex-specific effects of mGluR5 in ventromedial hypothalamus regulating estrogen signaling and glucose balance. *Proc. Natl. Acad. Sci. USA* **2020**, *117*, 19566–19577. [CrossRef] [PubMed]
55. Furigo, I.C.; de Souza, G.O.; Teixeira, P.D.S.; Guadagnini, D.; Frazão, R.; List, E.O.; Kopchick, J.J.; Prada, P.O.; Donato, J. Growth hormone enhances the recovery of hypoglycemia via ventromedial hypothalamic neurons. *FASEB J.* **2019**, *33*, 11909–11924. [CrossRef] [PubMed]
56. Seoane-Collazo, P.; Roa, J.; Rial-Pensado, E.; Liñares-Pose, L.; Beiroa, D.; Ruíz-Pino, F.; López-González, T.; Morgan, D.A.; Pardavila, J.Á.; Sánchez-Tapia, M.J.; et al. SF1-Specific AMPKα1 Deletion Protects Against Diet-Induced Obesity. *Diabetes* **2018**, db171538. [CrossRef]
57. Ramadori, G.; Fujikawa, T.; Anderson, J.; Berglund, E.D.; Frazao, R.; Michán, S.; Vianna, C.R.; Sinclair, D.A.; Elias, C.F.; Coppari, R. SIRT1 deacetylase in SF1 neurons protects against metabolic imbalance. *Cell Metab.* **2011**, *14*, 301–312. [CrossRef]
58. Felsted, J.A.; Chien, C.H.; Wang, D.; Panessiti, M.; Ameroso, D.; Greenberg, A.; Feng, G.; Kong, D.; Rios, M. Alpha2delta-1 in SF1+ Neurons of the Ventromedial Hypothalamus Is an Essential Regulator of Glucose and Lipid Homeostasis. *Cell Rep.* **2017**, *21*, 2737–2747. [CrossRef]
59. Felsted, J.A.; Meng, A.; Ameroso, D.; Rios, M. Sex-specific Effects of a2d-1 in the Ventromedial Hypothalamus of Female Mice Controlling Glucose and Lipid Balance. *Endocrinology* **2020**, *161*, bqaa068. [CrossRef]
60. Cardinal, P.; André, C.; Quarta, C.; Bellocchio, L.; Clark, S.; Elie, M.; Leste-Lasserre, T.; Maitre, M.; Gonzales, D.; Cannich, A.; et al. CB1cannabinoid receptor in SF1-expressing neurons of the ventromedial hypothalamus determines metabolic responses to diet and leptin. *Mol. Metab.* **2014**, *3*, 705–716. [CrossRef] [PubMed]
61. Coupé, B.; Leloup, C.; Asiedu, K.; Maillard, J.; Pénicaud, L.; Horvath, T.L.; Bouret, S.G. Defective autophagy in Sf1 neurons perturbs the metabolic response to fasting and causes mitochondrial dysfunction. *Mol. Metab.* **2021**, *47*, 101186. [CrossRef]
62. Toda, C.; Kim, J.D.; Impellizzeri, D.; Cuzzocrea, S.; Liu, Z.W.; Diano, S. UCP2 Regulates Mitochondrial Fission and Ventromedial Nucleus Control of Glucose Responsiveness. *Cell* **2016**, *164*, 872–883. [CrossRef]
63. Sun, J.S.; Yang, D.J.; Kinyua, A.W.; Yoon, S.G.; Seong, J.K.; Kim, J.; Moon, S.J.; Shin, D.M.; Choi, Y.H.; Kim, K.W. Ventromedial hypothalamic primary cilia control energy and skeletal homeostasis. *J. Clin. Investig.* **2021**, *131*. [CrossRef]
64. Minokoshi, Y.; Alquier, T.; Furukawa, N.; Kim, Y.-B.; Lee, A.; Xue, B.; Mu, J.; Foufelle, F.; Ferré, P.; Birnbaum, M.J.; et al. AMP-kinase regulates food intake by responding to hormonal and nutrient signals in the hypothalamus. *Nature* **2004**, *428*, 569–574. [CrossRef]
65. López, M.; Varela, L.; Vázquez, M.J.; Rodríguez-Cuenca, S.; González, C.R.; Velagapudi, V.R.; Morgan, D.A.; Schoenmakers, E.; Agassandian, K.; Lage, R.; et al. Hypothalamic AMPK and fatty acid metabolism mediate thyroid regulation of energy balance. *Nat. Med.* **2010**, *16*, 1001–1008. [CrossRef] [PubMed]
66. López, M.; Nogueiras, R.; Tena-Sempere, M.; Diéguez, C. Hypothalamic AMPK: A canonical regulator of whole-body energy balance. *Nat. Rev. Endocrinol.* **2016**, *12*, 421–432. [CrossRef] [PubMed]
67. Claret, M.; Smith, M.A.; Batterham, R.L.; Selman, C.; Choudhury, A.I.; Fryer, L.G.D.; Clements, M.; Al-Qassab, H.; Heffron, H.; Xu, A.W.; et al. AMPK is essential for energy homeostasis regulation and glucose sensing by POMC and AgRP neurons. *J. Clin. Investig.* **2007**, *117*, 2325–2336. [CrossRef] [PubMed]
68. Martínez-Sánchez, N.; Seoane-Collazo, P.; Contreras, C.; Varela, L.; Villarroya, J.; Rial-Pensado, E.; Buqué, X.; Aurrekoetxea, I.; Delgado, T.C.; Vázquez-Martínez, R.; et al. Hypothalamic AMPK-ER Stress-JNK1 Axis Mediates the Central Actions of Thyroid Hormones on Energy Balance. *Cell Metab.* **2017**, *26*, 212–229. [CrossRef]

69. Quenneville, S.; Labouèbe, G.; Basco, D.; Metref, S.; Viollet, B.; Foretz, M.; Thorens, B. Hypoglycemia-sensing neurons of the ventromedial hypothalamus require AMPK-induced TXN2 expression but are dispensable for physiological counterregulation. *Diabetes* **2020**, *69*, 2253–2266. [CrossRef] [PubMed]
70. Coppari, R. Metabolic actions of hypothalamic SIRT1. *Trends Endocrinol. Metab.* **2012**, *23*, 179–185. [CrossRef] [PubMed]
71. Orozco-Solis, R.; Ramadori, G.; Coppari, R.; Sassone-Corsi, P. SIRT1 relays nutritional inputs to the circadian clock through the Sf1 neurons of the ventromedial hypothalamus. *Endocrinology* **2015**, *156*, 2174–2184. [CrossRef] [PubMed]
72. Orozco-Solis, R.; Aguilar-Arnal, L.; Murakami, M.; Peruquetti, R.; Ramadori, G.; Coppari, R.; Sassone-Corsi, P. The circadian clock in the ventromedial hypothalamus controls cyclic energy expenditure. *Cell Metab.* **2016**, *23*, 467–478. [CrossRef]
73. Ramadori, G.; Fujikawa, T.; Fukuda, M.; Anderson, J.; Morgan, D.A.; Mostoslavsky, R.; Stuart, R.C.; Perello, M.; Vianna, C.R.; Nillni, E.A.; et al. SIRT1 deacetylase in POMC neurons is required for homeostatic defenses against diet-induced obesity. *Cell Metab.* **2010**, *12*, 78–87. [CrossRef]
74. Davies, A.; Hendrich, J.; Van Minh, A.T.; Wratten, J.; Douglas, L.; Dolphin, A.C. Functional biology of the $\alpha 2\delta$ subunits of voltage-gated calcium channels. *Trends Pharmacol. Sci.* **2007**, *28*, 220–228. [CrossRef]
75. Eroglu, Ç.; Allen, N.J.; Susman, M.W.; O'Rourke, N.A.; Park, C.Y.; Özkan, E.; Chakraborty, C.; Mulinyawe, S.B.; Annis, D.S.; Huberman, A.D.; et al. Gabapentin Receptor $\alpha 2\delta$-1 Is a Neuronal Thrombospondin Receptor Responsible for Excitatory CNS Synaptogenesis. *Cell* **2009**, *139*, 380–392. [CrossRef] [PubMed]
76. Garfield, A.S.; Shah, B.P.; Madara, J.C.; Burke, L.K.; Patterson, C.M.; Flak, J.; Neve, R.L.; Evans, M.L.; Lowell, B.B.; Myers, M.G.; et al. A parabrachial-hypothalamic cholecystokinin neurocircuit controls counterregulatory responses to hypoglycemia. *Cell Metab.* **2014**, *20*, 1030–1037. [CrossRef] [PubMed]
77. Kim, K.W.; Jo, Y.H.; Zhao, L.; Stallings, N.R.; Chua, S.C.; Parker, K.L. Steroidogenic factor 1 regulates expression of the cannabinoid receptor 1 in the ventromedial hypothalamic nucleus. *Mol. Endocrinol.* **2008**, *22*, 1950–1961. [CrossRef]
78. Fabelo, C.; Hernandez, J.; Chang, R.; Seng, S.; Alicea, N.; Tian, S.; Conde, K.; Wagner, E.J. Endocannabinoid signaling at hypothalamic steroidogenic factor-1/proopiomelanocortin synapses is sex-and diet-sensitive. *Front. Mol. Neurosci.* **2018**, *11*, 1–23. [CrossRef]
79. Ebato, C.; Uchida, T.; Arakawa, M.; Komatsu, M.; Ueno, T.; Komiya, K.; Azuma, K.; Hirose, T.; Tanaka, K.; Kominami, E.; et al. Autophagy Is Important in Islet Homeostasis and Compensatory Increase of Beta Cell Mass in Response to High-Fat Diet. *Cell Metab.* **2008**, *8*, 325–332. [CrossRef] [PubMed]
80. Zhang, Y.; Goldman, S.; Baerga, R.; Zhao, Y.; Komatsu, M.; Jin, S. Adipose-specific deletion of autophagy-related gene 7 (atg7) in mice reveals a role in adipogenesis. *Proc. Natl. Acad. Sci. USA* **2009**, *106*, 19860–19865. [CrossRef]
81. Coupé, B.; Ishii, Y.; Dietrich, M.O.; Komatsu, M.; Horvath, T.L.; Bouret, S.G. Loss of autophagy in pro-opiomelanocortin neurons perturbs axon growth and causes metabolic dysregulation. *Cell Metab.* **2012**, *15*, 247–255. [CrossRef]
82. Quan, W.; Kim, H.K.; Moon, E.Y.; Kim, S.S.; Choi, C.S.; Komatsu, M.; Jeong, Y.T.; Lee, M.K.; Kim, K.W.; Kim, M.S.; et al. Role of hypothalamic proopiomelanocortin neuron autophagy in the control of appetite and leptin response. *Endocrinology* **2012**, *153*, 1817–1826. [CrossRef]
83. Kaushik, S.; Rodriguez-Navarro, J.A.; Arias, E.; Kiffin, R.; Sahu, S.; Schwartz, G.J.; Cuervo, A.M.; Singh, R. Autophagy in hypothalamic agrp neurons regulates food intake and energy balance. *Cell Metab.* **2011**, *14*, 173–183. [CrossRef]
84. Sternson, S.M.; Shepherd, G.M.G.; Friedman, J.M. Topographic mapping of VMH $\rightarrow$ arcuate nucleus microcircuits and their reorganization by fasting. *Nat. Neurosci.* **2005**, *8*, 1356–1363. [CrossRef]
85. Raefsky, S.M.; Mattson, M.P. Adaptive responses of neuronal mitochondria to bioenergetic challenges: Roles in neuroplasticity and disease resistance. *Free Radic. Biol. Med.* **2017**, *102*, 203–216. [CrossRef] [PubMed]
86. Hass, D.T.; Barnstable, C.J. Mitochondrial uncoupling protein 2 knock-out promotes mitophagy to decrease retinal ganglion cell death in a mouse model of glaucoma. *J. Neurosci.* **2019**, *39*, 3582–3596. [CrossRef] [PubMed]
87. Cao, J.; Patisaul, H.B. Sexually dimorphic expression of hypothalamic estrogen receptors α and β and Kiss1 in neonatal male and female rats. *J. Comp. Neurol.* **2011**, *519*, 2954–2977. [CrossRef] [PubMed]
88. Santiago, A.M.; Clegg, D.J.; Routh, V.H. Estrogens modulate ventrolateral ventromedial hypothalamic glucose-inhibited neurons. *Mol. Metab.* **2016**, *5*, 823–833. [CrossRef] [PubMed]
89. Cheung, C.C.; Krause, W.C.; Edwards, R.H.; Yang, C.F.; Shah, N.M.; Hnasko, T.S.; Ingraham, H.A. Sex-dependent changes in metabolism and behavior, as well as reduced anxiety after eliminating ventromedial hypothalamus excitatory output. *Mol. Metab.* **2015**, *4*, 857–866. [CrossRef]
90. Canteras, N.S.; Simerly, R.B.; Swanson, L.W. Organization of projections from the ventromedial nucleus of the hypothalamus: A Phaseolus vulgaris-leucoagglutinin study in the rat. *J. Comp. Neurol.* **1994**, *348*, 41–79. [CrossRef] [PubMed]
91. Krieger, M.S.; Conrad, L.C.; Pfaff, D.W. An autoradiographic study of the efferent connections of the ventromedial nucleus of the hypothalamus. *J. Comp. Neurol.* **1979**, *183*, 785–815. [CrossRef] [PubMed]
92. Xu, Y.; O'Malley, B.W.; Elmquist, J.K. Brain nuclear receptors and body weight regulation. *J. Clin. Investig.* **2017**, *127*, 1172–1180. [CrossRef]
93. Yang, F.; Liu, Y.; Chen, S.; Dai, Z.; Yang, D.; Gao, D.; Shao, J.; Wang, Y.; Wang, T.; Zhang, Z.; et al. A GABAergic neural circuit in the ventromedial hypothalamus mediates chronic stress–induced bone loss. *J. Clin. Investig.* **2020**, *130*, 6539–6554. [CrossRef] [PubMed]

International Journal of *Molecular Sciences*

MDPI

Article

Angiopoietin-Like Growth Factor Involved in Leptin Signaling in the Hypothalamus

Yunseon Jang [1,2,3,†], Jun Young Heo [1,2,3,†], Min Joung Lee [1,2,3], Jiebo Zhu [1,2,3], Changjun Seo [1,2,3], Da Hyun Go [1,2,3], Sung Kyung Yoon [1,2,3], Date Yukari [4], Yuichi Oike [5], Jong-Woo Sohn [6], Minho Shong [7,*] and Gi Ryang Kweon [1,3,*]

1 Department of Biochemistry, Chungnam National University School of Medicine, Daejeon 35015, Korea; yunseonj@cnu.ac.kr (Y.J.); junyoung3@gmaiil.com (J.Y.H.); rmj1102@cnu.ac.kr (M.J.L.); zhujiebo2019@o.cnu.ac.kr (J.Z.); justin20@o.cnu.ac.kr (C.S.); goda1046@o.cnu.ac.kr (D.H.G.); 201950879@o.cnu.ac.kr (S.K.Y.)
2 Department of Medical Science, Chungnam National University School of Medicine, Daejeon 35015, Korea
3 Infection Control Convergence Research Center, Chungnam National University School of Medicine, Daejeon 35015, Korea
4 Frontier Science Research Center, University of Miyazaki, Miyazaki 889-1692, Japan; dateyuka@med.miyazaki-u.ac.jp
5 Department of Molecular Genetics, Graduate School of Medical Sciences, Kumamoto University, Kumamoto 860-8556, Japan; oike@gpo.kumamoto-u.ac.jp
6 Department of Biological Sciences, Korea Advanced Institute of Science and Technology, Daejeon 34141, Korea; jwsohn@kaist.ac.kr
7 Department of Internal Medicine, Chungnam National University School of Medicine, Daejeon 35015, Korea
* Correspondence: minhos@cnu.ac.kr (M.S.); mitochondria@cnu.ac.kr (G.R.K.); Tel.: +82-42-280-7161 (M.S.); +82-42-580-8226 (G.R.K.)
† Authors contributed equally to this work.

Citation: Jang, Y.; Heo, J.Y.; Lee, M.J.; Zhu, J.; Seo, C.; Go, D.H.; Yoon, S.K.; Yukari, D.; Oike, Y.; Sohn, J.-W.; et al. Angiopoietin-Like Growth Factor Involved in Leptin Signaling in the Hypothalamus. *Int. J. Mol. Sci.* **2021**, *22*, 3443. https://doi.org/10.3390/ijms22073443

Academic Editor: Rosalía Rodríguez-Rodríguez

Received: 26 February 2021
Accepted: 24 March 2021
Published: 26 March 2021

Abstract: The hypothalamic regulation of appetite governs whole-body energy balance. Satiety is regulated by endocrine factors including leptin, and impaired leptin signaling is associated with obesity. Despite the anorectic effect of leptin through the regulation of the hypothalamic feeding circuit, a distinct downstream mediator of leptin signaling in neuron remains unclear. Angiopoietin-like growth factor (AGF) is a peripheral activator of energy expenditure and antagonizes obesity. However, the regulation of AGF expression in brain and localization to mediate anorectic signaling is unknown. Here, we demonstrated that AGF is expressed in proopiomelanocortin (POMC)-expressing neurons located in the arcuate nucleus (ARC) of the hypothalamus. Unlike other brain regions, hypothalamic AGF expression is stimulated by leptin-induced signal transducers and activators of transcription 3 (STAT3) phosphorylation. In addition, leptin treatment to hypothalamic N1 cells significantly enhanced the promoter activity of AGF. This induction was abolished by the pretreatment of ruxolitinib, a leptin signaling inhibitor. These results indicate that hypothalamic AGF expression is induced by leptin and colocalized to POMC neurons.

Keywords: hypothalamus; AGF; leptin; POMC neuron

1. Introduction

The hypothalamus controls feeding behavior for the maintenance of whole-body energy balance, and dysregulation of energy intake leads to obesity [1]. Leptin is an anorectic hormone to regulate the hypothalamic neural network including proopiomelanocortin (POMC) neurons, producing α- melanocyte-stimulating hormone (MSH) [2]. Insensitivity to leptin signaling is observed in obese patients, despite the increase in leptin produced by adipose tissue [3,4]. It is reported that the combination of recombinant human leptin, metreleptin and pramlintide enhanced the reduction of body weight in a clinical trial [5]. Additionally, celastrol, originated from thunder god vine roots, increased brain leptin sensitivity to promote weight loss and food intake [6]. However, while considerable research on the beneficial effect of leptin on feeding control and energy expenditure has been done, downstream molecules of leptin signaling in the hypothalamic neuron still remain elusive.

Hypothalamic leptin signaling initiates Janus kinase 2-signal transducers and activators of transcription 3 (JAK2-STAT3) signaling and extracellular signal-regulated kinase (ERK) to regulate food intake [7] and adenosine monophosphate-activated protein kinase (AMPK) signaling for energy homeostasis [8]. The POMC neuronal population is located within the arcuate nucleus (ARC) of the hypothalamus. In postprandial state, a POMC neuron produces an anorectic molecule on leptin binding to leptin receptor (LEP-R) [2]. Leptin promotes POMC expression through the phosphorylation of STAT3. A significant increase in body weight was observed in *Pomc*-Cre, *Lepr*$^{flox/flox}$ mice, in which the leptin receptor was deleted in the POMC neuron [9]. Although leptin signaling in the POMC neuron is important for the regulation of appetite, the mediator of leptin signaling in the POMC neuron is only partially identified. In the current study, we identified angiopoietin-like growth factor (AGF) as a downstream of leptin, and AGF expression is enhanced by leptin in the hypothalamus.

AGF has been identified as a peripheral activator of energy expenditure [10,11] that antagonizes obesity and insulin resistance. Whole-body AGF-knockout (KO) mice exhibited severe obesity [12,13]. Adenovirus-mediated hepatic overexpression of AGF showed to reverse the obesity of AGF-KO mice [12]. AGF is a member of the angiopoietin-like proteins (ANGPTLs) family composed of eight members, all of which possess a coiled-coil domain at the N-terminus and a fibrinogen-like domain at the C-terminus [14]. ANGPTLs are orphan ligands that do not bind Tie1 (tyrosine kinase with immunoglobulin-like and epidermal growth factor (EGF)-like domains 1) or Tie2 receptors—targets of angiopoietin [14]. However, the existence of AGF and the regulating factor of AGF in brain is unknown. In the present study, we demonstrate that AGF is induced by the leptin hormone in the hypothalamus and expressed in POMC neurons.

2. Results

2.1. *Angptl6 Is Expressed in the Hypothalamus and Induced after Feeding*

Hypothalamic appetite control is critical for the maintenance of whole-body energy balance, and the impairment of anorectic signaling is associated with obesity [1]. AGF is known as an activator of energy expenditure in the liver and adipocytes [12,13]. Although AGF protein expression and function in the hypothalamus have not been investigated, AGF mRNA has been detected in the brain of C57BL/6 mice [15]. To determine whether AGF is expressed in the hypothalamus, we performed immunohistochemistry on human brain. As shown in Figure 1A, AGF immunoreactivity was detected in the hypothalamus of postmortem human brain. Because the hypothalamus is sensitive to nutritional status and governs whole-body energy homeostasis [1,2,16,17], we tested whether hypothalamic AGF expression is affected by changes in nutritional status. Mice were allowed to freely approach food, and the refed group was fed after overnight fasting. Fasting did not alter AGF mRNA level in the hypothalamus or epididymal white adipose tissue (eWAT), but did decrease hepatic AGF mRNA level (Figure S1A,B). In contrast, AGF mRNA level increased ~2-fold in the hypothalamus after refeeding (Figure 1B) and modestly increased in liver and eWAT, indicating that feeding induces AGF expression in metabolic organs. Fasting and refeeding status of mice were confirmed by measurements of blood glucose level and insulin concentration in plasma (Figure S1C,D).

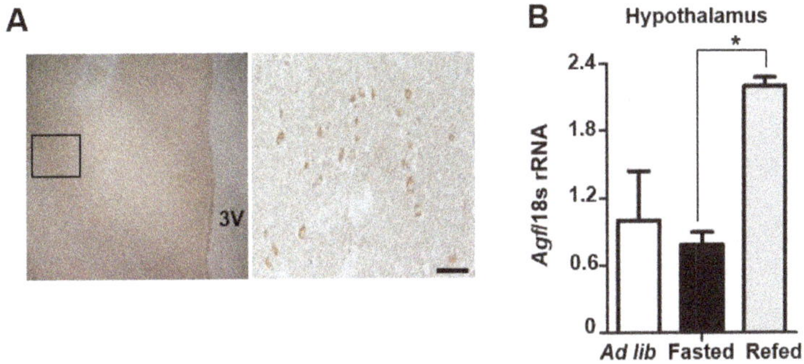

Figure 1. Angiopoietin-like growth factor (AGF) is expressed in the hypothalamus. (**A**) Immunohistochemical detection of AGF in the human brain. The boxed area within the image indicates a magnified field. (**B**) *Agf* mRNA expression in the hypothalamus of *ad libitum-fed*, overnight fasted or refed mice, confirmed by qPCR analysis ($n = 5$/group). Data represent means ± SEM (* $p < 0.05$). Scale bars: 50 μm.

2.2. *Angptl6 Is Expressed in POMC Neurons of the Hypothalamus*

The hypothalamus delivers pivotal anorectic signaling after feeding. Because AGF-positive cells were found in the human hypothalamus (Figure 1A), we performed immunofluorescence staining of AGF in mice to identify the localization and cell type in the hypothalamus. As shown in Figure 2A, AGF immunoreactivity was detected in the ARC where the feeding regulatory circuit is localized. AGF immunoreactivity exhibited a 91% overlap with the neuronal marker, NeuN. However, AGF did not colocalize with glial fibrillary acidic protein (GFAP) (Figure 2B), one of a marker of astrocytes (a type of glial cell). Leptin binds to the LEP-R of the POMC-expressing neuron, which is located in the ARC of the hypothalamus and activates anorectic signaling in a postprandial state [18,19]. POMC activation reduces food intake through the release of α-MSH, and *Pomc*-deficient mice in the ARC (Arc*Pomc*$^{-/-}$) showed obesity and hyperphagia [20]. Because AGF in the hypothalamus was induced by feeding (Figure 1B), we verified the spatial expression of AGF in relation to these major appetite-controlling neurons by performing immunofluorescence staining for AGF in the brain of POMC-mCherry reporter mice. We found that 45.0% of AGF-positive neurons in the ARC of the hypothalamus were also POMC-positive (Figure 2C,D). These results indicate that AGF localized to the POMC-expressing neuronal population and it is not colocalized to GFAP-positive cells.

2.3. *AGF Expression Is Associated with Leptin-Induced STAT3 Phosphorylation*

Insulin and leptin, which are secreted from peripheral organs and act on the hypothalamus, are induced during the postprandial state and serve to inhibit food intake [21,22]. Accordingly, we tested the possibility that this hormone might act through effects on AGF level to modulate feeding behavior. Treatment of the insulin receptor-expressing cell lines, SN4741 and C2C12, with insulin (100 nM) had no effect on AGF protein or mRNA level (Figure S2A–C). Next, we examined the relationship between AGF and leptin, the latter of which dominates feeding behavior through the activation of anorectic signaling in the hypothalamus [18,23]. The leptin receptor (LepRb) is expressed in hypothalamic neurons, and, upon binding leptin, induces the JAK2 (Janus kinase 2)-mediated phosphorylation of STAT3 (signal transducer and activator of transcription 3) [18,19,24]. To examine changes in AGF level induced by leptin in vivo, we injected mice intraperitoneally with leptin (3 mg/kg) and sacrificed mice after 1 h. Leptin induced an increase in AGF protein in the hypothalamus (Figure 3A,B), but not in other brain areas, including the cortex, striatum and midbrain (Figure S3A). Immunostaining for Tyr705–phosphorylated STAT3 (pSTAT3), a leptin signaling marker, was increased in hypothalamus tissue (Figure 3A,C) and the ARC by leptin injection (Figure S3B).

To demonstrate the response of AGF-expressing neurons to leptin, we immunostained for AGF in the ARC of mice injected with leptin or vehicle (Figure 3D). This analysis showed that leptin increased the number of AGF-positive neurons by 56% (Figure 3D,E). Collectively, these data suggest that leptin induces hypothalamic AGF level.

2.4. AGF Promoter Activity Is Enhanced by Leptin

Consistent with in vivo experiments, both AGF and pSTAT3 expression were increased after leptin treatment in N1 cells (Figure 4A–C). We further found that AGF mRNA level was 2.5-fold higher in the hypothalamus of leptin-injected mice (Figure 4D) and 2-fold higher in leptin-treated N1 cells compared with vehicle-treated controls (Figure 4E), suggesting that leptin induces AGF transcription. To test this, we first sought to identify transcription factor binding sites in the AGF promoter using an online search program (http://tfbind.hgc.jp/ accessed on 23 May 2020), which predicted three consensus STAT3 binding sequences in a 1020-bp (−1072 to −52 bp) 5′ untranslated region (Figure 4F). Using a promoter-reporter construct in which this 1020-bp region was placed upstream of the luciferase gene, we tested whether the promoter activity of AGF was affected by leptin treatment. In N1 cells transfected with this promoter-reporter construct, leptin (100 ng/ml) induced a 2.6-fold increase in reporter activity within 45 min (Figure 4G). Notably, induction of AGF promoter activity by exogenous leptin treatment was ablated by a 1 h pre-incubation with ruxolitinib (1 μM), a leptin signaling inhibitor that blocks JAK2-mediated phosphorylation (Figure 4G). These results indicate that AGF transcription is induced by leptin and mediated by phospho-STAT3, a downstream of leptin/JAK2 signaling.

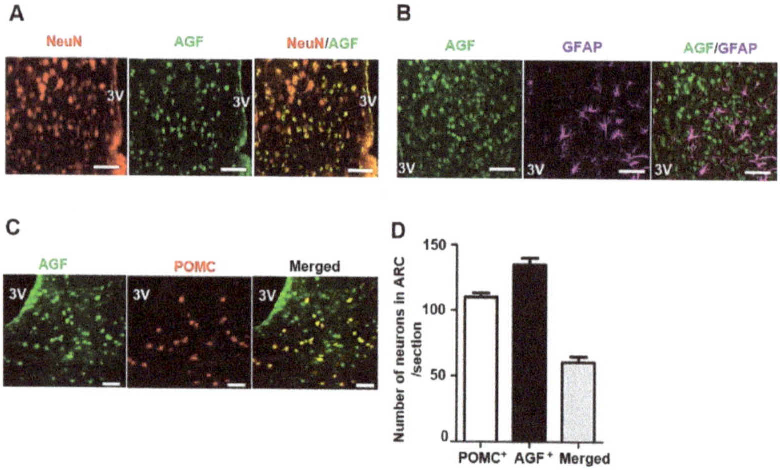

Figure 2. AGF is colocalized to proopiomelanocortin (POMC) neurons rather than glial fibrillary acidic protein (GFAP)-positive cells in the hypothalamus. (**A**,**B**) Immunofluorescence staining of AGF (green) and NeuN (red), a neuronal nuclei marker and GFAP (magenta), an astrocyte marker in the hypothalamus section. (**C**) POMC neurons in the hypothalamus, detected based on red fluorescence in POMC-specific tandem dimer Tomato (tdTomato) expressing reporter mice, and AGF immunofluorescence staining. Brain slices were prepared by cryosectioning (25 μm/section). Sections were incubated overnight at 4 °C with Alexa 488-conjugated mouse anti-AGF antibody (1:100). Fluorescence was visualized by confocal microscopy. A yellow-colored neuron represents the colocalization of AGF with POMC. (**D**) Number of AGF and POMC double-positive neurons in the ARC (between −1.7 and −3.4 mm from bregma) presented as means ± SD (n = 10/slide). Scale bars: 50 μm (upper).

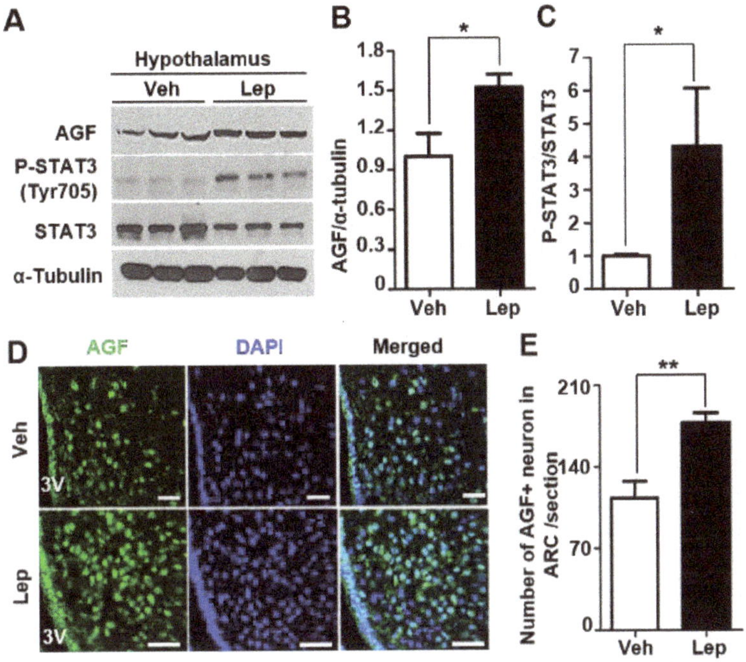

Figure 3. AGF expression is induced by leptin-stimulated signal transducer and activator of transcription 3 (STAT3) activation. (**A**) AGF level in the hypothalamus of C57BL/6 mice after intraperitoneal injection of recombinant mouse leptin (3 mg/kg) or vehicle, determined by Western blotting. AGF, pSTAT3 (Tyr705), STAT3 and α-tubulin were detected in tissue lysates containing equal amounts of protein from the hypothalamus of each group. (**B**,**C**) Band intensities of AGF and pSTAT3 measured by the ImageJ program; values are normalized to vehicle-injected samples. (**D**) AGF and DAPI (4′,6-diamidino-2-phenylindole) co-staining in brain sections from leptin- or vehicle-injected mice. Immunoreactivity was detected by confocal microscopy. (**E**) Number of AGF-positive neurons in the ARC. Data represent means ± SEM (* $p < 0.05$, ** $p < 0.01$).

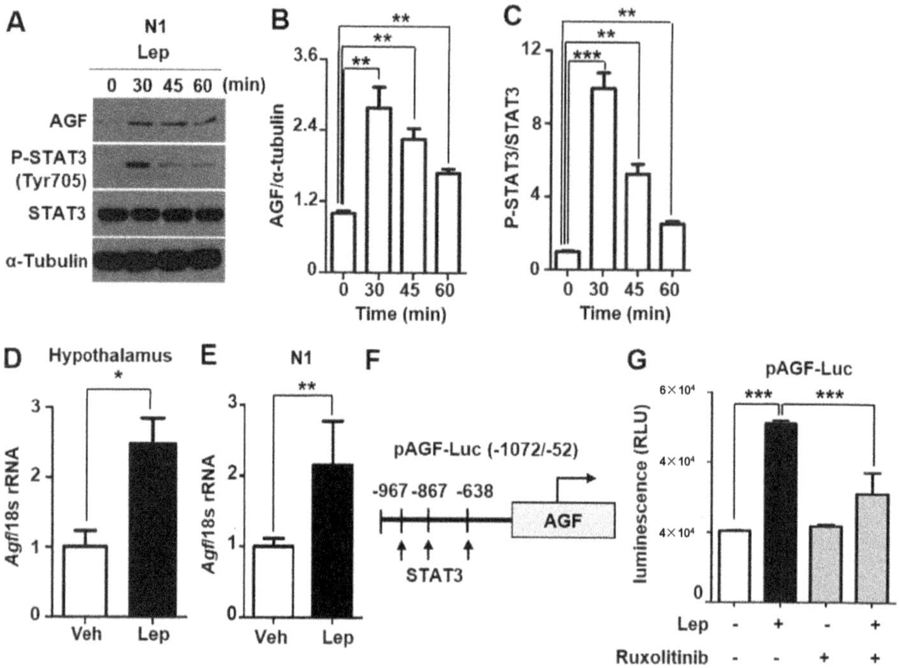

Figure 4. AGF promoter activity is increased by leptin. (**A**) AGF protein level in N1 cells, determined after treatment with 100 ng/ml leptin for 0, 30, 45 and 60 min. (**B**,**C**) AGF and pSTAT3 band intensities, measured using the ImageJ program. (**D**,**E**) AGF mRNA expression level in the hypothalamus (**D**) and in N1 cells (**E**) after leptin administration, normalized to 18s rRNA level, as determined by qPCR ($n = 5$/group). (**F**) Location of STAT3 binding sequences in the AGF promoter (−1072 to −52 bp). (**G**) AGF promoter activity, measured in N1 cells transfected with an AGF promoter-luciferase reporter plasmid after a 45 min incubation with leptin (100 ng/ml). Cells were pre-incubated in the presence or absence of the JAK2 inhibitor, ruxolitinib (1 μM) for 1 h. N1 cells were harvested and luminescence was measured. Data represent means ± SEM (* $p < 0.05$, ** $p < 0.01$, *** $p < 0.001$).

3. Discussion

Modulation of appetite by the hypothalamus is pivotal for energy homeostasis and is associated with the regulation of body weight [1,2,17]. Leptin binding to LEP-R on POMC neurons provokes anorectic signaling in the hypothalamus. It is reported that leptin activates ERK1/2 in the ARC for anorectic and sympathetic effect as confirmed by leptin-induced reduction of food intake was reversed by ERK inhibitor U0126 injection [7,8]. However, the downstream mediator of leptin signaling to modulate hypothalamic neuronal activity that is associated with appetite control is still elusive.

AGF, a known peripheral activator of energy expenditure, is determined by leptin in mice hepatocyte, and serum AGF level is paradoxically increased in human patients with metabolic disease due to compensation [25]. In the previous report, whole-body AGF KO mice showed extreme obesity due to a decrease in energy consumption with no change in food intake. The adenovirus-mediated reconstitution of AGF leads to a decrease in body weight and improved glucose tolerance [12], suggesting that peripheral AGF has a role in energy expenditure. However, alteration of brain AGF expression in this model was not examined and cell-type-specific AGF expression in the hypothalamus was unknown. In the current study, we address the fact that AGF is expressed in the hypothalamus and has leptin responsiveness. As we observed that AGF promoter activity was enhanced by the treatment of leptin and ablated this effect by ruxolitinib, JAK1/2 inhibitor in hypothalamic

N1 cells (Fig. 4G), it would be supportive to elucidate the involvement of AGF in leptin signaling if reproducible effects similar with cells occur in the hypothalamus of mice.

To assess whether AGF expression is dependent on leptin in the hypothalamus, *ob/ob* and *db/db* mice with leptin or leptin receptor deficiency could be useful. There is a possibility that AGF expression is reduced in ob/ob mice and exogenous leptin may restore AGF level in the hypothalamus. Additionally, feeding-induced hypothalamic increase in AGF expression may be ablated in these mice. Hypothalamic control of feeding by leptin is disrupted in obese subjects due to insensitivity to leptin signaling, despite the elevated level of circulating leptin [26–28]. The alteration of hypothalamic AGF in obesity remains to be elucidated.

AGF immunoreactivity was detectable in the hypothalamus of human postmortem brain around the third ventricle. Neuronal cell type and nucleus where AGF is localized in the human hypothalamus was not identified. The human protein atlas program (https://www.proteinatlas.org/, accessed on 23 May 2020) revealed AGF expression in the hypothalamus, but it does not present specific regions such as the paraventricular nucleus (PVN) and ARC. In the case of mice, AGF immunoreactivity predominantly colocalized to POMC neurons in the ARC (Figures 1A and 2C), not GFAP-positive cells. AGF expression was not restricted to the ARC but observed in dorsomedial hypothalamic nucleus (DMH) and PVN (data not shown). It is known that leptin administration activates NUCB2/nesfatin-1-expressing neurons in PVN to inhibit food intake [29]. Additionally, stimulation of the Thyrotropin-releasing hormone (TRH) neuron by leptin leads to an increase in pSTAT3 and neuronal activation, reducing appetite [30]. Therefore, to determine whether PVN AGF colocalize to neurons related to food intake regulation would give an insight into the role of AGF in appetite control and metabolism. Recently, it has been reported that tanycytes transport leptin into the hypothalamus and it is required for the regulation of normal hypothalamic leptin signaling [31]. To determine distinct neuronal populations which express AGF in the ARC and whether tanycytes co-express AGF could give an insight into AGF as a downstream mediator of leptin signaling.

A physiological role of hypothalamic AGF as a downstream effector of leptin in POMC neurons has not been determined in this study. Based on our finding that 45% of AGF-expressing neurons colocalized with POMC neurons in the ARC, the highly overlapped expression of AGF and POMC gives rise to the possibility that AGF has a role in POMC neuron-mediated metabolic changes such as appetite regulation and energy expenditure, which are promoted by leptin [32]. Both AGF expression and STAT3 phosphorylation were significantly induced after 30 min of leptin treatment. In contrast, pSTAT3 level declined after 60 min, whereas AGF level was sustained (Figure 4A). From this, we infer that, after induction by pSTAT3, AGF could act to regulate leptin downstream signaling, for example, by inducing JAK2 phosphatase, which is a negative regulator of this signaling [33]. It is also implied that leptin-induced AGF interacts with leptin signaling pathways such as ERK and AMPK in leptin receptor-expressing neurons to control appetite and energy balance. Further study for verification of the physiological role of hypothalamic AGF is including use of hypothalamic POMC neuron-specific AGF KO mice via assessment of an alteration of leptin signaling and α-MSH level. In addition, recording the neuronal activity of POMC in a condition of increasing AGF expression by leptin treatment using a patch-clamp study could be a useful approach. Our study suggests that AGF is a downstream molecule of leptin signaling in the hypothalamus.

4. Materials and Methods

4.1. Animals

Male C57BL/6 mice were purchased from Harlan Teklad (Indianapolis, IN, USA). Mice were maintained at 22 °C under a 12 h light–dark cycle. For diet-induced obesity, 5-week-old C57BL/6 mice were fed a high-fat diet (60% fat; Research Diets Inc., New Brunswick, NJ, USA). Leptin (3 mg/kg, R&D systems, Minneapolis, MN, USA) was dissolved in 0.9% saline and intraperitoneally injected into C57BL/6. Animal experiments were approved by the Institutional Animal Care and Use Committee of Chungnam National University (Ethical approval number, 201903A-CNU-46, approved on 1 June 2019).

4.2. Cell Culture

HypoE-N1 mice embryonic hypothalamus cells were maintained in Dulbecco's Modified Eagle Medium (DMEM) containing 10% fetal bovine serum (Gibco, Waltham, MA, USA), 1% penicillin/streptomycin at 37 °C under 5% CO_2/21% O_2 condition.

4.3. Human Brain Tissue

The hypothalamus was obtained from postmortem brains of otherwise healthy individuals at the University of Miyazaki, Japan, with the approval of the institutional review board (approval number, C-0037; 11 May 2018). Written informed consent was acquired from each outpatient, and the study was conducted according to provisions of the Declaration of Helsinki.

4.4. Immunofluorescence Staining and Immunohistochemistry

Postmortem hypothalamus was fixed in 4% paraformaldehyde for 24 h, embedded in paraffin, and cut into 4-μm–thick sections. Immunohistochemistry was performed using an anti-human AGF antibody, produced by immunizing rabbits with a synthetic peptide corresponding to amino acids 392–408 (NDKPESTVDRDRDSYSG) of the AGF protein. Sections were pretreated with periodic acid (Nichirei, Tokyo, Japan) and incubated with anti-human AGF antibody (1:100) overnight. Then, the sections were washed with phosphate-buffered saline (PBS) and immunostained using EnVision/horseradish peroxidase (Dako, Glostrup, Denmark). After incubating with 3,3-diaminobenzidine solution, images were acquired using an IX70 microscope (Olympus, Tokyo, Japan). Mice whole brain was dipped in the 4% PFA and moved to 30% sucrose solution. The brain section was cut into 25 μm and blocked for 1 hour with 3% donkey serum (Dako, Glostrup, Denmark) and 0.3% triton x-100. Then, brain sections, including the arcuate nucleus region, were obtained from −1.7 to −3.4 mm from bregma and were incubated with primary antibodies, including a mouse anti-AGF-488 alexa conjugated antibody (1:100, Bioss, MA, USA), anti-NeuN (1:500, Abcam, Cambridge, UK), a chicken anti-GFAP (1:1000, Abcam, Cambridge, UK), and a rabbit anti-POMC (1:300, Phoenix, CA, USA) antibody, overnight at 4 °C. Sections were washed with PBS and incubated in secondary fluorescence antibody. Fluorescence was visualized using a IX70 fluorescent microscope (Olympus, Tokyo, Japan).

4.5. RNA Iolation and Real-Time PCR

The hypothalamus was removed from mice brain and homogenized by TissueLyserII (Qiagen, Netherlands). Total RNA was isolated using Isol-RNA lysis reagent (5 PRIME, South San Francisco, CA, USA). cDNA was synthesized by an moloney murine leukemia virus (M-MLV) reverse transcriptase kit (Invitrogen, Carlsbad, CA, USA). For real-time PCR, after mixing cDNA, primers and 10X SYBR mix, mRNA expression was analyzed using a Rotor Gene 6000 system (Corbett Life Science, Venlo, Netherlands) and normalized to 18s rRNA. Gene-specific primers are listed in Table S1.

4.6. Western Blotting

Proteins of mice tissues were prepped and homogenized by TissueLyserII, and N1 cells were extracted using radioimmunoprecipitation assay (RIPA) buffer (1% Nonidet P-40, 0.1% SDS, 150 mM NaCl, 50 mM Tris–HCl pH 7.5 and 0.5% deoxycholate) with 10% of phosphatase and protease inhibitor (Roche, Basel, Switzerland). A total of 12 μg of protein samples was loaded on SDS-PAGE gel and run by electrophoresis; afterwards, it was transferred to polyvinylidene fluoride (PVDF) membrane, blocked by 5% skim milk. Membranes were incubated with primary antibody including a mouse anti-AGF (R&D systems, Minneapolis, MN, USA), a mouse anti-α-tubulin (Santa Cruz Biotechnology, Dallas, TX, USA), anti-phospho-STAT3 and anti-STAT3 (Cell Signaling, Danvers, MA,

USA) antibody at 4 °C overnight. Anti-Immunoglobulin G (IgG) horseradish peroxidase antibody (Pierce Biotechnology, MA, USA) corresponds with the host of the primary antibody and was used as a secondary antibody. A protein band was detected by the enhanced chemiluminescence (ECL) system (Thermo Scientific, Waltham, MA, USA)

4.7. Promoter-Luciferase Reporter Assay

N1 cells were grown in 6-well plates and transfected with 1 μg pAGF prom-Luc DNA mixture using lipofectamin reagents (Invitrogen, Carlsbad, CA, USA). The pLightSwitch-prom plasmid encodes RenSP luciferase sequence under control of 1020 bp AGF promoter. After transfection, cells were treated with 100 ng/ml leptin or vehicle for 45 min and harvested for luminiscence reading. Luciferase activity was measured using a Berthold LB9507 luminometer (Berthold Technologies, Black Forrest, Germany).

4.8. Statistical Analysis

All data are represented as mean ± standard error mean (SEM) from triplicate results. The statistical analysis was determined using Prizm version 5 software (Graphpad, San Diego, CA, USA). The significance of differences between two groups was analyzed by one-tailed student's *t*-test. p value < 0.05 was considered statistically significant.

Supplementary Materials: The following are available online at https://www.mdpi.com/article/10.3390/ijms22073443/s1, Figure S1: AGF mRNA expression is increased after feeding in mice, Figure S2: AGF expression is not affected by insulin, Figure S3: Leptin injection did not induce AGF expression in various brain region excluding hypothalamus.

Author Contributions: Y.J., J.Y.H., M.S. and G.R.K. contributed to conceptualization and study design. Y.J. and J.Y.H. were responsible for the data acquisition and performing animal experiment. M.S. and G.R.K. helped with data interpretation. Y.J., J.Y.H., M.J.L., M.S. and G.R.K. wrote the manuscript. J.Z., C.S., J.-W.S., S.K.Y. and D.H.G. contributed to discussion and article revision. D.Y. and Y.O. performed human tissue histology. M.S. and G.R.K. supervised, and all authors approved the final version of the manuscript. All authors have read and agreed to the published version of the manuscript.

Funding: This work was funded by Bio & Medical Technology Development Program of the National Research Foundation of Korea (NRF) grant and the Ministry of Science and ICT (MSIT) and the Ministry of Education (2014R1A6A1029617, 2019R1F1A1059586, 2019R1A2C2088345, 2019M3E5D1A02068575, 2017R1A5A201538, 2020R1I1A1A01061771, 2017R1E1A1A01075126), research fund of Chungnam National University and a grant of the Korea Health Technology R&D Project through the Korea Health Industry Development Institute (KHIDI), funded by the Ministry of Health & Welfare, Republic of Korea (HR20C0025).

Institutional Review Board Statement: The study was conducted according to the guidelines of the Declaration of Helsinki with the approval of institutional review board (approval number, C-0037; May 11, 2018). Animal experiments were approved by the Institutional Animal Care and Use Committee of Chungnam National University (Ethical approval number, 201903A-CNU-46, approved on 1 June 2019).

Informed Consent Statement: Written informed consent was acquired from each outpatient involved in the study.

Data Availability Statement: The datasets generated during and/or analyzed during the current study are available from the corresponding author on reasonable request. No applicable resources were generated or analyzed during the current study.

Conflicts of Interest: The authors declare no conflict of interest.

References

1. Suzuki, K.; Jayasena, C.N.; Bloom, S.R. Obesity and appetite control. *Exp. Diabetes Res.* **2012**, *2012*, 824305. [CrossRef]
2. Parker, J.A.; Bloom, S.R. Hypothalamic neuropeptides and the regulation of appetite. *Neuropharmacology* **2012**, *63*, 18–30. [CrossRef] [PubMed]

3. Barateiro, A.; Mahu, I.; Domingos, A.I. Leptin Resistance and the Neuro-Adipose Connection. *Front. Endocrinol.* **2017**, *8*, 45. [CrossRef]
4. Izquierdo, A.G.; Crujeiras, A.B.; Casanueva, F.F.; Carreira, M.C. Leptin, Obesity, and Leptin Resistance: Where Are We 25 Years Later? *Nutrients* **2019**, *11*, 2704. [CrossRef]
5. DePaoli, A.M. 20 years of leptin: Leptin in common obesity and associated disorders of metabolism. *J. Endocrinol.* **2014**, *223*, T71–T81. [CrossRef] [PubMed]
6. Greenhill, C. Celastrol identified as a leptin sensitizer and potential novel treatment for obesity. *Nat. Rev. Endocrinol.* **2015**, *11*, 444. [CrossRef] [PubMed]
7. Rahmouni, K.; Sigmund, C.D.; Haynes, W.G.; Mark, A.L. Hypothalamic ERK mediates the anorectic and thermogenic sympathetic effects of leptin. *Diabetes* **2009**, *58*, 536–542. [CrossRef] [PubMed]
8. Kwon, O.; Kim, K.W.; Kim, M.S. Leptin signalling pathways in hypothalamic neurons. *Cell Mol. Life Sci.* **2016**, *73*, 1457–1477. [CrossRef] [PubMed]
9. Balthasar, N.; Coppari, R.; McMinn, J.; Liu, S.M.; Lee, C.E.; Tang, V.; Kenny, C.D.; McGovern, R.A.; Chua, S.C., Jr.; Elmquist, J.K.; et al. Leptin receptor signaling in POMC neurons is required for normal body weight homeostasis. *Neuron* **2004**, *42*, 983–991. [CrossRef]
10. Abdul-Wahed, A.; Gautier-Stein, A.; Casteras, S.; Soty, M.; Roussel, D.; Romestaing, C.; Guillou, H.; Tourette, J.A.; Pleche, N.; Zitoun, C.; et al. A link between hepatic glucose production and peripheral energy metabolism via hepatokines. *Mol. Metab.* **2014**, *3*, 531–543. [CrossRef]
11. Oike, Y.; Tabata, M. Angiopoietin-like proteins–potential therapeutic targets for metabolic syndrome and cardiovascular disease. *Circ. J.* **2009**, *73*, 2192–2197. [CrossRef]
12. Oike, Y.; Akao, M.; Yasunaga, K.; Yamauchi, T.; Morisada, T.; Ito, Y.; Urano, T.; Kimura, Y.; Kubota, Y.; Maekawa, H.; et al. Angiopoietin-related growth factor antagonizes obesity and insulin resistance. *Nat. Med.* **2005**, *11*, 400–408. [CrossRef]
13. Kadomatsu, T.; Tabata, M.; Oike, Y. Angiopoietin-like proteins: Emerging targets for treatment of obesity and related metabolic diseases. *FEBS J.* **2011**, *278*, 559–564. [CrossRef]
14. Santulli, G. Angiopoietin-like proteins: A comprehensive look. *Front. Endocrinol.* **2014**, *5*, 4. [CrossRef] [PubMed]
15. Kim, H.K.; Shin, M.S.; Youn, B.S.; Kang, G.M.; Gil, S.Y.; Lee, C.H.; Choi, J.H.; Lim, H.S.; Yoo, H.J.; Kim, M.S. Regulation of energy balance by the hypothalamic lipoprotein lipase regulator Angptl3. *Diabetes* **2015**, *64*, 1142–1153. [CrossRef] [PubMed]
16. Minokoshi, Y.; Alquier, T.; Furukawa, N.; Kim, Y.B.; Lee, A.; Xue, B.; Mu, J.; Foufelle, F.; Ferre, P.; Birnbaum, M.J.; et al. AMP-kinase regulates food intake by responding to hormonal and nutrient signals in the hypothalamus. *Nature* **2004**, *428*, 569–574. [CrossRef]
17. Karatsoreos, I.N.; Thaler, J.P.; Borgland, S.L.; Champagne, F.A.; Hurd, Y.L.; Hill, M.N. Food for thought: Hormonal, experiential, and neural influences on feeding and obesity. *J. Neurosci.* **2013**, *33*, 17610–17616. [CrossRef]
18. Cowley, M.A.; Smart, J.L.; Rubinstein, M.; Cerdan, M.G.; Diano, S.; Horvath, T.L.; Cone, R.D.; Low, M.J. Leptin activates anorexigenic POMC neurons through a neural network in the arcuate nucleus. *Nature* **2001**, *411*, 480–484. [CrossRef]
19. Varela, L.; Horvath, T.L. Leptin and insulin pathways in POMC and AgRP neurons that modulate energy balance and glucose homeostasis. *EMBO Rep.* **2012**, *13*, 1079–1086. [CrossRef] [PubMed]
20. Yu, H.; Chhabra, K.H.; Thompson, Z.; Jones, G.L.; Kiran, S.; Shangguan, G.; Low, M.J. Hypothalamic POMC deficiency increases circulating adiponectin despite obesity. *Mol. Metab.* **2020**, *35*, 100957. [CrossRef]
21. Air, E.L.; Benoit, S.C.; Clegg, D.J.; Seeley, R.J.; Woods, S.C. Insulin and leptin combine additively to reduce food intake and body weight in rats. *Endocrinology* **2002**, *143*, 2449–2452. [CrossRef]
22. Baskin, D.G.; Figlewicz Lattemann, D.; Seeley, R.J.; Woods, S.C.; Porte, D., Jr.; Schwartz, M.W. Insulin and leptin: Dual adiposity signals to the brain for the regulation of food intake and body weight. *Brain Res.* **1999**, *848*, 114–123. [CrossRef]
23. Shimizu, H.; Inoue, K.; Mori, M. The leptin-dependent and -independent melanocortin signaling system: Regulation of feeding and energy expenditure. *J. Endocrinol.* **2007**, *193*, 1–9. [CrossRef] [PubMed]
24. Buettner, C.; Pocai, A.; Muse, E.D.; Etgen, A.M.; Myers, M.G., Jr.; Rossetti, L. Critical role of STAT3 in leptin's metabolic actions. *Cell Metab.* **2006**, *4*, 49–60. [CrossRef]
25. Kim, M.J.; Namkung, J.; Chang, J.S.; Kim, S.J.; Park, K.S.; Kong, I.D. Leptin regulates the expression of angiopoietin-like 6. *Biochem. Biophys. Res. Commun.* **2018**, *502*, 397–402. [CrossRef] [PubMed]
26. Ramirez, S.; Claret, M. Hypothalamic ER stress: A bridge between leptin resistance and obesity. *FEBS Lett.* **2015**, *589*, 1678–1687. [CrossRef] [PubMed]
27. Sainz, N.; Barrenetxe, J.; Moreno-Aliaga, M.J.; Martinez, J.A. Leptin resistance and diet-induced obesity: Central and peripheral actions of leptin. *Metabolism* **2015**, *64*, 35–46. [CrossRef]
28. Scarpace, P.J.; Matheny, M.; Tumer, N. Hypothalamic leptin resistance is associated with impaired leptin signal transduction in aged obese rats. *Neuroscience* **2001**, *104*, 1111–1117. [CrossRef]
29. Darambazar, G.; Nakata, M.; Okada, T.; Wang, L.; Li, E.; Shinozaki, A.; Motoshima, M.; Mori, M.; Yada, T. Paraventricular NUCB2/nesfatin-1 is directly targeted by leptin and mediates its anorexigenic effect. *Biochem. Biophys. Res. Commun.* **2015**, *456*, 913–918. [CrossRef]
30. Perello, M.; Stuart, R.C.; Nillni, E.A. The role of intracerebroventricular administration of leptin in the stimulation of prothyrotropin releasing hormone neurons in the hypothalamic paraventricular nucleus. *Endocrinology* **2006**, *147*, 3296–3306. [CrossRef]

31. Yoo, S.; Cha, D.; Kim, D.W.; Hoang, T.V.; Blackshaw, S. Tanycyte-Independent Control of Hypothalamic Leptin Signaling. *Front. Neurosci.* **2019**, *13*, 240. [CrossRef] [PubMed]
32. Dodd, G.T.; Decherf, S.; Loh, K.; Simonds, S.E.; Wiede, F.; Balland, E.; Merry, T.L.; Munzberg, H.; Zhang, Z.Y.; Kahn, B.B.; et al. Leptin and insulin act on POMC neurons to promote the browning of white fat. *Cell* **2015**, *160*, 88–104. [CrossRef] [PubMed]
33. Shintani, T.; Higashi, S.; Suzuki, R.; Takeuchi, Y.; Ikaga, R.; Yamazaki, T.; Kobayashi, K.; Noda, M. PTPRJ Inhibits Leptin Signaling, and Induction of PTPRJ in the Hypothalamus Is a Cause of the Development of Leptin Resistance. *Sci. Rep.* **2017**, *7*, 11627. [CrossRef] [PubMed]

International Journal of *Molecular Sciences*

MDPI

Article

Adiponectin Controls Nutrient Availability in Hypothalamic Astrocytes

Nuri Song [1,†], Da Yeon Jeong [1,†], Thai Hien Tu [1], Byong Seo Park [2], Hye Rim Yang [1], Ye Jin Kim [1], Jae Kwang Kim [1], Joon Tae Park [1], Jung-Yong Yeh [1], Sunggu Yang [3,*] and Jae Geun Kim [1,*]

1 Division of Life Sciences, College of Life Sciences and Bioengineering, Incheon National University, Incheon 406-772, Korea; sannudeul@naver.com (N.S.); dayeon@inu.ac.kr (D.Y.J.); thaihientu@gmail.com (T.H.T.); hr.yang0414@inu.ac.kr (H.R.Y.); 201721047@inu.ac.kr (Y.J.K.); kjkpj@inu.ac.kr (J.K.K.); joontae.park@inu.ac.kr (J.T.P.); yehjy@inu.ac.kr (J.-Y.Y.)

2 Department of Biological Science, University of Ulsan, Ulsan 44610, Korea; bbs0808@naver.com

3 Department of Nano-Bioengineering, Incheon National University, Incheon 406-772, Korea

* Correspondence: sungguyang@inu.ac.kr (S.Y.); jgkim@inu.ac.kr (J.G.K.); Tel.: +82-32-835-8257 (S.Y.); +82-32-835-8256 (J.G.K.)

† These authors contributed equally to this work.

Abstract: Adiponectin, an adipose tissue-derived hormone, plays integral roles in lipid and glucose metabolism in peripheral tissues, such as the skeletal muscle, adipose tissue, and liver. Moreover, it has also been shown to have an impact on metabolic processes in the central nervous system. Astrocytes comprise the most abundant cell type in the central nervous system and actively participate in metabolic processes between blood vessels and neurons. However, the ability of adiponectin to control nutrient metabolism in astrocytes has not yet been fully elucidated. In this study, we investigated the effects of adiponectin on multiple metabolic processes in hypothalamic astrocytes. Adiponectin enhanced glucose uptake, glycolytic processes and fatty acid oxidation in cultured primary hypothalamic astrocytes. In line with these findings, we also found that adiponectin treatment effectively enhanced synthesis and release of monocarboxylates. Overall, these data suggested that adiponectin triggers catabolic processes in astrocytes, thereby enhancing nutrient availability in the hypothalamus.

Keywords: adiponectin; astrocyte; energy metabolism; hypothalamus; glycolysis; metabolic diseases

Citation: Song, N.; Jeong, D.Y.; Tu, T.H.; Park, B.S.; Yang, H.R.; Kim, Y.J.; Kim, J.K.; Park, J.T.; Yeh, J.-Y.; Yang, S.; et al. Adiponectin Controls Nutrient Availability in Hypothalamic Astrocytes. *Int. J. Mol. Sci.* **2021**, *22*, 1587. https://doi.org/10.3390/ijms22041587

Academic Editor: Rosalía Rodríguez-Rodríguez

Received: 9 December 2020

Accepted: 1 February 2021

Published: 4 February 2021

1. Introduction

The brain constitutes a metabolically active organ that requires the highest energy demands in the human body. Although the adult brain represents approximately 2% of the total body weight, it consumes approximately a quarter of the total glucose used for its energy supply [1–4]. Therefore, nutrient availability in the central nervous system (CNS) is directly linked to the maintenance of life. As neurons expend high levels of energy resources, such as glucose and lactate, to initiate and propagate their action potentials [4,5], impairment of the energy supply can lead to perturbation of neuronal excitability. Consistent with these concepts, multiple brain disorders are also deeply associated with abnormalities of energy metabolism in the CNS.

Astrocytes, which comprise the most abundant cell type in the CNS, support normal neuronal functions by regulating the concentration of chemical substances in the synaptic cleft area and providing nutrients between blood vessels and neurons [4–6]. Although astrocytes are responsible for the metabolic processing of glucose absorbed by the brain [4,7], they do not require as much energy as they uptake. Rather, the primary driving factor underlying astrocyte participation in glucose uptake and utilization is the provision of energy sources from astrocytes to neurons.

Adiponectin, an adipokine predominantly secreted from adipocytes, has functional roles in regulating glucose and lipid metabolism along with insulin sensitivity. In particular, adiponectin facilitates systemic glucose and lipid homeostasis by regulating several major metabolic organs, such as adipose tissue, liver, and muscle [7–11]. Accordingly, adiponectin elicits beneficial effects in multiple metabolic diseases and their related secondary complications. For example, adiponectin improves hyperglycemia by alleviating glucose intolerance and insulin resistance [9,10]. Furthermore, it mitigates hepatic steatosis and dyslipidemia through regulation of lipid metabolism [10]. Specifically, adiponectin stimulates glucose uptake by skeletal and cardiac muscle and inhibits glucose production by the liver, consequently decreasing blood glucose levels [7,8]. However, although it is well established that adiponectin dynamically participates in the regulation of peripheral energy metabolism, its impact on nutrient metabolism in astrocytes of the hypothalamus, a central unit for the regulation of energy homeostasis, has not yet been clearly elucidated.

We hypothesized that central adiponectin regulates multiple metabolic processes in hypothalamic astrocytes including glucose uptake, glycolytic activity, fatty acid oxidation and metabolites secretion. To verify this hypothesis, in this study we mainly utilized primary astrocytes extracted from the mouse hypothalamus and determined the active roles of circulating adiponectin on hypothalamic astrocytes coupled to whole body energy homeostasis.

2. Results

2.1. Central Administration of Adiponectin Results in the Activation of Hypothalamic Astrocytes

Based on the evidence that astrocytes respond to metabolic alterations and reactive astrocytes display morphological changes [12], we evaluated the number of astrocytes and their pattern of interaction with blood vessels in the hypothalamus assessed by immunohistochemistry with an antibody against Gfap, a molecular maker for the astrocyte after central administration of adiponectin. Icv administration of recombinant adiponectin into the lateral ventricle of mice resulted in an elevated number of astrocytes in the hypothalamus (Figure 1A,B). We further examined the contact ratio between astrocytes and blood vessels by performing fluorescence immunohistochemistry combined with a visualization of blood vessels by cardiac infusion of lectin to speculate whether adiponectin participates in nutrient shuttling between astrocytes and blood vessels. Notably, icv administration of adiponectin led to an increase in astrocytes interaction with the blood vessel (Figure 1C). In order to further confirm whether adiponectin triggers reactive astrogliosis, we examined the mRNA expression involved in the astrocyte activation. Semi-quantitative RT-PCR results showed an elevation of *Gfap* and *Catenin beta 1(Ctnnb1)* transcripts involved in the processes of astrocyte activation (Figure 1D,E), indicating that adiponectin induced reactive astrogliosis.

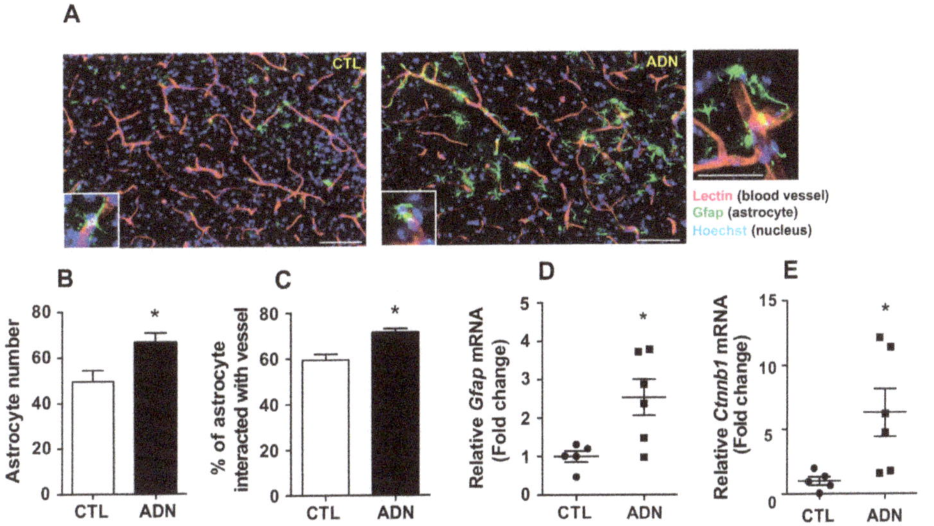

Figure 1. Central administration of adiponectin leads to activation of hypothalamic astrocytes. The whole brain was collected from mice that received an intracerebroventricular (icv) injection of adiponectin (ADN, 3 μg/2 μL) and intracardiac injection of lectin. The distribution of astrocytes was examined by immunohistochemistry using an antibody against glial fibrillary acidic protein (Gfap), a molecular marker for astrocytes. (**A**) Representative images of Gfap immunolabeling in the hypothalamus. Icv administration of adiponectin led to increased (**B**) number of astrocytes and (**C**) contact ratio between astrocytes and blood vessels in the hypothalamus of the mouse brain ($n = 5$ for each group). Icv administration of adiponectin elevated mRNA levels of (**D**) *Gfap* and (**E**) *Ctnnb1* as determined by qRT-PCR ($n = 5$ for CTL; $n = 6$ for ADN). Results are presented as mean ± SEM. * $p < 0.05$. Scale bar = 50 μm.

2.2. Adiponectin Enhances Glucose Uptake in Astrocytes

To validate the purification of primary astrocytes, we tested the enrichment of GFAP protein in cultured primary astrocytes compared to that in primary microglial cells. GFAP protein was predominantly present in primary astrocytes but not in primary microglia. Additionally, the Iba-1 protein was almost absent in primary astrocytes (Figure 2A). Given the well-known effect of adiponectin on phosphorylation of the AMPK protein, an evolutionarily conserved energy sensor and regulator of energy metabolism, we evaluated the induction of AMPK phosphorylation as determined by immunoblot assay to validate the cellular impact of recombinant adiponectin in primary astrocytes. A treatment of adiponectin resulted in a significant increase in AMPK phosphorylation (Figure 2B–D).

As multiple lines of evidence indicate that adiponectin enhances glucose uptake and utilization in metabolically active peripheral organs, such as muscle and adipose tissue [9], we interrogated whether adiponectin altered glucose uptake in astrocytes using mouse hypothalamic primary astrocytes. We first identified that a significant elevation in glucose uptake was observed in adiponectin-treated astrocytes compared to vehicle-treated astrocytes (Figure 2E). In support of this finding, we observed that exogenous treatment of adiponectin led to increased levels of glucose transporter-1 (Glut-1) protein (Figure 2F,G) and mRNA (Figure 2H), consistent with the effects of adiponectin on glucose metabolism in peripheral organs. To further verify the effect of adiponectin on glucose uptake, we quantified the mRNA expression of *Glut-1* after silencing the expression of adiponectin receptor 1 (AdipoR1) and adiponectin receptor 2 (AdipoR2). We validated the reduced levels of *AdipoR1* and *AdipoR2* mRNAs in response to transfection with siRNAs (Figure 2I,J). An increase of *Glut1* mRNA expression induced by adiponectin treatment was significantly reversed by AdipoR1 and AdipoR2 siRNAs (Figure 2K). These observations suggest that adiponectin has an active role for glucose uptake in hypothalamic astrocytes.

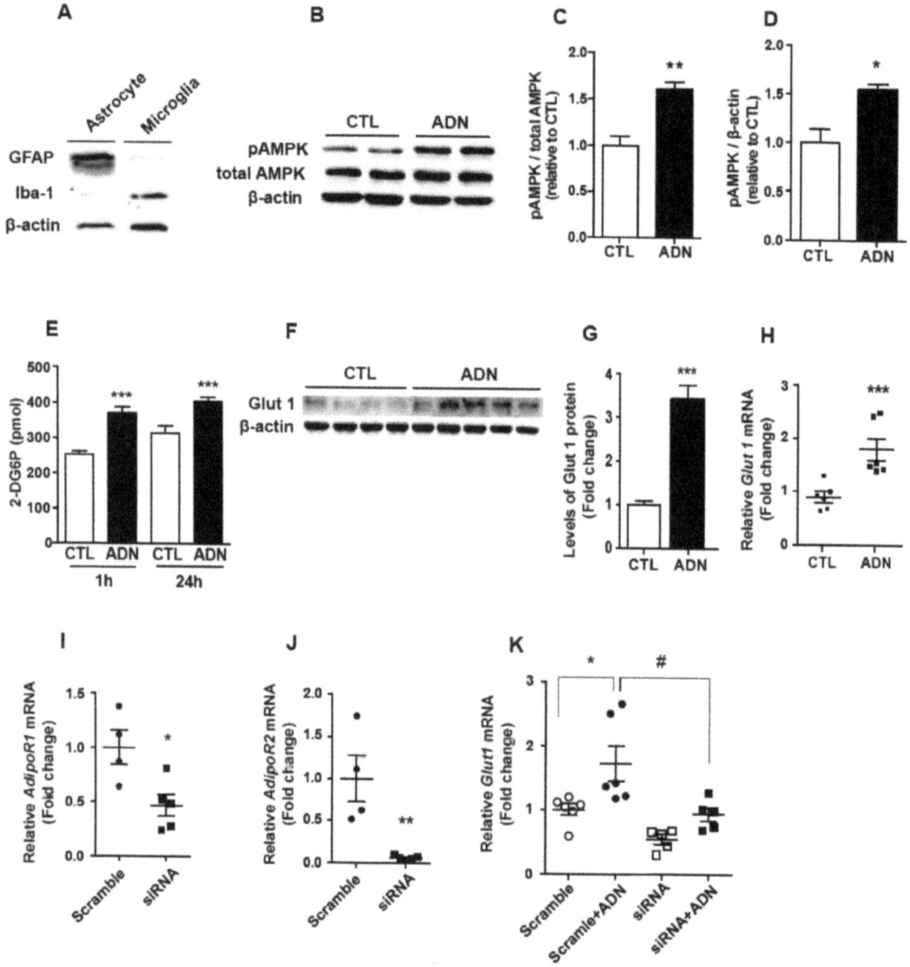

Figure 2. Adiponectin enhances glucose uptake in primary astrocytes. The primary astrocytes and microglia were set up at 5×10^5 cell/well and subjected to western blot analysis by using antibodies for GFAP and Iba-1 proteins. (**A**) GFAP protein was predominantly present in primary astrocytes but Iba-1 protein was almost absent in primary microglia. (**B–K**) Primary astrocytes were seeded at 5×10^5 cell/well, starved overnight, and treated with adiponectin (ADN, 1 μg/mL) for 24 h. Cell lysates were subjected to western blot analysis using antibodies against (**B–D**) pAMPK, total AMPK, β-actin and (**F,G**) glucose transporter-1 (GLUT-1) proteins and qRT-PCR using a primer set of (**H,K**) *Glut-1* gene. Exogenous treatment of adiponectin led to increased levels of (**F,G**) Glut-1 protein and (**H**) *Glut-1* mRNA in primary astrocytes. (**E**) Adiponectin treatment enhanced glucose uptake in cultured primary astrocytes as determined by 2DG-glucose uptake assay. AdipoR1 and AdipoR2-specific siRNA were transfected into the primary astrocytes to confirm the mRNA expression of *Glut1* induced by adiponectin treatment. qPCR data showed a significant decrease in (**I**) *AdipoR1* and (**J**) *AdipoR2* mRNAs. Elevated (**K**) *Glut1* mRNA induced by adiponectin was effectively rescued by transfection of AdipoR1 and AdipoR2-specific siRNAs. Results are presented as the means ± SEM. n= 4 for (**B–D**); n= 3 for (**E**); n= 4–5 for (**F,G**); n= 6 for (**H**); n= 4–5 for (**I,J**); n= 5–6 for (**K**). All experiments were performed from at least three different preparations of astrocytes. * $p < 0.05$; ** $p < 0.01$; *** $p < 0.001$; # $p < 0.05$.

2.3. Adiponectin Enhances Glycolytic Activities in Astrocytes

We next examined the effect of adiponectin on the glycolytic activities in primary astrocytes. Semi-quantitative RT-PCR results showed that adiponectin treatment elevated the expression of *hexokinase1 (Hk1)*, which catalyzes glucose phosphorylation during glycolysis (Figure 3A). In addition, we further evaluated glycolytic activity utilizing a seahorse XF-24 extracellular flux analyzer. The ECAR in primary astrocytes was significantly elevated by adiponectin treatment (Figure 3B). To verify that enhanced glycolytic activity was triggered by adiponectin, we performed GC-MS to measure the metabolites produced in the glycolytic processes in the hypothalamus of mice injected with adiponectin. Central administration of adiponectin led to an elevation of pyruvate and lactate levels, which correlated with glycolytic activity (Figure 3C,D). Furthermore, icv injection of adiponectin induced increased in levels of fumarate, malate, and citrate generated during the TCA cycle (Figure 3E–G). Notably, elevated glucose levels were observed in the hypothalamus of adiponectin-treated mice (Figure 3H). These data indicated that adiponectin facilitated brain glucose utilization by reinforcing glycolytic activity in astrocytes.

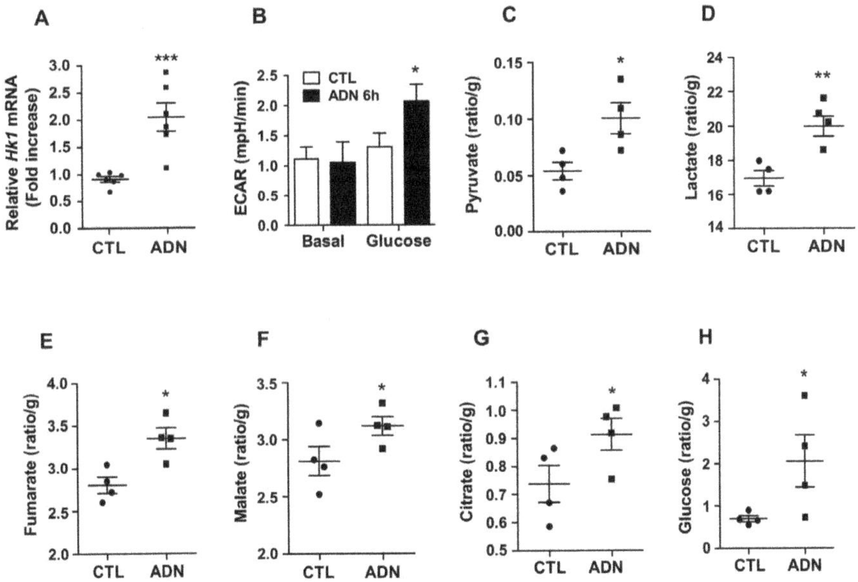

Figure 3. Adiponectin enhances glycolytic activity in primary astrocytes. Primary astrocytes were seeded at 5×10^5 cell/well, starved overnight, and treated with adiponectin (ADN, 1 μg/mL) for 24 h. (**A**) The level of mRNA encoding *hexokinase 1 (Hk1)* was upregulated in adiponectin-treated primary astrocytes compared to that in vehicle-treated primary astrocytes as determined by qRT-PCR (n = 6 for each group). (**B**) The extracellular acidification rates (ECAR) were elevated by adiponectin treatment (n = 3 for each group). GC-MS results showed increased levels of (**C**) pyruvate, (**D**) lactate, (**E**) fumarate, (**F**) malate, (**G**) citrate, and (**H**) glucose in the hypothalamus of adiponectin-treated mice compared with that of vehicle-treated mice (n = 4 for each group). Results are presented as the means ± SEM. qPCR and ECAR experiments were performed from at least three different preparations of astrocytes. * $p < 0.05$; ** $p < 0.01$; *** $p < 0.001$.

2.4. Adiponectin Promotes Synthesis and Release of Monocarboxylates in Primary Astrocytes

As the production and release of monocarboxylates from astrocytes are coupled to glycolytic processes [4,13], we next interrogated whether adiponectin alters production of monocarboxylates by performing the analysis of mRNA expression in primary astrocytes treated with recombinant adiponectin. We observed that adiponectin treatment led to increased mRNA level of *lactate dehydrogenase (Ldh)*, a catalytic enzyme that reversibly induces conversion of lactate to pyruvate (Figure 4A), and upregulated *monocarboxylate transporter-1 (Mct-1)* (Figure 4B). In support of these molecular observations, an exogenous treatment of recombinant adiponectin led to an elevation of lactate release in cultured primary astrocytes (Figure 4C). In addition, adiponectin-treated astrocytes showed increased mRNA levels of genes involved in the biosynthesis of ketone bodies including *hydroxymethylglutaryl-CoA synthase (Hmgcs)* (Figure 4D) and *hydroxymethylglutaryl-CoA lyase (Hmgcl)* (Figure 4E). In accordance with the patterns of mRNA expression, the medium level of β-hydroxybutyrate was significantly elevated in adiponectin-treated primary astrocytes (Figure 4F). To confirm the increased synthesis and release of ketone bodies, we identified that adiponectin-treated astrocytes showed increased mRNA levels of fatty acid transport protein *(FATP)*, a fatty acid transporter and genes involved in fatty acid oxidation, such as peroxisome proliferator-activated receptor-α *(PPAR-α)* and carnitine palmitoyltransferase 1-α *(CTP1-α)* (Figure 4G–I). These findings indicate that adiponectin promotes synthesis and release of monocarboxylates including lactate and β-hydroxybutyrate in the hypothalamic astrocyte through promoting glucose and lipid utilization.

2.5. Central Administration of Adiponectin Leads to an Elevation of Catabolic Processes in the Hypothalamic Astrocytes

To further specifically confirmed the impacts of adiponectin on the metabolic process in hypothalamic astrocyte, we used Ribo-Tag technique with *Gfap-Cre;Rpl22*^{HA} mice that expressed HA-tagged ribosomal protein Rpl22 in astrocytes. The immunohistochemistry experiment confirmed astrocyte-specific Cre recombination by identifying the immunosignals of HA protein in the Gfap-positive hypothalamic astrocytes (Figure 5A). In addition, we validated the purification of mRNA extracted from the hypothalamic astrocyte by confirming a predominant expression of *Gfap* mRNA, a molecular maker for the astrocytes and minor expression of *Iba-1* mRNA, a molecular marker for microglia and *NeuN* mRNA, a maker for neuron in purified sample compared with the input control sample (Figure 5B). In consistent with mRNA expression data obtained from mouse total hypothalamus and primary astrocytes, we observed that icv administration of adiponectin effectively elevated mRNA levels of *Glut-1* (Figure 5C) and *Hk1* (Figure 5D), indicating enhanced glucose utilization. In addition, Ribo-tag results revealed that central administration of adiponectin resulted in an increase of mRNA levels of genes involved in the synthesis of monocarboxylates including *Ldh* (Figure 5E), *Hmgcl* (Figure 5F) and *Hmgcs* (Figure 5G), enzymes for the synthesis of lactate and ketone body as well as *Mct-1 gene* (Figure 5H). In support of these observations and in vitro findings, we also identified that central administration adiponectin resulted in elevated levels of mRNAs involved in the fatty acid utilization including *FATP* (Figure 5I), *PPAR-α* (Figure 5J) and *CTP1-α* (Figure 5K). From these data, we successfully confirmed that adiponectin triggers enhanced the catabolic process of nutrients such as glucose and fatty acids, and monocarboxylates production in hypothalamic astrocytes utilizing a mouse model that enabled the purification of astrocyte-specific mRNA.

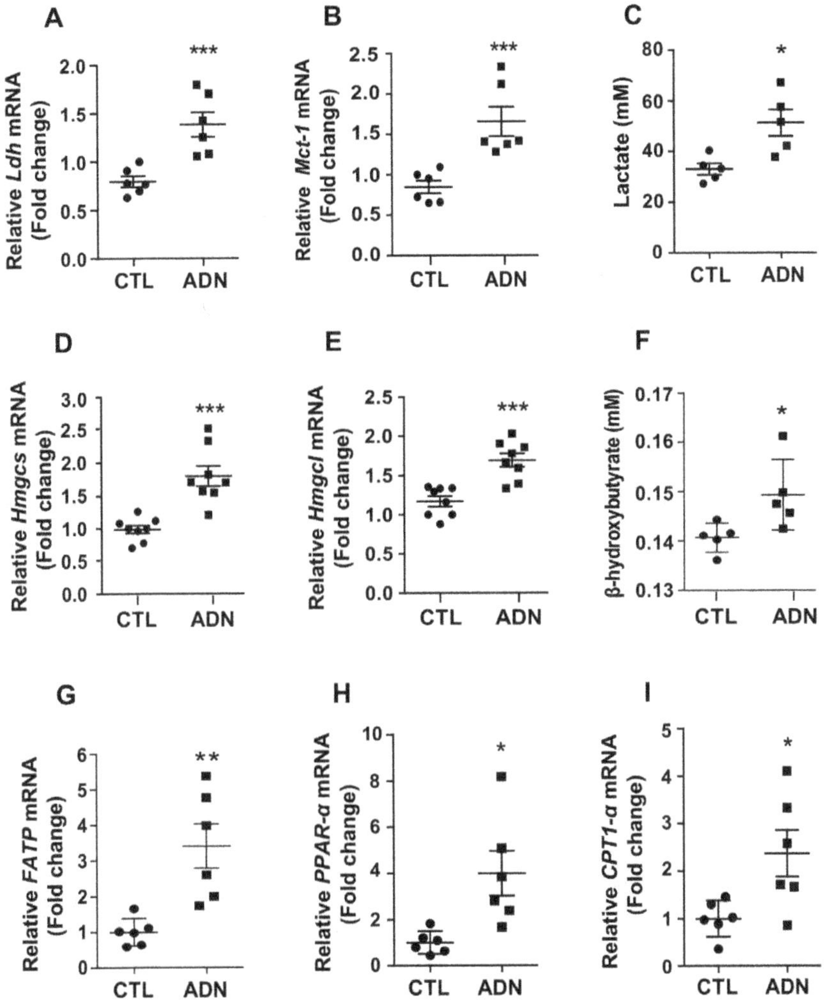

Figure 4. Adiponectin promotes production of lactate and ketone body in primary astrocytes. Primary astrocytes were seeded at 5×10^5 cell/well, starved overnight, and treated with adiponectin (1 μg/mL) for 24 h. The elevated levels of mRNA encoding (**A**) *lactate dehydrogenase (Ldh)* and (**B**) *monocarboxylate transporter-1 (Mct-1)* genes were observed in adiponectin-treated primary astrocyte as determined by qRT-PCR ($n = 6$ for each group). (**C**) The concentration of medium lactate was increased 24 h after adiponectin treatment ($n = 5$ for each group). The elevated mRNA levels of (**D**) *hydroxymethylglutaryl-CoA synthase (Hmgcs)* and (**E**) *hydroxymethylglutaryl-CoA lyase (Hmgcl)* were observed in adiponectin-treated primary astrocytes as determined by qRT-PCR ($n = 8$ for each group). (**F**) The medium concentration of β-hydroxybutyrate was increased 24 h after adiponectin treatment ($n = 5$ for each group). The increased mRNA levels of (**G**) *fatty acid transport protein (FATP)*, (**H**) *peroxisome proliferator-activated receptor-α (PPAR-α)* and (**I**) *carnitine palmitoyltransferase 1-α* (*CTP1-α*) were observed in the adiponectin-treated astrocytes compared with the that in vehicle-treated astrocytes ($n = 6$ for each group). Results are presented as the means ± SEM. All experiments were performed from at least three different preparations of astrocytes. * $p < 0.05$; ** $p < 0.01$; *** $p < 0.001$.

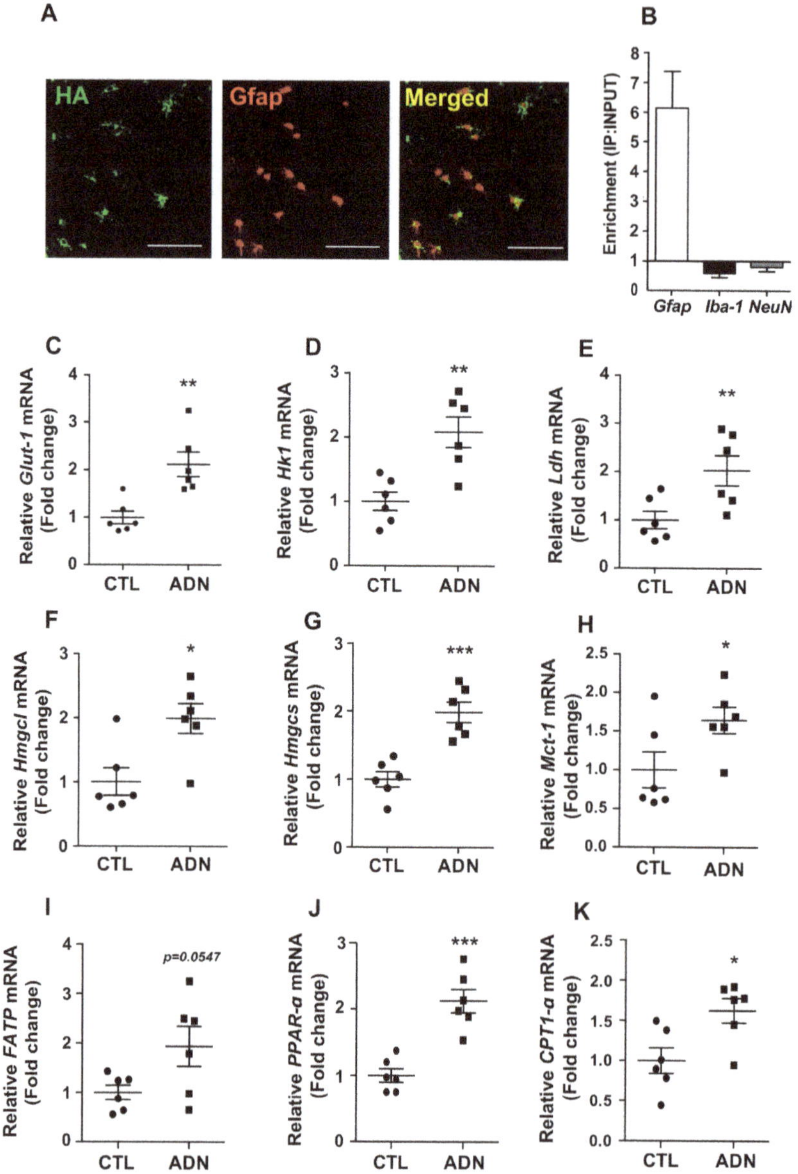

Figure 5. Central administration of adiponectin leads to enhanced glucose utilization and monocarboxylates production in hypothalamic astrocytes of the mice. Ribo-tag analysis were performed to determine alterations in astrocyte-specific mRNA expression in the hypothalamus of *Gfap-CreER*^{T2}*:Rpl22*^{HA} mice. (**A**) Representative images revealing co-expression of HA and Gfap immunosignals in the hypothalamus of *Gfap-CreER*^{T2}*:Rpl22*^{HA} mice. (**B**) qRT-PCR data showing enrichment of *Gfap* mRNA, a molecular marker for astrocyte (but not *Iba-1*, a molecular maker for microglia and *NeuN*, a molecular marker for neuron) in the purified RNA immunoprecipitated with HA antibody compared with the input RNA extracted from hypothalamus. Elevated mRNA levels of (**C**) *Glut-1*, (**D**) *Hk1*, (**E**) *Ldh*, (**F**) *Hmgcl*, (**G**) *Hmgcs*, (**H**) *Mct-1*, (**I**) *FATP*, (**J**) *PPAR-α* and (**K**) *CTP1-α* were observed in hypothalamic astrocytes from *Gfap-CreER*^{T2}*:Rpl22*^{HA} mice that received an icv injection of adiponectin (ADN, 3 μg/2 μL) compared with vehicle-treated control group (CTL). Results are presented as mean ± SEM. $n = 6$ for each group. * $p < 0.05$, ** $p < 0.01$, *** $p < 0.001$. Scale bar = 50 μm.

2.6. Adiponectin Rescued 2-DG-Induced Hyperphagia

Central administration of 2-deoxy-D-glucose (2-DG) led to an increase in appetite, suggesting that lower glucose availability drives feeding behavior in association with hypothalamic circuit activity [14]. Thus, we further evaluated the effect of adiponectin on the hyperphagic response induced by 2-DG treatment to identify the physiological relevance of adiponectin-induced alteration in cellular metabolism of hypothalamic astrocytes. In accordance with previous findings, icv administration of 2-DG led to an increase in food intake compared to that in vehicle-treated mice (Figure 6A,B). Increased 2 h and 18 h cumulative food intake induced by 2-DG administration was effectively rescued by adiponectin treatment (Figure 6A,B). These observations suggest that the enhanced catabolic processes of nutrients in hypothalamic astrocytes triggered by adiponectin may be coupled with the homeostatic feeding behavior under the hypoglycemic conditions.

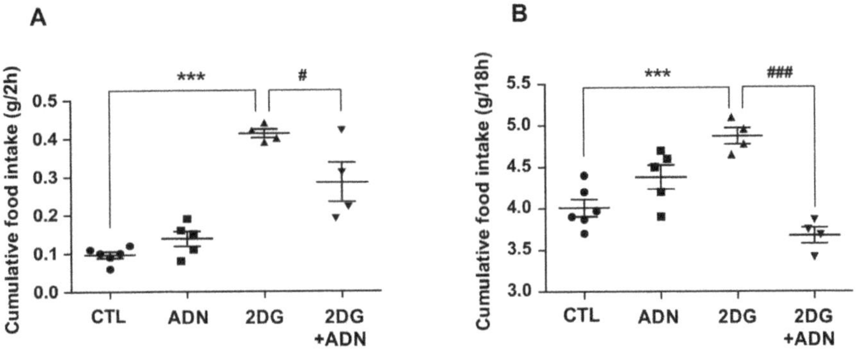

Figure 6. Central administration of adiponectin reversed increased food intake in response to icv injection of 2-DG. Icv injection of adiponectin curbed the increase of (**A**) 2 h and (**B**) 18 h cumulative food intake induced by icv injection of 2-DG. Results are presented as the mean ± SEM. n = 4–6 for each group. *** $p < 0.001$ for the 2-DG-treated versus the CTL groups; # $p < 0.05$, ### $p < 0.001$ for the 2-DG + ADN-treated versus 2-DG-treated groups.

3. Discussion

The current study highlights an active role of adiponectin in the regulation of nutrient availability in hypothalamic astrocytes. In turn, this suggests that the alteration of nutrient availability between the hypothalamic neuron-astrocyte-blood vessel axis controlled by adiponectin might be tightly coupled to the function of the hypothalamic circuit that controls whole-body energy metabolism [4,15].

Consistent with the concretized evidence that adiponectin affects glucose and lipid utilization in peripheral organs, we here verified the active contributions of adiponectin in glucose and lipid metabolism of hypothalamic astrocytes. We identified that adiponectin facilitates the availability of nutrients in the hypothalamus by promoting the utilization of glucose and fatty acids through the multiple in vitro and in vivo strategies. The hypothalamus constitutes the master organ that controls energy homeostasis by mediating afferent signals derived from metabolically active peripheral organs [16]. In particular, adipose tissue dynamically communicates with the hypothalamus through its own chemical messengers termed adipokines [17,18]. Notably, the circulating levels of most adipokines are proportional to adiposity and long-term elevation of adipokines elicits adverse effects, such as inflammation, oxidative stress, and endoplasmic reticulum stress, which might comprise potential pathogenic elements for the development of metabolic disorders [19].

However, the effects of adiponectin on metabolic controls are distinct from those of general adipokines. In particular, it exerts beneficial effects against multiple cellular stresses and the development of metabolic disorders [15]. In accordance with previous studies, which indicate that adiponectin improves inflammatory responses, [20,21], our results validated the anti-inflammatory properties of adiponectin in hypothalamic astrocytes (Supplementary Figure S1). Moreover, as most adipokines participate in the operation of hypothalamic circuit activity, central adiponectin also acts as an appetite regulator by targeting hypothalamic neurons associated with the brain glucose concentration [22]. Together with our data, these observations suggest that adiponectin-controlled glucose metabolism in hypothalamic astrocytes might be linked to the hypothalamic control of energy homeostasis. Consistent with this, a growing body of evidence has suggested that the nutrient availability between hypothalamic astrocytes and neurons determines the operation of the circuit activity triggering homeostatic feeding behavior [6,23]. Furthermore, recent reports have shown that the metabolic shift induced by fasting or high-fat diet treatment results in alteration of hypothalamic metabolites, such as lactate and ketone bodies [24]. In the present study, we also identified elevated synthesis and release of the monocarboxylates such as lactate and ketone body in primary astrocytes in response to adiponectin treatment. Thus, it is reasonable to hypothesize that adiponectin may affect the excitability of hypothalamic neurons by modulating astrocyte-derived substances, such as gliotransmitters and metabolites. Recently, considerable effort has been paid to unmask the direct contributions of hypothalamic glial cells in regulation of the hypothalamic neuronal circuit [4,25]. However, there is still insufficient information regarding the astrocyte-derived tropic factors that give rise to homeostatic behaviors and responses directly coupled to hypothalamic neuronal functions. In the present study, we proposed that the hypothalamic astrocyte responds to fluctuations in circulating adiponectin and presumably participates in operation of the hypothalamic circuit by modulating the nutrient availability linked to various cellular metabolic processes, including nutrient uptake and release between the extracellular environment and hypothalamic neurons. Notably, a recent study showed that adiponectin evokes the excitation of hypothalamic proopiomelanocortin neurons, which promotes satiety signals under low glucose conditions [22]. In this regard, we identified that central administration of adiponectin effectively curbed hyperphagic behavior induced by 2-DG-mediated limited glucose utilization. As neuron-glia metabolic coupling constitutes a critical cellular event to preserve normal brain functions and the availability of oxygen and energy resources is tightly coupled to neuronal excitability and functions [26], the impairment of nutrient shuttling between astrocytes and neurons may serve as a primary pathological event in the development of multiple neurodegenerative diseases. Thus, our findings raised an open question regarding the potential beneficial effects of central adiponectin in the development of metabolic diseases caused by the degeneration of hypothalamic neurons coupled with the impairment of the neuron-glia metabolic interaction in the hypothalamic circuitry. Therefore, further studies are required to clarify whether disruptions of neuronal functions linked to metabolic disorders can be reversed by adiponectin treatment. Additionally, further investigations are warranted to elucidate whether adiponectin-controlled astrocytes control the activity of hypothalamic neurons linked to energy homeostasis via tropic factors or modulating synaptic inputs. Nevertheless, the current findings collectively suggest that central adiponectin may support normal hypothalamic functions by promoting nutrient availability in hypothalamic astrocytes.

4. Materials and Methods

4.1. Animals

Eight-week-old male C57BL/6 mice (Dae Han Bio Link, Seoul, Korea) were housed in a 12-h light-dark cycle at 25 °C and 55 ± 5% humidity. The mice were allowed access to normal diet and tap water *ad libitum*. All animal care and experimental procedures were performed in accordance with a protocol approved by the Institutional Animal Care and Use Committee (IACUC) at the Incheon National University (permission number: INU-2016-001).

4.2. Cannula Implantation for Intracerebroventricular (Icv) Injection

The mice were anesthetized with an intraperitoneal injection of tribromoethanol (250 mg/kg, Sigma-Aldrich, St. Louis, MO, USA) and placed in a stereotaxic apparatus (Stoelting, Wood Dale, IL, USA). The 2.5-mm cannula (26 gauge) was implanted into the lateral ventricle (X: 1 mm, Y: 0.4 mm to the bregma) and secured to the skull with dental cement. Animals were kept warm until they recovered from the anesthesia and then placed in individual cages. After surgery, a recovery period of seven days was allowed prior to the initiation of experiments.

4.3. Icv Injection of Adiponectin

Mice were matched based on body weight and food intake during the adaptation period and divided into adiponectin and phosphate-buffered saline (PBS)-injected groups. Recombinant globular adiponectin (Lugen Sci, Bucheon, Korea) was dissolved in PBS to a concentration of 1.5 mg/mL. Each solution was freshly prepared on the day of administration and free of any contaminants, such as endotoxin. Mice were administered the first icv injection of adiponectin (3 μg/2 μL) 24 h before sacrifice and hypothalamic tissues were harvested 1 h after the second injection of adiponectin (3 μg/2 μL). For the test of feeding behavior, mice were injected with recombinant globular adiponectin (3 μg/2 μL) 1 h before icv injection of 2-deoxy-D-glucose (2.5 mg/2 μL, Sigma-Aldrich, St. Louis, MO, USA), and their food intake was measured.

4.4. Immunohistochemistry

Mice were anesthetized, their thoracic cavities were opened, and 30 μg of tomato lectin (Vectorlabs, Burlingame, CA, USA) in 100 μL volume was injected directly into the left ventricle of the heart, over a period of approximately 30 s. The heart continued to beat for approximately 1 min following injection. Subsequently, the animals were transcardially perfused with 0.9% saline (wt/vol), followed by fresh fixative of 4% paraformaldehyde in phosphate buffer (PB, 0.1 M, pH 7.4). Brains were collected and post-fixed overnight before coronal sections (50 μm thickness) were taken by vibratome (5100 mz Campden Instruments, Leicestershire, England). After washing in PB several times, the sections were pre-incubated with 0.3% Triton X-100 (Sigma-Aldrich) in PB for 30 min at room temperature (RT) and incubated overnight with rabbit anti-GFAP antibody (1:1000 dilution, ab7260, abcam, Cambridge, UK) or mouse anti-HA antibody (1:1000; MMS-101R, BioLegend, San Diego, CA, USA) at RT. Immunofluorescence was performed with the secondary antibodies (Alexa Fluor 488-labeled anti-mouse antibody, 1: 500; A11001, Invitrogen, Carlsbad, CA, USA or Alexa Fluor 594-labeled anti-rabbit antibody, 1: 500; A21209, Invitrogen, Carlsbad, CA, USA) for 2 h at RT. The sections were then mounted onto glass slides and covered by coverslips with a drop of mounting medium (Dako North America Inc, Carpinteria, CA, USA). The coverslips were sealed with nail polish to prevent desiccation and movement of the samples under the microscope. The images were recorded using fluorescence microscopy (Axioplan2 Imaging, Carl Zeiss Microimaging Inc., Thornwood, NY, USA) and subjected to analyses.

4.5. IHC Image Capture and Analyses

Images were acquired by fluorescence microscopy (Axioplan2 Imaging; Carl Zeiss Microimaging Inc.). For IHC analyses, sections were anatomically matched with the mouse brain using atlas50 (hypothalamic region: between 1.46 and −1.82 mm from bregma). Both sides of the bilateral hypothalamic region were analyzed for two brain sections per mouse. The number of GFAP-positive astrocytes was counted using ImageJ 1.47 v software (National Institutes of Health, Bethesda, MD, USA; http://rsbweb.nih.gov/ij/). Hoechst (Sigma-Aldrich, St. Louis, MO, USA) staining was performed to identify cell nuclei. In sections double-labeled with GFAP antibody and tomato lectin, the number of astrocytes in contact with blood vessels was calculated as a percentage of the GFAP-positive cells in contact with blood vessels per total number of GFAP-positive cells.

4.6. Primary Astrocyte Culture

Following decapitation of five C57BL6 mice (5 days old), the hypothalamic tissues were removed, combined in a sterile dish and triturated in Dulbecco's modified Eagle medium (DMEM) F-12 containing 1% penicillin-streptomycin. The cell suspension was filtered through a 100-μm sterile cell strainer to remove debris and fibrous layers. The suspension was centrifuged and the pellet was resuspended in DMEM F-12 containing 10% fetal bovine serum (FBS) and 1% penicillin-streptomycin. Cells were grown in this culture medium in 75-cm^3 culture flasks at 37 °C and 5% CO_2. When cells grew to confluence (about 9 days), the flasks were placed in a 37 °C shaking incubator at 240 rpm for 16 h. The cells were then harvested using 0.05% trypsin-ethylenediamine tetraacetic acid, resuspended in DMEM F-12 containing 10% FBS and 1% penicillin-streptomycin, and centrifuged for 5 min at 1000 rpm. Cells were seeded at a concentration of 5×10^5 cells/mL in culture plates previously treated with poly-L-lysine hydrobromide (50 μg/mL) and grown for 24 h. The medium was changed to DMEM F-12 containing 1% antibiotics without FBS; 24 h later, the same medium plus either 1 μg/mL of recombinant globular adiponectin or vehicle was added. Cells and/or medium were collected 1 or 24 h after treatment. For siRNA transfection, primary astrocytes were seeded (5×10^5 cells/mL) in 6-well culture plates and transfected with siRNAs specific for AdipoR1 (# 72674–1, Bioneer, Daejeon, South Korea) and AdipoR2 (# 68465–1, Bioneer, Daejeon, South Korea) using Lipofectamine 3000 reagent (Invitrogen, Carlsbad, CA, USA) according to the manufacturer's protocols. A scrambled siRNA (Bioneer) was used as a negative control.

4.7. Measurement of Glucose Uptake

Mouse primary astrocytes were seeded at a concentration of 5×10^5 cells/well in 6-well plates. After 24 h treatment of globular adiponectin (1 μg/mL), the cells were processed for a glucose uptake assay following the glucose uptake assay kit instructions (Bio Vision Research Products, Milpitas, CA, USA).

4.8. Measurement of Monocarboxylates

The mouse primary astrocytes were seeded at a concentration of 1×10^5 cells/well in 12-well plates and treated with PBS (vehicle control) or globular adiponectin (1 μg/mL) for 24 h. After treatment, the medium were collected and centrifuged to eliminate remaining cells and debris, and processed to measure the concentration of monocarboxylates. For the lactate release assay, the concentration of lactate in the collected medium was examined by using the lactate assay kit (Biomedical Research Service Center, Buffalo, NY, USA) according to the manufacturer's instructions. The concentration of β-hydroxybutyrate (ketone body) in the collected medium was measured by using β-hydroxybutyrate colorimetric assay kit (Cayman, 700190, MI, USA) according to the manufacturer's instructions.

4.9. Extraction and Analysis of Hydrophilic Metabolites in the Hypothalamus

The extraction and analysis of hydrophilic metabolites in the hypothalamus was performed as previously described [27]. Each sample was extracted with 1 mL of methanol: water:chloroform solution (5:2:2, *v*/*v*/*v*) and 0.03 mL of L-2-chlorophenylalanine in distilled water (0.3 mg/mL) as an internal standard (IS). A mixture of approximately 300 mg glass beads (acid-washed, 425–600 μm, G8772, Sigma-Aldrich, St. Louis, MO, USA) was homogenized for 20 s using a bead beater (Mini Beadbeater-96, BioSpec Products, Bartlesville, OK, USA) and sonicated for 10 min. Thereafter, the samples were incubated in a thermomixer (model 5355, Eppendorf AG, Hamburg, Germany) at 1200 rpm for 30 min at 37 °C and centrifuged at 16,000× *g* at 4 °C for 5 min. The liquid from upper layer (0.8 mL) was transferred into a clean tube and mixed with 0.4 mL distilled water. After centrifugation under the same conditions, 0.9 mL was separated in a new tube and dried using a centrifugal concentrator (VS-802F, Visionbionex, Gyeonggi, Korea) for at least 3 h. The sample was placed in a freeze-dryer (MCFD8512, Ilshin, Gyeonggi-do, Korea) for 16 h to achieve complete concentration. Thereafter, 0.08 mL methoxyamine hydrochloride (20 mg/mL) in pyridine was added for derivatization and incubated at 1200 rpm at 30 °C for 90 min. Next, 0.08 mL *N*-methyl-*N*-(trimethylsilyl) trifluoroacetamide was added and allowed to react at 1200 rpm at 37 °C for 30 min. The derivatized hydrophilic metabolites (1 µL) were analyzed using gas chromatography–mass spectrometry (GC–MS). The GC–MS system consisted of an auto-sampler AOC-20i and GCMS-QP2010 Ultra system (both Shimadzu, Kyoto, Japan) equipped with a DB-5 column (30 m × 0.25 mm id, film thickness 1.0 µm, 122–5033, Agilent, Santa Clara, CA, USA). The flow rate of helium as carrier gas was 1.1 mL/min, and split ratio was 1:10. The injector temperature was 280 °C. The column oven program was as follows: 100 °C for 4 min, increased at a rate of 10 °C/min to 320 °C, and then maintained at 320 °C for 11 min. The ion source temperature was 200 °C, and the interface was set at 280 °C. The scanned mass range was 45–600 m/z. Lab solutions GCMS solution software (version 4.11; Shimadzu, Kyoto, Japan) was used to identify the hydrophilic metabolites in the hypothalamus. The results were filtered with their retention times and mass spectra with reference to standard compounds and the in-house library. Quantitative analysis was conducted using the ratio of the analyte peak area to the IS peak area.

4.10. Ribo Tag Analysis

To specifically evaluated mRNA expression from hypothalamic astrocytes, we utilized the Ribo-Tag translational profiling system [28,29]. In order to generate Ribo-tag mice (*Gfap-CreER*^{T2}: *Rpl22*^{HA}), *Rpl22*^{HA} mice (Stock No. 011029, Jackson Laboratory) were crossbred with *glial fibrillary acidic protein (Gfap)-CreER*^{T2} mice (Stock No.012849), which specifically expresses *Cre* recombinase in astrocytes. Since the *Gfap-CreER*^{T2} mice expressed *Cre* recombinase under the control of the tamoxifen inducible GFAP promoter, 8-week-old *Gfap-CreER*^{T2}: *Rpl22*^{HA} mice received daily intraperitoneal injections for 5 days of tamoxifen (100 mg/kg,T5648, Sigma-Aldrich, St. Louis, MO, USA) dissolved in corn oil (C8267, Sigma-Aldrich). RNA isolation with the Ribo-Tag system was performed as described by a previous reporters [29,30]. Briefly, the hypothalamus was harvested and homogenized before RNA extraction. RNA was extracted from 10% of the cleared lysate and used as an input control. The remaining lysate was incubated with mouse anti-HA antibody for 4 h at 4 °C followed by the addition of protein G agarose beads (LGP-1018B, Lugen, Gyeonggi-Do, South Korea) and overnight incubation at 4 °C. The beads were washed three times in high-salt solution. The bound ribosomes and RNA were separated from the beads with 30 s of vortexing. Total RNA was extracted using a QIAGEN RNeasy Micro Kit (74034, Qiagen, Hilden, Germany), according to the manufacturer's instructions and quantified with NanoDrop Lite (Thermo Scientific, Waltham, MA, USA). To evaluate the levels of ribosome-associated mRNA in astrocytes, we synthesized cDNA using a high-capacity cDNA reverse transcription kit (Applied Biosystems, Foster City, CA, USA) and performed quantitative real-time PCR (qRT-PCR).

4.11. Quantitative Real-Time Reverse Transcription-Polymerase Chain Reaction (qRT-PCR)

Total RNA was extracted from the hypothalamus or cultured cells according to the Tri-Reagent protocol, and cDNA was then synthesized from total RNA using a high-capacity cDNA reverse transcription kit. Real-time PCR amplification of the cDNA was detected using the SYBR Green Real-time PCR Master Mix (Toyobo Co., Ltd., Osaka, Japan) in a Bio-Rad CFX 96 Real-Time Detection System (Bio-Rad Laboratories, Hercules, CA, USA). The results were analyzed by the CFX Manager software and normalized to the levels of *β-actin* and *L19*, housekeeping genes. The primer sequences used are shown in Table 1. All reactions were performed under the following conditions: initial denaturation at 95 °C for 3 min, followed by 40 cycles of 94 °C for 15 s, 60°C for 20 s, and 72 °C for 40 s.

Table 1. Real-time PCR primer sequences.

Target Gene	Direction of Primer	Sequence (5′→3′)
Gfap	Forward	TCA ATG ACC GCT TTG CTA GC
	Reverse	ACT CGT GCA GCC TTA CAC AG
Iba-1	Forward	TCT GCC GTC CAA ACT TGA AG
	Reverse	TCT AGG TGG GTC TTG GGA AC
NeuN	Forward	ATG GTG CTG AGA TTT ATG GAG G
	Reverse	CGA TGG TGT GAT GGT AAG GAT C
Ctnnb1	Forward	ATC CAA AGA GTA GCT GCA GG
	Reverse	TCA TCC TGG CGA TAT CCA AG
Glut-1	Forward	CTT CAT TGT GGG CAT GTG CTT C
	Reverse	AGG TTC GGC CTT TGG TCT CAG
Hk1	Forward	AGA GGC CTA GAC CAC CTG AAT GTA A
	Reverse	ACT GTT TGG TGC ATG ATT CTG GAG
Ldh	Forward	AGC CCT GAC TGC ACC ATC ATC
	Reverse	CGG AAT CGA GCA GAA TCC AGA
Mct-1	Forward	AAT GAT CGC TGG TGG TTG TC
	Reverse	TTG AAA GCA AGC CCA AGA CC
Hmgcs	Forward	TTT GAT GCA GCT GTT TGA GG
	Reverse	CCA CCT GTA GGT CTG GCA TT
Hmgcl	Forward	CCA GCT TTG TTT CTC CCA AG
	Reverse	TCA GAC ACA GCA CCG AAG AC
L19	Forward	GGT GAC CTG GAT GAG AAG GA
	Reverse	TTC AGC TTG TGG ATG TGC TC
FATP	Forward	GCA GCA TTG CCA ACA TGG AC
	Reverse	GTG TCC TCA TTG ACC TTG ACC AGA
PPAR-α	Forward	ACG CTC CCG ACC CAT CTT TAG
	Reverse	TCC ATA AAT CGG CAC CAG GAA
CPT1- α	Forward	CCA GGC TAC AGT GGG ACA TT
	Reverse	GAA CTT GCC CAT GTC CTT GT
AdipoR1	Forward	TGA CTG GCT GAA AGA CAA CG
	Reverse	TTG GTC TCA GCA TCG TCA AG
AdipoR2	Forward	ATC CCT CAC GAT GTG CTA CC
	Reverse	TAA AAG ATC CCC AGG CAC AG
IL-1β	Forward	ATA CTG CCT GCC TGA AGC TCT TGT
	Reverse	AGG GCT GCT TCC AAA CCT TTG AC
IL-6	Forward	TGG TCT TCT GGA GTA CCA TAG C
	Reverse	TCT GAA GGA CTC TGG CTT TGT C
β-actin	Forward	GAT CTG GCA CCA CAC CTT CT
	Reverse	GGG GTG TTG AAG GTC TCA AA

4.12. Immunoblotting

Primary astrocytes were seeded at 5×10^5 cells/well in 6-well plates, allowed to attach overnight, and then incubated in DMEM containing 1 μg/mL adiponectin for 24 h. The cells were washed twice with PBS, followed by scraping and resuspension of the cell pellet in RIPA lysis buffer containing protease inhibitors and centrifugation to remove debris, unbroken cells, and cellular nuclei. Protein content was determined by Bradford's method using bovine serum albumin as a standard. Samples containing 10 μg of total protein were separated using 12% sodium dodecyl sulfate acrylamide gels. After electrophoresis, proteins were transferred to a nitrocellulose membrane and then blocked with 5% skim milk in TBS (Tris-buffered saline, 0.1% Tween20) buffer for 2 h and incubated in the primary antibodies, including anti-GFAP (1:1000 dilution, Sigma-Aldrich, USA), anti-Iba-1 (1:1000 dilution, Wako, Osaka, Japan), anti-Glut1 (1:500 dilution, Santa Cruz Biotechnology, Dallas, TX, USA), anti-phosphorylated (p)AMPK, anti-total AMPK (1:1000 dilution, Cell Signaling, Danvers, MA, USA), and anti-β-actin (1:10,000 dilution, Sigma-Aldrich, USA).

4.13. Measurement of Extracellular Acidification Rates (ECAR)

ECAR measurements were performed using the XF24 Extracellular Flux analyzer (Seahorse Bioscience, North Billerica, MA, USA). Briefly, primary astrocytes were plated into XF24 (V7) polystyrene cell culture plates (Seahorse Bioscience). Primary astrocytes were seeded at 5×10^4 cells/well in poly-L-lysine hydrobromide-coated XF24 plates. The cells were attached overnight and then incubated with globular adiponectin (1 μg/mL) for 6 h in a humidified 37 °C incubator with 5% CO_2. Prior to performing an assay, growth medium in the wells of an XF cell plate was exchanged with the appropriate assay medium to achieve a minimum 1:1000 dilution of growth medium. Subsequently, 560 μL of the assay medium was added to cells for an XF assay. While sensor cartridges were calibrated, cell plates were incubated in a 37 °C/non-CO_2 incubator for 60 min prior to the start of an assay. All experiments were performed at 37 °C. Each measurement cycle consisted of a mixing time of 3 min and a data acquisition period of 3 min (12 data points) for the XF24. All compounds were prepared at appropriate concentrations in desired assay medium and adjusted to pH 7.4. For XF24, 80 μL of compound was added to each injection port. In a typical experiment, three baseline measurements were taken prior to the addition of any compound, and three response measurements were taken after the addition of each compound. ECAR were reported as absolute rates mph/min for ECAR.

4.14. Statistical Analysis

Data were analyzed using analysis of variance (ANOVA) followed by Student's *t*-test using Prism GraphPad. p value ≤ 0.05 was considered statistically significant. The values are represented as the means ± standard error of the mean (SEM).

Supplementary Materials: The following are available online at https://www.mdpi.com/1422-0067/22/4/1587/s1, Figure S1: Adiponectin leads to decreased mRNA levels of proinflammatory cytokines in primary astrocytes.

Author Contributions: Conceptualization, N.S., D.Y.J., S.Y. and J.G.K.; investigation, N.S., D.Y.J., T.H.T., B.S.P., H.R.Y., Y.J.K., J.K.K., J.T.P. and S.Y.; data curation, B.S.P., J.T.P. and J.-Y.Y.; writing—original draft preparation, N.S., D.Y.J., T.H.T., S.Y. and J.G.K.; writing—review and editing, S.Y. and J.G.K. All authors have read and agreed to the published version of the manuscript.

Funding: This work was supported by an Incheon National University (Foreign post-doctor program-2016) Research Grant for Thai Hien Tu and supported by a grant of the Korea Health Technology R&D Project through the Korea Health Industry Development Institute (KHIDI), funded by the Ministry of Health & Welfare, Republic of Korea (grant number: HI17C1600).

Institutional Review Board Statement: This study was approved by the Institutional Animal Care and Use Committee (IACUC) at the Incheon National University (permission number: INU-2016-001).

Informed Consent Statement: Not applicable.

Data Availability Statement: All data reported in the manuscript and in the Supplementary materials.

Conflicts of Interest: The authors declare no competing financial interest.

References

1. Mergenthaler, P.; Lindauer, U.; Dienel, G.A.; Meisel, A. Sugar for the brain: The role of glucose in physiological and pathological brain function. *Trends Neurosci.* **2013**, *36*, 587–597. [CrossRef]
2. Attwell, D.; Laughlin, S.B. An energy budget for signaling in the grey matter of the brain. *J. Cereb. Blood Flow Metab.* **2001**, *21*, 1133–1145. [CrossRef]
3. Goyal, M.S.; Hawrylycz, M.; Miller, J.A.; Snyder, A.Z.; Raichle, M.E. Aerobic glycolysis in the human brain is associated with development and neotenous gene expression. *Cell Metab.* **2014**, *19*, 49–57. [CrossRef] [PubMed]
4. Bélanger, M.; Allaman, I.; Magistretti, P.J. Brain Energy Metabolism: Focus on Astrocyte-Neuron Metabolic Cooperation. *Cell Metab.* **2011**, *14*, 724–738. [CrossRef]
5. Magistretti, P.J.; Allaman, I. A cellular perspective on brain energy metabolism and functional imaging. *Neuron* **2015**, *86*, 883–901. [CrossRef]
6. Falkowska, A.; Gutowska, I.; Goschorska, M.; Nowacki, P.; Chlubek, D.; Baranowska-Bosiacka, I. Energy Metabolism of the Brain, Including the Cooperation between Astrocytes and Neurons, Especially in the Context of Glycogen Metabolism. *Int. J. Mol. Sci.* **2015**, *16*, 25959–25981. [CrossRef] [PubMed]
7. Karbowska, J.; Kochan, Z. Role of adiponectin in the regulation of carbohydrate and lipid metabolism. *J. Physiol. Pharmacol.* **2006**, *57* (Suppl. 6), 103–113.
8. Kadowaki, T.; Yamauchi, T.; Kubota, N.; Hara, K.; Ueki, K.; Tobe, K. Adiponectin and adiponectin receptors in insulin resistance, diabetes, and the metabolic syndrome. *J. Clin. Investig.* **2006**, *116*, 1784–1792. [CrossRef] [PubMed]
9. Berg, A.H.; Combs, T.P.; Scherer, P.E. ACRP30/adiponectin: An adipokine regulating glucose and lipid metabolism. *Trends Endocrinol. Metab.* **2002**, *13*, 84–89. [CrossRef]
10. Tao, C.; Sifuentes, A.; Holland, W.L. Regulation of glucose and lipid homeostasis by adiponectin: Effects on hepatocytes, pancreatic β cells and adipocytes. *Best Pract. Res. Clin. Endocrinol. Metab.* **2014**, *28*, 43–58. [CrossRef] [PubMed]
11. Qiao, L.; Kinney, B.; Yoo, H.S.; Lee, B.; Schaack, J.; Shao, J. Adiponectin increases skeletal muscle mitochondrial biogenesis by suppressing mitogen-activated protein kinase phosphatase-1. *Diabetes* **2012**, *61*, 1463–1470. [CrossRef]
12. Kim, J.G.; Suyama, S.; Koch, M.; Jin, S.; Argente-Arizon, P.; Argente, J.; Liu, Z.-W.; Zimmer, M.R.; Jeong, J.K.; Szigeti-Buck, K.; et al. Leptin signaling in astrocytes regulates hypothalamic neuronal circuits and feeding. *Nat. Neurosci.* **2014**, *17*, 908–910. [CrossRef]
13. Rogatzki, M.J.; Ferguson, B.S.; Goodwin, M.L.; Gladden, L.B. Lactate is always the end product of glycolysis. *Front. Neurosci.* **2015**, *9*, 22. [CrossRef] [PubMed]
14. Miselis, R.R.; Epstein, A.N. Feeding induced by intracerebroventricular 2-deoxy-D-glucose in the rat. *Am. J. Physiol.* **1975**, *229*, 1438–1447. [CrossRef] [PubMed]
15. Qi, Y.; Takahashi, N.; Hileman, S.M.; Patel, H.R.; Berg, A.H.; Pajvani, U.B.; Scherer, P.E.; Ahima, R.S. Adiponectin acts in the brain to decrease body weight. *Nat. Med.* **2004**, *10*, 524–529. [CrossRef]
16. Sainsbury, A.; Cooney, G.J.; Herzog, H. Hypothalamic regulation of energy homeostasis. *Best Pract. Res. Clin. Endocrinol. Metab.* **2002**, *16*, 623–637. [CrossRef] [PubMed]
17. Ahima, R.S.; Lazar, M.A. Adipokines and the peripheral and neural control of energy balance. *Mol. Endocrinol.* **2008**, *22*, 1023–1031. [CrossRef] [PubMed]
18. Ahima, R.S.; Qi, Y.; Singhal, N.S. Adipokines that link obesity and diabetes to the hypothalamus. *Prog. Brain Res.* **2006**, *153*, 155–174.
19. Tu, T.H.; Nam-Goong, I.S.; Lee, J.; Yang, S.; Kim, J.G. Visfatin Triggers Anorexia and Body Weight Loss through Regulating the Inflammatory Response in the Hypothalamic Microglia. *Mediat. Inflamm.* **2017**, *2017*, 1958947. [CrossRef]
20. Lee, H.; Tu, T.H.; Park, B.S.; Yang, S.; Kim, J.G. Adiponectin Reverses the Hypothalamic Microglial Inflammation during Short-Term Exposure to Fat-Rich Diet. *Int. J. Mol. Sci.* **2019**, *20*, 5738. [CrossRef] [PubMed]
21. Ouchi, N.; Walsh, K. Adiponectin as an anti-inflammatory factor. *Clin. Chim. Acta* **2007**, *380*, 24–30. [CrossRef] [PubMed]
22. Suyama, S.; Maekawa, F.; Maejima, Y.; Kubota, N.; Kadowaki, T.; Yada, T. Glucose level determines excitatory or inhibitory effects of adiponectin on arcuate POMC neuron activity and feeding. *Sci. Rep.* **2016**, *6*, 1–9. [CrossRef] [PubMed]
23. Blouet, C.; Schwartz, G.J. Hypothalamic nutrient sensing in the control of energy homeostasis. *Behav. Brain Res.* **2010**, *209*, 1–12. [CrossRef]
24. Thevenet, J.; De Marchi, U.; Domingo, J.S.; Christinat, N.; Bultot, L.; Lefebvre, G.; Sakamoto, K.; Descombes, P.; Masoodi, M.; Wiederkehr, A. Medium-chain fatty acids inhibit mitochondrial metabolism in astrocytes promoting astrocyte-neuron lactate and ketone body shuttle systems. *FASEB J.* **2016**, *30*, 1913–1926. [CrossRef] [PubMed]
25. Gao, Y.; Layritz, C.; Legutko, B.; Eichmann, T.O.; Laperrousaz, E.; Moullé, V.S.; Cruciani-Guglielmacci, C.; Magnan, C.; Luquet, S.; Woods, S.C.; et al. Disruption of Lipid Uptake in Astroglia Exacerbates Diet-Induced Obesity. *Diabetes* **2017**, *66*, 2555–2563. [CrossRef] [PubMed]
26. Barres, B.A. The mystery and magic of glia: A perspective on their roles in health and disease. *Neuron* **2008**, *60*, 430–440. [CrossRef] [PubMed]

27. Tu, T.H.; Kim, H.; Yang, S.; Kim, J.K.; Kim, J.G. Linoleic acid rescues microglia inflammation triggered by saturated fatty acid. *Biochem. Biophys. Res. Commun.* **2019**, *513*, 201–206. [CrossRef]
28. Kang, S.S.; Ebbert, M.T.W.; Baker, K.E.; Cook, C.; Wang, X.; Sens, J.P.; Kocher, J.-P.; Petrucelli, L.; Fryer, J.D. Microglial translational profiling reveals a convergent APOE pathway from aging, amyloid, and tau. *J. Exp. Med.* **2018**, *215*, 2235–2245. [CrossRef]
29. Sanz, E.; Yang, L.; Su, T.; Morris, D.R.; McKnight, G.S.; Amieux, P.S. Cell-type-specific isolation of ribosome-associated mRNA from complex tissues. *Proc. Natl. Acad. Sci. USA* **2009**, *106*, 13939–13944. [CrossRef]
30. Jin, S.; Kim, K.K.; Park, B.S.; Kim, D.H.; Jeong, B.; Kang, D.; Lee, T.H.; Park, J.W.; Kim, J.G.; Lee, B.J. Function of astrocyte MyD88 in high-fat-diet-induced hypothalamic inflammation. *J. Neuroinflamm.* **2020**, *17*, 195. [CrossRef]

International Journal of
Molecular Sciences

MDPI

Article

A Negative Energy Balance Is Associated with Metabolic Dysfunctions in the Hypothalamus of a Humanized Preclinical Model of Alzheimer's Disease, the 5XFAD Mouse

Antonio J. López-Gambero [1,2,3], Cristina Rosell-Valle [1], Dina Medina-Vera [1,2,3,4,5], Juan Antonio Navarro [1,2,4], Antonio Vargas [1,2], Patricia Rivera [1,2], Carlos Sanjuan [6], Fernando Rodríguez de Fonseca [1,2,*] and Juan Suárez [1,4,7,*]

1 Instituto de investigación Biomédica de Málaga-IBIMA, 29010 Málaga, Spain; antonio.lopez@ibima.eu (A.J.L.-G.); cristina.rosell@ibima.eu (C.R.-V.); dina.medina@ibima.eu (D.M.-V.); juan_naga@hotmail.es (J.A.N.); antonio.vargas@ibima.eu (A.V.); patricia.rivera@ibima.eu (P.R.)
2 UGC Salud Mental, Hospital Regional Universitario de Málaga, 29010 Málaga, Spain
3 Departamento de Biología Celular, Genética y Fisiología, Campus de Teatinos s/n, Universidad de Málaga, Andalucia Tech, 29071 Málaga, Spain
4 Facultad de Medicina, Campus de Teatinos s/n, Universidad de Málaga, Andalucia Tech, 29071 Málaga, Spain
5 UGC Corazón, Hospital Universitario Virgen de la Victoria, 29010 Málaga, Spain
6 EURONUTRA S.L, Parque Tecnológico de Andalucía, 29590 Campanillas, Spain; euronutra@euronutra.eu
7 Departamento de Anatomía Humana, Medicina Legal e Historia de la Ciencia, Universidad de Málaga, 29071 Málaga, Spain
* Correspondence: fernando.rodriguez@ibima.eu (F.R.d.F.); juan.suarez@uma.es (J.S.); Tel.: +34-952614012 (F.R.d.F.); +34-952133433 (J.S.)

Citation: López-Gambero, A.J.; Rosell-Valle, C.; Medina-Vera, D.; Navarro, J.A.; Vargas, A.; Rivera, P.; Sanjuan, C.; Rodríguez de Fonseca, F.; Suárez, J. A Negative Energy Balance Is Associated with Metabolic Dysfunctions in the Hypothalamus of a Humanized Preclinical Model of Alzheimer's Disease, the 5XFAD Mouse. *Int. J. Mol. Sci.* **2021**, *22*, 5365. https://doi.org/10.3390/ijms22105365

Academic Editors: Rosalía Rodríguez-Rodríguez and Cristina Miralpeix

Received: 30 April 2021
Accepted: 16 May 2021
Published: 20 May 2021

Abstract: Increasing evidence links metabolic disorders with neurodegenerative processes including Alzheimer's disease (AD). Late AD is associated with amyloid (Aβ) plaque accumulation, neuroinflammation, and central insulin resistance. Here, a humanized AD model, the 5xFAD mouse model, was used to further explore food intake, energy expenditure, neuroinflammation, and neuroendocrine signaling in the hypothalamus. Experiments were performed on 6-month-old male and female full transgenic ($Tg^{5xFAD/5xFAD}$), heterozygous ($Tg^{5xFAD/-}$), and non-transgenic (Non-Tg) littermates. Although histological analysis showed absence of Aβ plaques in the hypothalamus of 5xFAD mice, this brain region displayed increased protein levels of GFAP and IBA1 in both $Tg^{5xFAD/-}$ and $Tg^{5xFAD/5xFAD}$ mice and increased expression of IL-1β in $Tg^{5xFAD/5xFAD}$ mice, suggesting neuroinflammation. This condition was accompanied by decreased body weight, food intake, and energy expenditure in both $Tg^{5xFAD/-}$ and $Tg^{5xFAD/5xFAD}$ mice. Negative energy balance was associated with altered circulating levels of insulin, GLP-1, GIP, ghrelin, and resistin; decreased insulin and leptin hypothalamic signaling; dysregulation in main metabolic sensors (phosphorylated IRS1, STAT5, AMPK, mTOR, ERK2); and neuropeptides controlling energy balance (NPY, AgRP, orexin, MCH). These results suggest that glial activation and metabolic dysfunctions in the hypothalamus of a mouse model of AD likely result in negative energy balance, which may contribute to AD pathogenesis development.

Keywords: Alzheimer's disease; 5xFAD; insulin signaling; energy expenditure; hypothalamus; neuroinflammation

1. Introduction

Physical decline and impairment of metabolic homeostasis are the main features of the human aging process [1]. The dysregulation of the metabolic network leads to an age-related elevated risk of suffering from chronic metabolic disorders, especially insulin resistance-related pathologies. In addition to the well-known peripheral role for insulin on glucose and energy storage, insulin also regulates a series of cognitive processes such as memory formation through its effects on glial–neuronal metabolic coupling. Central insulin resistance is a common feature linked to premature aging and observed in neurological disorders, including early stages of Alzheimer's disease (AD) [2].

AD dementia is the main form of dementia in humans. AD is a progressive neurological disorder with a mostly sporadic origin that is characterized by the loss of cognitive functions such as memory, reasoning, or language, leading to death at about 3–9 years after diagnosis. Common features of AD are aberrant production and deposition of β-amyloid (Aβ) peptides, either in Aβ40 or Aβ42 fragments and Tau protein hyperphosphorylation aggregates (neurofibrillary tangles), disruption of normal autophagic processes, generation of reactive oxygen species (ROS) and reactive nitrogen species (RNS), and cell death [3,4]. The combination of the above factors is seen be cause for cognitive impairment and memory loss. Recent studies suggest that unhealthy dietary habits and microbiota changes mediate neuroinflammation, modify neurovascular coupling, and create metabolic disturbances deriving on oxidate/nitrosative stress and insulin resistance. All these factors have been identified to contribute to the cognitive decline in AD patients.

Studies have shown AD progression is linked to dysregulated insulin signaling in the frontal cortex and hippocampus, and postmortem analysis of human hippocampal tissue shows correlation between high serine-inhibitory phosphorylation of insulin receptor substrate 1 (IRS1) and those of oligomeric Aβ plaques, which were negatively associated with working memory and episodic memory [5]. Further studies have shown that early hyperactivation of insulin signaling may cause negative feedback mediated by energy sensors such as mTOR, thus leading to impaired metabolic sensing in the neuronal population [6,7].

Aside from metabolic impairment, epidemiological studies have long appreciated weight loss associated with development of neurodegenerative diseases such as AD in the elderly [8]. Clinical data shows that weight loss precedes development of dementia in AD and may be a preclinical indicator of the disease [9]. Increased presence of AD markers (Aβ and total Tau) in cerebrospinal fluid has been correlated to decreased body mass index (BMI), which reinforces the link between body weight and metabolic disturbances with development of dementia and AD [10]. This negative energy status in AD patients with disturbed insulin signaling in the brain may exacerbate AD pathology. However, despite the age-associated frailty, the molecular and physiological mechanisms deriving from weight loss and metabolic impairment in AD patients are still poorly described, and data on neuroendocrine changes preceding AD are scarce.

The hypothalamus is the key regulator of metabolic processes in the organism. Food intake and energy expenditure are controlled by neuronal populations secreting orexigenic neuropeptides NPY and AgRP (responsible for promoting appetite and decreasing metabolic rate) and anorexigenic neuropeptides POMC and CART (deriving on satiety and promoting energy expenditure). Studies in AD mouse model 3xtg, which enhances Tau hyperphosphorylation, have shown inflammation and loss of neuronal population controlling energy expenditure and food intake such as POMC- and NPY-expressing neurons in 6-month-old mice [11]. Other studies have shown a hypermetabolic state in 3xtg mice, as observed by increased food intake and energy expenditure in young mice, whereas energy expenditure is significantly decreased in 3xtg mice at 18 months old, accompanied by decreased body weight [12]. NPY electrophysiological responses were also observed in another mouse model of AD overexpressing the Swedish mutation of APP (Tg2576) [13]. Studies in mice have helped gain insight into how hypothalamic dysfunction and metabolic disturbances play an early role in AD development. However, exact changes and molecular mechanisms underlying these events are still to be discerned.

Changes in brain metabolism in humans have been observed to be distinct from males and females, as the female brain undergoes earlier signals of hypometabolic status preceding and stating susceptibility in developing cognitive decline and AD [14]. Given differences in brain metabolic rate between males and females, sex differences should also be regarded as a main factor in the development of AD. However, there is a lack of studies on sex differences regarding the use of humanized preclinical models of AD.

A relevant issue on the clinical approach to AD and related pathologies that lead to cognitive impairment is the fact that most of the research effort on therapeutics have focused on either fighting the symptoms by boosting certain deteriorated transmission pathways (as it is the example of anti-acetylcholinesterase drugs to enhance cholinergic transmission) or reducing Aβ load via immunotherapy. However, there is a clear lack of therapeutic development designed to restore metabolic impairments associated with these neurodegenerative disorders. In our study, we aimed to investigate food intake, energy expenditure, and neuroinflammation and the mechanisms regulating these processes in the hypothalami of 6-month-old male and female 5xFAD (FAD: family Alzheimer's disease) transgenic mouse model of AD, which is one of the most early onset AD models of neurodegeneration and amyloid pathology, where cognitive decline starts by age of 4 months increases through age [15–17]. Biomarkers of neuroinflammation, metabolic-regulating hormones, insulin signaling, energy sensors, and neuropeptide levels were assessed in an attempt to gain insight into how energy balance is disrupted, contributing to development of AD pathology.

2. Results

2.1. Heterozygous ($Tg^{5xFAD/-}$) and Homozygous ($Tg^{5xFAD/5xFAD}$) Transgenic 5xFAD Mice Showed Decreased Body Weight, Food Intake, and Energy Expenditure at 6 Months

Since the first signs of glucose metabolism decline in 5xFAD mice brain regions are observed at 6 months [18], we assessed body weight and metabolic parameters in 5xFAD at this age. At 6 months old, two-way ANOVA showed no interactions between genotype and sex ($F_{2,43} = 2.766$; $p = 0.0741$), but a main effect of sex ($F_{1,43} = 120.6$; $p < 0.001$) and genotype $F_{2,43} = 5.182$; $p = 0.0096$) (Figure 1A). Multiple comparisons by Tukey's test showed decreased body weight in male $Tg^{5xFAD/5xFAD}$ as compared to non-Tg ($p < 0.01$) and $Tg^{5xFAD/-}$ ($p < 0.01$). Both female $Tg^{5xFAD/5xFAD}$ and $Tg^{5xFAD/-}$ also showed significant decrease in body weight as compared to non-Tg females ($p < 0.05$) (Figure 1A).

Next, we aimed to determine if these defects in body weight were related to changes in food intake patterns and energy expenditure (EE) for 48 h as they were placed in metabolic cages. Two-way ANOVA showed the main genotype effect on food intake per body weight during light phase (day) ($F_{2,47} = 4.096$; $p = 0.0229$) (Figure 1B) and dark phase (night) ($F_{2,48} = 15.44$; $p < 0.001$) (Figure 1C). Male $Tg^{5xFAD/5xFAD}$ showed a slight increase in food intake as compared to $Tg^{5xFAD/-}$ mice during the day (Tukey's test: $p < 0.05$) (Figure 1B). However, during the night, male $Tg^{5xFAD/5xFAD}$ showed a great decrease in food intake as compared to non-Tg (Tukey's test: $p < 0.001$) (Figure 1C). Both female $Tg^{5xFAD/-}$ and $Tg^{5xFAD/5xFAD}$ mice also showed decreased food intake as compared to non-Tg females (Tukey's test: $p < 0.01$ and $p < 0.001$, respectively) (Figure 1C).

EE is calculated by indirect calorimetry, using VO_2 and VCO_2 flow rates and normalized to body weight (BW) and raised to the power 0.75, since small animals such as mice show greater basal energy expenditure than larger animals such as humans (Figure 1D–G). No interaction effect was observed by two-way ANOVA, but there was a main effect of genotype during the day ($F_{2,44} = 3.227$; $p = 0.0492$) and night ($F_{2,44} = 3.661$; $p = 0.0338$), as well as a sex main effect during the day ($F_{1,44} = 17.41$; $p = 0.001$) and night ($F_{1,44} = 17.42$; $p = 0.001$), as female mice usually have a greater metabolic rate. Tukey's test showed only $Tg^{5xFAD/5xFAD}$ mice had slight decreases in EE, as shown by male mice during the day ($p < 0.05$ vs. non-Tg group), as well as being more pronounced in female mice during the day ($p < 0.05$ vs. Non-Tg and $Tg^{5xFAD/-}$ groups) and night ($p < 0.05$ vs. Non-Tg and $Tg^{5xFAD/-}$ groups). Pattern of activity and rearing was also different between males and females during the night (Figure S1A–D), as shown by main sex effect ($F_{1,64} = 8.305$, $p = 0.0054$ for night activity; $F_{1,64} = 16.41$, $p = 0.001$ for night rearing), showing decreased night activity in $Tg^{5xFAD/5xFAD}$ females compared to non-Tg females ($p < 0.05$) (Figure S1B).

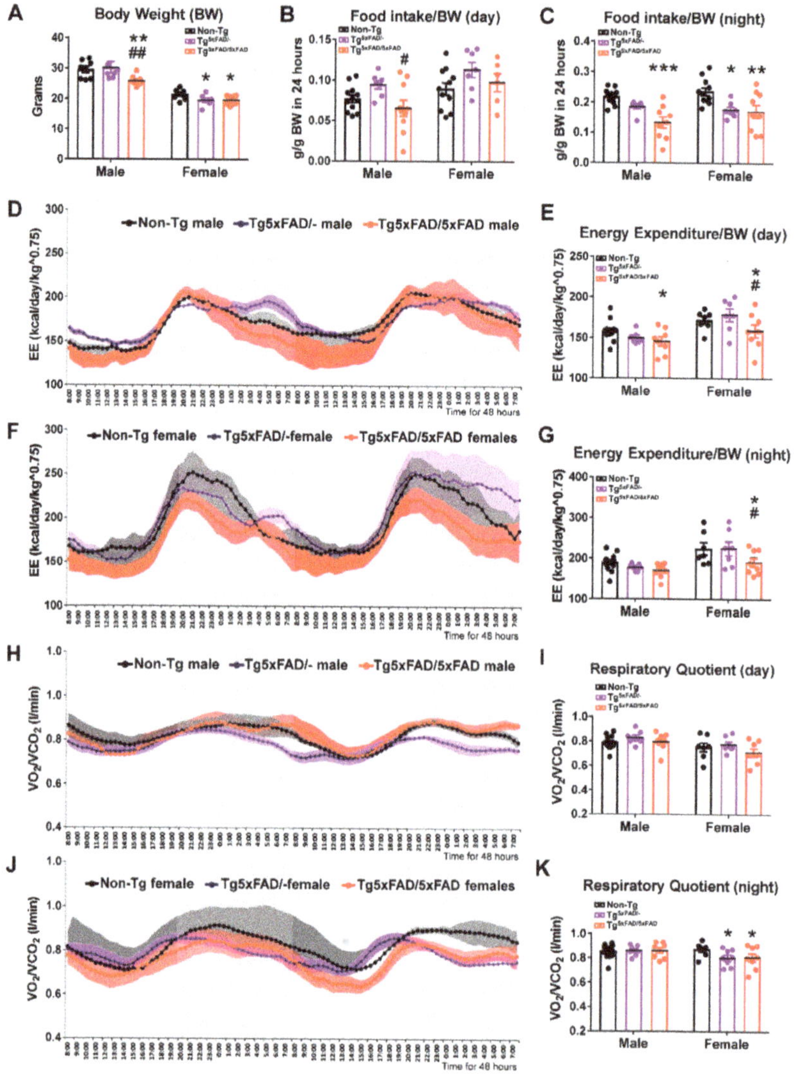

Figure 1. Negative energy balance in 5xFAD mice was associated with decreased food intake and energy expenditure for 48 h. (**A**) Body weight showing significantly decreased body weight in $Tg^{5xFAD/5xFAD}$ males, $Tg^{5xFAD/-}$ females, and $Tg^{5xFAD/5xFAD}$ females at 6 months of age. (**B**,**C**) Decreased food intake per body weight in $Tg^{5xFAD/5xFAD}$ males during the day (light phase, 8 a.m. to 8 p.m.) and night (dark phase, 8 p.m. to 8 a.m.). Decreased food intake also occurred in $Tg^{5xFAD/-}$ females and $Tg^{5xFAD/5xFAD}$ females during the night. (**D**–**G**) Energy expenditure normalized per body weight (EE/BW) in males and females, showing decreased mean EE/BW in $Tg^{5xFAD/5xFAD}$ males during the day, and in $g^{5xFAD/5xFAD}$ females during the day and night. (**H**–**K**) Respiratory quotient showing decreased ratio of vO2/vCO2 in $Tg^{5xFAD/-}$ females and $Tg^{5xFAD/5xFAD}$ females during the night, indicating decreased glucose utilization/increased fatty acid oxidation as energy source. n = 7–15 per group. Two-way ANOVA analysis with Tukey's post hoc test: * = $p < 0.05$, ** = $p < 0.01$, *** = $p < 0.001$ versus same-sex non-Tg group; # = $p < 0.05$, ## = $p < 0.01$ versus same-sex $Tg^{5xFAD/-}$ group.

Respiratory quotient (RQ) as calculated by VCO_2/VO_2 indicates rate of glucose or fat fuel utilization, since lower RQ measurements correlate to higher O_2 consumption in lipid β-oxidation (Figure 1H–K). Two-way ANOVA analysis determined only sex effect during the day ($F_{1,44} = 7.176$; $p = 0.0104$). However, both female $Tg^{5xFAD/-}$ and $Tg^{5xFAD/5xFAD}$ mice presented lower RQ during the night (Tukey's test; $p < 0.05$ vs. non-Tg group) (Figure 1K).

2.2. Heterozygous ($Tg^{5xFAD/-}$) and Homozygous ($Tg^{5xFAD/5xFAD}$) Transgenic 5xFAD Mice Showed Increased Hypothalamic Tau Phosphorylation and Inflammation Despite Absence of Aβ Plaques in the Hypothalamus

We investigated if changes in body weight and metabolic parameters were related to abnormalities in the hypothalamus, as it is the main brain region responsible for control of energy homeostasis [19]. There were no visible Aβ plaques in the hypothalamus of none of the three genotypes (Figure 2A–F). However, the presence of Aβ plaques was significantly found in the hippocampus of both $Tg^{5xFAD/-}$ and $Tg^{5xFAD/5xFAD}$ male and female mice (Figure 2G), indicating amyloid pathology affecting the brain region responsible for learning and memory, exhibiting a projection to the hypothalamus via fimbria-fornix bundle.

Despite the absence of Aβ plaques, as observed by immunohistochemical sections, we aimed to determine by Western blotting if there were changes in Tau protein activity in the hypothalamus, as its hyperphoshorylation that leads to formation of toxic microtubule neurofibrillary tangles (NFTs), a hallmark of AD [5]. We also evaluated changes in protein levels of Tau kinases (Figure 2H–N). Tau is seen to be a target of the protein kinase CDK5, a member of the cyclin-dependent kinases (Cdks). CDK5 is activated physiologically via p35. However, in pathological conditions, as observed in AD, CDK5 is hyperactivated by p25, a truncated form of p35, leading to hyperphosphorylation [20].

Two-way ANOVA analysis revealed a significant main effect of genotype in Tau-AT8 phosphorylation ($F_{2,28} = 5.22$; $p = 0.0118$), CDK5 protein levels ($F_{2,28} = 6.353$; $p = 0.0053$), and p25 protein levels ($F_{2,28} = 10.72$; $p < 0.001$), but also main sex effect in CDK5 ($F_{1,28} = 32.17$; $p < 0.001$) and p25 protein levels ($F_{1,28} = 17.04$; $p < 0.001$). Interaction was also observed in CDK5 ($F_{2,28} = 6.438$; $p = 0.005$). Multiple comparison showed increased Tau-AT8 hyperphosphorylation in both male and female $Tg^{5xFAD/5xFAD}$ compared to Non-Tg ($p < 0.05$) (Figure 2G). Main sex effect was observed in kinase activity in females, as both $Tg^{5xFAD/-}$ and $Tg^{5xFAD/5xFAD}$ female mice showed increased CDK5 protein levels (Tukey's test; $p < 0.001$ and $p < 0.05$, respectively) and p25 ($p < 0.001$ and $p < 0.05$, respectively) as compared to Non-Tg females (Figure 2I,L). These results suggest male hyperactivation of Tau may be regulated by different kinase activities. Intriguingly, despite high CDK5 and p25 protein levels in heterozygous $Tg^{5xFAD/-}$ females, no significant increase in Tau activity or protein levels were observed (Figure 2H).

We also evaluated protein expression of LRP1, a receptor that mediates Aβ internalization and degradation in the brain, as well as Aβ clearance across the blood–brain barrier (BBB) [21]. Intriguingly, LRP1 levels were significantly increased in $Tg^{5xFAD/-}$ (Figure 2M) but not $Tg^{5xFAD/5xFAD}$ male mice (Tukey's test; $p < 0.01$), and non-significantly in $Tg^{5xFAD/-}$ female mice ($p = 0.146$). These results suggest heterozygous 5xFAD mice could exhibit a more marked Aβ clearance as compared to homozygous 5xFAD mice.

As a means to assess pathogenicity in the hypothalamus, we determined mRNA and protein expression of inflammatory cytokine neuroinflammatory markers (Figure 3A–G). Two-way ANOVA analysis showed the main genotype effect on *Tnf* ($F_{2,31} = 6.73$; $p = 0.0037$) and *Il1b* ($F_{2,31} = 7.198$; $p = 0.0027$) mRNA levels, but no significant changes in expression of *Il6*. It should be noted that inflammatory response observed was different between genotypes, since the expression of *Tnf* was significantly elevated in $Tg^{5xFAD/-}$ mice but not in $Tg^{5xFAD/5xFAD}$ and, on the contrary, that of *Il1b* was augmented in $Tg^{5xFAD/5xFAD}$ but not in $Tg^{5xFAD/-}$ as compared to Non-Tg mice (Figure 3A,B). Protein expression of glia (GFAP) and microglia/macrophage (IBA1) markers was also significantly elevated in $Tg^{5xFAD/-}$ and $Tg^{5xFAD/5xFAD}$ mice (Figure 3D,F). However, Il-1β protein levels were only significantly increased in female $Tg^{5xFAD/5xFAD}$ mice (Figure 3E). These results showcase that both heterozygous $Tg^{5xFAD/-}$ and homozygous $Tg^{5xFAD/5xFAD}$ showed an inflammatory response in the hypothalamus, despite absence of Aβ immunoreactivity, with little differences between males and females and a different pattern of cytokine release profile.

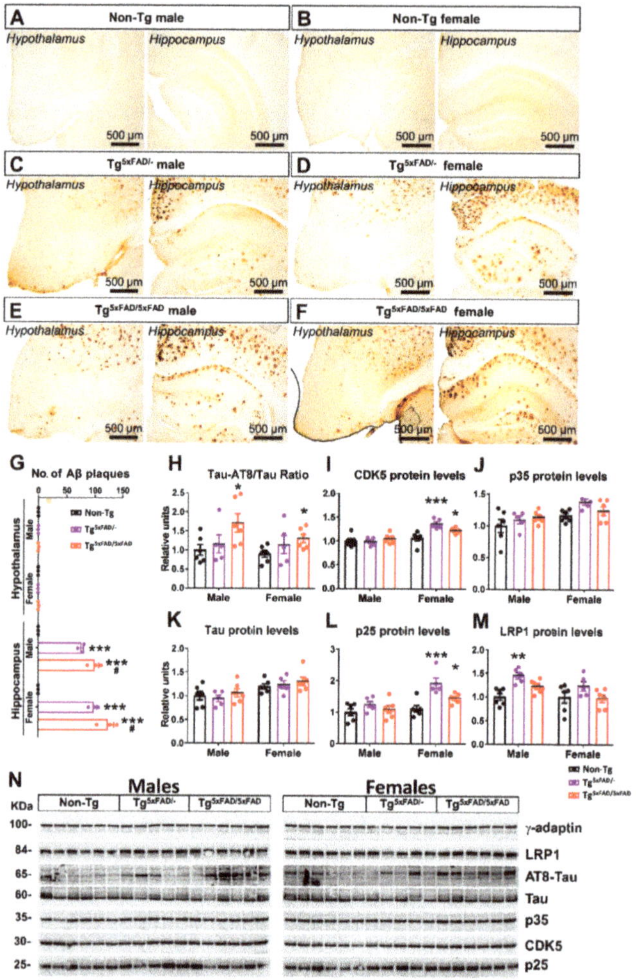

Figure 2. Absence of amyloid-β 42 fragments in hypothalamus of 5xFAD mice at 6 months of age but increased hypothalamic Tau phosphorylation. (**A–F**) Immunohistochemical sections of Aβ_{42} in the hypothalamus and hippocampus of 5xFAD mice, showing the presence of Aβ_{42} plaques in hippocampus of Tg$^{5xFAD/-}$ and Tg$^{5xFAD/-}$ males and females, but in the absence in the hypothalamus in all the groups. (**G**) Quantification of total Aβ plaques in the hypothalamus and hippocampus of 5xFAD mice, showing the absence of Aβ in hypothalamic sections and higher total Aβ content in both male and female Tg$^{5xFAD/5xFAD}$ mice with respect to Tg$^{5xFAD/+}$ mice. (**H–M**) Densitometric evaluation of Tau activation (Tau-AT8/Tau), CDK5 protein levels, p35 protein levels, Tau protein levels, p25 protein levels, and LRP1 protein levels (n = 5–6 per group). Protein levels were normalized with γ-adaptin. Associated phosphorylations were normalized with respective total protein levels. Non-Tg males were set as 1 for protein relative units. (**N**): representative Western blot images. Two-way ANOVA analysis with Tukey's post hoc test: * = $p < 0.05$, ** = $p < 0.01$, *** = $p < 0.001$ versus same-sex non-Tg group; # = $p < 0.05$ versus same-sex Tg$^{5xFAD/-}$ group.

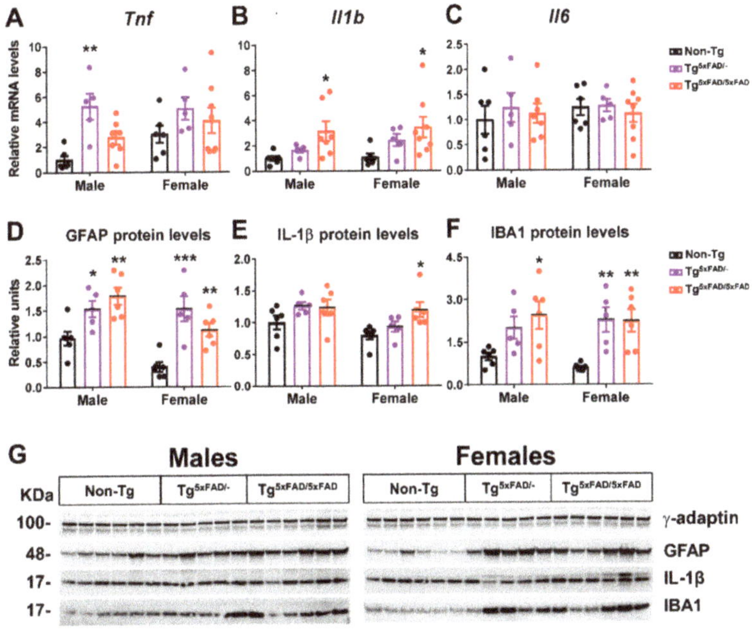

Figure 3. Inflammatory markers were present in the hypothalamus of 5xFAD mice at 6 months of age. (**A–C**) Representative quantification of expression of cytokine genes (TNF-α, IL-1β, IL-6) in hypothalamus assessed by qPCR (n = 5–8 per group). (**D–F**) Representation of Western blot membranes and (**G**) densitometric evaluation of GFAP, IL-1β, and IBA1 protein levels (n = 5–6 per group). Protein levels were normalized with γ-adaptin. Non-Tg males were set as 1 for relative mRNA units and protein relative units. Two-way ANOVA analysis with Tukey's post hoc test: * = $p < 0.05$, ** = $p < 0.01$, *** = $p < 0.001$ versus same-sex non-Tg group.

Correlation test was run to assess if neuroinflammation was directly associated with decreased body weight in 5xFAD mice (Table 1). We found no correlation between body weight and neuroinflammatory marker levels (IL-1β, TNF-α, GFAP, and IBA1), which were significantly elevated in male mice of either genotype. In Non-Tg females, body weight was negatively correlated to TNF-α ($p < 0.05$) and IBA1 ($p < 0.05$). In $Tg^{5xFAD/-}$ females, IL-1β ($p < 0.05$) and IBA1 ($p < 0.05$) were also related to body weight. $Tg^{5xFAD/5xFAD}$ female body weights showed more associations with IL-1β ($p < 0.05$), TNF-α ($p < 0.05$), and IBA1 ($p < 0.05$).

Table 1. Correlations between body weight and hypothalamic neuroinflammatory markers in 5xFAD mice at 6 months of age.

	Males			Females		
Body Weight Versus:	**Non-Tg**	**$Tg^{5xFAD/-}$**	**$Tg^{5xFAD/5xFAD}$**	**Non-Tg**	**$Tg^{5xFAD/-}$**	**$Tg^{5xFAD/5xFAD}$**
Il-1β (hyp)	ns [1]	ns	ns	ns	R^2 = 0.820 p = 0.034	R^2 = 0.593 p = 0.025
TNF-α (hyp)	ns	ns	ns	R^2 = 0.799 p = 0.016	R^2 = 0.713 p = 0.071	R^2 = 0.548 p = 0.035
GFAP (hyp)	ns	ns	ns	ns	ns	ns
IBA1 (hyp)	ns	ns	ns	R^2 = 0.769 p = 0.021	R^2 = 0.783 p = 0.046	R^2 = 0.783 p = 0.019

[1] ns, not significant.

2.3. Transgenic 5xFAD Mice Had Low Circulating Levels of Insulin, GLP-1, Ghrelin, and Resistin, as Well as Altered Activity and Expression of Hormone Receptors in the Hypothalamus

Brain insulin resistance is a pattern observed in mice models of AD. Previous studies have observed low plasma levels of leptin in a different mouse model of AD [13]. In addition, the pancreas expresses amyloid precursor protein (APP), and there is a suspicion of a contribution of Aβ-induced damage in the pancreas to insulin resistance-linked alterations in AD [22]. Thus, we wanted to determine if metabolic dysfunction was associated with either increased or decreased plasma levels of insulin, insulin secretagogues, or adipokines, as well as altered hormone receptor activity in the hypothalamus, in order to explain metabolic changes in these animals. We found significant genotype differences in basal plasma levels of insulin (two-way ANOVA; $F_{2,33} = 10.26$; $p < 0.001$), with a marked decrease in plasma insulin $Tg^{5xFAD/-}$ and $Tg^{5xFAD/5xFAD}$ mice (Figure 4A). These results were accompanied by decreased mRNA expression of insulin receptor (*Insr*) in female $Tg^{5xFAD/-}$ and $Tg^{5xFAD/5xFAD}$ mice (Figure 4D). Activity of insulin receptor as measured by tyrosine phosphorylation (p-IR/IR protein levels ratio) was not significantly different between genotypes, although a non-significant decrease was observed again in female $Tg^{5xFAD/-}$ and $Tg^{5xFAD/5xFAD}$ mice (Figure 4G,J).

GLP-1 is known to promote glucose-stimulated insulin release, but its effects in absence of glucose bolus are under discussion [23]. Nevertheless, basal insulin levels correlate to those of basal GLP-1, and decreased GLP-1 plasma is observed after weight loss [24]. Interestingly, plasma levels of GLP-1 were decreased in female $Tg^{5xFAD/5xFAD}$ mice (Figure 4B). Regarding GLP-1 activity, according to insulin and insulin receptor levels, we also observed decreased GLP-1 receptor (*Glp1r*) mRNA expression in the hypothalamus of $Tg^{5xFAD/-}$ and $Tg^{5xFAD/5xFAD}$ mice (Figure 4E) with a significant genotype main effect ($F_{1,33} = 8.042$; $p = 0.0078$) and a significant decrease in GLP1-R protein levels in male $Tg^{5xFAD/5xFAD}$ mice as compared to $Tg^{5xFAD/-}$ and non-Tg male mice (Figure 4H,J). These results over GLP-1 activity in the hypothalamus may agree with an impaired insulin response. Despite this, we also observed decreased plasma levels of ghrelin in female $Tg^{5xFAD/5xFAD}$ mice (Figure 4C). Ghrelin reduces glucose-stimulated insulin secretion, and thus ghrelin levels would be expected to be either unchanged or slightly increased [25]. Since ghrelin is secreted in an autocrine fashion in the hypothalamus, we measured ghrelin (*Ghrl*) mRNA levels and found a slight increase in male $Tg^{5xFAD/5xFAD}$ mice as compared to $Tg^{5xFAD/-}$ mice (Figure 4F). However, ghrelin receptor (*Ghsr*) mRNA levels were significantly decreased in male but not female $Tg^{5xFAD/-}$ and $Tg^{5xFAD/5xFAD}$ mice (Figure 4I).

Since $Tg^{5xFAD/-}$ and $Tg^{5xFAD/5xFAD}$ mice show a leaner phenotype, we also measured profile of adipokines leptin and resistin. Basal levels of both hormones are normally elevated in obese mice. Sexual dimorphism is also seen, with plasma leptin being increased in male mice, whereas plasma resistin is higher in female mice [26]. In accordance with this, we found significant sex differences in plasma leptin (two-way ANOVA; $F_{1,33} = 8.826$; $p = 0.0055$) and resistin ($F_{1,33} = 29.36$; $p < 0.001$) and also genotype differences in resistin ($F_{2,33} = 9.612$; $p < 0.001$) (Figure 5A,B).

Despite a leaner phenotype, no changes in leptin receptor mRNA expression or protein activity (as determined by Tyr1077 phosphorylation) were observed, although there was a non-significant trend towards lower receptor signaling in the hypothalamus of $Tg^{5xFAD/-}$ and $Tg^{5xFAD/5xFAD}$ female mice (Figure 5D,G,H). Resistin is regarded as a promoter of obesity-mediated hypothalamic inflammation via TLR4 receptor, being correlated to insulin resistance [27]. It was surprising to find decreased plasma levels of resistin in both male and female $Tg^{5xFAD/-}$ mice as compared to non-Tg mice (Figure 5B). We also measured central resistin (*Retn*) mRNA expression, which showed a marked decrease in female $Tg^{5xFAD/-}$ and $Tg^{5xFAD/5xFAD}$ mice as compared to non-Tg ($p < 0.05$ and $p < 0.05$, respectively) (Figure 5E). These results were not accompanied by changes on TLR4 (*Tlr4*) mRNA, but increased TLR4 protein levels were also observed in female $Tg^{5xFAD/-}$ ($p < 0.001$) and $Tg^{5xFAD/5xFAD}$ ($p < 0.05$) mice (Figure 5C,F,H). It is seen that autocrine or central resistin infusion interferes in food intake and peripheral insulin sensitivity fatty acid synthesis via modulation of neuropeptide NPY expression [28]. Changes in hypothalamic resistin expression and TLR4 levels may be more in accordance with a metabolic rather than inflammatory role in whole-body energy homeostasis.

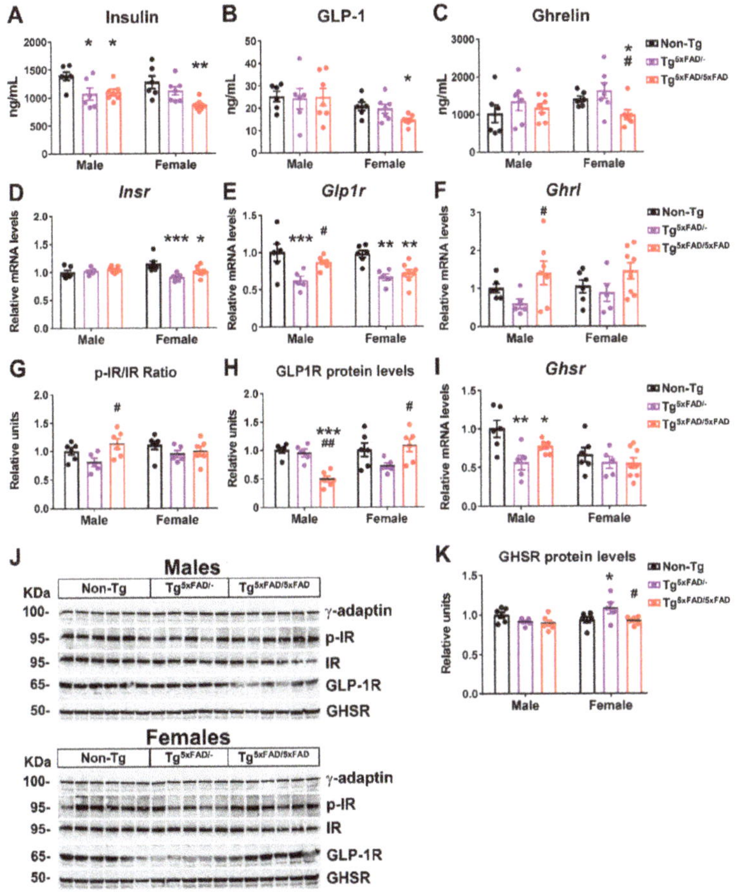

Figure 4. Altered insulin, leptin, and GLP-1 plasma levels and hypothalamic receptors were genotype and sex-specific in 5xFAD mice at 6 months of age. (**A–C**) Plasma levels of insulin, leptin, and GLP-1 (n = 6–7 per group). (**D–F**) Representative quantification of expression of insulin receptor (*Insr*), leptin receptor (*Lepr*), and GLP-1 receptor (*Glp1r*) in hypothalamus assessed by qPCR (n = 5–8 per group). (**G–I**) Representation of Western blot membranes, and (**J**) densitometric evaluation of insulin receptor (IR) activation (p-IR/IR), leptin receptor (LepR) activation (p-LepR/LepR), and GLP-1 receptor (GLP1R) protein levels (n = 5–6 per group). Protein levels were normalized with γ-adaptin. Associated phosphorylations were normalized with respective total protein levels. Non-Tg males were set as 1 for relative mRNA units and protein relative units. Two-way ANOVA analysis with Tukey's post hoc test: * = $p < 0.05$, ** = $p < 0.01$, *** = $p < 0.001$ versus same-sex non-Tg group; # = $p < 0.05$, ## = $p < 0.01$ versus same-sex $Tg^{5xFAD/-}$ group.

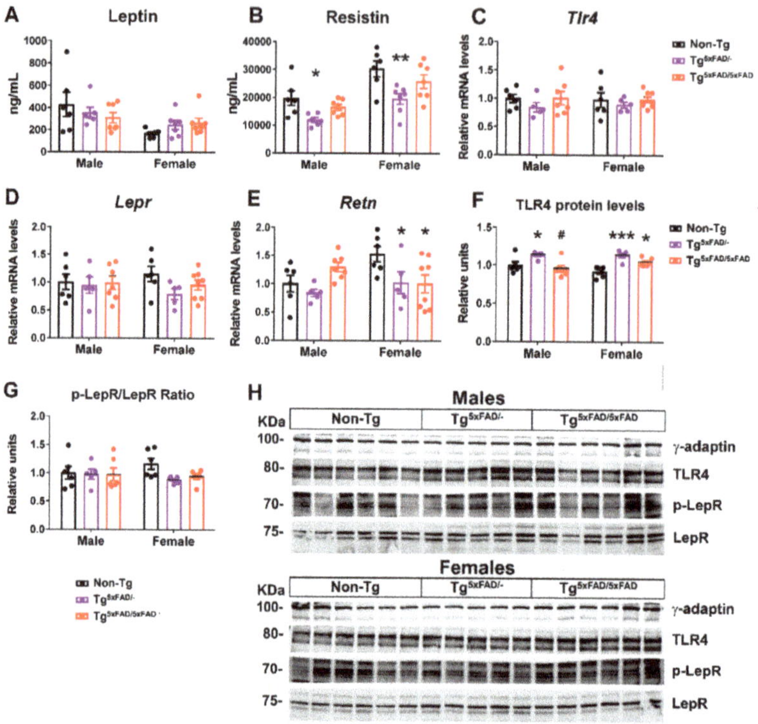

Figure 5. Decreased ghrelin and resistin plasma levels, with hypothalamic receptors being more specific in 5xFAD female mice at 6 months of age. (**A**,**B**) Plasma levels of ghrelin and resistin (n = 6–7 per group). (**C–F**) Representative quantification of expression of hypothalamic ghrelin (*Ghrl*), resistin (*Retn*), ghrelin receptor (*Ghsr*), and resistin putative receptor (toll-like receptor 4, *Tlr4*) assessed by qPCR (n = 5–8 per group). (**G–H**) Representation of Western blot membranes and (**I**) densitometric evaluation of ghrelin receptor (GHSR) and resistin putative receptor (TLR4) protein levels (n = 5–6 per group). Protein levels were normalized with γ-adaptin. Non-Tg males were set as 1 for relative mRNA units and protein relative units. Two-way ANOVA analysis with Tukey's post hoc test: * = $p < 0.05$, ** = $p < 0.01$, *** = $p < 0.001$ versus same-sex non-Tg group; # = $p < 0.05$ versus same-sex $Tg^{5xFAD/-}$ group.

Along with GLP-1, GIP also promotes glucose-stimulated insulin release. We found a main genotype effect for decreased GIP plasma levels ($F_{2,33} = 3.394$; $p = 0.0457$). GIP levels were lower in male $Tg^{5xFAD/-}$ and $Tg^{5xFAD/5xFAD}$ mice compared to non-Tg, but changes did not reach statistical significance in $Tg^{5xFAD/-}$ mice ($p = 0.010$ and $p < 0.05$, respectively) (Figure S2A). Glucagon and PAI-1 levels were, however, not significantly different between groups (Figure S2B,C).

Correlation test was run to assess if abnormal plasma and hypothalamic levels of hormones were related to decreased body weight in 5xFAD mice (Table 2). We found insulin levels were also directly related to body weight in $Tg^{5xFAD/-}$ males ($p < 0.05$), $Tg^{5xFAD/5xFAD}$ males ($p < 0.01$), $Tg^{5xFAD/-}$ females ($p < 0.05$), and $Tg^{5xFAD/5xFAD}$ females ($p < 0.05$). Leptin was also positively related to body weight only in $Tg^{5xFAD/-}$ males ($p < 0.001$) and $Tg^{5xFAD/5xFAD}$ males ($p < 0.05$), but not females, indicating low leptin levels were effectively lower with decreased body weight. Plasma levels of resistin were positively related to body weight in $Tg^{5xFAD/5xFAD}$ males ($p < 0.05$), $Tg^{5xFAD/-}$ females ($p < 0.01$), and $Tg^{5xFAD/5xFAD}$ females ($p < 0.05$). Hypothalamic resistin expression was also related to decreased body weight in $Tg^{5xFAD/-}$ females ($p < 0.01$) and $Tg^{5xFAD/5xFAD}$

females ($p < 0.05$). As for hormones regulating insulin secretion, ghrelin and GLP-1 were positively associated with body weight in $Tg^{5xFAD/5xFAD}$ females ($p < 0.05$) ($p < 0.05$ for both hormones), whereas GIP was positively associated with body weight in $Tg^{5xFAD/-}$ males ($p < 0.05$) and $Tg^{5xFAD/5xFAD}$ males ($p < 0.05$). Plasma levels of glucagon were not associated with body weight in any cases.

Table 2. Correlations between body weight and hypothalamic and plasma hormones in 5xFAD mice at 6 months of age.

	Males			**Females**		
Body Weight Versus:	**Non-Tg**	**$Tg^{5xFAD/-}$**	**$Tg^{5xFAD/5xFAD}$**	**Non-Tg**	**$Tg^{5xFAD/-}$**	**$Tg^{5xFAD/5xFAD}$**
Resistin (hyp)	$R^2 = 0.584$ $p = 0.076$	ns [1]	ns	ns	$R^2 = 0.920$ $p = 0.009$	$R^2 = 0.546$ $p = 0.036$
Ghrelin (hyp)	ns	ns	ns	ns	ns	ns
Resistin (plasma)	$R^2 = 0.594$ $p = 0.072$	ns	$R^2 = 0.704$ $p = 0.018$	ns	$R^2 = 0.894$ $p = 0.001$	$R^2 = 0.755$ $p = 0.011$
Ghrelin (plasma)	ns	ns	ns	ns	ns	$R^2 = 0.580$ $p = 0.046$
Insulin (plasma)	ns	$R^2 = 0.760$ $p = 0.023$	$R^2 = 0.851$ $p = 0.003$	ns	$R^2 = 0.708$ $p = 0.017$	$R^2 = 0.578$ $p = 0.047$
Leptin (plasma)	ns	$R^2 = 0.942$ p =0.001	$R^2 = 0.612$ $p = 0.037$	ns	ns	ns
GLP-1 (plasma)	ns	ns	ns	ns	ns	$R^2 = 0.622$ $p = 0.034$
GIP (plasma)	ns	$R^2 = 0.813$ $p = 0.014$	$R^2 = 0.644$ $p = 0.029$	ns	ns	ns
PAI-1 (plasma)	ns	ns	ns	ns	$R^2 = 0.628$ $p = 0.033$	$R^2 = 0.708$ $p = 0.017$
Glucagon (plasma)	ns	ns	ns	ns	ns	ns

[1] ns, not significant.

Correlation tests were also run to assess if decreased insulin levels were related to decreased incretins (GLP-1 and GIP) plasma levels or increased ghrelin and glucagon levels (Table 3). Interestingly, decreased plasma levels of insulin were differently associated in males and females, with insulin being positively associated with GIP in $Tg^{5xFAD/-}$ males ($p < 0.05$) and $Tg^{5xFAD/5xFAD}$ males ($p < 0.05$), whereas it was associated with plasma levels of GLP-1 in $Tg^{5xFAD/5xFAD}$ females ($p < 0.01$). No correlation was observed for plasma levels of glucagon or ghrelin.

Table 3. Correlations between plasma insulin and insulin release-regulating hormones in 5xFAD mice at 6 months of age.

	Males			**Females**		
Insulin (Plasma) Versus:	**Non-Tg**	**$Tg^{5xFAD/-}$**	**$Tg^{5xFAD/5xFAD}$**	**Non-Tg**	**$Tg^{5xFAD/-}$**	**$Tg^{5xFAD/5xFAD}$**
GIP	ns [1]	$R^2 = 0.712$ $p = 0.034$	$R^2 = 0.572$ $p = 0.048$	$R^2 = 0.659$ $p = 0.049$	ns	ns
GLP-1	ns	ns	ns	ns	ns	$R^2 = 0.846$ $p = 0.003$
Ghrelin	ns	ns	ns	ns	ns	ns
Glucagon	ns	ns	ns	$R^2 = 0.582$ $p = 0.077$	ns	$R^2 = 0.495$ $p = 0.077$

[1] ns, not significant.

2.4. Insulin and Leptin Signaling Was Decreased in Hypothalamus of Transgenic 5xFAD Mice at 6 Months of Age

The classical mechanisms of action in insulin and leptin signaling have been extensively described and reviewed [29,30]. The binding of insulin to its receptor (IR) in target tissues promotes tyrosine autophosphorylation, recruiting IR substrates as IRS1, whereas hypothalamic leptin receptor (LepR) actions modulating metabolism include the JAK-STAT3 and STAT5 signaling.

Since IRS1 activity is controlled by inhibitory and activating phosphorylations on serine and tyrosine residues, respectively, we determined the ratio of activity by determining the pTyr892/pSer612 ratio and total protein levels (Figure 6A,D,G). There was a main genotype effect to decrease IRS phosphorylation activity in 5xFAD mice (two-way ANOVA; $F_{2,28} = 8.865$; $p = 0.001$). Tukey's test also revealed significant decreased pTyr892/pSer612 ratio in $Tg^{5xFAD/-}$ and $Tg^{5xFAD/5xFAD}$ male mice compared to non-Tg males ($p < 0.05$; $p < 0.01$, respectively) and also in $Tg^{5xFAD/5xFAD}$ female mice compared to both $Tg^{5xFAD/-}$ and non-Tg females ($p < 0.05$; $p < 0.05$, respectively) (Figure 6A). There was also a main genotype effect observed in decreased IRS1 protein levels ($F_{2,28} = 6.404$; $p = 0.0051$) (Figure 6D).

We also determined activity and total protein level LepR signaling elements STAT5 and STAT3 (Figure 6B,C,E–G). Phosphorylation of Tyr1077 on LepR, which we measured, regulates primarily STAT5 [31]. In a similar trend to what we observed with pTyr1077 on LepR, there was a main genotype effect in a decreased STAT5 activating phosphorylation ($F_{2,28} = 4.752$; $p = 0.0167$). Again, Tukey's test revealed particular significant STAT5 phosphorylation in both $Tg^{5xFAD/-}$ and $Tg^{5xFAD/5xFAD}$ female mice compared to non-Tg females ($p < 0.05$; $p < 0.01$, respectively) (Figure 6B). Intriguingly, there were no changes in STAT3 phosphorylation (Figure 6C,G), but we observed a main sex and genotype effect and interaction in increased STAT3 protein levels (two-way ANOVA; $F_{1,28} = 20.78$, $p < 0.001$; $F_{2,28} = 14.32$, $p < 0.001$; $F_{2,28} = 14.78$, $p < 0.001$, respectively). Tukey's test showed $Tg^{5xFAD/5xFAD}$ males had higher STAT3 protein levels than $Tg^{5xFAD/-}$ ($p < 0.001$) and non-Tg ($p < 0.01$) males, whereas $Tg^{5xFAD/-}$ and $Tg^{5xFAD/5xFAD}$ females had lower STAT3 protein levels compared to non-Tg females ($p < 0.001$; $p < 0.001$, respectively) (Figure 6F,G).

2.5. Hypothalamic Energy Sensors Were Dysregulated in Transgenic 5xFAD Mice at 6 Months of Age

Energy sensors in the hypothalamus, such as AMPK, mTOR, and ERK proteins, interplay in hormone receptor signaling pathways and regulate processes of food intake, glucose uptake, mitochondrial metabolism, and whole-body energy homeostasis [32–34]. Normally, mTOR acts as a pro-anabolic factor in response to insulin signaling. AMPK, in contrast, counteracts mTOR activity in fasting state. Changes in these energy sensors in the hypothalamus could help interpretation of negative energy balance in transgenic 5xFAD mice (Figure 7).

Two-way ANOVA analysis showed a main effect of genotype for increased phosphorylation of AMPK in 5xFAD mice ($F_{2,28} = 5.317$; $p = 0.011$), but also a main sex effect for increased pAMPK in females ($F_{1,28} = 14.9$; $p < 0.001$) and interaction "sex x genotype" ($F_{2,28} = 4.275$; $p = 0.024$). This is resumed in a particular significant increase in pAMPK only on $Tg^{5xFAD/5xFAD}$ males compared to $Tg^{5xFAD/-}$ ($p < 0.001$) and non-Tg males ($p < 0.001$) (Figure 7A,I). There were also significant main effects of sex ($F_{1,14} = 5.665$; $p = 0.0321$), genotype ($F_{2,14} = 4.392$; $p = 0.0331$), and interaction "sex x genotype" in AMPK total protein levels ($F_{2,14} = 6.305$; $p = 0.0112$). Tukey's test showed $Tg^{5xFAD/5xFAD}$ female mice had increased AMPK protein levels compared to Tg ($p < 0.05$) and non-Tg females ($p < 0.001$) (Figure 7D,I). As expected, phosphorylation of mTOR showed a contrary trend, with a main genotype effect for decreased pmTOR ($F_{2,28} = 3.399$; $p = 0.0477$). In particular, female $Tg^{5xFAD/5xFAD}$ had decreased pmTOR compared to non-Tg females ($p < 0.05$) (Figure 7B,I). mTOR total protein levels were generally increased in 5xFAD mice, as observed with main genotype effect (two-way ANOVA; $F_{2,28} = 10.49$; $p < 0.001$). This was also shown in significant differences between mTOR total protein in $Tg^{5xFAD/5xFAD}$ males compared to $Tg^{5xFAD/-}$ ($p < 0.05$) and non-Tg males ($p < 0.05$), and also in Tg$^{5xFAD/5xFAD}$ females compared to Tg$^{5xFAD/-}$ ($p < 0.05$) and non-Tg females ($p < 0.05$) (Figure 7E,I).

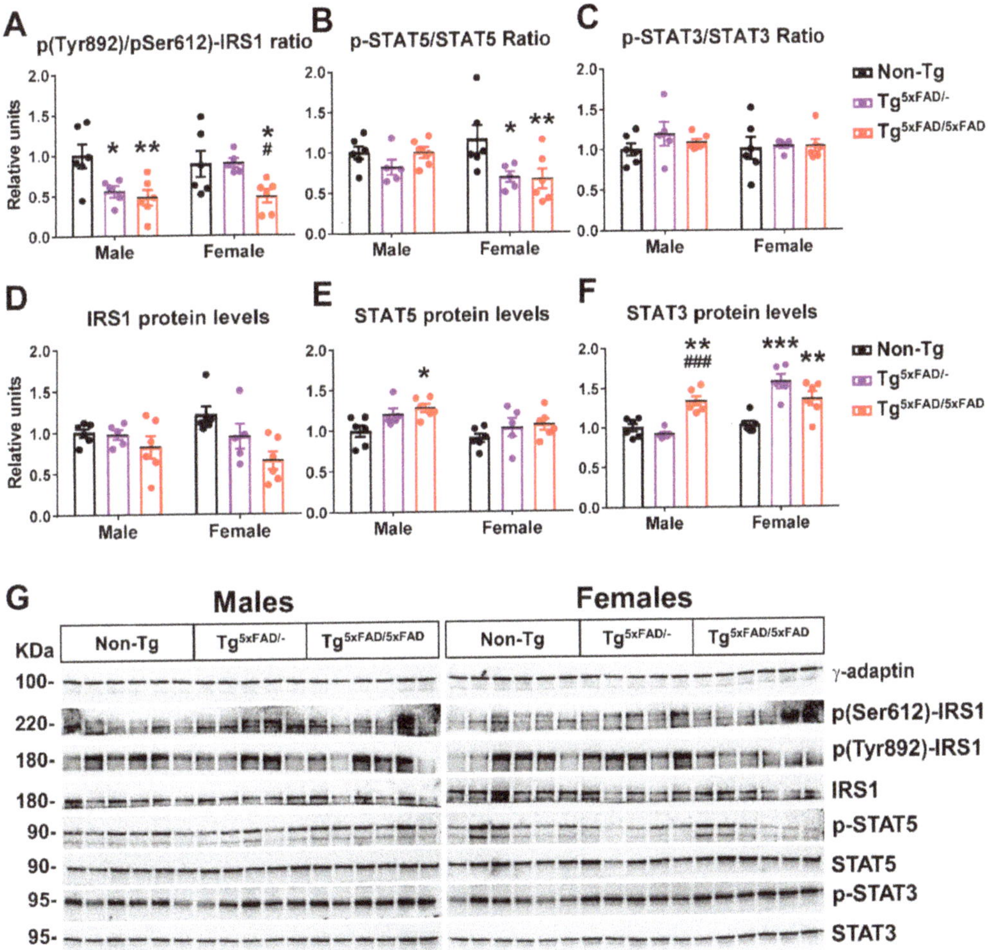

Figure 6. Insulin and leptin hypothalamic signaling were decreased in 5xFAD mice at 6 months of age, as shown by secondary messengers IRS1 and STAT5. (**A–F**) Representation of Western blot membranes and (**G**) densitometric evaluation of insulin receptor substrate 1 (IRS1) activation ratio (pTyr692-IRS1/pSer612-IRS1), IRS1 protein levels, signal transducer and activator of transcription 5 (STAT5) activation (p-STAT5/STAT5), STAT5 protein levels, STAT3 activation (p-STAT3/STAT3), and STAT3 protein levels (n = 5–6 per group). Protein levels were normalized with γ-adaptin. Associated phosphorylations were normalized per activation/inhibition phosphorylation ratio (IRS1) or respective total protein levels (STAT5 and STAT3). Non-Tg males were set as 1 for protein relative units. Two-way ANOVA analysis with Tukey's post hoc test: * = $p < 0.05$, ** = $p < 0.01$, *** = $p < 0.001$ versus same-sex non-Tg group; # = $p < 0.05$, ### = $p < 0.001$ versus same-sex Tg$^{5xFAD/-}$ group.

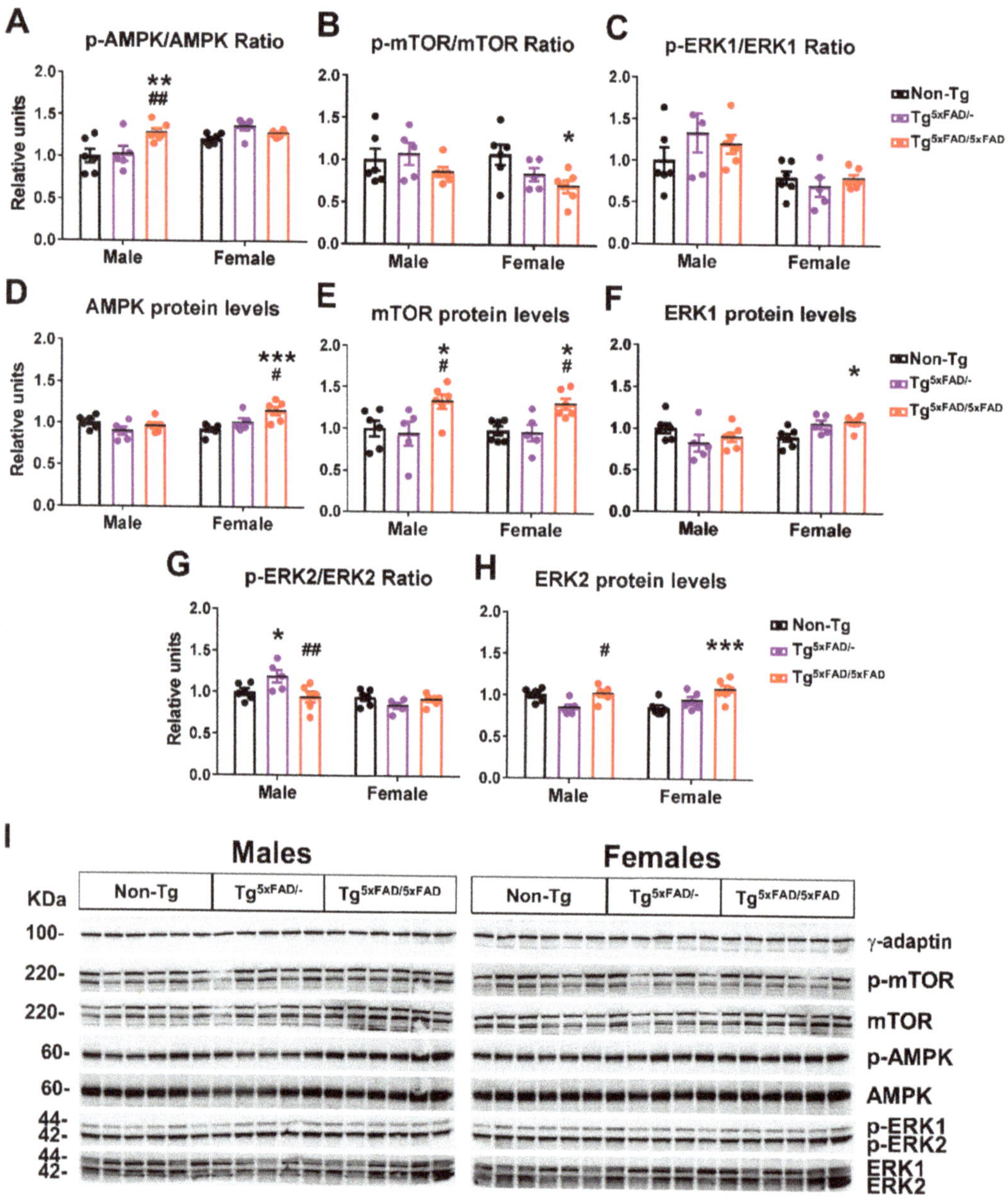

Figure 7. Metabolic sensors AMPK, mTOR, and ERK were altered as a consequence of impaired insulin signaling in the hypothalamus of 5xFAD mice at 6 months of age. (**A–H**) Representation of Western blot membranes and (**I**) densitometric evaluation of AMP kinase (AMPK) activation (p-AMPK/AMPK), AMPK protein levels, mammalian target of rapamycin (mTOR) activation (p-mTOR/mTOR), extracellular-regulated kinase 1 (ERK1) activation (p-ERK1/ERK1), ERK1 protein levels, ERK2 activation (p-ERK2/ERK2), and ERK2 protein levels (n = 5–6 per group). Protein levels were normalized with γ-adaptin. Associated phosphorylations were normalized with respective total protein levels. Non-Tg males were set as 1 for protein relative units. Two-way ANOVA analysis with Tukey's post hoc test: * = $p < 0.05$, ** = $p < 0.01$, *** = $p < 0.001$ versus same-sex non-Tg group; # = $p < 0.05$, ## = $p < 0.01$ versus same-sex $Tg^{5xFAD/-}$ group.

As for ERK activity, there was a significant main effect of sex showing increased pERK1 ($F_{1,28} = 14.95$; $p < 0.001$) and pERK2 ($F_{1,28} = 11.77$; $p = 0.0019$) in males. Tukey's test only showed a particular increase in pERK2 in $Tg^{5xFAD/-}$ males compared to $Tg^{5xFAD/5xFAD}$ males ($p < 0.01$) and non-Tg males ($p < 0.05$) (Figure 7G,I). There was an interaction of "sex x genotype" in ERK1 and ERK2 total protein levels ($F_{2,28} = 4.661$, $p = 0.0179$; $F_{2,28} = 4.88$, $p = 0.0152$, respectively), showing a trend towards decreased total ERK1 and ERK2 in 5xFAD males and increase in 5xFAD females. Tukey's test also showed increased ERK1 protein levels in $Tg^{5xFAD/5xFAD}$ females compared to non-Tg females ($p < 0.05$) and the same trend for ERK2 protein levels in $Tg^{5xFAD/5xFAD}$ females compared to non-Tg females ($p < 0.001$) (Figure 7F,H).

2.6. A Decrease in Orexigenic Neuropeptides Was Observed in Male Transgenic 5xFAD Mice at 6 Months of Age

Control of energy homeostasis in the hypothalamus resides ultimately in the interplay of distinct neuropeptides NPY and AgRP, to promote body mass increase and suppress appetite; POMC and CART, which cause weight loss by inhibiting food intake and stimulating energy expenditure; and others such as orexin and MCH, linking appetite patterns to circadian rhythms, with orexin increasing appetite and wakefulness in a food-seeking context, and MCH stimulating consumption of palatable food and promoting paradoxical sleep [19,35–37]. Balance in the secretion of these neuropeptides is essential in controlling behavior intake and whole-energy homeostasis. We determined mRNA expression of these neuropeptides in the whole hypothalamus.

Two-way ANOVA showed that there was a main effect of genotype to decrease NPY hypothalamic mRNA expression (*Npy*) ($F_{2,31} = 4.511$; $p = 0.0191$) and also an interaction in "sex x genotype", revealing decreased NPY primarily in males ($F_{2,31} = 3.479$; $p = 0.0434$). Tukey's test showed this decrease was more pronounced in $Tg^{5xFAD/-}$ males than $Tg^{5xFAD/5xFAD}$ males as compared to non-Tg males ($p < 0.01$; $p < 0.05$, respectively) (Figure 8A). AgRP mRNA expression (*AgRP*) showed a similar trend as NPY, but no significant main effects nor particular differences were observed (Figure 8B). Contrary to orexigenic neuropeptides, no overall changes in mRNA expression of anorexigenic neuropeptides POMC (*Pomc*) and CART (*Cartpt*) were observed (Figure 8C,D). Intriguingly, as opposed to NPY, orexin mRNA expression (*Hcrt*) was significantly increased in $Tg^{5xFAD/-}$ males compared to non-Tg males ($p < 0.01$) (Figure 8E). There was also a main genotype effect in an increase in MCH mRNA expression (*Pmch*) in 5xFAD mice ($F_{2,31} = 3.317$; $p = 0.0495$) with a significant increase in $Tg^{5xFAD/5xFAD}$ females compared to non-Tg females ($p < 0.05$) (Figure 8F).

Correlation tests were run to assess if neuropeptide levels were directly associated with cytokines that were significantly increased in 5xFAD mice (IL-1β and TNF-α) (Table 4). POMC levels were negatively associated with IL-1β in $Tg^{5xFAD/5xFAD}$ females ($p < 0.05$). MCH was associated with IL-1β in non-Tg males ($p < 0.01$), but not in transgenic 5xFAD mice. Interestingly, hypothalamic resistin was negatively associated with IL-1β and TNF-α differently in males and females. Resistin was negatively correlated to IL-1β in $Tg^{5xFAD/-}$ females ($p < 0.01$) and $Tg^{5xFAD/5xFAD}$ females ($p < 0.05$), and also to TNF-α in $Tg^{5xFAD/5xFAD}$ females ($p < 0.01$). In male mice, resistin was not significantly associated with IL-1β, but correlation was observed with TNF-α in $Tg^{5xFAD/-}$ males ($p < 0.05$) and $Tg^{5xFAD/5xFAD}$ males ($p < 0.01$). Since NPY dysfunction has been observed in AD mice [13], we also ran a correlation test to determine if NPY expression was accompanied with expression of other neuropeptides. NPY levels were associated to AgRP in all groups but Non-Tg males, as they are expressed in the same neuronal population in the arcuate nucleus of the hypothalamus (ARC). Interestingly, NPY levels were also related to hypothalamic POMC in in $Tg^{5xFAD/-}$ females ($p < 0.05$) and $Tg^{5xFAD/5xFAD}$ females ($p < 0.05$). NPY levels were also related similarly to orexin (HCRT) and MCH in $Tg^{5xFAD/-}$ males, $Tg^{5xFAD/5xFAD}$ males, and $Tg^{5xFAD/5xFAD}$ females ($p < 0.05$ for all).

2.7. Lipid Plasma Profile Was Altered in Female 5xFAD Mice

Plasma lipid profile in 5xFAD mice was determined in order to assess if AD metabolic dysfunction was related to altered lipid metabolism as previously observed in AD patients [38,39]. We determined levels of plasma PAI-1, as it is associated with increased risk of thrombolysis (Figure S2C). There was a significant sex effect in PAI-1 levels ($F_{1,33} = 8.537$; $p = 0.0062$), with a tendency to increased PAI-1 plasma levels in female 5xFAD mice that did not reach statistical significance. Biochemistry showed altered total cholesterol, HDL, LDL, and triglyceride plasma levels and also hepatic glutamic oxaloacetic transaminase (GOT) (Figure S3B–E). Two-way ANOVA showed interaction in "sex x genotype" in total cholesterol ($F_{2,42} = 3.39$; $p = 0.0432$). Breaking down cholesterol results, we observed for HDL a main effect of genotype ($F_{2,42} = 5.653$; $p = 0.0067$), sex ($F_{1,42} = 16.64$; $p = 0.0002$), and interaction sex x genotype ($F_{2,42} = 3.577$; $p = 0.0368$), while in LDL there was also a main effect in genotype ($F_{2,42} = 3.331$; $p = 0.0454$) and sex ($F_{1,42} = 5.56$; $p = 0.0231$). Triglyceride levels showed an interaction "sex x genotype" ($F_{2,42} = 3.275$; $p = 0.0477$). GOT was distinctly affected by genotype ($F_{2,42} = 6.773$; $p = 0.0028$). Total cholesterol was significantly decreased in $Tg^{5xFAD/-}$ females compared to non-Tg females ($p < 0.01$). Conversely, HDL levels were significantly increased in $Tg^{5xFAD/5xFAD}$ females compared to both non-Tg and $Tg^{5xFAD/-}$ females ($p < 0.01$ for both) (Figure S3C), whereas LDL was significantly decreased in $Tg^{5xFAD/5xFAD}$ females also compared to both non-Tg and $Tg^{5xFAD/-}$ females ($p < 0.05$ for both) (Figure S3D). GOT levels were significantly increased in $Tg^{5xFAD/5xFAD}$ females compared to $Tg^{5xFAD/-}$ females ($p < 0.05$) and non-Tg females ($p < 0.01$) (Figure S3H).

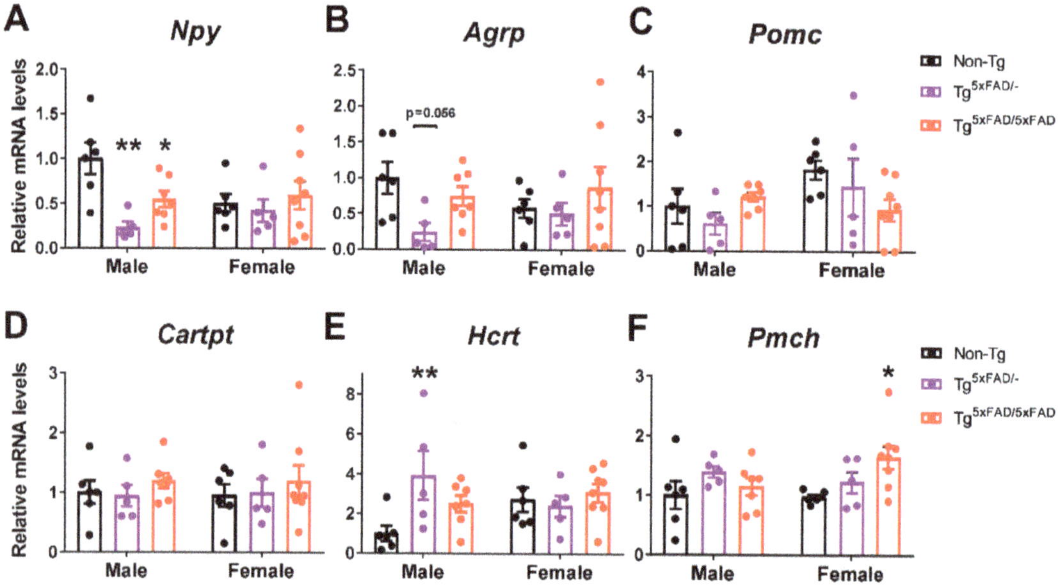

Figure 8. Hypothalamic orexigenic neuropeptide expression was decreased in 5xFAD males accompanied by altered orexin expression in males and MCH expression in females. (**A**–**F**) Representative quantification of expression of hypothalamic neuropeptide Y (*Npy*), agouti-related peptide (*Agrp*), pro-opiomelanocortin (*Pomc*), cocaine and amphetamine-regulated transcript prepropeptide (*Cartpt*), orexin (hipocretin; *Hcrt*), and ppre-melanin concentrating hormone (*Pmch*) assessed by qPCR (n = 5–8 per group). Non-Tg males were set as 1 for relative mRNA units. Two-way ANOVA analysis with Tukey's post hoc test: * = $p < 0.05$, ** = $p < 0.01$ versus same-sex non-Tg group.

Table 4. Correlations between food intake-regulating hypothalamic neuropeptides and inflammatory markers IL-1β and TNF-α, as well as NPY in 5xFAD mice at 6 months of age.

	Males			Females		
	Non-Tg	**$Tg^{5xFAD/-}$**	**$Tg^{5xFAD/5xFAD}$**	**Non-Tg**	**$Tg^{5xFAD/-}$**	**$Tg^{5xFAD/5xFAD}$**
Il-1β versus						
NPY	ns [1]	ns	ns	ns	ns	ns
AgRP	ns	ns	ns	ns	ns	ns
POMC	ns	ns	ns	ns	$R^2 = 0.211$ $p = 0.009$	$R^2 = 0.546$ $p = 0.036$
CART	ns	ns	$R^2 = 0.566$ $p = 0.051$	ns	ns	ns
HCRT	ns	ns	ns	ns	$R^2 = 0.765$ $p = 0.052$	ns
MCH	$R^2 = 0.868$ $p = 0.006$	ns	ns	ns	ns	ns
Resistin (hyp)	ns	ns	ns	ns	$R^2 = 0.947$ $p = 0.005$	$R^2 = 0.513$ $p = 0.045$
Ghrelin (hyp)	ns	ns	ns	ns	ns	ns
TNF-α versus						
NPY	ns	ns	ns	$R^2 = 0.831$ $p = 0.011$	ns	ns
AgRP	ns	ns	ns	ns	ns	ns
POMC	ns	ns	ns	ns	ns	ns
CART	ns	$R^2 = 0.829$ $p = 0.031$	ns	ns	ns	ns
HCRT	ns	ns	ns	ns	ns	ns
MCH	ns	ns	ns	ns	ns	$R^2 = 0.436$ $p = 0.074$
Resistin (hyp)	ns	$R^2 = 0.902$ $p = 0.013$	$R^2 = 0.833$ $p = 0.004$	ns	ns	$R^2 = 0.724$ $p = 0.007$
Ghrelin (hyp)	ns	ns	ns	ns	ns	ns
NPY versus						
AgRP	ns	$R^2 = 0.945$ $p = 0.005$	$R^2 = 0.896$ $p = 0.001$	$R^2 = 0.707$ $p = 0.035$	$R^2 = 0.952$ $p = 0.004$	$R^2 = 0.985$ $p < 0.0001$
POMC	ns	$R^2 = 0.707$ $p = 0.074$	ns	ns	$R^2 = 0.797$ $p = 0.041$	$R^2 = 0.640$ $p = 0.017$
CART	ns	ns	ns	ns	$R^2 = 0.7987$ $p = 0.0409$	ns
HCRT	ns	$R^2 = 0.780$ $p = 0.046$	$R^2 = 0.692$ $p = 0.020$	ns	ns	$R^2 = 0.511$ $p = 0.046$
MCH	ns	$R^2 = 0.790$ $p = 0.043$	$R^2 = 0.701$ $p = 0.018$	ns	ns	$R^2 = 0.672$ $p = 0.012$

[1] ns, not significant.

To assess if altered lipid plasma levels could be affected by white adipose tissue (WAT) metabolism, we determined uncoupling protein 1 (UCP1) levels in WAT. We found a main genotype effect in UCP1 ($F_{2,28} = 4.995$; $p = 0.014$) and interaction "sex x genotype" ($F_{2,28} = 3.625$; $p = 0.0398$). UCP1 protein levels were significantly elevated in $Tg^{5xFAD/5xFAD}$ males compared to non-Tg males ($p < 0.05$), and also in $Tg^{5xFAD/-}$ females compared to non-Tg females ($p < 0.05$) (Figure S3J,K).

We ran a correlation test to assess if altered triglyceride and GOT levels in female 5xFAD mice were related to altered total cholesterol and cholesterol particle HDL and LDL in plasma (Table 5). The triglyceride levels were positively correlated with total cholesterol in females ($p < 0.001$ for non-Tg; $p < 0.05$ for $Tg^{5xFAD/-}$ and $Tg^{5xFAD/5xFAD}$ females). Triglyceride levels were also related to HDL in $Tg^{5xFAD/-}$ females ($p < 0.05$) and $Tg^{5xFAD/5xFAD}$ females ($p < 0.05$), and LDL in $Tg^{5xFAD/5xFAD}$ females ($p < 0.01$). Increased GOT levels were negatively associated with LDL levels in $Tg^{5xFAD/5xFAD}$ females ($p < 0.05$).

Table 5. Correlations between cholesterol levels and triglyceride levels or glutamic oxaloacetic transaminase (GOT) in the plasma of 5xFAD mice at 6 months of age.

	Males			Females		
	Non-Tg	**$Tg^{5xFAD/-}$**	**$Tg^{5xFAD/5xFAD}$**	**Non-Tg**	**$Tg^{5xFAD/-}$**	**$Tg^{5xFAD/5xFAD}$**
Triglycerides versus						
Total cholesterol	ns [1]	ns	ns	$R^2 = 0.912$ $p = 0.0002$	$R^2 = 0.737$ $p = 0.013$	$R^2 = 0.582$ $p = 0.010$
HDL	ns	ns	ns	ns	$R^2 = 0.746$ $p = 0.012$	$R^2 = 0.570$ $p = 0.011$
LDL	ns	ns	ns	ns	ns	$R^2 = 0.761$ $p = 0.001$
GOT versus						
Total cholesterol	ns	ns	ns	ns	ns	ns
HDL	ns	ns	ns	ns	ns	ns
LDL	ns	ns	ns	ns	ns	$R^2 = 0.426$ $p = 0.040$

[1] ns, not significant.

3. Discussion

AD is the most prevalent form of dementia, affecting 10% of the population over 65 years of age. Most of the approaches in neurodegenerative diseases such as AD are carried out from a cognitive perspective, focusing on structural and molecular disturbances in the hippocampus, prefrontal cortex, or motor cortex, since these brain regions are responsible for the control of memory and locomotor capacities that are the most characteristic features of AD. However, changes in body mass index (BMI) and weight loss occur in the prodromal phases and predict AD development [40,41]. Extensive research links both peripheral and brain insulin resistance with metabolic abnormalities and synaptic dysfunction, contributing to late-onset development of AD [42]. In the present study, we examined the hypothesis that early metabolic deficits observed in AD are a result of hypothalamic dysfunction and a dysregulation in metabolic hormonal signaling derived in a negative energy balance. We used both heterozygous ($Tg^{5xFAD/-}$) and full transgenic ($Tg^{5xFAD/5xFAD}$) mice, aiming to determine how Aβ burden could affect peripheral metabolism. Since women are at greater risk for development of AD [43], we also investigated sex differences between male and female mice. We found that 5xFAD mice had a decreased body weight, which was associated with a decreased food intake rather than changes in energy expenditure, although 5xFAD female mice exhibited a decreased energy expenditure and respiratory quotient, suggesting a sex-specific change in lipid oxidation resulting from Aβ overexpression. As revealed by glial activation, 5xFAD mice had a neuroinflammatory status in the hypothalamus despite absence of immunohistochemically traces of Aβ deposition. This phenotype was more pronounced in the full transgenic phenotype and females. Energy deficit was accompanied by low levels of plasma insulin in 5xFAD mice. Sex differences were observed as insulin signaling was defective in both male and female 5xFAD mice, but leptin signaling in the hypothalamus was impaired in females, albeit leptin levels were unaltered, as observed by secondary messengers IRS1 and STAT5, and metabolic sensors mTOR and AMPK. Intriguing results showed low orexigenic NPY levels in males but not females, which may partially explain the decreased appetite observed. All together, these observations provide evidence that hypothalamic dysfunction occurs in early–mid stages of AD, before severe neuropathology and associated cognitive deficits are evident. They also suggest that decreased food intake along with insulin impairment contribute to a negative energy balance that may explain metabolic phenotypes observed in AD patients in the prodromal stages (Figure 9).

3.1. Body Weight Deficit in 5xFAD Mice

Our data in the 5xFAD model of AD shows that, in the early–mid stages of cognitive decline at 6 months of age, there was an evident decrease in body weight that was more prominent in females, as both $Tg^{5xFAD/-}$ and $Tg^{5xFAD/5xFAD}$ showed this feature. Our results are in accordance with previous studies in 5xFAD mice of 6, 7.5, 9, and 12 months of age, showing weight loss in female full transgenic mice [44,45]. However, we also observed male weight loss in $Tg^{5xFAD/5xFAD}$ males, but not in the $Tg^{5xFAD/-}$ mice. Studies in different mouse models of AD such as the 3xTg have shown differences in body weight at early age, showing decreased body weight in male mice but, conversely, increased in female mice when other studies showed decreased body weight in both male and female 3xTG mice at 3 months of age [13,46]. According to previous studies in mouse models of AD, this body weight deficit seems to be mainly associated with white fat mass loss [13,43,45]. Weight loss is usually observed in AD patients, and BMI is also a marker for AD cognitive decline as changes in stages of AD are related to cumulative weight loss [47–49]. Moreover, it has been found that weight loss precedes one or two decades the appearance of cognitive symptoms in patients with autosomal dominant AD (ADAD) [50]. The exact causes for weight loss are, however, still under discussion and seem to be fairly related to decreased food intake and disruptions in the sleep pattern [48,49]. In our study, we found that body weight decrease was significantly associated with decreased food intake, more prominently during the active phase (night), in accordance with previous studies in the same mouse model of AD [45]. It has also been reported that behavioral factors such as anhedonia and depression in AD patients are correlated to decreased food intake and weight loss [51]. Poor food intake may lead to malnutrition in 5xFAD mice. Malnutrition is observed in AD and mild cognitive deficit (MCI) patients and correlates with cognitive performance [52]. This result supports a role of appetite in the manifestation of metabolic disturbances and a negative energy balance in AD patients and may precede cognitive decline. We cannot discard that transgenesis might direct or indirectly affect intestinal transit, nutrient absorption, and microbiota composition. Current ongoing research will address this important aspect to clarify whether weight loss has a neglected gastrointestinal component that has to be controlled to identify early metabolic dysfunctions in AD.

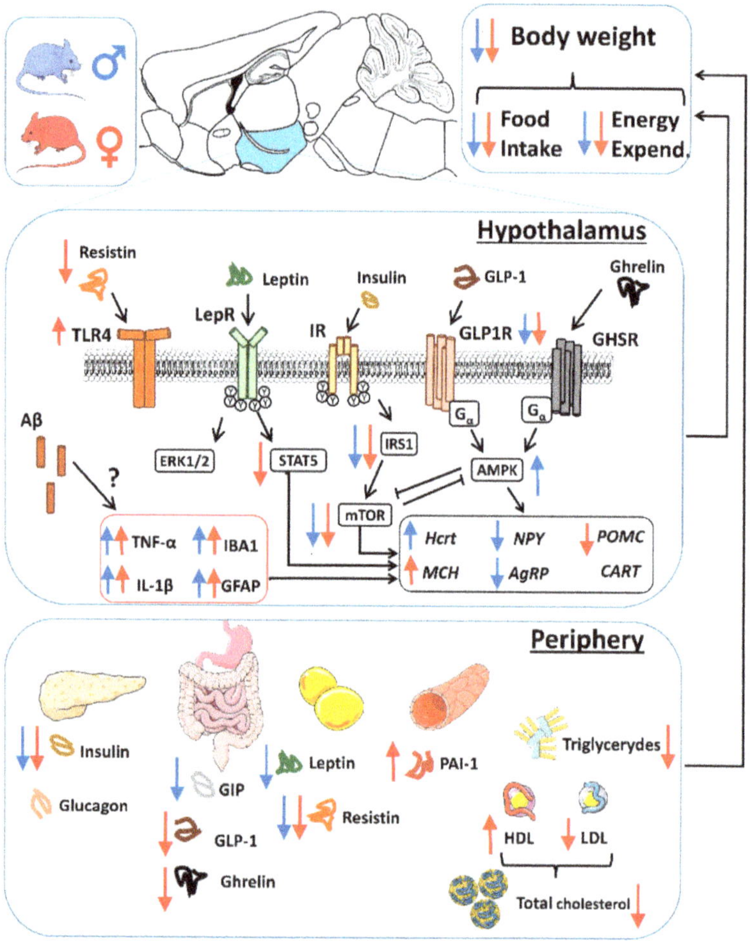

Figure 9. Schematic view of peripheral and hypothalamic alterations derived in negative energy balance and decreased body weight in 5xFAD mice. Increased or decreased observed levels are represented in blue for males and red for females. In the periphery, decreased insulin levels are associated with decreased GIP in males and GLP-1 in females, which are related to decreased body weight pattern. Low leptin levels were observed in males, whereas decreased ghrelin and resistin were more pronounced in females. Females also showed lower total triglyceride and total cholesterol levels, with increased HDL and decreased LDL, contributing to peripheral metabolic impairment. In the hypothalamus, insulin signaling was decreased in 5xFAD mice, accompanied by decreased leptin signaling and resistin hypothalamic levels in females. Despite absence of amyloid plaques, hypothalamic neuroinflammation was observed and contributed to hypothalamic dysfunction and lower body weight. Alterations in NPY/AgRP and orexin (Hcrt) in males seemed to contribute to decreased food intake, whereas females showed a tendency towards decreased overall neuropeptide expression but increased MCH levels. Hypothalamic and peripheral neuroendocrine dysfunction, which are sex-specific and aggravated in the full transgenic (increased Aβ burden) mice, are proposed as contributors to whole negative energy balance in 5RExFAD mice.

3.2. Alterations in Food Intake and Energy Expenditure in 5xFAD Mice

Along with decreased food intake, it has been proposed that hypermetabolism may also be a cause for decreased body weight in AD. There is some controversy over contribution of energy expenditure on metabolic deficits in AD patients. While some studies show increased metabolism in AD [53], others indicate decreased or non-significant changes in energy expenditure (EE) in AD patients [54]. As previously seen, the Tg4510 and 3xTg mouse models of AD show a hypermetabolic state at an early age [13,46,55]. We, in contrast, did not find increased EE when normalized per body weight and quantified by indirect calorimetry. Instead, we also observed a decreased EE in full 5xFAD transgenic mice, more preeminently in female mice during active phase (night), in accordance with the food intake pattern of activity we observed. These results show that weight loss in 5xFAD mice at 6 months of age are not related to hyperactivity nor energy expenditure, as other studies have also shown decreased motility and locomotor activity in male and female 5xFAD mice, which were associated with increased frailty [45,56]. The main differences between energy metabolism in 5xFAD and other mouse models of AD reside in either mutation leading to Tau hyperphosphorylation or behavioral traits, which may be a cause for observed hypermetabolism [55]. This seems to be consistent as suppression of transgene overexpression of Tau in the Tg4510 model of AD attenuates observed hyperactivity [57]. Surprisingly, we also found a decreased respiratory quotient (RQ) in female 5xFAD mice. This observation indicates a decline in glucose/fat fuel utilization ratio, since lower RQ measurements correlate with higher O_2 consumption in lipid β-oxidation. This tendency in fatty acid utilization as fuel may be a triggering risk factor for the development of AD. One of the most indicative risk factors for late-onset development of AD is the presence of the APOE4 allele, causing a higher prevalence in female carriers. Previous studies have shown that mice carrying APOE4 mutation exhibit decreased body weight and fat mass, and increased fatty acid utilization as fuel [58]. A preliminary study in humans also observed that young female E4 carriers displayed lower resting EE and a redirected flux towards aerobic glycolysis instead of glucose oxidative phosphorylation. Overall, the phenomena occurring in 5xFAD mice provided supportive evidence for the fact that negative energy balance occurring in AD patients is likely related to alterations in appetite and energy utilization, and that these are drifted away from normal ranges with more intensity in females, making them more susceptible to aggravation of their metabolic dysfunctions. These data support nutritional and metabolic interventions as a useful tool for management of AD from a gender perspective.

3.3. Aβ Pathology and Neuroinflammation in the Hypothalamus of 5xFAD Mice

Because the hypothalamus has a marked role in integrating peripheral metabolic and hormonal signals and controlling whole-body energy homeostasis [19], we examined the Aβ pathology in the hypothalamus of 5xFAD mice in order to gain insight into the mechanisms derived in negative energy balance that we previously observed. Several studies have associated hypothalamic abnormalities in AD patients, showing decreased hypothalamus volume [59] and atrophy [60], as well as the presence of amyloid deposits and NFTs in all hypothalamic nuclei [61]. Although we did not find immunohistochemical traces of Aβ plaques in the hypothalamus of 5xFAD mice at 6 months of age, we shall not discard the possibility of influence of other forms of Aβ or APP fragments in hypothalamic neuronal status. Other mouse models of AD have stated metabolic abnormalities before detection of Aβ plaques in the brain [13]. In our case, other areas such as the hippocampus and prefrontal cortex had widespread and large Aβ plaques at 6 months of age, which can be fairly related to deficits in memory and decision tasks. There has been extensive research suggesting that soluble oligomers, which we did not observe by immunohistochemistry, can lead to activation of microglia, brain insulin resistance, and Tau hyperphosphorylation [62]. This should be highlighted as the possible cause for neuroinflammatory status observed in 5xFAD mice. This is especially relevant since glial cells are main regulators of Aβ levels in the brain and mediate Aβ clearance in the brain, and hence activation of microglia response is likely triggered by soluble Aβ [63].

Inflammation in the hypothalamus is a common cause in metabolic diseases. Neuroinflammation is observed in several hypothalamic nuclei in obesity, and cytokine infiltration worsens the disease [64]. We evaluated the expression of inflammatory markers and found an increase expression of IL-1β in full transgenic and TNF-α in heterozygous 5xFAD mice, but no changes on IL-6 expression. Although crucial information on inflammatory transducers such as NF-kB, cyclooxygenase, and nitric oxide synthase is lacking, data clearly suggest that inflammation is triggered in the hypothalamus. The exact mechanisms derived in this distinct pattern of cytokine release may depend on Aβ burden difference between heterozygous and full transgenic genotypes. However, both cytokines are part of microglia-activated response, and elevation in their expression is observed in AD patients and the hippocampus, as well as the brain cortex, in mouse models of AD. Despite the absence of Aβ plaques, Aβ oligomers may be present in the hypothalamus and bind to microglial cells, triggering a pro-inflammatory response. The immune response was accompanied by increased protein levels of GFAP (indicative of astrogliosis) and IBA1 (microgliosis). There is evidence that neuroinflammation derived from astrocytes and microglia in a context of Aβ deposition triggers Tau protein misfolding and hyperphosphorylation [65]. We observed increased Tau phosphorylation in AD led by higher levels of p25 and CDK5 in 5xFAD mice. As pointed out previously, Tau hyperphosphorylation in the hypothalamus of mice models of AD leads to increased locomotor activity and EE. Since the 5xFAD mouse does not contain Tau-specific mutations, this increased phosphorylation we observed as part of neuroinflammatory status in the hypothalamus may not be sufficient for lead and/or second hypermetabolism. The placement of the hypothalamus next to the fourth ventricle makes it highly sensitive to leakage of peripheral inflammatory molecules when disruptions on the blood–brain barrier (BBB) occur, which is also a feature of neurodegenerative processes in course with a chronic neuroinflammatory status. This inflammation of peripheral immune cells and cytokines may contribute to worsening the hypothalamic neurodegeneration.

Inflammation in the hypothalamus is thought to induce several changes in metabolism. Previous studies have also shown increased expression of IL-1β, TNF-α, and IL-6 in the hypothalamus of the 3xTg mouse model of AD [11,46]. IL-1β in the arcuate nucleus of the hypothalamus (ARC) is thought to play a major role in food intake, inducing hypophagia and reducing body weight [66]. This mechanism may account for the detrimental phenotype of the full transgenic 5xFAD mice as compared to the heterozygous 5xFAD mice in energy balance. In turn, hypothalamic TNF-α reduces thermogenesis and energy expenditure and leads to hypothalamic desensitization of insulin response [67].

3.4. Alterations in Plasma Hormones and Hypothalamic Signaling in 5xFAD Mice

The metabolic phenotype of 5xFAD mice was accompanied by decreased insulin, leptin, and GLP-1 hypothalamic signaling, as shown by low activity of insulin and leptin receptors and the secondary messengers IRS1 and STAT5, respectively. Moreover, low plasma levels of insulin were observed in 5xFAD mice, as well as low GLP-1 in female 5xFAD mice. It is well known that hypothalamic inflammation impairs neuronal response to neuroendocrine signals, such as insulin and leptin [68], and is observed in AD mice [69]. Both insulin are important neuroprotective growth factors, and impaired signaling of insulin, leptin, and GLP-1 in the hypothalamus are seen to contribute to worsening of Aβ pathology [8]. Recently, it has been demonstrated that cognitive impairment and AD progression are related to a defective insulin signaling in the hippocampus and frontal cortex. Postmortem analysis of human hippocampal tissue shows a correlation between high serine-inhibitory phosphorylation of IRS1 and oligomeric Aβ plaques, which were negatively associated with working memory and episodic memory [5]. The low insulin levels are likely a major contribution to the negative metabolic balance phenotype observed in 5xFAD mice. Supporting this, low-insulin plasma levels were positively correlated to body weight in 5xFAD mice. It is commonly thought that high-insulin plasma levels in pre-diabetic to type 2 diabetes mellitus (T2DM) patients contributes to AD development, with increasing evidence suggesting an exacerbation of cognitive impairment, neuroinflammation, Aβ aggregation, and tau hyperphosphorylation in AD [70–72]. Moreover, diabetic patients show a high prevalence and increased risk of developing AD [73]. However, low insulin levels in plasma are also a predictor for late-onset dementia and AD. Two follow-up studies have shown that low-insulin plasma levels during fasting were higher risk predictors for dementia and AD as compared to high insulin levels, and this association was independent of preclinical T2DM [74,75]. This describes a U-shaped association between fasting plasma insulin levels and risk for development of AD, as stated by studies in aging men and women [74,76]. These studies support a role of low insulin levels observed in 5xFAD mice in the early onset of AD. Other authors have shown low insulin levels in the Tg2576 mouse model of AD at 14 months of age [13]. This observation is, however, concomitant with low leptin levels in earlier age, as also seen in the 3xTg mice with weight loss [46]. Hypoleptinemia is also described in AD patients and seems to be related to low fat adipose tissue mass in weight loss and defective adipokyne section [77,78]. We also observed positive correlation between plasma leptin in male 5xFAD mice and body weight, but not in females, suggesting a difference in the mechanism of leptin secretion between both sexes. Despite the fact that we did not observe significant decreased plasma leptin levels, impaired leptin hypothalamic signaling in 5xFAD mice, which is likely related to neuroinflammation, may reproduce these mechanisms of metabolic dysfunction.

Sex bias is preeminently observed in other hormones previously described to contribute to metabolic dysfunctions aside from insulin and leptin. Along with leptin and insulin, both GIP and GLP-1 signaling in the hypothalamus are thought to contribute to insulin sensitivity and also provide a neuroprotective effect and modulate food intake and energy expenditure [79,80]. We observed that low insulin levels in male and female 5xFAD mice were related to a deficiency in both incretins. Male 5xFAD mice showed decreased basal GIP plasma levels that were also positively correlated with insulin plasma levels and body weight. Studies with GIP analogs in AD models have shown improved cognitive performance [80]. Moreover, dual GIP/GLP-1 agonists are arising as a promising therapeutic for AD as they show neuroprotective effects in AD models [81]. Specific GLP-1R stimulation reduces Aβ aggregation in vitro and in 3xTg mice in vivo. The fact that we also observed low levels of plasma GLP-1 in female mice that were positively correlated with low inulin plasma levels and body weight, but also low GLP-1R levels in the hypothalamus, suggest a worsening pattern of hypothalamic control of energy balance, especially in female mice. Ghrelin levels were also lower in female 5xFAD mice. Ghrelin also exerts a neuroprotective effect in AD independently of insulin signaling and promotes food intake [82,83]. Treatment of 5xFAD mice with ghrelin agonist MK-0677 has been shown to decrease Aβ burden, neuroinflammation, and neurodegeneration in the hippocampus [84]. Although there is no evidence of altered levels of GLP-1 and ghrelin in AD patients, the mechanisms involving both signaling pathways in the hypothalamus may account for cumulative deficits in hypothalamic control on energy balance, resulting in metabolic dysfunction. Deficiencies in neuroendocrine signaling show an early age pattern of metabolic dysfunction and weight loss.

We should highlight the finding of low levels of the adipokine resistin in the plasma of 5xFAD mice. This hormone is secreted in fatty tissue in mice and macrophages in humans and is involved in fatty acid synthesis in adipose tissue and the triggering of an inflammatory response in the brain and hypothalamic insulin resistance through its binding to TLR4 receptors [28,85,86]. Moreover, association between resistin levels and increased inflammatory markers in AD patients suggests a pro-inflammatory role of resistin in AD [87]. We found that low resistin plasma levels were positively related to low body weight in both males and females, which could also have been due to reduced fat mass in 5xFAD mice. Moreover, low hypothalamic localized resistin mRNA expression was also observed in heterozygous and full transgenic 5xFAD mice. Since central resistin infusion modulates hypothalamic neuronal activity and promotes food intake, the low levels of resistin may imply a major role in appetite regulation rather than neuroinflammatory process in 5xFAD mice. Moreover, association between resistin levels and neuroinflammatory process in the hypothalamus seems to be in the opposite direction, as we observed a negative correlation between resistin and the cytokine IL-1β in female $Tg^{5xFAD/-}$ and $Tg^{5xFAD/5xFAD}$ mice as well as TNF-α in female $Tg^{5xFAD/5xFAD}$ mice. This association was also observed with body weight in female mice, suggesting an important interplay between hypothalamic neuroinflammation, resistin levels, and weight loss specifically observed in females. Our results seem to be in accordance with previous results showing a reduction in hypothalamic resistin fibers in both obese and food-deprived underweight young mice [88]. In accordance, damage induced to the ARC, where resistin co-localizes with POMC neurons, decreases resistin immunoreactivity [88]. Since metabolic dysfunction appears to be accompanied by poor endocrine signaling involving a greater range of hormones in females, this may be a cause that explains why they develop a more pronounced underweight phenotype.

3.5. Alterations in Energy Sensors in 5xFAD Mice

Energy sensors are constitutive of hypothalamic response to metabolites and hormones, playing a major role on hypothalamic control of energy balance. AMPK is considered the main energy sensor in the organism and its activity depends on cell energy status, having an important role on metabolic homeostasis, autophagy, cell growth, and inflammation [32]. AMPK activity is characteristic of energy requirements in the cell, and the higher phosphorylation levels observed in $Tg^{5xFAD/5xFAD}$ mice might indicate low energy availability. The main mechanism derived from hypothalamic AMPK activation is the drive for an orexigenic response and increased appetite, implying release of NPY and AgRP mainly in the ARC [89]. The fact that we observed counter-intuitive low levels of NPY and AgRP in 5xFAD male mice shows that this response might be blunted. In fact, studies in other mouse models of AD have shown the presence of Aβ fragments 1–42 blunts response from NPY neurons [8]. Lower levels of both orexigenic and anorexigenic neuropeptides in the hypothalamus have also been associated with neuronal loss in the aged 3xTg mice of AD [11]. AMPK is also a negative regulator of the mTOR pathway, which is mainly involved in the control cell cycle, autophagy, and neuronal plasticity [90,91]. Activity of mTOR is also defined by energy requirements of the cell and is a key element of the insulin signaling pathway, deriving an anabolic response. It has been defined that brain insulin resistance occurs primarily with early hyperactivation of insulin signaling, including mTOR hyperactivity, which causes negative feedback on the insulin receptor, IRS1 and mTOR itself [6]. The fact that we observed low mTOR activity in 5xFAD mice measured by its phosphorylation on the residue Ser2448 is in accordance with low insulin levels and decreased pTyr/pSer ratio on IRS1, which translates into decreased insulin responsiveness in the hypothalamus. mTOR opposes AMPK-mediated autophagy. Hence, low mTOR activity and increased autophagy may help Aβ clearance, as previously seen [92]. However, decreased mTOR activity has been observed in 5xFAD mice, APP/PS1 AD mice, and AD patients, being impaired as a result of exposure to Aβ peptides, contributing to synaptic and cognitive deficiencies [93–95]. Moreover, low mTOR signaling has been related to increased cytokine expression in AD rats [96]. Hence, both impaired response to hormonal cues and enhanced cytotoxic activity of Aβ may exacerbate hypothalamic insulin signaling response and metabolic sensing through AMPK and mTOR, contributing to the negative energy balance observed in 5xFAD mice. Along with AMPK and mTOR, ERK's role in the hypothalamic control of energy homeostasis is fairly involved in growth factor signaling as a second messenger in the Ras–ERK pathway, with some events of crosstalk with the mTOR pathway as an anabolic response [33]. However, we observed no influence of total ERK activity as measured by its phosphorylation, but total protein levels tended to be decreased in males as opposed to higher levels in females. We also observed higher levels of total mTOR in both $Tg^{5xFAD/5xFAD}$ males and females. Higher total mTOR levels and lower mTOR relative activity as measured by p-mTOR/mTOR ratio has been observed in normal aging process in the hippocampi of mice [6]. Because insulin signaling in the hypothalamus was decreased, we also evaluated and observed low hypothalamic GSK3-β activity in 5xFAD mice (Figure S4). Both mTOR and GSK3-β have been previously related to Tau hyperphosphorylation [97]. However, our results suggesting increased Tau phosphorylation in 5xFAD mice seem to be related to CDK5 and neuroinflammation [98].

3.6. Alterations in Neuropeptides in 5xFAD Mice

Ultimately, hypothalamic control of food intake resides on the release of orexigenic/anorexigenic neuropeptides. NPY/AgRP are co-localized in neurons in the ARC, promoting food intake and lowering energy expenditure, whereas POMC/CART neurons do the opposite [19,35–37]. Previous studies based on other mouse models of AD with hypoleptinemia assumed hypothalamic neurons had abnormal or absent responses to hormonal cues, causing the metabolic deficits observed [13,99]. In our study, we confirmed that the hypothalamus of the 5xFAD mouse model of AD showed a blunted response to insulin and leptin in both male and female mice, as well as low plasma levels of ghrelin, GLP-1, and resistin in female mice, promoting an exacerbated status of negative energy balance. The heterozygous model resulted in a less pronounced phenotype, especially in male mice. However, distinct patterns of food intake and activity were observed in accordance with hypothalamic neuropeptide expression. We found decreased NPY and AgRP levels in heterozygous and full transgenic 5xFAD male mice after a night of food deprivation. This result agrees with the decreased food intake pattern in the full transgenic $Tg^{5xFAD/5xFAD}$ mice. However, lower expression of orexigenic neuropeptides in the heterozygous $Tg^{5xFAD/-}$ mice did not account for a significant decrease in food intake. Nevertheless, we did find a peak in EE in the last hours of nighttime (active phase) and first hours of daytime (resting phase) in heterozygous $Tg^{5xFAD/-}$ mice. Heterozygous mice showed a tendency towards increased intake during the day, which, in the case of males, was likely related to increased expression of orexin in the hypothalamus. Orexin promotes wakefulness, appetite, and EE, and its expression is inhibited by NPY projections [100]. This is in accordance with the negative correlation we observed between NPY and orexin expression in the hypothalamus. The fact that mice were sacrificed at the very first hours of daytime, when energy expenditure was slightly increased in heterozygous mice, might imply why orexin levels were higher. Both male and female 5xFAD mice have shown decreased sleep at 4–6.5 months of age, and sleep disruption was more prominent during late hours of night time and early hours of day time, as matched with our results of EE in full transgenic mice and more pronounced in heterozygous 5xFAD mice at the same age [101]. The reason why we observed this pattern in a more pronounced way in heterozygous mice may rely upon a possible inability of orexin to compensate for low NPY levels in full transgenic $Tg^{5xFAD/5xFAD}$ mice.

As previously reported in the Tg2576 mouse model of AD, hypothalamic NPY neurons lose response to leptin, which has a hyper-polarizing effect, and ghrelin, which in turn depolarizes NPY neurons and promotes neuropeptide release. This effect is mediated by the presence of Aβ peptides [13]. The fact that we did not observe changes in NPY/AgRP mRNA expression in female mice could have been a result of small fasting time before sacrifice. Changes in NPY expression in the hypothalamus of Tg2576 mice were observed after 48h of food deprivation, but not in the fed state [13]. However, male Tg2576 mice showed low POMC and CART expression without food deprivation, which resembles the non-significant tendency we observed in POMC mRNA expression in 5xFAD female mice after a short period of food deprivation. In our study, NPY and POMC expressions were positively correlated in female 5xFAD mice, as it would be expected to be inversely correlated. This observation could imply a mechanism of generalized neuronal dysfunction in the hypothalamus, leading to impaired food intake or energy expenditure response during fasting.

3.7. Plasma Lipid Profile and Increased PAI-1 Levels in Female 5xFAD Mice

PAI-1 is an inhibitor of tissue (tPA) and urokinase plasminogen activator (uPA), playing an important role in inhibiting fibrinolysis, hence decreasing blood clot clearance. Because of this, increased levels of plasma PAI-1 are observed in obesity and metabolic syndrome, leading to increased risk of atherosclerosis development. Our results showed a sex-biased increment in plasma PAI-1 levels in female $Tg^{5xFAD/-}$ and $Tg^{5xFAD/5xFAD}$ mice and a negative correlation between body weight and PAI-1 plasma levels in the aforementioned groups. Interestingly, altered PAI-1 levels have been observed in the brain of APP/PS1 mice and AD patients, being increased in correlation to age and progress of dementia. Studies in mice have showed inhibition of PAI-1 [102]. Addition of PAI-1 inhibitors in APP/PS1 and Tg2576 AD mice improve clearance and reduce levels of plasma and brain Aβ and restore memory function [103,104]. The fact that plasma levels of PAI-1 are altered and seem to relate to prodromal weight loss suggest that systemic metabolic imbalance may be related to worsening of AD pathology in females in a mechanism different from obesity-related increments of PAI-1 levels.

We investigated plasma lipid profile of 5xFAD AD mice because several studies relate altered plasma lipid levels in AD patients. We observed again a sex-specific decay in triglycerides and total cholesterol of female 5xFAD mice that were positively correlated. From a metabolic point of view, depletion of triglycerides in female 5xFAD mice may be indicative of a shift in lipid utilization as energy source, in accordance with the results observed and previously discussed on the low RQ observed and increased lipid utilization. In fact, increased expression of WAT UCP1 and UCP2 was also observed in $Tg^{5xFAD/-}$ and $Tg^{5xFAD/5xFAD}$ female mice. Previously, it has been observed patients with probable AD have abnormally low levels of total cholesterol, triglycerides, and LDL cholesterol when compared to control groups, and these low levels were inversely but not significantly correlated with cognitive performance. The specific reason why blood lipids are altered in AD patients is not fully understood. Normally, higher levels of total cholesterol, triglycerides, and LDL are related to metabolic syndrome and increased risk of AD, which implies a specific relationship between obesity-derived problems and AD. Some other studies have assessed increased risk of dementia in patients with higher mid-life and late-life levels of total cholesterol [38,39]. However, a drastic decrease in total cholesterol occurs before dementia diagnosis and is predictive of dementia appearance [39]. Cholesterol is a key element for several processes such as maintenance and fluidity of cell membrane, synaptic transmission, and synthesis of steroid hormones, all of which may experience a negative outcome in AD patients. It should be noted that full transgenic $Tg^{5xFAD/5xFAD}$ female mice had significantly low LDL and high HDL levels in plasma. It has been shown that non-carriers of APOE4 allele AD patients manifest an improvement in cognitive performance with lipid therapy increasing LDL levels, whereas an increment in HDL levels were associated with a poorer score in cognitive tests [105]. Our results indicate that mechanisms of dysregulation of lipid metabolism and altered lipid profile may contribute to worsening of AD pathology, specifically in the higher incidence in females as compared to males in a different pattern non-related to AD prevalence in obesity and diabetic patients.

3.8. Limitations of the Study

Some differences between AD patients and the 5XFAD mouse model of AD must be taken into consideration when pointing out the results hereby obtained. 5xFAD mice contain five mutations, two of the in presenilin-1 (PSEN-1) and three in APP, leading to early production of Aβ fragments. 5xFAD mice hence develop early cognitive impairments similar to those of AD patients. However, 5xFAD mice are unable to develop NFTs and do not develop severe Tau hyperphosphorylation [106]. Although we observe increased Tau phosphorylation in the hypothalamus of 5xFAD mice, this does not seem to reproduce the hypermetabolic status observed in AD mice bearing mutations in Tau protein such as 3xTg [12,13]. However, the higher similarities in neurodegeneration and cognitive impairment in the 5xFAD with AD patients with respect to other AD mice may imply Aβ pathology linked to metabolic dysfunction could be reproduced with more accuracy in our mouse model of AD. As our mice were fed standard chow diet, we could not ascertain whether a high-fat diet (HFD) (a very common nutritional problem in western countries) has or does not have an impact in the progression of metabolic and cognitive decline in AD. Since 5xFAD mice presented impairments in insulin, leptin, and resistin signaling, this could lead to a hyper-sensitive and detrimental response to exposure to fat-enriched nutrition, as this has been previously observed in other mouse models of AD [46]. A previous study has shown detrimental effects of HFD in 5xFAD glucose tolerance, lipid profile, and microbiome composition [44]. However, we are still lacking information as to whether HFD could differently affect the heterozygous non-full transgenic bearing less Aβ burden and also if 5xFAD mice, which have been proven to be more susceptible to metabolic impairments, could exert a more deteriorated phenotype. Moreover, the results observed did not focus on the neuroendocrine system and hypothalamic regulation of energy balance. Future studies should address these questions thoroughly and comparisons should be made between 5xFAD and other AD mice.

In addition, the expression of either Aβ oligomers or plaque deposition in peripheral tissues (including gut mucosa and myenteric plexus, autonomic ganglia, pancreatic islets, liver, and adipose tissue) has to be addressed in order of clarify the contribution of peripheral tissues when amyloid deposition is boosted by mutations.

4. Conclusions

In conclusion, we found a negative energy balance in 5xFAD mice leading to weight loss at 6 months of age that was related to alterations in food intake, hypothalamic dysfunction despite absence of observable Aβ plaques, and neuroendocrine dysregulation. Decreases in body weight were differently regulated and more severe in female 5xFAD mice and worsened with Aβ burden in the full transgenic genotype. The existence of clear differences regarding the heterozygous versus homozygous conditions is remarkable, a fact that might help in the search for models that boost amyloid deposition where there is not a complete penetration mutation. Thus, heterozygous animals have to be included in future studies as a way of analyzing non-genetic AD contributing factors.

In addition, we observed low plasma levels of insulin and insulin-releasing hormones in both male and female 5xFAD mice that were related to decreased body weight and decreased hypothalamic insulin and leptin signaling. Neuroinflammation and low resistin levels in the hypothalamus were specifically related to weight loss in female mice, which also exhibited altered plasma lipid profile, likely worsening the whole-body energy homeostasis. This study provides useful information for detection of possible early metabolic markers indicative of the outcome of AD and serve as possible therapeutic targets in a sex-specific point of view. This strategy should help in stratification of AD patients. Our work also points out a different focus on ageing-related prodromal weight loss as a comorbid state of AD and focus on less-explored areas in the brain AD pathology such as the hypothalamus, opening the scope of further research in AD.

5. Materials and Methods

5.1. Ethics Statement

The research procedures were approved by the Research and Clinical Ethics Committee of the Regional University Hospital of Malaga and the University of Malaga. All experimental procedures with animals were carried out in strict accordance with the guidelines of Royal Decree 1201/2005 of 21 October 2005 (BOE no. 252), and in compliance with Directive 86/609/ECC of the European Community (24 November 1986) in relation to the regulation of research with animals. All efforts were made to minimize the suffering of the animals, as well as to reduce the number of animals used.

5.2. Animals

Animals used were 6-month-old male and female non-transgenic (non-Tg) mice and heterozygous ($Tg^{5xFAD/-}$) and homozygous ($Tg^{5xFAD/5xFAD}$) transgenic 5xFAD mice. 5xFAD mice co-express and co-inherit familial Alzheimer disease (FAD) mutant forms of human APP (the Swedish mutation: K670N, M671L; the Florida mutation: 1716V; the London mutation: V7171) and PSI (M146L; L286V) transgenes under transcriptional control of the neuron-specific mouse Thy-1 promotor (Tg6799 line) [15]. 5XFAD lines (B6/SJL genetic background) were maintained by crossing heterozygous transgenic mice with B6/SJL F1 breeders (The Jackson Laboratory, Bar Harbor, ME, USA). Non-Tg wild-type littermate mice served as a control.

The rodents were housed in the Animal Center for Experimentation at the University of Malaga. This center complies with all current regulations for breeding and housing. The animals were housed individually with free access to food and water under standardized conditions: 20 ± 2 °C of room temperature, relative humidity of 40 ± 5%, and light/dark cycle of 12 h with dawn/sunset effect. The mice were fed on a standard pellet diet (STD) (3.02 Kcal/g with 30 Kcal% protein, 55 Kcal% carbohydrate, and 15 Kcal% fat; purchased from Harlam (Tecklad, Madison WI, USA)). All study animals were sacrificed at 6 months of age with sodium pentobarbital (150 mg/kg, i.p.). Blood was drawn directly from the right atrium and perfused with 0.1M PBS.

5.3. Sample Collection

Blood was drawn directly from the right atrium and perfused with 0.1M PBS. Blood was centrifuged (2100× *g* for 10 min, 4 °C), and the plasma was kept at −80 °C for a biochemical analysis. Histology samples were obtained from the left hemisphere of the brain and kept in 4% paraformaldehyde for 48 h. Samples for biochemical analysis were also obtained from the right hemisphere of the brain and brown fat. Samples from right hemisphere and brown fat were flash frozen in liquid nitrogen, then stored at −80 °C until analysis.

5.4. Measurement of Metabolites in Plasma

The following plasma metabolites were measured: glucose, urea, creatinine, triglycerides, total cholesterol, HDL cholesterol, LDL cholesterol and the hepatic enzymes glutamic oxaloacetic transaminase (GOT) and glutamate pyruvate transaminase (GPT). These metabolites were analyzed using commercial kits, according to the manufacturer's instructions, and a Hitachi 737 Automatic Analyzer (Hitachi Ltd., Tokyo, Japan). Each metabolite concentration was expressed in mg/dL. Hepatic enzyme levels were expressed in IU/L.

5.5. Bio-Plex Pro Multiplex Diabetes Assay

Plasma levels of hormones insulin, glucagon, ghrelin, leptin, glucagon-like peptide-1 (GLP-1), plasminogen activator inhibitor-1 (PAI-1), gastric inhibitory polypeptide (GIP), and resistin were determined by multiplex immunoassay system using commercial kits: Bio-Plex Pro™ mouse diabetes 8-plex immunoassay (Bio-Rad, Hercules, CA, USA, cat. number: #171F7001M). Plates were run on a Bio-Plex MAGPIX™ Multiplex Reader with Bio-Plex anager™ MP Software (Luminex, Austin, TX, USA). Hormone concentrations were expressed in pg/mL, and detection limits were 68.29 (insulin), 0.50 (glucagon), 0.64 (ghrelin), 5.07 (leptin), 0.59 (GLP-1), 2.98 (PAI-1), 4.31 (GIP), and 184.89 (resistin) pg/mL.

5.6. RNA Isolation and RT-qPCR Analysis

We performed real-time PCR (TaqMan, ThermoFisher Scientific, Waltham, MA, USA) as described previously [107] using specific sets of primer probes from TaqMan® Gene Expression Assays, as shown in Table S1. Total RNA was extracted from tibia samples using the Trizol® method according to the manufacturer's instructions (ThermoFisher Scientific). RNA samples were isolated with RNAeasy minelute cleanup-kit including digestion with DNase I column (Qiagen) and quantified using a spectrophotometer to ensure

A260/280 ratios of 1.8–2.0. After the reverse transcript reaction from 1 μg of mRNA, a quantitative real-time reverse transcription polymerase chain reaction (qPCR) was performed in a CFX96TM Real-Time PCR Detection System (Bio-Rad, Hercules, CA, USA) and the FAM dye labeled format for the TaqMan® Gene Expression Assays (ThermoFisher Scientific). A melting curve analysis was performed to ensure that only a single product was amplified. After analyzing several reference genes, we normalized values obtained from the tibia samples in relation to Actb levels (Mm02619580_g1, amplicon length: 143; ThermoFisher Scientific), which were found not to vary significantly between experimental groups.

5.7. Protein Extraction and Western Blot Analysis

Total protein from 5–15 mg of hypothalamic samples was extracted using 500 μL ice-cold cell lysis buffer for 30 min, as previously described [108,109]. A quantity of 50 μg of protein was resolved on a 4–12% (Bis-Tris) Criterion XT Precast Gel (Bio-Rad Laboratories, Inc., Hercules, CA, USA, cat. number: 3450124), and then transferred onto nitrocellulose membranes (Bio-Rad Laboratories, Inc., Hercules, CA, USA). Total protein content was visualized after staining with Ponceau red. Membranes were blocked in TBS-T (50 mM Tris-HCl (pH 7.6), 200 mM NaCl, and 0.1% Tween 20) with 2% albumin fraction V from BSA (Roche, Mannheim, Germany) for 1h at room temperature. For specific protein detection, the membrane was incubated overnight at 4 °C in TBS-T containing 2% BSA and the corresponding primary antibody (Table S2). Mouse γ-adaptin was used as the reference protein. After several washes in TBS-T containing 1% Tween 20, an HRP-conjugated anti-rabbit or anti-mouse IgG (H+L) secondary antibody (Promega, Madison, WI, USA) diluted 1:10,000 was added, followed by incubation for 1 h at room temperature. After extensive washing in TBS-T, the membranes were incubated for 1 min with the Western Blotting Luminol Reagent kit (Santa Cruz Biotechnology, Santa Cruz, CA, USA), and the specific protein bands were visualized and quantified by chemiluminescence using a Chemi-Doc TM MP Imaging System (Bio-Rad, Barcelona, Spain). For specific detection of phosphorylated form of proteins, after measuring phosphorylation proteins, the specific antibodies were removed from membrane by incubation with stripping buffer (2% SDS, 62.5 mM Tris HCl (pH 6.8), 0.8% β-mercaptoethanol) for 30 min at 50 °C. Membranes were extensively washed in ultrapure water, and then re-incubated with the corresponding antibody specific for the total protein. Quantification of results was performed using ImageJ software (http://imagej.nih.gov/ij, accessed on 5 May 2020). The results are expressed as either the phosphorylated form of target protein/target protein ratios or target protein/γ-adaptin ratios. Results for Non-Tg male protein levels were arbitrarily set as 1.

5.8. Immunohistochemistry

Left hemisphere of the brains were post-fixed in 4% paraformaldehyde for 48h and cryopreserved in 30% sucrose in 0.1M PBS solution for at 4 °C until processing. Free-floating coronal sections of mouse hypothalamus and hippocampus were selected from −1.22 to −1.94 mm of Bregma levels [110]. Serial sections were blocked with 5% donkey serum, 0.5% Triton X-100 in 0.1M PBS for 45 min at room temperature, as previously described [111]. For plaque amyloid-β analysis, we used rabbit anti-Aβ (1:500, Abcam). For neuroinflammation analysis, we used rabbit anti-glial fibrillary acidic protein (GFAP; 1:1000; Dako) and rabbit anti-Iba1 (1:500, Abcam). Primary antibodies were incubated overnight at room temperature. After rinsing, the sections were incubated with secondary antibody biotinylated goat anti-rabbit (1:500, GE Healthcare) for 2 h at room temperature. All antibodies were diluted in PBS, 0.5% Triton X-100, and 2.5% donkey serum (Sigma-Aldrich, St. Louis, MO, USA). We used the peroxidase-conjugated ExtraAvidin method and diaminobenzidine as the chromogen to visualize the reaction product. Quantification was performed using ImageJ software (http://imagej.nih.gov/ij, accessed on 5 May 2020). Total amyloid plaques were counted using three binarized sections of hypothalamus and hippocampus per animal.

5.9. Measurement of Energy Expenditure and Respiratory Quotient

For 48 h, mice were analyzed for energy expenditure (EE, kcal/kg lean mass), respiratory quotient (RQ, VCO_2/VO_2), and food intake using a calorimetric system (LabMaster, TSE System, Bad Homburg, Germany) as previously described [112]. This system is an open-circuit instrument that determines (1) the energy consumed by the amount of caloric intake (kilocalories) along time (hours) and normalized by the lean mass (kilograms) and (2) the ratio between the CO_2 production and O_2 consumption (VCO_2/VO_2). Activity was measured by infrared system counting mice movement and rearing. Previously, all rats were acclimated to the experimental room and habituated to the system for 48 h before starting the measurements.

5.10. Statistical Analysis

Graph-Pad Prism 7.0 software was used to analyze the data. Values are represented as mean ± standard error of the mean (SEM) for each in vivo experimental group, according to the assay. The significance of differences within and between groups was evaluated by a two-way analysis of variance (ANOVA), factors: "genotype" x "sex", followed by Tukey post-hoc test for multiple comparisons. Alternatively, for comparisons between two groups, Student *t*-test was also used. A *p*-value ≤ 0.05 was considered to be statistically significant. (* = $p < 0.05$; ** = $p < 0.01$; *** = $p < 0.001$) versus same-sex non-Tg group. (# = $p < 0.05$; ## = $p < 0.01$; ### = $p < 0.001$) versus same-sex $Tg^{5xFAD/-}$ group.

Supplementary Materials: The following are available online at https://www.mdpi.com/article/10.3390/ijms22105365/s1, Figure S1: Activity (A,B) and rearing (C,D) measured as counts by infrared in 5xFAD mice placed in metabolic cages for 48 h, showing decreased activity in $Tg^{5xFAD/5xFAD}$ females during the night (n = 8–15 per group). Two-way ANOVA analysis with Tukey's post hoc test: * = $p < 0.05$ versus same-sex non-Tg group. Figure S2: Plasma levels of grastric inhibitory polypeptide ((A), GIP), glucagon (B), and plasminogen activator inhibitor-1 ((C), PAI-1) plasma levels showing a tendency to decreased GIP in males and increased PAI-1 in females (n = 6–7 per group). Figure S3: Plasma biochemical analysis of glucose (A), lipids (B–E), urea (F), creatinine (G), and hepatic enzyme (H,I) levels showing decreased trigyceride, total cholesterol, and LDL, and increased HDL and GOT in 5xFAD female mice. (J) Representation of Western blot membranes and (K) densitometric evaluation of UCP1 levels in white adipose tissue of 5xFAD mice at 6 months of age, showing increased UCP1 levels in transgenic 5xFAD male and female mice (n = 5–6 per group). Protein levels were normalized with γ-adaptin. Non-Tg males were set as 1 for protein relative units. Two-way ANOVA analysis with Tukey's post hoc test: * = $p < 0.05$, ** = $p < 0.01$, *** = $p < 0.001$ vs. same-sex non-Tg group; # = $p < 0.05$, ## = $p < 0.01$, ### = $p < 0.001$ versus same-sex $Tg^{5xFAD/-}$ group. Figure S4: Densitometric evaluation of the ratio of serine (inhibitory) and tyrosine (activating) phosphorylation of GSK3-β (A) and GSK3-β protein levels (B) in the hypothalamus of 5xFAD mice at 6 months of age, showing increased Ser/Tyr ratio in transgenic 5xFAD male and female mice (n = 5–6 per group). Protein levels were normalized with γ-adaptin. Non-Tg males were set as 1 for protein relative units. Two-way ANOVA analysis with Tukey's post hoc test: * = $p < 0.05$, ** = $p < 0.01$, *** = $p < 0.001$ vs. same-sex non-Tg group; # = $p < 0.05$, ## = $p < 0.01$, ### = $p < 0.001$ versus same-sex $Tg^{5xFAD/-}$ group. Figure S5: Unedited blots. Table S1: Primer references for TaqMan® Gene Expression Assays (Applied Biosystems). Table S2: Antibody references for Western blot assays.

Author Contributions: Conceptualization, A.J.L.-G., F.R.d.F. and J.S.; methodology, A.J.L.-G., C.R.-V., D.M.-V., J.A.N. and A.V.; software, A.J.L.-G.; validation, A.J.L.-G., P.R., C.S., F.R.d.F. and J.S.; formal analysis, A.J.L.-G., C.R.-V. and D.M.-V.; investigation, A.J.L.-G., F.R.d.F. and J.S.; resources, P.R., F.R.d.F. and J.S.; data curation, A.J.L.-G. and J.S.; writing—original draft preparation, A.J.L.-G. and J.S; writing—review and editing, A.J.L.-G., P.R., F.R.d.F. and J.S.; supervision, F.R.d.F., P.R. and J.S.; project administration, F.R.d.F., P.R. and J.S.; funding acquisition, F.R.d.F., P.R. and J.S. All authors have read and agreed to the published version of the manuscript.

Funding: This research was funded by the European Regional Development Funds-European Union (ERDF-EU) and Fatzheimer project EULAC-HEALTH H2020, grant number EULACH16/T010131; Ministerio de Economía, Industria y Competitividad, Gobierno de España, grant number RTC-2016-4983-1; EU-ERDF and Instituto de Salud Carlos III (ISCIII), grant numbers PI19/01577 and PI19/00343; Ministerio de Sanidad, Delegación de Gobierno para el Plan Nacional sobre Drogas, grant numbers 2019/040 and 2020/048; Consejería de Transformación Económica, Industria, Conocimiento y Universidades, Junta de Andalucía, grant number P18-TP-5194. J.S. (CPII17/00024) holds a "Miguel Servet II" research contract from the National System of Health, EU-ERDF-ISCIII. P.R. (CP19/00068) holds a "Miguel Servet I" research contract from the National System of Health, EU-ERDF-ISCIII. D.M-V. (FI20/00227) holds a "PFIS" predoctoral contract from the National System of Health, EU-ERDF-ISCIII. A.J.L.-G. (IFI18/00042) holds an "iPFIS" predoctoral contract from the National System of Health, EU-ERDF-ISCIII.

Institutional Review Board Statement: The study was conducted according to the principles of laboratory animal care (National Research Council, Neuroscience CoGftUoAi, Research B, 2003) following the European Community Council Directive (86/609/EEC) and approved by the Ethics Committee for Animal Research of the University of Málaga, Spain (Code: CTS-8221; date: July 2016).

Informed Consent Statement: Not applicable.

Data Availability Statement: The data that support the findings of this study are available on reasonable request from the corresponding author.

Conflicts of Interest: C.S. declares he receives a salary from and has shares in the Euronutra company. The remaining authors declare that they have no known competing financial interests or personal relationships that could have appeared to influence the work reported in this paper. The funders had no role in the design of the study; in the collection, analyses, or interpretation of data; in the writing of the manuscript; or in the decision to publish the results.

References

1. Azzu, V.; Valencak, T.G. Energy Metabolism and Ageing in the Mouse: A Mini-Review. *Gerontology* **2017**, *63*, 327–336. [CrossRef] [PubMed]
2. Milstein, J.L.; Ferris, H.A. The brain as an insulin-sensitive metabolic organ. *Mol. Metab.* **2021**, 101234. [CrossRef]
3. López-Gambero, A.J.; Sanjuan, C.; Serrano-Castro, P.J.; Suárez, J.; Rodríguez de Fonseca, F. The Biomedical Uses of Inositols: A Nutraceutical Approach to Metabolic Dysfunction in Aging and Neurodegenerative Diseases. *Biomedicines* **2020**, *8*, 295. [CrossRef]
4. Pitt, J.; Thorner, M.; Brautigan, D.; Larner, J.; Klein, W.L. Protection against the synaptic targeting and toxicity of Alzheimer's-associated Aβ oligomers by insulin mimetic chiro-inositols. *FASEB J.* **2013**, *27*, 199–207. [CrossRef] [PubMed]
5. Talbot, K.; Wang, H.Y.; Kazi, H.; Han, L.Y.; Bakshi, K.P.; Stucky, A.; Fuino, R.L.; Kawaguchi, K.R.; Samoyedny, A.J.; Wilson, R.S.; et al. Demonstrated brain insulin resistance in Alzheimer's disease patients is associated with IGF-1 resistance, IRS-1 dysregulation, and cognitive decline. *J. Clin. Investig.* **2012**, *122*, 1316–1338. [CrossRef]
6. Barone, E.; Di Domenico, F.; Cassano, T.; Arena, A.; Tramutola, A.; Lavecchia, M.A.; Coccia, R.; Butterfield, D.A.; Perluigi, M. Impairment of biliverdin reductase-A promotes brain insulin resistance in Alzheimer disease: A new paradigm. *Free Radic. Biol. Med.* **2016**, *91*, 127–142. [CrossRef] [PubMed]
7. Barone, E.; Di Domenico, F.; Cenini, G.; Sultana, R.; Cini, C.; Preziosi, P.; Perluigi, M.; Mancuso, C.; Butterfield, D.A. Biliverdin reductase-A protein levels and activity in the brains of subjects with Alzheimer disease and mild cognitive impairment. *Biochim. Biophys. Acta Mol. Basis Dis.* **2011**, *1812*, 480–487. [CrossRef] [PubMed]
8. McGuire, M.J.; Ishii, M. Leptin Dysfunction and Alzheimer's Disease: Evidence from Cellular, Animal, and Human Studies. *Cell. Mol. Neurobiol.* **2016**, *36*, 203–217. [CrossRef]
9. Johnson, D.K.; Wilkins, C.H.; Morris, J.C. Accelerated weight loss may precede diagnosis in Alzheimer disease. *Arch. Neurol.* **2006**, *63*, 1312–1317. [CrossRef]
10. Vidoni, E.D.; Townley, R.A.; Honea, R.A.; Burns, J.M. Alzheimer disease biomarkers are associated with body mass index. *Neurology* **2011**, *77*, 1913–1920. [CrossRef]
11. Do, K.; Laing, B.T.; Landry, T.; Bunner, W.; Mersaud, N.; Matsubara, T.; Li, P.; Yuan, Y.; Lu, Q.; Huang, H. The effects of exercise on hypothalamic neurodegeneration of Alzheimer's disease mouse model. *PLoS ONE* **2018**, *13*, e0190205. [CrossRef] [PubMed]
12. Knight, E.M.; Verkhratsky, A.; Luckman, S.M.; Allan, S.M.; Lawrence, C.B. Hypermetabolism in a triple-transgenic mouse model of Alzheimer's disease. *Neurobiol. Aging* **2012**, *33*, 187–193. [CrossRef] [PubMed]
13. Ishii, M.; Wang, G.; Racchumi, G.; Dyke, J.P.; Iadecola, C. Transgenic mice overexpressing amyloid precursor protein exhibit early metabolic deficits and a pathologically low leptin state associated with hypothalamic dysfunction in arcuate neuropeptide Y neurons. *J. Neurosci. Off. J. Soc. Neurosci.* **2014**, *34*, 9096–9106. [CrossRef]
14. Zhao, L.; Mao, Z.; Woody, S.K.; Brinton, R.D. Sex differences in metabolic aging of the brain: Insights into female susceptibility to Alzheimer's disease. *Neurobiol. Aging* **2016**, *42*, 69–79. [CrossRef] [PubMed]

15. Oakley, H.; Cole, S.L.; Logan, S.; Maus, E.; Shao, P.; Craft, J.; Guillozet-Bongaarts, A.; Ohno, M.; Disterhoft, J.; Van Eldik, L.; et al. Intraneuronal beta-amyloid aggregates, neurodegeneration, and neuron loss in transgenic mice with five familial Alzheimer's disease mutations: Potential factors in amyloid plaque formation. *J. Neurosci.* **2006**, *26*, 10129–10140. [CrossRef]
16. Ohno, M.; Chang, L.; Tseng, W.; Oakley, H.; Citron, M.; Klein, W.L.; Vassar, R.; Disterhoft, J.F. Temporal memory deficits in Alzheimer's mouse models: Rescue by genetic deletion of BACE1. *Eur. J. Neurosci.* **2006**, *23*, 251–260. [CrossRef] [PubMed]
17. Kimura, R.; Devi, L.; Ohno, M. Partial reduction of BACE1 improves synaptic plasticity, recent and remote memories in Alzheimer's disease transgenic mice. *J. Neurochem.* **2010**, *113*, 248–261. [CrossRef]
18. Xiao, N.A.; Zhang, J.; Zhou, M.; Wei, Z.; Wu, X.L.; Dai, X.M.; Zhu, Y.G.; Chen, X.C. Reduction of Glucose Metabolism in Olfactory Bulb is an Earlier Alzheimer's Disease-related Biomarker in 5XFAD Mice. *Chin. Med. J.* **2015**, *128*, 2220–2227. [CrossRef]
19. López-Gambero, A.J.; Martínez, F.; Salazar, K.; Cifuentes, M.; Nualart, F. Brain Glucose-Sensing Mechanism and Energy Homeostasis. *Mol. Neurobiol.* **2019**, *56*, 769–796. [CrossRef]
20. Kimura, T.; Ishiguro, K.; Hisanaga, S.-I. Physiological and pathological phosphorylation of tau by Cdk5. *Front. Mol. Neurosci.* **2014**, *7*, 65. [CrossRef]
21. Van Gool, B.; Storck, S.E.; Reekmans, S.M.; Lechat, B.; Gordts, P.L.S.M.; Pradier, L.; Pietrzik, C.U.; Roebroek, A.J.M. LRP1 Has a Predominant Role in Production over Clearance of Aβ in a Mouse Model of Alzheimer's Disease. *Mol. Neurobiol.* **2019**, *56*, 7234–7245. [CrossRef] [PubMed]
22. Kulas, J.A.; Puig, K.L.; Combs, C.K. Amyloid precursor protein in pancreatic islets. *J. Endocrinol.* **2017**, *235*, 49–67. [CrossRef] [PubMed]
23. Meloni, A.R.; DeYoung, M.B.; Lowe, C.; Parkes, D.G. GLP-1 receptor activated insulin secretion from pancreatic β-cells: Mechanism and glucose dependence. *Diabetes Obes. Metab.* **2013**, *15*, 15–27. [CrossRef] [PubMed]
24. de Luis, D.A.; Gonzalez Sagrado, M.; Conde, R.; Aller, R.; Izaola, O. Decreased basal levels of glucagon-like peptide-1 after weight loss in obese subjects. *Ann. Nutr. Metab.* **2007**, *51*, 134–138. [CrossRef] [PubMed]
25. Ueno, H.; Nakazato, M. Is ghrelin a glucagon-like peptide-1 secretagogue? *J. Diabetes Investig.* **2016**, *7*, 466–467. [CrossRef]
26. Gui, Y.; Silha, J.V.; Murphy, L.J. Sexual dimorphism and regulation of resistin, adiponectin, and leptin expression in the mouse. *Obes. Res.* **2004**, *12*, 1481–1491. [CrossRef]
27. Benomar, Y.; Taouis, M. Molecular Mechanisms Underlying Obesity-Induced Hypothalamic Inflammation and Insulin Resistance: Pivotal Role of Resistin/TLR4 Pathways. *Front. Endocrinol.* **2019**, *10*. [CrossRef]
28. Singhal, N.S.; Lazar, M.A.; Ahima, R.S. Central Resistin Induces Hepatic Insulin Resistance via Neuropeptide Y. *J. Neurosci.* **2007**, *27*, 12924–12932. [CrossRef]
29. Boucher, J.; Kleinridders, A.; Kahn, C.R. Insulin receptor signaling in normal and insulin-resistant states. *Cold Spring Harb. Perspect. Biol.* **2014**, *6*, a009191. [CrossRef]
30. Kwon, O.; Kim, K.W.; Kim, M.S. Leptin signalling pathways in hypothalamic neurons. *Cell. Mol. Life Sci. CMLS* **2016**, *73*, 1457–1477. [CrossRef]
31. Gong, Y.; Ishida-Takahashi, R.; Villanueva, E.C.; Fingar, D.C.; Münzberg, H.; Myers, M.G., Jr. The long form of the leptin receptor regulates STAT5 and ribosomal protein S6 via alternate mechanisms. *J. Biol. Chem.* **2007**, *282*, 31019–31027. [CrossRef] [PubMed]
32. Williams, T.; Courchet, J.; Viollet, B.; Brenman, J.E.; Polleux, F. AMP-activated protein kinase (AMPK) activity is not required for neuronal development but regulates axogenesis during metabolic stress. *Proc. Natl. Acad. Sci. USA* **2011**, *108*, 5849–5854. [CrossRef]
33. Hua, Z.; Gu, X.; Dong, Y.; Tan, F.; Liu, Z.; Thiele, C.J.; Li, Z. PI3K and MAPK pathways mediate the BDNF/TrkB-increased metastasis in neuroblastoma. *Tumour Biol. J. Int. Soc. Oncodevelopmental Biol. Med.* **2016**. [CrossRef]
34. Inoki, K.; Zhu, T.; Guan, K.L. TSC2 mediates cellular energy response to control cell growth and survival. *Cell* **2003**, *115*, 577–590. [CrossRef]
35. Schwartz, M.W.; Woods, S.C.; Porte, D.; Seeley, R.J.; Baskin, D.G. Central nervous system control of food intake. *Nature* **2000**, *404*, 661–671. [CrossRef] [PubMed]
36. Cone, R.; Cowley, M.A.; Butler, A.; Fan, W.; Marks, D.; Low, M. The arcuate nucleus as a conduit for diverse signals relevant to energy homeostasis. *Int. J. Obes.* **2001**, *25*, S63–S67. [CrossRef]
37. Barson, J.R.; Morganstern, I.; Leibowitz, S.F. Complementary roles of orexin and melanin-concentrating hormone in feeding behavior. *Int. J. Endocrinol.* **2013**, *2013*, 983964. [CrossRef]
38. Reitz, C.; Tang, M.X.; Luchsinger, J.; Mayeux, R. Relation of plasma lipids to Alzheimer disease and vascular dementia. *Arch. Neurol.* **2004**, *61*, 705–714. [CrossRef]
39. Notkola, I.L.; Sulkava, R.; Pekkanen, J.; Erkinjuntti, T.; Ehnholm, C.; Kivinen, P.; Tuomilehto, J.; Nissinen, A. Serum total cholesterol, apolipoprotein E epsilon 4 allele, and Alzheimer's disease. *Neuroepidemiology* **1998**, *17*, 14–20. [CrossRef]
40. Gu, Y.; Scarmeas, N.; Cosentino, S.; Brandt, J.; Albert, M.; Blacker, D.; Dubois, B.; Stern, Y. Change in body mass index before and after Alzheimer's disease onset. *Curr. Alzheimer Res.* **2014**, *11*, 349–356. [CrossRef]
41. Cova, I.; Clerici, F.; Rossi, A.; Cucumo, V.; Ghiretti, R.; Maggiore, L.; Pomati, S.; Galimberti, D.; Scarpini, E.; Mariani, C.; et al. Weight Loss Predicts Progression of Mild Cognitive Impairment to Alzheimer's Disease. *PLoS ONE* **2016**, *11*, e0151710. [CrossRef] [PubMed]

42. Folch, J.; Olloquequi, J.; Ettcheto, M.; Busquets, O.; Sánchez-López, E.; Cano, A.; Espinosa-Jiménez, T.; García, M.L.; Beas-Zarate, C.; Casadesús, G.; et al. The Involvement of Peripheral and Brain Insulin Resistance in Late Onset Alzheimer's Dementia. *Front. Aging Neurosci.* **2019**, *11*, 236. [CrossRef] [PubMed]
43. Laws, K.R.; Irvine, K.; Gale, T.M. Sex differences in Alzheimer's disease. *Curr. Opin. Psychiatry* **2018**, *31*, 133–139. [CrossRef]
44. Reilly, A.M.; Tsai, A.P.; Lin, P.B.; Ericsson, A.C.; Oblak, A.L.; Ren, H. Metabolic Defects Caused by High-Fat Diet Modify Disease Risk through Inflammatory and Amyloidogenic Pathways in a Mouse Model of Alzheimer's Disease. *Nutrients* **2020**, *12*, 2977. [CrossRef]
45. Gendron, W.H.; Fertan, E.; Pelletier, S.; Roddick, K.M.; O'Leary, T.P.; Anini, Y.; Brown, R.E. Age Related Weight Loss in Female 5xFAD Mice from 3 to 12 Months of Age. *Behav. Brain Res.* **2021**, 113214. [CrossRef]
46. Robison, L.S.; Gannon, O.J.; Thomas, M.A.; Salinero, A.E.; Abi-Ghanem, C.; Poitelon, Y.; Belin, S.; Zuloaga, K.L. Role of sex and high-fat diet in metabolic and hypothalamic disturbances in the 3xTg-AD mouse model of Alzheimer's disease. *J. Neuroinflammation* **2020**, *17*, 285. [CrossRef]
47. White, H.; Pieper, C.; Schmader, K. The association of weight change in Alzheimer's disease with severity of disease and mortality: A longitudinal analysis. *J. Am. Geriatr. Soc.* **1998**, *46*, 1223–1227. [CrossRef]
48. Gillette Guyonnet, S.; Abellan Van Kan, G.; Alix, E.; Andrieu, S.; Belmin, J.; Berrut, G.; Bonnefoy, M.; Brocker, P.; Constans, T.; Ferry, M.; et al. IANA (International Academy on Nutrition and Aging) Expert Group: Weight loss and Alzheimer's disease. *J. Nutr. Health Aging* **2007**, *11*, 38–48.
49. Soto, M.E.; Secher, M.; Gillette-Guyonnet, S.; Abellan van Kan, G.; Andrieu, S.; Nourhashemi, F.; Rolland, Y.; Vellas, B. Weight loss and rapid cognitive decline in community-dwelling patients with Alzheimer's disease. *J. Alzheimer Dis. JAD* **2012**, *28*, 647–654. [CrossRef]
50. Müller, S.; Preische, O.; Sohrabi, H.R.; Gräber, S.; Jucker, M.; Dietzsch, J.; Ringman, J.M.; Martins, R.N.; McDade, E.; Schofield, P.R.; et al. Decreased body mass index in the preclinical stage of autosomal dominant Alzheimer's disease. *Sci. Rep.* **2017**, *7*, 1225. [CrossRef]
51. Volicer, L.; Frijters, D.H.; van der Steen, J.T. Apathy and weight loss in nursing home residents: Longitudinal study. *J. Am. Med. Dir. Assoc.* **2013**, *14*, 417–420. [CrossRef]
52. Doorduijn, A.S.; Visser, M.; van de Rest, O.; Kester, M.I.; de Leeuw, F.A.; Boesveldt, S.; Fieldhouse, J.L.P.; van den Heuvel, E.; Teunissen, C.E.; Scheltens, P.; et al. Associations of AD Biomarkers and Cognitive Performance with Nutritional Status: The NUDAD Project. *Nutrients* **2019**, *11*, 1161. [CrossRef] [PubMed]
53. Doorduijn, A.S.; de van der Schueren, M.A.E.; van de Rest, O.; de Leeuw, F.A.; Hendriksen, H.M.A.; Teunissen, C.E.; Scheltens, P.; van der Flier, W.M.; Visser, M. Energy intake and expenditure in patients with Alzheimer's disease and mild cognitive impairment: The NUDAD project. *Alzheimer Res. Ther.* **2020**, *12*, 116. [CrossRef]
54. Poehlman, E.T.; Dvorak, R.V. Energy expenditure, energy intake, and weight loss in Alzheimer disease. *Am. J. Clin. Nutr.* **2000**, *71*, 650S–655S. [CrossRef]
55. Joly-Amado, A.; Serraneau, K.S.; Brownlow, M.; Marín de Evsikova, C.; Speakman, J.R.; Gordon, M.N.; Morgan, D. Metabolic changes over the course of aging in a mouse model of tau deposition. *Neurobiol. Aging* **2016**, *44*, 62–73. [CrossRef]
56. O'Leary, T.P.; Mantolino, H.M.; Stover, K.R.; Brown, R.E. Age-related deterioration of motor function in male and female 5xFAD mice from 3 to 16 months of age. *Genes Brain Behav.* **2020**, *19*, e12538. [CrossRef]
57. Jul, P.; Volbracht, C.; de Jong, I.E.; Helboe, L.; Elvang, A.B.; Pedersen, J.T. Hyperactivity with Agitative-Like Behavior in a Mouse Tauopathy Model. *J. Alzheimer Dis. JAD* **2016**, *49*, 783–795. [CrossRef]
58. Huebbe, P.; Dose, J.; Schloesser, A.; Campbell, G.; Glüer, C.C.; Gupta, Y.; Ibrahim, S.; Minihane, A.M.; Baines, J.F.; Nebel, A.; et al. Apolipoprotein E (APOE) genotype regulates body weight and fatty acid utilization-Studies in gene-targeted replacement mice. *Mol. Nutr. Food Res.* **2015**, *59*, 334–343. [CrossRef]
59. Loskutova, N.; Honea, R.A.; Brooks, W.M.; Burns, J.M. Reduced limbic and hypothalamic volumes correlate with bone density in early Alzheimer's disease. *J. Alzheimer Dis. JAD* **2010**, *20*, 313–322. [CrossRef]
60. Callen, D.J.; Black, S.E.; Gao, F.; Caldwell, C.B.; Szalai, J.P. Beyond the hippocampus: MRI volumetry confirms widespread limbic atrophy in AD. *Neurology* **2001**, *57*, 1669–1674. [CrossRef]
61. Braak, H.; Braak, E. Neuropathological stageing of Alzheimer-related changes. *Acta Neuropathol.* **1991**, *82*, 239–259. [CrossRef] [PubMed]
62. Ettcheto, M.; Cano, A.; Busquets, O.; Manzine, P.R.; Sánchez-López, E.; Castro-Torres, R.D.; Beas-Zarate, C.; Verdaguer, E.; García, M.L.; Olloquequi, J.; et al. A metabolic perspective of late onset Alzheimer's disease. *Pharmacol. Res.* **2019**, *145*, 104255. [CrossRef]
63. Ries, M.; Sastre, M. Mechanisms of Aβ Clearance and Degradation by Glial Cells. *Front. Aging Neurosci.* **2016**, *8*. [CrossRef] [PubMed]
64. Thaler, J.P.; Yi, C.X.; Schur, E.A.; Guyenet, S.J.; Hwang, B.H.; Dietrich, M.O.; Zhao, X.; Sarruf, D.A.; Izgur, V.; Maravilla, K.R.; et al. Obesity is associated with hypothalamic injury in rodents and humans. *J. Clin. Investig.* **2012**, *122*, 153–162. [CrossRef] [PubMed]
65. Cortés, N.; Andrade, V.; Guzmán-Martínez, L.; Estrella, M.; Maccioni, R.B. Neuroimmune Tau Mechanisms: Their Role in the Progression of Neuronal Degeneration. *Int. J. Mol. Sci.* **2018**, *19*, 956. [CrossRef] [PubMed]
66. Chaskiel, L.; Bristow, A.D.; Bluthé, R.M.; Dantzer, R.; Blomqvist, A.; Konsman, J.P. Interleukin-1 reduces food intake and body weight in rat by acting in the arcuate hypothalamus. *Brain Behav. Immun.* **2019**, *81*, 560–573. [CrossRef]

67. Arruda, A.P.; Milanski, M.; Coope, A.; Torsoni, A.S.; Ropelle, E.; Carvalho, D.P.; Carvalheira, J.B.; Velloso, L.A. Low-grade hypothalamic inflammation leads to defective thermogenesis, insulin resistance, and impaired insulin secretion. *Endocrinology* **2011**, *152*, 1314–1326. [CrossRef]
68. Velloso, L.A.; Araújo, E.P.; de Souza, C.T. Diet-induced inflammation of the hypothalamus in obesity. *Neuroimmunomodulation* **2008**, *15*, 189–193. [CrossRef]
69. Lee, Y.-H.; Hsu, H.-C.; Kao, P.-C.; Shiao, Y.-J.; Yeh, S.H.-H.; Shie, F.-S.; Hsu, S.-M.; Yeh, C.-W.; Liu, H.-K.; Yang, S.-B.; et al. Augmented Insulin and Leptin Resistance of High Fat Diet-Fed APPswe/PS1dE9 Transgenic Mice Exacerbate Obesity and Glycemic Dysregulation. *Int. J. Mol. Sci.* **2018**, *19*, 2333. [CrossRef]
70. Holscher, C. Diabetes as a risk factor for Alzheimer's disease: Insulin signalling impairment in the brain as an alternative model of Alzheimer's disease. *Biochem. Soc. Trans.* **2011**, *39*, 891–897. [CrossRef]
71. Chatterjee, S.; Mudher, A. Alzheimer's Disease and Type 2 Diabetes: A Critical Assessment of the Shared Pathological Traits. *Front. Neurosci.* **2018**, *12*, 383. [CrossRef]
72. de la Monte, S.M.; Wands, J.R. Alzheimer's disease is type 3 diabetes-evidence reviewed. *J. Diabetes Sci. Technol.* **2008**, *2*, 1101–1113. [CrossRef]
73. Wong, C.H.Y.; Wanrooy, B.J.; Bruce, D.G. Chapter 10-Neuroinflammation, Type 2 Diabetes, and Dementia. In *Type 2 Diabetes and Dementia*; Srikanth, V., Arvanitakis, Z., Eds.; Academic Press: Cambridge, MA, USA, 2018; pp. 195–209. [CrossRef]
74. Mehlig, K.; Lapidus, L.; Thelle, D.S.; Waern, M.; Zetterberg, H.; Björkelund, C.; Skoog, I.; Lissner, L. Low fasting serum insulin and dementia in nondiabetic women followed for 34 years. *Neurology* **2018**, *91*, e427–e435. [CrossRef]
75. Rönnemaa, E.; Zethelius, B.; Sundelöf, J.; Sundström, J.; Degerman-Gunnarsson, M.; Berne, C.; Lannfelt, L.; Kilander, L. Impaired insulin secretion increases the risk of Alzheimer disease. *Neurology* **2008**, *71*, 1065–1071. [CrossRef] [PubMed]
76. Peila, R.; Rodriguez, B.L.; White, L.R.; Launer, L.J. Fasting insulin and incident dementia in an elderly population of Japanese-American men. *Neurology* **2004**, *63*, 228–233. [CrossRef] [PubMed]
77. Ma, J.; Zhang, W.; Wang, H.F.; Wang, Z.X.; Jiang, T.; Tan, M.S.; Yu, J.T.; Tan, L. Peripheral Blood Adipokines and Insulin Levels in Patients with Alzheimer's Disease: A Replication Study and Meta-Analysis. *Curr. Alzheimer Res.* **2016**, *13*, 223–233. [CrossRef] [PubMed]
78. Freeman, L.R.; Zhang, L.; Dasuri, K.; Fernandez-Kim, S.O.; Bruce-Keller, A.J.; Keller, J.N. Mutant amyloid precursor protein differentially alters adipose biology under obesogenic and non-obesogenic conditions. *PLoS ONE* **2012**, *7*, e43193. [CrossRef]
79. Tramutola, A.; Arena, A.; Cini, C.; Butterfield, D.A.; Barone, E. Modulation of GLP-1 signaling as a novel therapeutic approach in the treatment of Alzheimer's disease pathology. *Expert Rev. Neurother.* **2017**, *17*, 59–75. [CrossRef]
80. Zhang, Z.Q.; Hölscher, C. GIP has neuroprotective effects in Alzheimer and Parkinson's disease models. *Peptides* **2020**, *125*, 170184. [CrossRef]
81. Hölscher, C. Novel dual GLP-1/GIP receptor agonists show neuroprotective effects in Alzheimer's and Parkinson's disease models. *Neuropharmacology* **2018**, *136*, 251–259. [CrossRef]
82. Kunath, N.; van Groen, T.; Allison, D.B.; Kumar, A.; Dozier-Sharpe, M.; Kadish, I. Ghrelin agonist does not foster insulin resistance but improves cognition in an Alzheimer's disease mouse model. *Sci. Rep.* **2015**, *5*, 11452. [CrossRef]
83. Al Massadi, O.; López, M.; Tschöp, M.; Diéguez, C.; Nogueiras, R. Current Understanding of the Hypothalamic Ghrelin Pathways Inducing Appetite and Adiposity. *Trends Neurosci.* **2017**, *40*, 167–180. [CrossRef] [PubMed]
84. Jeong, Y.O.; Shin, S.J.; Park, J.Y.; Ku, B.K.; Song, J.S.; Kim, J.J.; Jeon, S.G.; Lee, S.M.; Moon, M. MK-0677, a Ghrelin Agonist, Alleviates Amyloid Beta-Related Pathology in 5XFAD Mice, an Animal Model of Alzheimer's Disease. *Int. J. Mol. Sci.* **2018**, *19*, 1800. [CrossRef] [PubMed]
85. Park, H.K.; Ahima, R.S. Resistin in rodents and humans. *Diabetes Metab. J.* **2013**, *37*, 404–414. [CrossRef]
86. Benomar, Y.; Amine, H.; Crépin, D.; Al Rifai, S.; Riffault, L.; Gertler, A.; Taouis, M. Central Resistin/TLR4 Impairs Adiponectin Signaling, Contributing to Insulin and FGF21 Resistance. *Diabetes* **2016**, *65*, 913–926. [CrossRef] [PubMed]
87. Demirci, S.; Aynalı, A.; Demirci, K.; Demirci, S.; Arıdoğan, B.C. The Serum Levels of Resistin and Its Relationship with Other Proinflammatory Cytokines in Patients with Alzheimer's Disease. *Clin. Psychopharmacol. Neurosci.* **2017**, *15*, 59–63. [CrossRef] [PubMed]
88. Wilkinson, M.; Wilkinson, D.; Wiesner, G.; Morash, B.; Ur, E. Hypothalamic resistin immunoreactivity is reduced by obesity in the mouse: Co-localization with alpha-melanostimulating hormone. *Neuroendocrinology* **2005**, *81*, 19–30. [CrossRef] [PubMed]
89. Minokoshi, Y.; Alquier, T.; Furukawa, N.; Kim, Y.B.; Lee, A.; Xue, B.; Mu, J.; Foufelle, F.; Ferré, P.; Birnbaum, M.J.; et al. AMP-kinase regulates food intake by responding to hormonal and nutrient signals in the hypothalamus. *Nature* **2004**, *428*, 569–574. [CrossRef]
90. López-Gambero, A.J.; Rodríguez de Fonseca, F.; Suárez, J. Energy sensors in drug addiction: A potential therapeutic target. *Addict. Biol.* **2021**, *26*, e12936. [CrossRef] [PubMed]
91. Cota, D.; Proulx, K.; Smith, K.A.; Kozma, S.C.; Thomas, G.; Woods, S.C.; Seeley, R.J. Hypothalamic mTOR signaling regulates food intake. *Science* **2006**, *312*, 927–930. [CrossRef]
92. Shibuya, Y.; Chang, C.C.; Huang, L.H.; Bryleva, E.Y.; Chang, T.Y. Inhibiting ACAT1/SOAT1 in microglia stimulates autophagy-mediated lysosomal proteolysis and increases Aβ1-42 clearance. *J. Neurosci. Off. J. Soc. Neurosci.* **2014**, *34*, 14484–14501. [CrossRef] [PubMed]

93. Lafay-Chebassier, C.; Paccalin, M.; Page, G.; Barc-Pain, S.; Perault-Pochat, M.C.; Gil, R.; Pradier, L.; Hugon, J. mTOR/p70S6k signalling alteration by Abeta exposure as well as in APP-PS1 transgenic models and in patients with Alzheimer's disease. *J. Neurochem.* **2005**, *94*, 215–225. [CrossRef]
94. Avrahami, L.; Farfara, D.; Shaham-Kol, M.; Vassar, R.; Frenkel, D.; Eldar-Finkelman, H. Inhibition of glycogen synthase kinase-3 ameliorates β-amyloid pathology and restores lysosomal acidification and mammalian target of rapamycin activity in the Alzheimer disease mouse model: In vivo and in vitro studies. *J. Biol. Chem.* **2013**, *288*, 1295–1306. [CrossRef]
95. Zeng, Y.; Zhang, J.; Zhu, Y.; Zhang, J.; Shen, H.; Lu, J.; Pan, X.; Lin, N.; Dai, X.; Zhou, M.; et al. Tripchlorolide improves cognitive deficits by reducing amyloid β and upregulating synapse-related proteins in a transgenic model of Alzheimer's Disease. *J. Neurochem.* **2015**, *133*, 38–52. [CrossRef] [PubMed]
96. Wang, X.; Li, G.-J.; Hu, H.-X.; Ma, C.; Ma, D.-H.; Liu, X.-L.; Jiang, X.-M. Cerebral mTOR signal and pro-inflammatory cytokines in Alzheimer's disease rats. *Transl. Neurosci.* **2016**, *7*, 151–157. [CrossRef] [PubMed]
97. Mueed, Z.; Tandon, P.; Maurya, S.K.; Deval, R.; Kamal, M.A.; Poddar, N.K. Tau and mTOR: The Hotspots for Multifarious Diseases in Alzheimer's Development. *Front. Neurosci.* **2018**, *12*, 1017. [CrossRef]
98. Czapski, G.A.; Gąssowska, M.; Wilkaniec, A.; Chalimoniuk, M.; Strosznajder, J.B.; Adamczyk, A. The mechanisms regulating cyclin-dependent kinase 5 in hippocampus during systemic inflammatory response: The effect on inflammatory gene expression. *Neurochem. Int.* **2016**, *93*, 103–112. [CrossRef] [PubMed]
99. Ishii, M.; Hiller, A.J.; Pham, L.; McGuire, M.J.; Iadecola, C.; Wang, G. Amyloid-Beta Modulates Low-Threshold Activated Voltage-Gated L-Type Calcium Channels of Arcuate Neuropeptide Y Neurons Leading to Calcium Dysregulation and Hypothalamic Dysfunction. *J. Neurosci. Off. J. Soc. Neurosci.* **2019**, *39*, 8816–8825. [CrossRef] [PubMed]
100. Fu, L.Y.; Acuna-Goycolea, C.; van den Pol, A.N. Neuropeptide Y inhibits hypocretin/orexin neurons by multiple presynaptic and postsynaptic mechanisms: Tonic depression of the hypothalamic arousal system. *J. Neurosci. Off. J. Soc. Neurosci.* **2004**, *24*, 8741–8751. [CrossRef]
101. Sethi, M.; Joshi, S.S.; Webb, R.L.; Beckett, T.L.; Donohue, K.D.; Murphy, M.P.; O'Hara, B.F.; Duncan, M.J. Increased fragmentation of sleep-wake cycles in the 5XFAD mouse model of Alzheimer's disease. *Neuroscience* **2015**, *290*, 80–89. [CrossRef]
102. Liu, R.M.; van Groen, T.; Katre, A.; Cao, D.; Kadisha, I.; Ballinger, C.; Wang, L.; Carroll, S.L.; Li, L. Knockout of plasminogen activator inhibitor 1 gene reduces amyloid beta peptide burden in a mouse model of Alzheimer's disease. *Neurobiol. Aging* **2011**, *32*, 1079–1089. [CrossRef] [PubMed]
103. Akhter, H.; Huang, W.T.; van Groen, T.; Kuo, H.C.; Miyata, T.; Liu, R.M. A Small Molecule Inhibitor of Plasminogen Activator Inhibitor-1 Reduces Brain Amyloid-β Load and Improves Memory in an Animal Model of Alzheimer's Disease. *J. Alzheimer Dis. JAD* **2018**, *64*, 447–457. [CrossRef] [PubMed]
104. Jacobsen, J.S.; Comery, T.A.; Martone, R.L.; Elokdah, H.; Crandall, D.L.; Oganesian, A.; Aschmies, S.; Kirksey, Y.; Gonzales, C.; Xu, J.; et al. Enhanced clearance of Aβ in brain by sustaining the plasmin proteolysis cascade. *Proc. Natl. Acad. Sci. USA* **2008**, *105*, 8754–8759. [CrossRef]
105. de Oliveira, F.F.; Chen, E.S.; Smith, M.C.; Bertolucci, P.H.F. Longitudinal lipid profile variations and clinical change in Alzheimer's disease dementia. *Neurosci. Lett.* **2017**, *646*, 36–42. [CrossRef] [PubMed]
106. Maarouf, C.L.; Kokjohn, T.A.; Whiteside, C.M.; Macias, M.P.; Kalback, W.M.; Sabbagh, M.N.; Beach, T.G.; Vassar, R.; Roher, A.E. Molecular Differences and Similarities Between Alzheimer's Disease and the 5XFAD Transgenic Mouse Model of Amyloidosis. *Biochem. Insights* **2013**, *6*, 1–10. [CrossRef]
107. Rivera, P.; Ramírez-López, M.T.; Vargas, A.; Decara, J.; Vázquez, M.; Arco, R.; Gómez de Heras, R.; Argente, J.; Rodríguez de Fonseca, F.; Chowen, J.A.; et al. Perinatal free-choice of a high-calorie low-protein diet affects leptin signaling through IRS1 and AMPK dephosphorylation in the hypothalami of female rat offspring in adulthood. *Acta Physiol.* **2019**, *226*, e13244. [CrossRef]
108. Rivera, P.; Guerra-Cantera, S.; Vargas, A.; Díaz, F.; García-Úbeda, R.; Tovar, R.; Ramírez-López, M.T.; Argente, J.; de Fonseca, F.R.; Suárez, J.; et al. Maternal hypercaloric diet affects factors involved in lipid metabolism and the endogenous cannabinoid systems in the hypothalamus of adult offspring: Sex-specific response of astrocytes to palmitic acid and anandamide. *Nutr. Neurosci.* **2020**, 1–14. [CrossRef]
109. Navarro, J.A.; Decara, J.; Medina-Vera, D.; Tovar, R.; Suarez, J.; Pavón, J.; Serrano, A.; Vida, M.; Gutierrez-Adan, A.; Sanjuan, C.; et al. D-Pinitol from Ceratonia siliqua Is an Orally Active Natural Inositol that Reduces Pancreas Insulin Secretion and Increases Circulating Ghrelin Levels in Wistar Rats. *Nutrients* **2020**, *12*, 2030. [CrossRef]
110. Paxinos, G.; Franklin, K.B. *Paxinos and Franklin's the Mouse Brain in Stereotaxic Coordinates*; Academic Press: Cambridge, MA, USA, 2019.
111. Medina-Vera, D.; Rosell-Valle, C.; López-Gambero, A.J.; Navarro, J.A.; Zambrana-Infantes, E.N.; Rivera, P.; Santín, L.J.; Suarez, J.; Rodríguez de Fonseca, F. Imbalance of Endocannabinoid/Lysophosphatidylinositol Receptors Marks the Severity of Alzheimer's Disease in a Preclinical Model: A Therapeutic Opportunity. *Biology* **2020**, *9*, 377. [CrossRef]
112. Suárez, J.; Rivera, P.; Arrabal, S.; Crespillo, A.; Serrano, A.; Baixeras, E.; Pavón, F.J.; Cifuentes, M.; Nogueiras, R.; Ballesteros, J.; et al. Oleoylethanolamide enhances β-adrenergic-mediated thermogenesis and white-to-brown adipocyte phenotype in epididymal white adipose tissue in rat. *Dis. Models Mech.* **2014**, *7*, 129–141. [CrossRef]

International Journal of *Molecular Sciences*

MDPI

Article

Effects of Overexpression of Neurosecretory Protein GL-Precursor Gene on Glucose Homeostasis and Insulin Sensitivity in Mice

Keisuke Fukumura, Yuki Narimatsu, Shogo Moriwaki, Eiko Iwakoshi-Ukena, Megumi Furumitsu and Kazuyoshi Ukena *

Laboratory of Neurometabolism, Graduate School of Integrated Sciences for Life, Hiroshima University, Higashi-Hiroshima, Hiroshima 739-8521, Japan; kfuku@hiroshima-u.ac.jp (K.F.); d214243@hiroshima-u.ac.jp (Y.N.); m203300@hiroshima-u.ac.jp (S.M.); iwakoshi@hiroshima-u.ac.jp (E.I.-U.); mfurumi@hiroshima-u.ac.jp (M.F.)

* Correspondence: ukena@hiroshima-u.ac.jp

Abstract: A high-fat diet (HFD) quickly induces obesity with insulin resistance and hyperglycemia. We previously reported that a novel hypothalamic small protein, named neurosecretory protein GL (NPGL), stimulates feeding and fat accumulation in mice. However, the effects of NPGL on insulin sensitivity and glucose homeostasis remain unknown. Hence, we subjected NPGL-precursor gene (*Npgl*)-overexpressing mice to the oral glucose tolerance test (OGTT) and intraperitoneal insulin tolerance test (IPITT) under normal chow (NC) and HFD conditions. *Npgl* overexpression promoted body mass gain and tended to increase food intake of NC-fed mice, whereas it had little effect on HFD-fed mice. The OGTT showed elevated blood glucose and insulin levels in *Npgl*-overexpressing NC-fed mice 15 min after glucose administration. Both the OGTT and IPITT demonstrated that *Npgl* overexpression decreased blood glucose levels in HFD-fed mice 60 min after glucose and insulin treatments. Notably, *Npgl* overexpression increased adipose tissue masses only in NC-fed mice, and it decreased blood glucose and insulin levels in HFD-fed mice at the experimental end point. It also increased the mRNA expression of galanin, one of the feeding and metabolic regulatory neuropeptides, in the hypothalamus of HFD-fed mice. Therefore, NPGL may alleviate HFD-induced hyperglycemia and insulin resistance in mice.

Keywords: neurosecretory protein GL; hypothalamus; neuropeptide; obesity; glucose homeostasis; insulin sensitivity

Citation: Fukumura, K.; Narimatsu, Y.; Moriwaki, S.; Iwakoshi-Ukena, E.; Furumitsu, M.; Ukena, K. Effects of Overexpression of Neurosecretory Protein GL-Precursor Gene on Glucose Homeostasis and Insulin Sensitivity in Mice. *Int. J. Mol. Sci.* **2021**, 22, 4681. https://doi.org/10.3390/ijms22094681

Academic Editor: Rosalía Rodríguez-Rodríguez

Received: 27 March 2021
Accepted: 26 April 2021
Published: 28 April 2021

1. Introduction

Obesity is a disease that has reached epidemic proportions. It is a major health-related concern worldwide because this condition and its comorbidities, such as depression, type 2 diabetes, cardiovascular disease, and certain cancers, have created a huge economic burden [1–3]. The increase in the prevalence of obesity has emphasized the need for research focusing on its biological causative factors [3]. During obesity development, excess fat accumulation promotes insulin resistance and glucose intolerance, which is generally accompanied by chronic inflammation in adipose tissue [4–6]. As overfeeding and/or biased feeding, such as continuous feeding of a high-fat diet (HFD), quickly leads to metabolic disorders, the regulatory mechanisms of feeding behavior and metabolism have been investigated [7–9]. To date, several hypothalamic neuropeptides involved in feeding behavior have been identified in the arcuate nucleus of the hypothalamus, for instance, potent orexigenic factors such as neuropeptide Y (NPY) and agouti-related peptide (AgRP), and the anorexigenic factor, proopiomelanocortin (POMC)-derived α-melanocyte-stimulating hormone [7–9]. Among peripheral factors, ghrelin and leptin are well-known, feeding-regulatory peptides. Ghrelin, an orexigenic peptide secreted by the stomach, stimulates feeding behavior via NPY/AgRP neurons [10–12]. Leptin, an anorexigenic polypeptide secreted from white adipose tissue (WAT) in proportion to body fat reserves, influences NPY/AgRP and POMC neuron activities [13–16]. Moreover, insulin, a key regulator secreted from pancreatic β-cells, converts dietary carbohydrates into fat deposits and maintains systemic glucose homeostasis [17,18]. Although many factors involved in the regulation of energy homeostasis have been identified in the last few decades, the hormonal controls of insulin sensitivity and glucose homeostasis that underlie obesity development are not fully understood.

To elucidate the regulatory mechanism of energy homeostasis, we investigated previously unknown bioactive substances and their modes of action in animals. The search for novel neuropeptides and peptide hormone precursors in the hypothalamus led to the identification of a novel cDNA in the chick hypothalamus [19]. As the deduced precursor protein contains a small secretory protein of 80 amino acids with Gly-Leu-NH_2 at the C-terminus, the novel neuropeptide has been named neurosecretory protein GL (NPGL) [19]. Furthermore, homologous NPGL proteins have been discovered in mammals, including humans, rats, and mice, suggesting that the primary structure of NPGL is highly conserved in mammals and avian species [20]. In addition, its paralogous neuropeptide has been identified and named neurosecretory protein GM [21]. Similar to the effects of NPGL infusion on energy metabolism in avian species [22,23], acute intracerebroventricular (i.c.v.) infusion of NPGL stimulates feeding behavior in mice fed with normal chow (NC) [24]. Chronic i.c.v. infusion of NPGL decreases locomotor activity during the dark period in mice [25]. In addition, chronic i.c.v. infusion of NPGL increases food intake with considerable fat accumulation in mice fed with a medium fat/medium sucrose diet (MFSD), whereas it induces moderate fat accumulation without changing food intake in NC-fed mice [25]. We previously showed that chronic i.c.v. infusion of NPGL elicits food intake and subsequent fat accumulation through de novo lipogenesis in rats [26]. Notably, overexpression of the NPGL-precursor gene (*Npgl*) in the mouse hypothalamus increases food intake and fat deposits under NC and MFSD conditions and elevates blood insulin levels without changing the blood glucose levels [27]. However, the effects of NPGL on insulin sensitivity and glucose homeostasis have not been elucidated.

In this study, we subjected *Npgl*-overexpressing mice to the oral glucose tolerance test (OGTT) and intraperitoneal insulin tolerance test (IPITT) under a series of NC and HFD feedings to investigate whether NPGL affects insulin sensitivity and glucose tolerance. Additionally, we measured body mass gain, food intake, body composition, blood parameters, and mRNA expression of feeding and metabolic regulatory genes at the experimental end point to determine the effects of HFD feeding on NPGL action in mice.

2. Results

2.1. *Effects of NPGL-Precursor Gene Overexpression on Food Intake and Body Mass Gain under NC and HFD Conditions*

To determine the effects of NPGL on feeding behavior, metabolism, glucose homeostasis, and insulin sensitivity under NC and HFD conditions, we conducted a series of experiments, as shown in Figure 1A. Briefly, mice were fed NC for 28 days for the first OGTT and IPITT, and subsequently fed an HFD for 28 days for the second OGTT and IPITT. When NC feeding was started, control mice and *Npgl*-overexpressing mice were weighed at 23.6 ± 0.2 g and 23.4 ± 0.3 g, respectively. When subsequent HFD feeding was started, control mice and *Npgl*-overexpressing mice were weighed at 25.6 ± 0.4 g and 26.7 ± 0.3 g, respectively. Quantitative RT-PCR (qRT-PCR) showed chronic adeno-associated virus (AAV)-induced *Npgl* overexpression in the mediobasal hypothalamus (MBH) of mice at the experimental end point (Figure S1). *Npgl* overexpression significantly increased body mass gain from day 12 and tended to increase cumulative food intake in mice fed with NC for 28 days (Figure 1B,C). In contrast, *Npgl* overexpression did not affect body mass gain or cumulative food intake under HFD conditions (Figure 1B,C).

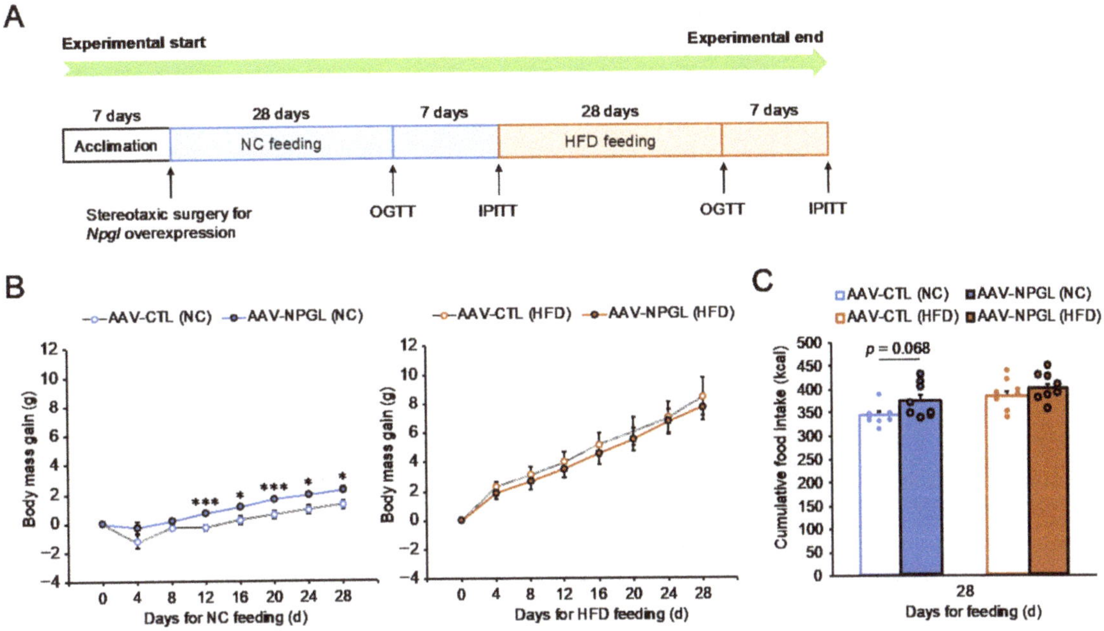

Figure 1. Effects of *Npgl* overexpression on body mass gain and food intake in normal chow (NC)-fed and high-fat diet (HFD)-fed mice. The panels show the data obtained upon injection of the AAV-based control vector (AAV-CTL) or the AAV-based NPGL-precursor gene vector (AAV-NPGL) in NC-fed and HFD-fed mice. (**A**) Experimental procedure. After animal surgery, mice were fed NC for 28 days until the first oral glucose tolerance test (OGTT) and intraperitoneal insulin tolerance test (IPITT). Thereafter, the diet was changed to an HFD and mice were fed for 28 days until the second OGTT and IPITT. (**B**) Body mass gain and (**C**) cumulative food intake. Circles in **C** represent individual data points. Each value represents the mean ± standard error of the mean ($n = 8$; * $p < 0.05$, *** $p < 0.005$ for Student's t-test).

2.2. *Effects of NPGL-Precursor Gene Overexpression on Glucose Homeostasis and Insulin Sensitivity under NC Conditions*

After 28 days of *Npgl* overexpression under NC conditions, we performed the OGTT. Blood glucose and insulin levels were significantly higher 15 min after oral glucose administration in *Npgl*-overexpressing mice (Figure 2A,C). In contrast, a calculation of the area under the curve (AUC) above the glucose baseline showed no significant difference between control and *Npgl*-overexpressing mice (Figure 2B). After one week of recovery from OGTT damage, we performed the IPITT. There was no significant difference in the blood glucose level and inverse AUC below the glucose baseline after intraperitoneal insulin administration (Figure 2D,E).

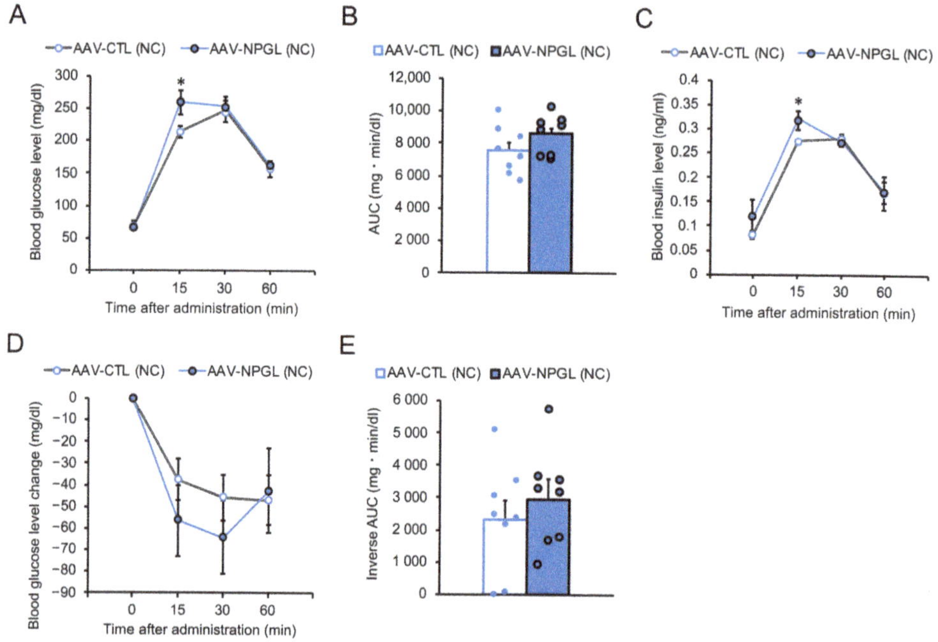

Figure 2. Effects of *Npgl* overexpression on glucose tolerance and insulin sensitivity in normal chow (NC)-fed mice. The panels show the data obtained upon injection of the AAV-based control vector (AAV-CTL) or the AAV-based NPGL-precursor gene vector (AAV-NPGL) in NC-fed mice. (**A–C**) Results of oral glucose tolerance test (OGTT) (**A**: blood glucose levels, **B**: area under the curve (AUC) for blood glucose levels, **C**: corresponding blood insulin secretion curves) for NC-fed mice at multiple time points. (**D**,**E**) Results of intraperitoneal insulin tolerance test (IPITT) (**D**: changes in blood glucose levels compared with those at time 0, **E**: inverse AUC for blood glucose levels) for NC-fed mice at multiple time points. Circles in **B** and **E** represent individual data points. Each value represents the mean ± standard error of the mean ($n = 8$; * $p < 0.05$ for Student's *t*-test).

2.3. Effects of NPGL-Precursor Gene Overexpression on Glucose Homeostasis and Insulin Sensitivity under HFD Conditions

After the OGTT and IPITT under NC conditions, we changed the diet from NC to an HFD and performed the same tests after 28 days of rearing under HFD conditions (Figure 1A). In the OGTT, the blood glucose level was significantly lower 60 min after oral glucose administration in *Npgl*-overexpressing mice (Figure 3A). In addition, a slight decrease in the AUC was observed for *Npgl*-overexpressing mice (Figure 3B), whereas there was no significant difference in the blood insulin levels (Figure 3C). After one week of recovery from OGTT damage, we performed the IPITT. In *Npgl*-overexpressing mice, the blood glucose level was significantly lower 60 min after intraperitoneal insulin administration, and the inverse AUC below the glucose baseline tended to be higher but without a significant difference (Figure 3D,E).

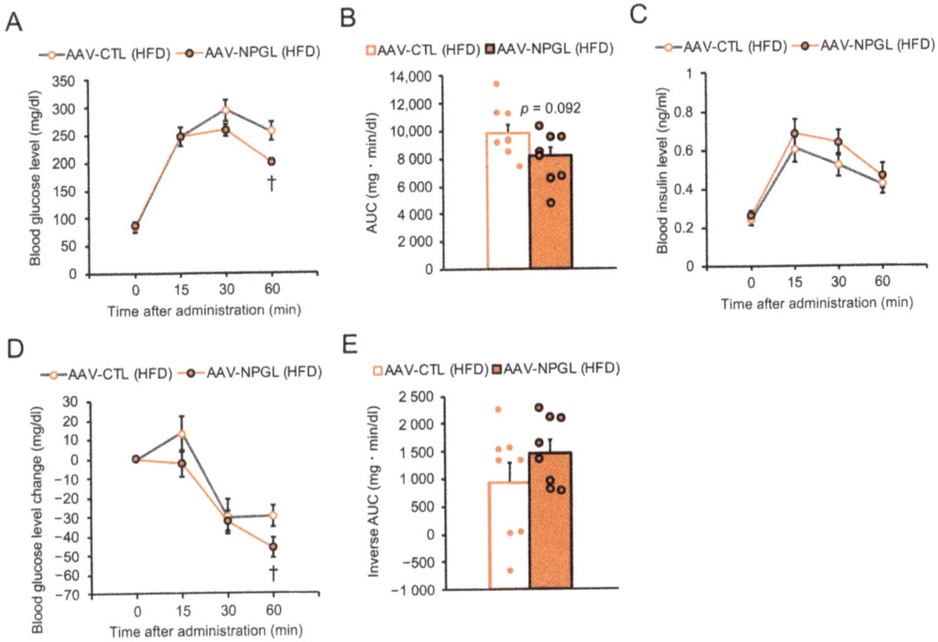

Figure 3. Effects of *Npgl* overexpression on glucose tolerance and insulin sensitivity in high-fat diet (HFD)-fed mice. The panels show the data obtained upon injection of the AAV-based control vector (AAV-CTL) or the AAV-based NPGL-precursor gene vector (AAV-NPGL) in HFD-fed mice. (**A–C**) Results of oral glucose tolerance test (OGTT) ((**A**): blood glucose levels, (**B**): area under the curve (AUC) for blood glucose levels, (**C**): corresponding blood insulin secretion curves) for HFD-fed mice at multiple time points. (**D**,**E**) Results of intraperitoneal insulin tolerance test (IPITT) ((**D**): changes in blood glucose levels compared with those at time 0, (**E**): inverse AUC for blood glucose levels) for HFD-fed mice at multiple time points. Circles in (**B**,**E**) represent individual data points. Each value represents the mean ± standard error of the mean (n = 8; † $p < 0.05$ for Student's t-test).

2.4. Effects of NPGL-Precursor Gene Overexpression on Body Composition and Blood Parameters

To examine the effects of nutrition on NPGL action with respect to body composition and blood parameters, we measured the masses of adipose tissues, muscle, and several organs under NC and HFD conditions. Independent of the series of experiments described in Figure 1A, we overexpressed *Npgl* in the hypothalamus for 40 days to analyze body composition under NC conditions. Under NC, the masses of interscapular brown adipose tissue (BAT) and WAT were significantly higher in *Npgl*-overexpressing mice than in control mice (Figure S2A,B). In contrast, there were no significant differences in the tissue masses of the control and *Npgl*-overexpressing mice under HFD conditions (Figure S2A,B). In addition, under both feeding conditions, *Npgl* overexpression did not affect the masses of the gastrocnemius muscle, liver, testis, kidney, and heart (Figure S3). Notably, the blood glucose and insulin levels at the experimental end point without fasting under HFD conditions were significantly lower in *Npgl*-overexpressing mice, whereas there were no changes in blood triglyceride (TG) and free fatty acid (FFA) levels (Figure 4).

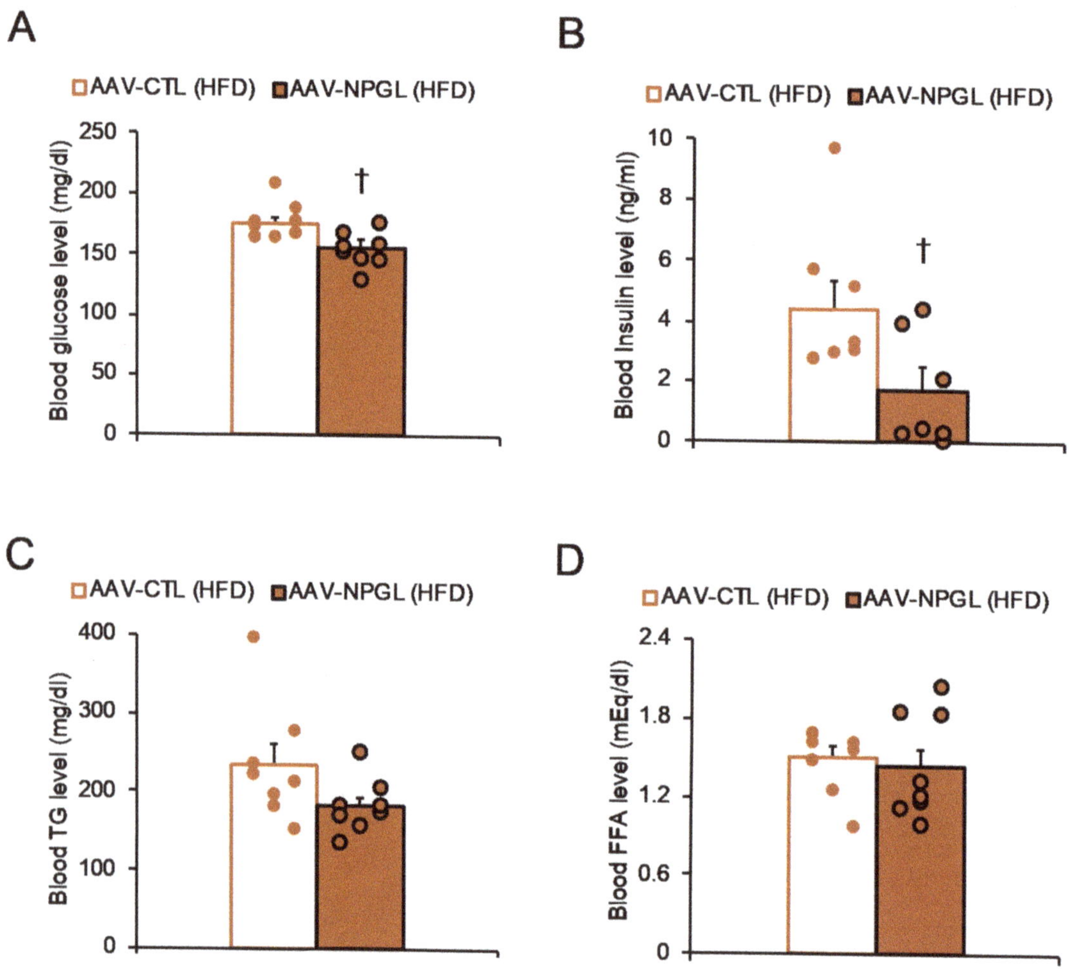

Figure 4. Effects of *Npgl* overexpression on blood parameters in high-fat diet (HFD)-fed mice. The panels show the data obtained upon injection of the AAV-based control vector (AAV-CTL) or the AAV-based NPGL-precursor gene vector (AAV-NPGL) in HFD-fed mice. (**A**) Blood glucose, (**B**) insulin, (**C**) triglyceride (TG), and (**D**) free fatty acid (FFA) levels. Circles represent individual data points. Each value represents the mean ± standard error of the mean ((**A**,**C**,**D**), $n = 8$; **B**, $n = 7$; † $p < 0.05$ for Student's t-test).

2.5. Effects of NPGL-Precursor Gene Overexpression on mRNA Expression of Neuropeptides and Genes Related to Gluconeogenesis and Glucose Uptake

Since we observed decreases in blood glucose levels both in the OGTT and IPITT under HFD conditions and at the end point of a series of experiments, we measured the mRNA expression levels of neuropeptides, which are involved in whole-body energy metabolism, in the MBH. The qRT-PCR showed that *Npgl* overexpression increased the mRNA expression of galanin (*Gal*), whereas it had no effect on the expression of *Npy*, *Agrp*, or *Pomc* under HFD conditions (Figure 5). On the other hand, we measured mRNA expression of genes related to lipid metabolism, glycolysis, glucose and lipid uptake, and browning in the inguinal WAT (iWAT), such as acetyl-CoA carboxylase (*Acc*), fatty acid synthase (*Fas*), carbohydrate-responsive element-binding protein α (*Chrebpα*), carnitine palmitoyltransferase 1a (*Cpt1a*), adipose triglyceride lipase (*Atgl*), hormone-sensitive lipase (*Hsl*), glyceraldehyde-3-phosphate dehydrogenase (*Gapdh*), solute carrier family 2 member 4 (*Slc2a4*), cluster of differentiation 36 (*Cd36*), peroxisome proliferator-activated receptor γ coactivator 1α (*Pgc1α*), uncoupling protein 1 (*Ucp1*), and type II iodothyronine deiodinase (*Dio2*). However, *Npgl* overexpression did not affect the expression of these genes (Figure S4). In addition, when we measured mRNA expression of genes related to gluconeogenesis and glucose uptake in the liver, such as glucose-6-phosphatase (*G6pase*), phosphoenolpyruvate carboxykinase (*Pepck*), solute carrier family 2 member 2 (*Slc2a2*), and fibroblast growth factor 21 (*Fgf21*), *Npgl* overexpression induced no changes in these genes (Figure S5).

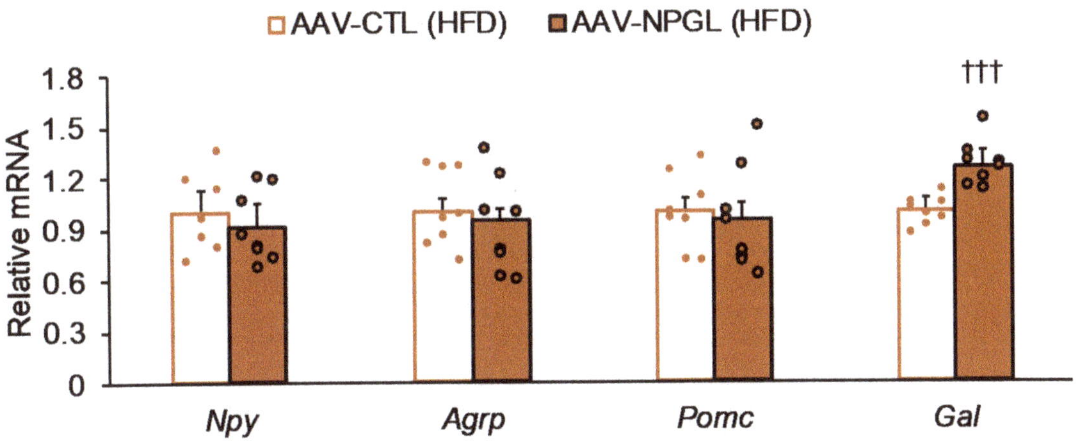

Figure 5. Effects of *Npgl* overexpression on mRNA expression of neuropeptides in high-fat diet (HFD)-fed mice. The graph shows the data obtained upon injection of the AAV-based control vector (AAV-CTL) or the AAV-based NPGL-precursor gene vector (AAV-NPGL) in HFD-fed mice. The mRNA expression levels of neuropeptide Y (*Npy*), agouti-related peptide (*Agrp*), proopiomelanocortin (*Pomc*), and galanin (*Gal*) in the mediobasal hypothalamus. Circles represent individual data points. Each value represents the mean ± standard error of the mean ($n = 8$; ††† $p < 0.005$ for Student's t-test).

3. Discussion

Hypothalamic neuropeptides predominantly control feeding behavior and metabolism and are closely linked to obesity development [28,29]. We recently demonstrated that hypothalamic overexpression of *Npgl*, a novel small secretory protein precursor gene, elicits food intake and fat accumulation in mice [27]. However, the effect of NPGL on insulin sensitivity and glucose homeostasis has not been elucidated. In this study, we subjected *Npgl*-overexpressing mice to the OGTT and IPITT under both NC and HFD conditions. Our data showed that *Npgl* overexpression under HFD conditions restrained an increase in blood glucose level in the OGTT, whereas it promoted a decrease in blood glucose level in the IPITT. Furthermore, *Npgl* overexpression reduced blood glucose and insulin levels in HFD-fed mice at the experimental end point.

Our data highlighted the effects of *Npgl* overexpression, which mitigated glucose intolerance in the OGTT and insulin resistance in the IPITT under HFD conditions. In contrast, we observed increases in blood glucose and insulin levels 15 min after oral glucose administration in *Npgl*-overexpressing mice under NC conditions. Several studies have indicated that an increase in blood glucose level immediately after glucose administration can reflect glucose absorption into the circulation through the intestine [30–32]. In addition, our previous study has suggested that *Npgl* overexpression promotes the absorption of glucose, as a substrate of de novo lipogenesis, from the circulation into WAT in NC-fed mice [27]. These data imply that NPGL affects multiple tissues, such as the intestine and WAT, to orchestrate glucose absorption and carbohydrate use for efficient de novo lipogenesis under NC conditions. In contrast to our previous study demonstrating that *Npgl* overexpression is unable to affect the blood glucose level in NC-fed mice [27], we again emphasize the finding that *Npgl* overexpression alleviated glucose intolerance, insulin resistance, and hyperglycemia in HFD-fed mice at the experimental end point of the present study. Over the past 30 years, novel molecular mechanisms linking obesity and its related disorders have been deciphered. Although the field has primarily focused on the direct impact of obesity-associated alterations in peripheral tissues such as the liver, skeletal muscle, and adipose tissue, the role of the central nervous system as a regulator of energy homeostasis among different organs has not received the same attention [33]. In this study, we found that *Npgl* overexpression upregulated the mRNA expression of *Gal*, a 29-amino-acid peptide, in the hypothalamus. Several studies have recently uncovered new aspects of neuropeptides in energy metabolism. For instance, GAL ameliorates insulin resistance and improves glucose metabolism by activating the trafficking of glucose transporter 4 and glucose uptake in the skeletal muscle and adipose tissue [34]. In addition, we have shown the co-localization of GAL and NPGL in the same neurons of the hypothalamic arcuate nucleus in mice [25], raising the possibility that these neuropeptides modulate each other at the transcriptional level, perhaps via the autocrine system. In contrast, the receptor for NPGL and its intracellular signaling remain unidentified. Although our data are limited to transcriptional changes of the neuropeptides, future studies to analyze central hormonal relay and peripheral insulin signaling will open up new avenues for the hypothalamic regulation of insulin sensitivity and glucose homeostasis.

In this study, we observed differences in the effects of *Npgl* overexpression on body mass gain and fat accumulation under NC and HFD conditions. *Npgl* overexpression increased adipose tissue mass under NC conditions, whereas it had little effect under HFD conditions. We previously demonstrated that NPGL stimulates fat accumulation in WAT through de novo lipogenesis using dietary carbohydrates in rats [26]. In addition, we recently revealed that NPGL promotes fat accumulation in high-sucrose diet (HSD)-fed rats, although it does not induce an increase in food intake [35]. In contrast, it is well known that an HFD suppresses de novo lipogenesis in rodents [36,37]. Moreover, the present data indicated that *Npgl* overexpression hardly increased WAT mass in HFD-fed mice. In addition, we could not observe activated de novo lipogenesis at the transcriptional level in the WAT of *Npgl*-overexpressing HFD-fed mice (data not shown). Dietary carbohydrates and fat regulate de novo lipogenesis partially via certain transcriptional factors in adipose tissue. Among them, carbohydrate response element binding protein (ChREBP) is a critical transcriptional factor of systemic lipid metabolism, including de novo lipogenesis, in various peripheral tissues [38]. ChREBP is activated and promotes de novo lipogenesis in response to carbohydrate intake, whereas dietary fat, such as polyunsaturated fatty acids, inhibits ChREBP-induced de novo lipogenesis [39]. It has been shown that *Npgl* overexpression upregulates mRNA expression of *ChREBP* in mice fed with NC and an MFSD, which include a large amount of carbohydrates [27]. Hence, we speculate that dietary carbohydrates and fat have opposite effects on NPGL action in fat accumulation, perhaps via the transcriptional factors in peripheral tissues, such as WAT. Since these transcriptional factors are regulated by post-translational modification as well as at the transcriptional and translational levels, detailed analysis is required to understand the regulatory mechanisms of lipid metabolism by NPGL. On the other hand, BAT mass was also increased in NC-fed mice, but not HFD-fed mice in the present study. However, the regulatory mechanisms of fat accumulation in the BAT and the functional relationship between the WAT and BAT remain unknown at this time.

Our data supported the finding that *Npgl* overexpression stimulates food intake in NC-fed mice [27], whereas it did not affect feeding behavior under HFD conditions. To date, a considerable amount of research has revealed that dietary nutrients and metabolic status influence neuropeptide functions in feeding behavior and metabolism [40,41]. For instance, fasting and long-term HFD feeding evoke neuronal activation, including increased spike frequency in NPY/AgRP neurons via peripheral signaling in mice [42,43]. Similarly, we demonstrated that NPGL stimulates food intake at different intensities under feeding with different nutritional compositions, based on the species. Under MFSD conditions,

NPGL significantly stimulates food intake, whereas it has little effect under NC conditions in rats [26]. In addition, we recently showed that NPGL cannot induce an increase in food intake of an HSD in rats [35]. Since both HFD and HSD are highly unbalanced nutrient diets, NPGL might stimulate feeding behavior only toward MFSD. A recent report demonstrated that the hypothalamic corticotropin-releasing hormone promotes the intake of carbohydrate over fat in mice [44]. Peripheral FGF21 suppresses simple sugar intake as a negative feedback in response to dietary carbohydrates in mice [45,46]. In contrast, although we observed that NPGL-like immunoreactive fibers contact the anorexigenic POMC neurons in the arcuate nucleus in mice [24], the molecular mechanisms by which NPGL influences feeding behavior remain unclear. Therefore, further study is needed to elucidate the molecular mechanisms of feeding behavior, including feeding preferences.

In summary, this study revealed that *Npgl* overexpression exerted different effects on feeding behavior and fat accumulation under NC and HFD conditions. The results of a series of previous studies [25–27] strongly suggest that NPGL plays multiple roles in energy homeostasis according to dietary nutrients. The present data showing the differences of NPGL action in body mass gain and fat accumulation under NC and HFD conditions require future study to analyze other metabolic parameters, including locomotor activity and energy expenditure under different nutrition. Notably, under HFD conditions, the OGTT and IPITT revealed a novel function of NPGL associated with insulin sensitivity and glucose homeostasis. Based on the reduced blood glucose and insulin levels in *Npgl*-overexpressing HFD-fed mice at the experimental end point, we propose that NPGL prevents HFD-induced glucose intolerance, insulin resistance, and hyperglycemia in mice. Further research on the NPGL action, including analysis of loss of function as well as *Npgl* overexpression, will help understand the complicated mechanisms of the central and peripheral energy metabolisms, including nutrient selection, obesity development, and related disorders.

4. Material and Methods

4.1. Animals

Male C57BL/6J mice (7 weeks old) were purchased from SLC (Hamamatsu, Japan) and housed individually in general cages (l: 26 cm, w: 18 cm, h: 13 cm, CL-0103-2; CLEA Japan, Tokyo, Japan) under standard conditions (25 ± 1 °C under a 12-h light/dark cycle) with ad libitum access to water and NC (CE-2; CLEA Japan) until animal surgery for *Npgl* overexpression. Thereafter, the mice were fed NC followed by an HFD (45% of calories from fat/17.5% of calories from sucrose, D12451; Research Diets, New Brunswick, NJ, USA), as described below. Animals were operated on under isoflurane anesthesia.

4.2. Production of AAV-Based Vectors

AAV-based vectors were produced following a previously reported method [26]. In the present study, the primers for mouse *Npgl* were 5′-CGATCGATACCATGGCTGATCCTG GGC-3′ (sense primer) and 5′-CGGAATTCTTATTTTCTCTTTACTTCCAGC-3′ (antisense primer). The AAV-based vectors were prepared at a concentration of 1×10^9 particles/µL and stored at −80 °C until use.

4.3. Npgl Overexpression

For *Npgl* overexpression, mice were bilaterally injected with 0.5 µL/site (5.0×10^8 particles/site) of AAV-based vectors (AAV-NPGL or AAV-CTL), using a Neuros Syringe (7001 KH; Hamilton, Reno, NV, USA), at the mediobasal hypothalamic region with the following coordinates: 2.2 mm caudal to the bregma, 0.25 mm lateral to the midline, and 5.8 mm ventral to the skull surface. *Npgl* overexpression was maintained for 70 days during a series of experiments, as shown in Figure 1A, or for 40 days in mice fed with NC alone. *Npgl* overexpression was confirmed by qRT-PCR at the experimental end point. Food intake and body mass were measured every morning (9:00 a.m.). Body composition and blood parameters were measured at the experimental end point of *Npgl* overexpression.

4.4. OGTT and IPITT

The OGTT and IPITT were performed according to a previously reported method [47]. Briefly, mice were fasted for 16 h (overnight fasting) for the OGTT and 4 h (morning fasting) for the IPITT at weekly intervals. Using the GLUCOCARD G+ (Arkray, Kyoto, Japan), blood glucose levels were measured 0, 15, 30, and 60 min after oral glucose (1 g/kg body weight) administration and intraperitoneal insulin (0.75 units/kg) injection. A 35-μL blood sample was collected from the tail vein using a heparinized plastic hematocrit tube (Drummond Scientific Company, Broomall, PA, USA), and plasma was separated by centrifugation at 2500× *g* for 30 min. After centrifugation, the plasma was stored at −80 °C for future insulin measurement. The Rebis Insulin-mouse U ELISA kit (Shibayagi, Gunma, Japan) was used to measure insulin levels. The AUC for blood glucose was calculated using the linear trapezoidal method for both the OGTT and IPITT.

4.5. Quantitative RT-PCR

The MBH was dissected using fine forceps and small scissors, according to the mouse brain atlas [48], and snap frozen in liquid nitrogen for RNA processing. The regions included the supraoptic nucleus, dorsomedial hypothalamus, ventromedial hypothalamus, arcuate nucleus, lateral hypothalamic area, and mammillary nucleus. Total RNA was extracted using TRIzol reagent (Invitrogen, Carlsbad, CA, USA) for the MBH and liver, and QIAzol lysis reagent (QIAGEN, Venlo, Netherlands) for the iWAT in accordance with the manufacturer's instructions. First-strand cDNA was synthesized from total RNA using a PrimeScript RT reagent Kit with gDNA Eraser (Takara Bio, Shiga, Japan).

The primer sequences used in this study are listed in Table 1. The qRT-PCR was conducted following previously reported methods [25,26]. Relative quantification of each gene was performed by the $2^{-\Delta\Delta Ct}$ method using beta-actin (*Actb*) for the MBH and liver, and ribosomal protein S18 (*Rps18*) for the iWAT as an internal control.

Table 1. Sequences of oligonucleotide primers for quantitative RT-PCR.

Gene	Sense Primer (5′ to 3′)	Antisense Primer (5′ to 3′)
Npgl	GGAACCATGGCTTAGGAAGG	TCTAAGGAGCTGAGAATATGCA
Npy	TATCTCTGCTCGTGTGTTTG	GATTGATGTAGTGTCGCAGA
Agrp	TGTTCCCAGAGTTCCCAGGTC	GCATTGAAGAAGCGGCAGTAGCAC
Pomc	AGCTGCCTTTCCGCGACA	ATCTATGGAGGTCTGAAGCA
Gal	GAGCCTTGATCCTGCACTGA	AGTGGCTGACAGGGTCACAA
Acc	TCCGCACTGACTGTAACCACAT	TGCTCCGCACAGATTCTTCA
Fas	AGGGGTCGACCTGGTCCTCA	GCCATGCCCAGAGGGTGGTT
Chrebpα	CGACACTCACCCACCTCTTC	TTGTTCAGCCGGATCTTGTC
Cpt1a	CCTGGGCATGATTGCAAAG	GGACGCCACTCACGATGTT
Atgl	AACACCAGCATCCAGTTCAA	GGTTCAGTAGGCCATTCCTC
Hsl	GCTGGGCTGTCAAGCACTGT	GTAACTGGGTAGGCTGCCAT
Gapdh	AAGGTCATCCCAGAGCTGAA	CTGCTTCACCACCTTCTTGA
Slc2a4	GTAACTTCATTGTCGGCATGG	AGCTGAGATCTGGTCAAACG
Cd36	TCCTCTGACATTTGCAGGTCTATC	AAAGGCATTGGCTGGAAGAA
Pgc1α	GCAACATGCTCAAGCCAAAC	TGCAGTTCCAGAGAGTTCCA
Ucp1	CAAAAACAGAAGGATTGCCGAAA	TCTTGGACTGAGTCGTAGAGG
Dio2	CCACCTTCTTGACTTTGCCA	GGTGAGCCTCATCAATGTATAC
G6pase	ACTGTGGGCATCAATCTCCTC	CGGGACAGACAGACGTTCAGC
Pepck	GTGCTGGAGTGGATGTTCGG	CTGGCTGATTCTCTGTTTCAGG
Slc2a2	GGCTAATTTCAGGACTGGTT	TTTCTTTGCCCTGACTTCCT
Fgf21	CCTCTAGGTTTCTTTGCCAACAG	AAGCTGCAGGCCTCAGGAT
Actb	GGCACCACACCTTCTACAAT	AGGTCTCAAACATGATCTGG
Rps18	CCTGAGAAGTTCCAGCACAT	TTCTCCAGCCCTCTTGGTG

4.6. Blood Biochemical Analysis

Blood biochemicals were analyzed at the experimental end point following previously reported methods [25,26]. Briefly, the GLUCOCARD G+ meter was used to measure glucose content (Arkray). NEFA C-Test Wako (Wako Pure Chemical Industries, Osaka, Japan) was used to measure FFA levels. Triglyceride E-Test Wako (Wako Pure Chemical Industries) was used to measure TG levels. The Rebis Insulin-mouse T ELISA kit (Shibayagi) was used to measure insulin levels.

4.7. Statistical Analysis

Group differences between AAV-NPGL- and AAV-CTL-injected animals were assessed using the unpaired two-tailed Student's *t*-test and Mann–Whitney *U* test. p values < 0.05 were considered significant. Statistical comparisons between every two groups at each time point were conducted with the unpaired two-tailed Student's *t*-test in Figure 1B and the results of OGTT and IPITT.

Supplementary Materials: Supplementary materials can be found at https://www.mdpi.com/article/10.3390/ijms22094681/s1.

Author Contributions: Conceptualization, K.F. and K.U.; methodology, K.F., Y.N., S.M., E.I-U. and M.F.; investigation, K.F., Y.N., S.M., E.I.-U., M.F. and K.U.; writing—original draft preparation, K.F.; writing—review and editing, K.F. and K.U.; visualization, K.F.; project administration, K.U.; funding acquisition, K.F., E.I.-U. and K.U. All authors have read and agreed to the published version of the manuscript.

Funding: This work was supported by JSPS KAKENHI Grant (JP20K22741 to K.F., JP19K06768 to E.I.-U., and JP19H03258 to K.U.), the Takeda Science Foundation (K.U.), the Uehara Memorial Foundation (K.U.), and the Ono Medical Research Foundation (K.U.).

Institutional Review Board Statement: All animal experiments were performed according to the Guide for the Care and Use of Laboratory Animals prepared by Hiroshima University (Higashi-Hiroshima, Japan), and these procedures were approved by the Institutional Animal Care and Use Committee of Hiroshima University (permit numbers: 30-92-2, 4 June 2019; and C19-8, 30 August 2019).

Informed Consent Statement: Not applicable.

Data availability Statement: No big data repositories needed. The raw data supporting the findings of this manuscript will be made available by the corresponding author, K.U., to any qualified researchers upon reasonable request.

Acknowledgments: We are grateful to Atsuki Kadota (Hiroshima University) for the experimental support.

Conflicts of Interest: The authors declare no conflict of interest.

References

1. Haslam, D.W.; James, W.P.T. Obesity. *Lancet* **2005**, *366*, 1197–1209. [CrossRef]
2. Hruby, A.; Hu, F.B. The epidemiology of obesity: A big picture. *Pharmacoeconomics* **2015**, *33*, 673–689. [CrossRef] [PubMed]
3. Ghosh, S.; Bouchard, C. Convergence between biological, behavioural and genetic determinants of obesity. *Nat. Rev. Genet.* **2017**, *18*, 731–748. [CrossRef]
4. Hotamisligil, G.S.; Shargill, N.S.; Spiegelman, B.M. Adipose expression of tumor necrosis factor-α: Direct role in obesity-linked insulin resistance. *Science* **1993**, *259*, 87–91. [CrossRef] [PubMed]
5. Hotamisligil, G.S.; Peraldi, P.; Budavari, A.; Ellis, R.; White, M.F.; Spiegelman, B.M. IRS-1-mediated inhibition of insulin receptor tyrosine kinase activity in TNF-alpha- and obesity-induced insulin resistance. *Science* **1996**, *271*, 665–668. [CrossRef] [PubMed]
6. Weisberg, S.P.; McCann, D.; Desai, M.; Rosenbaum, M.; Leibel, R.L.; Ferrante, A.W. Obesity is associated with macrophage accumulation in adipose tissue. *J. Clin. Investig.* **2003**, *112*, 1796–1808. [CrossRef]
7. Schwartz, M.W.; Porte, D. Diabetes, obesity, and the brain. *Science* **2005**, *307*, 375–379. [CrossRef]
8. Morton, G.J.; Cummings, D.E.; Baskin, D.G.; Barsh, G.S.; Schwartz, M.W. Central nervous system control of food intake and body weight. *Nature* **2006**, *443*, 289–295. [CrossRef]
9. Morton, G.J.; Meek, T.H.; Schwartz, M.W. Neurobiology of food intake in health and disease. *Nat. Rev. Neurosci.* **2014**, *15*, 367–378. [CrossRef]
10. Kamegai, J.; Tamura, H.; Shimizu, T.; Ishii, S.; Sugihara, H.; Wakabayashi, I. Chronic central infusion of ghrelin increases hypothalamic neuropeptide Y and Agouti-related protein mRNA levels and body weight in rats. *Diabetes* **2001**, *51*, 2438–2443. [CrossRef]
11. Nakazato, M.; Murakami, N.; Date, Y.; Kojima, M.; Matsuo, H.; Kangawa, K.; Matsukura, S. A role for ghrelin in the central regulation of feeding. *Nature* **2001**, *409*, 194–198. [CrossRef]

12. Cowley, M.A.; Smith, R.G.; Diano, S.; Tschöp, M.; Pronchuk, N.; Grove, K.L.; Strasburger, C.J.; Bidlingmaier, M.; Esterman, M.; Heiman, M.L.; et al. The distribution and mechanism of action of ghrelin in the CNS demonstrates a novel hypothalamic circuit regulating energy homeostasis. *Neuron* **2003**, *37*, 649–661. [CrossRef]
13. Beutler, L.R.; Chen, Y.; Ahn, J.S.; Lin, Y.C.; Essner, R.A.; Knight, Z.A. Dynamics of gut-brain communication underlying hunger. *Neuron* **2017**, *96*, 461–475. [CrossRef] [PubMed]
14. Cowley, M.A.; Smart, J.L.; Rubinstein, M.; Cerdán, M.G.; Diano, S.; Horvath, T.L.; Cone, R.D.; Low, M.J. Leptin activates anorexigenic POMC neurons through a neural network in the arcuate nucleus. *Nature* **2001**, *411*, 480–484. [CrossRef] [PubMed]
15. Vong, L.; Ye, C.; Yang, Z.; Choi, B.; Chua, S.; Lowell, B.B. Leptin action on GABAergic neurons prevents obesity and reduces inhibitory tone to POMC neurons. *Neuron* **2011**, *71*, 142–154. [CrossRef] [PubMed]
16. Zhang, Y.; Proenca, R.; Maffei, M.; Barone, M.; Leopold, L.; Friedman, J.M. Positional cloning of the mouse obese gene and its human homologue. *Nature* **1994**, *372*, 425–432. [CrossRef]
17. Koch, L.; Wunderlich, F.T.; Seibler, J.; Könner, A.C.; Hampel, B.; Irlenbusch, S.; Brabant, G.; Kahn, C.R.; Schwenk, F.; Brüning, J.C. Central insulin action regulates peripheral glucose and fat metabolism in mice. *J. Clin. Investig.* **2008**, *118*, 2132–2147. [CrossRef]
18. Petersen, M.C.; Shulman, G.I. Mechanisms of insulin action and insulin resistance. *Physiol. Rev.* **2018**, *98*, 2133–2223. [CrossRef]
19. Ukena, K.; Iwakoshi-Ukena, E.; Taniuchi, S.; Bessho, Y.; Maejima, S.; Masuda, K.; Shikano, K.; Kondo, K.; Furumitsu, M.; Tachibana, T. Identification of a cDNA encoding a novel small secretory protein, neurosecretory protein GL, in the chicken hypothalamic infundibulum. *Biochem. Biophys. Res. Commun.* **2014**, *446*, 298–303. [CrossRef]
20. Ukena, K. Avian and murine neurosecretory protein GL participates in the regulation of feeding and energy metabolism. *Gen. Comp. Endocrinol.* **2018**, *260*, 164–170. [CrossRef]
21. Shikano, K.; Bessho, Y.; Kato, M.; Iwakoshi-Ukena, E.; Taniuchi, S.; Furumitsu, M.; Tachibana, T.; Bentley, G.E.; Kriegsfeld, L.J.; Ukena, K. Localization and function of neurosecretory protein GM, a novel small secretory protein, in the chicken hypothalamus. *Sci. Rep.* **2018**, *8*, 704. [CrossRef]
22. Shikano, K.; Kato, M.; Iwakoshi-Ukena, E.; Furumitsu, M.; Matsuura, D.; Masuda, K.; Tachibana, T.; Bentley, G.E.; Kriegsfeld, L.J.; Ukena, K. Effects of chronic intracerebroventricular infusion of neurosecretory protein GL on body mass and food and water intake in chicks. *Gen. Comp. Endocrinol.* **2018**, *256*, 37–42. [CrossRef]
23. Shikano, K.; Iwakoshi-Ukena, E.; Kato, M.; Furumitsu, M.; Bentley, G.E.; Kriegsfeld, L.J.; Ukena, K. Neurosecretory protein GL induces fat accumulation in chicks. *Front. Endocrinol.* **2019**, *10*, 392. [CrossRef]
24. Matsuura, D.; Shikano, K.; Saito, T.; Iwakoshi-Ukena, E.; Furumitsu, M.; Ochi, Y.; Sato, M.; Bentley, G.E.; Kriegsfeld, L.J.; Ukena, K. Neurosecretory protein GL, a hypothalamic small secretory protein, participates in energy homeostasis in male mice. *Endocrinology* **2017**, *158*, 1120–1129. [CrossRef]
25. Shikano, K.; Iwakoshi-Ukena, E.; Saito, T.; Narimatsu, Y.; Kadota, A.; Furumitsu, M.; Bentley, G.E.; Kriegsfeld, L.J.; Ukena, K. Neurosecretory protein GL induces fat accumulation in mice. *J. Endocrinol.* **2020**, *244*, 1–12. [CrossRef] [PubMed]
26. Iwakoshi-Ukena, E.; Shikano, K.; Kondo, K.; Taniuchi, S.; Furumitsu, M.; Ochi, Y.; Sasaki, T.; Okamoto, S.; Bentley, G.E.; Kriegsfeld, L.J.; et al. Neurosecretory protein GL stimulates food intake, de novo lipogenesis, and onset of obesity. *Elife* **2017**, *6*, e28527. [CrossRef]
27. Narimatsu, Y.; Fukumura, K.; Shikano, K.; Iwakoshi-Ukena, E.; Furumitsu, M.; Bentley, G.E.; Kriegsfeld, L.J.; Ukena, K. Hypothalamic overexpression of neurosecretory protein GL leads to obesity in Mice. *bioRxiv* **2021**. [CrossRef]
28. Berthoud, H.R.; Morrison, C. The brain, appetite, and obesity. *Annu. Rev. Psychol.* **2008**, *59*, 55–92. [CrossRef] [PubMed]
29. Andermann, M.L.; Lowell, B.B. Toward a wiring diagram understanding of appetite control. *Neuron* **2017**, *95*, 757–778. [CrossRef]
30. Andrikopoulos, S.; Blair, A.R.; Deluca, N.; Fam, B.C.; Proietto, J. Evaluating the glucose tolerance test in mice. *Am. J. Physiol. Endocrinol. Metab.* **2008**, *295*, E1323–E1332. [CrossRef] [PubMed]
31. Ahrén, B.; Winzell, M.S.; Pacini, G. The augmenting effect on insulin secretion by oral versus intravenous glucose is exaggerated by high-fat diet in mice. *J. Endocrinol.* **2008**, *197*, 181–187. [CrossRef]
32. Horakova, O.; Kroupova, P.; Bardova, K.; Buresova, J.; Janovska, P.; Kopecky, J.; Rossmeisl, M. Metformin acutely lowers blood glucose levels by inhibition of intestinal glucose transport. *Sci. Rep.* **2019**, *9*, 6156. [CrossRef]
33. Ruud, J.; Steculorum, S.M.; Bruning, J.C. Neuronal control of peripheral insulin sensitivity and glucose metabolism. *Nat. Commun.* **2017**, *8*, 15259. [CrossRef]
34. Fang, P.; Yu, M.; Shi, M.; Bo, P.; Zhang, Z. Galanin peptide family regulation of glucose metabolism. *Front. Neuroendocrinol.* **2020**, *56*, 100801. [CrossRef] [PubMed]
35. Fukumura, K.; Shikano, K.; Narimatsu, Y.; Iwakoshi-Ukena, E.; Furumitsu, M.; Naito, M.; Ukena, K. Effects of neurosecretory protein GL on food intake and fat accumulation under different dietary nutrient compositions in rats. *Biosci. Biotechnol. Biochem.* **2021**, zbab064. [CrossRef] [PubMed]
36. Delgado, T.C.; Pinheiro, D.; Caldeira, M.; Castro, M.M.; Geraldes, C.F.; López-Larrubia, P.; Cerdán, S.; Jones, J.G. Sources of hepatic triglyceride accumulation during high-fat feeding in the healthy rat. *NMR Biomed.* **2009**, *22*, 310–317. [CrossRef] [PubMed]
37. Duarte, J.A.; Carvalho, F.; Pearson, M.; Horton, J.D.; Browning, J.D.; Jones, J.G.; Burgess, S.C. A high-fat diet suppresses de novo lipogenesis and desaturation but not elongation and triglyceride synthesis in mice. *J. Lipid Res.* **2014**, *55*, 2541–2553. [CrossRef]
38. Abdul-Wahed, A.; Guilmeau, S.; Postic, C. Sweet sixteenth for ChREBP: Established roles and future goals. *Cell Metab.* **2017**, *26*, 324–341. [CrossRef] [PubMed]

39. Dentin, R.; Benhamed, F.; Pégorier, J.P.; Foufelle, F.; Viollet, B.; Vaulont, S.; Girard, J.; Postic, C. Polyunsaturated fatty acids suppress glycolytic and lipogenic genes through the inhibition of ChREBP nuclear protein translocation. *J. Clin. Investig.* **2005**, 2843–2854. [CrossRef]
40. Caron, A.; Richard, D. Neuronal systems and circuits involved in the control of food intake and adaptive thermogenesis. *Ann. N. Y. Acad. Sci.* **2017**, *1391*, 35–53. [CrossRef]
41. Lieu, L.; Chau, D.; Afrin, S.; Dong, Y.; Alhadeff, A.L.; Betley, J.N.; Williams, K.W. Effects of metabolic state on the regulation of melanocortin circuits. *Physiol. Behav.* **2020**, *224*, 113039. [CrossRef]
42. Takahashi, K.A.; Cone, R.D. Fasting induces a large, leptin-dependent increase in the intrinsic action potential frequency of orexigenic arcuate nucleus neuropeptide Y/Agouti-related protein neurons. *Endocrinology* **2005**, *146*, 1043–1047. [CrossRef] [PubMed]
43. Baver, S.B.; Hope, K.; Guyot, S.; Bjørbaek, C.; Kaczorowski, C.; O'Connell, K.M.S. Leptin modulates the intrinsic excitability of AgRP/NPY neurons in the arcuate nucleus of the hypothalamus. *J. Neurosci.* **2014**, *34*, 5486–5496. [CrossRef] [PubMed]
44. Okamoto, S.; Sato, T.; Tateyama, M.; Kageyama, H.; Maejima, Y.; Nakata, M.; Hirako, S.; Matsuo, T.; Kyaw, S.; Shiuchi, T.; et al. Activation of AMPK-regulated CRH neurons in the PVH is sufficient and necessary to induce dietary preference for carbohydrate over fat. *Cell Rep.* **2018**, *22*, 706–721. [CrossRef] [PubMed]
45. Von Holstein-Rathlou, S.; BonDurant, L.D.; Peltekian, L.; Naber, M.C.; Yin, T.C.; Claflin, K.E.; Urizar, A.I.; Madsen, A.N.; Ratner, C.; Holst, B.; et al. FGF21 mediates endocrine control of simple sugar intake and sweet taste preference by the liver. *Cell Metab.* **2016**, *23*, 335–343. [CrossRef] [PubMed]
46. Talukdar, S.; Owen, B.M.; Song, P.; Hernandez, G.; Zhang, Y.; Zhou, Y.; Scott, W.T.; Paratala, B.; Turner, T.; Smith, A.; et al. FGF21 regulates sweet and alcohol preference. *Cell Metab.* **2016**, *23*, 344–349. [CrossRef]
47. Nagy, C.; Einwallner, E. Study of in vivo glucose metabolism in high-fat diet-fed mice using oral glucose tolerance test (OGTT) and insulin tolerance test (ITT). *J. Vis. Exp.* **2018**, *131*, 56672. [CrossRef]
48. Franklin, K.B.J.; Paxinos, G. *The Mouse Brain in Stereotaxic Coordinates*; Elsevier: Amsterdam, The Netherlands; Academic Press: Boston, MA, USA, 1997.

MDPI
St. Alban-Anlage 66
4052 Basel
Switzerland
Tel. +41 61 683 77 34
Fax +41 61 302 89 18
www.mdpi.com

International Journal of Molecular Sciences Editorial Office
E-mail: ijms@mdpi.com
www.mdpi.com/journal/ijms

www.ingramcontent.com/pod-product-compliance
Lightning Source LLC
Chambersburg PA
CBHW040803250926
51850CB00007BA/1070